A TUTORIAL GUIDE TO

AUTOCAD®

RELEASE 13

FOR WINDOWS®

Includes Coverage of
AutoVision™ and
AutoCAD® Designer

A TUTORIAL GUIDE TO

AUTOCAD®
RELEASE 13
FOR WINDOWS®

Includes Coverage of
AutoVision™ and
AutoCAD® Designer

Shawna D. Lockhart

Addison-Wesley Publishing Company

Reading, Massachusetts • Menlo Park, California • New York • Don Mills, Ontario
Harlow, U.K. • Amsterdam • Bonn • Paris • Milan • Madrid • Sydney • Singapore
Tokyo • Seoul • Taipei • Mexico City • San Juan, Puerto Rico

The Publishing Team

Acquisitions Editor: Denise Penrose
Executive Editor: Dan Joraanstad
Marketing Manager: Mary Tudor
Assistant Editor: MaryLynne Wrye
Project Manager: Cindy Johnson
Development Editor: Kathleen Habib
Senior Production Editor: Teri Holden

Manager of Visual Communications: Don Kesner
Text Designer: Jean Hammond
Copyeditor: Pamela Mayne
Technical Validators: MaryLynne Wrye, Angela Chadwick
Compositor: Gex, Inc.
Cover Designer: Ark Stein
Manufacturing Coordinator: Janet Weaver

Library of Congress Cataloging-in-Publication Data

Lockhart, Shawna D., 1957-

A tutorial guide to AutoCAD release 13 for Windows: includes coverage of AutoVision and AutoCAD Designer/Shawna D. Lockhart.

p.   cm.

Includes index.

ISBN 0-201-82373-X

1. Mechanical drawing. 2. AutoCAD for Windows. 3. AutoVision. 4. AutoCAD designer. I. Title.

T353.L854   1996

620'.0042'02855369—dc20   95-39483
                                                                                        CIP

Autodesk Trademarks

The following are registered trademarks of Autodesk, Inc.: ADI, Advanced Modeling Extension, AME, Animator Pro, ATC, AutoCAD, AutoCAD Development System, Autodesk, Autodesk Animator, the Autodesk logo, AutoLISP, AutoShade, AutoSketch, AutoSolid, AutoSurf, Geodyssey, HOOPS, Multimedia Explorer, Office Series, TinkerTech, World-Creating Toolkit, 3D Plan, and 3D Studio.

The following are trademarks of Autodesk, Inc.: ACAD, Advanced User Interface, AEMULUS, AEMULUS(mf), AME Link, Animation Partner, Animation Player, Animation Pro Player, A Studio in Every Computer, ATLAST, AUI, AutoCAD Data Extension, AutoCAD Simulator, AutoCAD SQL Extension, AutoCAD SQL Interface, Autodesk Animator Clips, Autodesk Animator Theatre, Autodesk Device Interface, Autodesk Software Developer's Kit, Autodesk WorkCenter, AutoCDM, AutoEDM, AutoFlix, AutoLathe, AutoVision, DXF, FLI, Flic, Generic 3D, SketchTools, SmartCursor, Supportdesk, Texture Universe, Transforms ideas Into Reality, and Visual Link.

The following are service marks of Autodesk, Inc.: Autodesk Strategic Developer, Autodesk Strategic Developer logo, Autodesk Registered Developer, Autodesk Registered Developer logo, and the Autodesk University logo.

Third Party Trademarks

Renderman is a registered trademark of Pixar used by Autodesk under license from Pixar.
Microsoft, Windows, MS-DOS, Windows NT, and the Windows logo are either registered trademarks or trademarks of Microsoft Corporation.
All other trademarks are trademarks of their respective holders.

ISBN 0-201-82373-X

1 2 3 4 5 6 7 8 9 10—CRW—99 98 97 96 95

AutoCAD® is the most widely used design and drafting software for desktop computers in the world. AutoCAD® Release 13 for Windows® provides anyone with a personal computer with the ability to create complex and accurate drawings. Its position as the industry standard makes it an essential tool for anyone preparing for a career in engineering, design, or technology.

Because it is the industry standard, AutoCAD is the ideal cornerstone of your design and drafting skill set. With a knowledge of AutoCAD, you will find it easy to add any number of a wide range of applications to create a complete design environment suited to your needs. AutoVision™, photo-realistic rendering software, and AutoCAD® Designer, parametric solid modeling software, are two popular applications that operate within AutoCAD to expand your design capability.

This tutorial-based manual will teach you step-by-step to use AutoCAD Release 13 for Windows, AutoVision, and AutoCAD Designer to create 2D, 3D, and parametric models and the engineering drawings that describe them. Written for the novice AutoCAD user, the manual uses a proven tutorial approach that guides you through the creation of actual drawings and models. Information about AutoCAD is presented in a need-to-know fashion that makes it easy to remember. Tips and shortcuts are included where appropriate to help you become an efficient and proficient AutoCAD user. A Command Summary and Glossary make it easy to review what you've learned and to use the manual as a reference.

This manual may be used in conjunction with a basic engineering graphics course, introductory engineering, introductory architecture and/or design courses, or independently for self-study.

Features

To facilitate your study of AutoCAD Release 13 for Windows, this tutorial guide includes:

- Step-by-step tutorials written for the novice user
- A complete chapter on configuring AutoCAD for performance and use with these tutorials
- Configuration instructions for Windows 95 users
- Tutorials organized to parallel an introductory engineering graphics course
- "TIP" boxes that offer suggestions and warnings to students as they progress through the tutorials.
- Challenging end-of-tutorial drawing exercises, with applications in mechanical, civil, and electrical engineering, as well as architecture
- Key Terms and Key Commands summaries to recap important topics and commands learned in each tutorial

Organization

The tutorials proceed in a logical fashion to guide you from drawing basic shapes to building three-dimensional models. You will learn to create 3D models first in AutoCAD, then with

AutoCAD Designer's parametric modeling tools. Later tutorials introduce special drawing techniques and advanced dimensioning, and the final tutorial shows you how to create photo-realistic renderings of your models with AutoVision.

Part 1, Getting Started, first helps you configure AutoCAD's environment and menus for the step-by-step instructions in this manual, then takes you on a guided tour of the AutoCAD Release 13 screen display, help facility, and keyboard/mouse usage conventions. Both of the Getting Started chapters address AutoCAD operations under Windows 3.1 and Windows 95.

Part 2, Tutorials, introduces AutoCAD Release 13 for Windows, AutoCAD Designer, and AutoVision in the context of technical drawing.

- Tutorials 1 and 2 introduce AutoCAD's basic drawing commands and build proficiency with the menus, toolbars, and drawing aids.
- Tutorials 3 and 4 introduce editing commands and geometric constructions and provide instruction on plotting a drawing.
- Tutorial 5 focuses on good drawing management with prototype drawings, layers, and plotting from paper space.
- Tutorial 6 is devoted to the concepts of drawing orthographic views.
- Tutorial 7 introduces basic dimensioning and the use of dimension styles.
- Tutorial 8 shows how to use blocks and to customize toolbars.
- Tutorials 9 and 10 teach 3D solid modeling from a single part to an assembly drawing.
- Tutorials 11, 12, and 13 show how to use AutoCAD Designer to create 3D parametric models, update them, and generate 2D drawings from them, including an intelligent assembly using global parameters. The AutoCAD Designer chapters may be covered at any point in the manual after Tutorial 5.

- Tutorials 14 and 15 introduce section and auxiliary views and show how to create them in 2D and from a 3D model. These tutorials may be undertaken after solid modeling, Tutorials 9 and 10.
- Tutorial 16 is devoted to advanced dimensioning topics such as geometric tolerances. It may be used immediately after Tutorial 7, if desired.
- Tutorial 17 shows how to use AutoVision to render and shade a 3D model and export it to 3D Studio. This tutorial may be completed any time after solid modeling (Tutorial 10) if you wish to use AutoVision to render models created in later tutorials.

The Glossary defines key terms used in the tutorials. The Command Summaries show where the key commands used in the tutorials can be found. The Command Summaries are divided into three parts for ease of use: for AutoCAD, AutoVision, and AutoCAD Designer.

Acknowledgments

I wish to acknowledge the individuals who contributed to the conceptualization and implementation of these tutorials. First, thanks to the many teaching colleagues who responded to inquiries and helped to shape the tutorials in this book. We cannot list all the individuals with whom we spoke in the course of our research, but we are grateful to the reviewers listed below for their help.

Reuben R. Aronovitz
Delaware County Community College

Adrian G. Baird
Ricks College

Karen Coen-Brown
University of Nebraska-Lincoln

Michael H. Gregg
Virginia Polytechnic Institute and State University

Gary J. Hordemann
Gonzaga University

Henry Horwitz
Dutchess Community College

Craig L. Miller
Purdue University

Jim Raschka
Delaware County Community College

Gordon F. Snyder Jr.
Springfield Technical Community College

Gary Sobczak
Purdue University

Kyle Tage
Montana State University

Our thanks go to Doug Baese, Steve Brackman, Craig Bradley, Tom Bryson, Karen Coen-Brown, Joe Evers, Mary Ann Koen, D. Krall, Kim Manner, Shawn Murphy, Torian Roesch, Kyle Tage, John Walker, and Wendy Warren for submitting exercises and creating art files for us; to Shannon Kyles and James Bethune, authors of previous AutoCAD tutorial guides, from which some material was adapted for this manual; and to James Earle, for permission to reprint several exercises from his text.

I wish to thank Richard Cuneo, Vice President U.S. Sales; Jim Purcell, Director of Education; Mark Sturges, Education Sales Manager; Alan Jacobs, Manager of Virtual Publishing; Jimm Meloy, Education Programs Manager; Lisa Senauke, Developer Marketing Programs Coordinator; Mary Vance, Corporate Counsel; Carrie Costamagna Bustillos, Manager Education Marketing America; Maureen Barrow, Contracts Administrator; and Daniel Vinson, Consultant, Education Department, who lent their talents to the partnership between Addison-Wesley and Autodesk, Inc. that supported the development of this manual.

Editing, testing, and producing step-by-step tutorials requires a tremendous amount of publishing expertise, and I would be remiss if I did not acknowledge those who worked to make this book complete, technically accurate, and lovely to look at, especially Denise Penrose, MaryLynne Wrye, Kathleen Habib, and Cindy Johnson for their enthusiasm, efficiency, and professionalism throughout this project. Last but not least, I wish to thank the many great students I have had the pleasure to teach at Montana State University and from whom I have learned a great deal.

Shawna D. Lockhart

CONTENTS

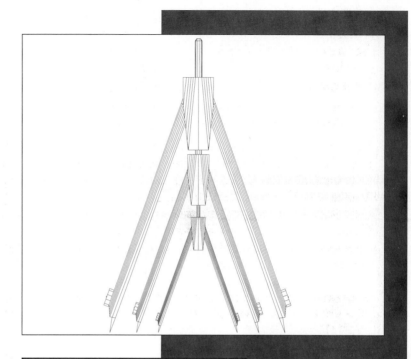

PART ONE

Getting Started

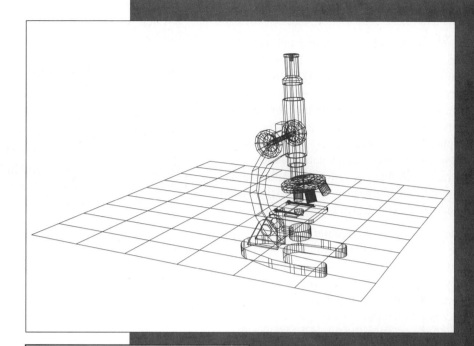

CHAPTER 1

Preparing AutoCAD for the Tutorials

Objectives

This chapter describes how to prepare your computer system and
AutoCAD Release 13 for Windows for use with the tutorials in this manual.
As you read the chapter, you will

1. Understand how to use a mouse to complete these tutorials.

2. Recognize the typographical conventions used in this book.

3. Configure Windows for maximum AutoCAD performance and support.

4. Create a separate AutoCAD configuration and icon to use for these tutorials.

5. Create and prepare datafile and working directories for use with tutorials.

6. Compile the AutoCAD menu to include the AutoVision and Designer menus.

7. Configure your AutoCAD, AutoVision, and AutoCAD Designer for use with these
   tutorials.

Introduction to the Tutorials

The 17 tutorials in this manual will teach you to use AutoCAD Release 13 for Windows, AutoVision 2.0, and AutoCAD Designer 1.2 through a series of step-by-step exercises. You will learn the fundamental operating procedures and how to use the drawing tools of each application. As you progress through the tutorials, you will employ some of the more advanced features of AutoCAD, AutoVision, and AutoCAD Designer to create, dimension, shade, and print drawings.

The tutorials in this manual assume that you are using AutoCAD's default settings and that AutoVision and AutoCAD Designer are properly installed. If your system has been customized for other uses, some of the settings may not match those assumed by the step-by-step instruction in the tutorials. Please check your system against the configuration in this chapter so that you can work through the tutorials as instructed.

If you are using AutoCAD on a network at your school, ask your professor or system administrator about how the software is configured. You should not try to reconfigure AutoCAD unless instructed to do so by network personnel. Read about mouse techniques, typographical conventions, and end-of-tutorial exercises in this chapter, then go on to Chapter 2, AutoCAD Basics.

To complete these tutorials, your system must be running Microsoft Windows Version 3.1, or a later release.

Basic Mouse Techniques

The tutorials in this manual assume that you will be using a mouse or pointing device to work with AutoCAD Release 13 for Windows.

The following terms will be used to streamline the instructions you are given for using the mouse.

Term	Meaning
Pick or Click	To quickly press and release the left mouse button
Double-click	To click the left mouse button twice in rapid succession
Drag	To press and hold down the left mouse button while you move the mouse
Point	To move the mouse until the mouse pointer on the screen is positioned above the item you want

> ■ *TIP* Your mouse may have more than one button. To click (or pick) in Windows and in AutoCAD, use the left mouse button. This button is also referred to as the *pick button.* ■

Recognizing Typographical Conventions

When you work with AutoCAD, AutoVision, and AutoCAD Designer, you'll use your keyboard and your pointing device to input information. As you read this manual, you'll encounter special type styles that will help you determine the commands required and the information you must input.

Special symbols illustrate certain computer keys. For example, the Enter or Return key is represented by the symbol ⏎.

> ■ *TIP* The right button on your mouse performs the same function as the ⏎ key. This button is also referred to as the *return button*. ■

The manual also employs special typefaces when it presents instructions for performing computer operations. Instructions are set off from the main text to indicate a series of actions or AutoCAD command prompts. Boldface type in all capital letters is used for the letters and numbers to be input by you. For example,

Command: **LINE**

instructs you to type the Line command.

From point: ***(pick point A)***

instructs you to select point A on your screen by clicking it with the mouse. The words "Command" and "From point" represent the AutoCAD prompts you would see on your screen. The AutoCAD prompt line is always in regular type.

Instruction words, such as "pick," "type," "select," and "press," appear in italic type. For example, "type" instructs you to strike or press several keys in sequence. For the following instruction,

Type: **R(4,4)**

you would type *R(4,4)* in that order on your keyboard. Remember to type exactly what you see, including spaces, if any.

"Pick" tells you to pick an object or an icon or to choose commands from the menu. For example,

Pick: **Line icon**

instructs you to use your pointing device to select the Line icon from the toolbar on your screen. To select an icon, move the pointer until the cursor is over the command and press the pick button.

"Press" means to strike a key once. For example,

Press: ⏎

means strike the Enter or Return key once. Sometimes "press" is followed by two keys, such as

Press: ALT-F1

When this occurs, press and hold down the first key, press the second key once and release it, and then release the first key.

Default values that are displayed as part of a command or prompt are represented in angle brackets: < >. For example,

Trace width<0.0500>:

is the command prompt when the Trace command has a default value of .05.

New terms are in italics in the text when they are introduced, and are defined in the Glossary at the end of the manual.

End-of-Tutorial Exercises

Exercises at the end of the tutorials are divided into four types, designated by the following icons:

Mechanical Engineering—Exercises in the design of machines and tools

Electrical Engineering—Exercises in the design of electrical systems

Civil Engineering—Exercises in the design of roads, bridges, and other public and private works

Architecture—Exercises in the design of buildings

Configuring Windows for AutoCAD

AutoCAD Release 13 for Windows takes advantage of common Windows file operations, and allows you to run AutoCAD while you are running other applications.

Windows 3.1, Windows NT, and Windows 95

AutoCAD Release 13 is a 32-bit program, designed to take full advantage of today's computers. In order to run AutoCAD under Windows 3.1x, a 16-bit platform, your system uses the Win32s subsystem to emulate 32-bit operation. If you are running Windows NT or Windows 95, both of which are 32-bit programs, some of the tips in this manual may not apply to your system. For example, AutoCAD Release 13 for Windows does not support multiple drawing sessions under Windows 3.1, but it does under NT. The Win32s subsystem is included with the AutoCAD software. You should have installed Win32s when you installed AutoCAD.

AutoCAD can be installed with Windows 3.1, Windows NT, or Windows 95. To get the best performance, you may want to change your Windows settings as described in this section.

The figures in the tutorials were captured in Windows 3.1. The AutoCAD screen in Windows 95 is very similar to the screens you will see in this manual, however, and contains all the elements you need to complete the tutorials. There may be instances where step-by-step instructions for working with files and the operating system need to be adjusted slightly for Windows 95, but the operations are described fully enough to make it easy to substitute Windows 95 operations. Differences between Windows 3.1x and Windows 95 operation are spelled out in this chapter and the next.

Managing Memory for Performance

Windows uses a lot of your computer's memory. If you have terminate-and-stay-resident applications or applications running in the background, your Windows and AutoCAD performance may suffer.

AutoCAD is designed to adjust memory usage as necessary. Systems with 16 MB of memory, however, may require fine-tuning. If you are running Windows 3.1x, AutoCAD Release 13 recommends that you use a 64-MB Windows permanent swap file. To check this setting, open the Main program group, then open the Control Panel.

■ *Warning:* The minimum requirement for the Windows NT version of AutoCAD is 32 MB of RAM. ■

Double-click: **386 Enhanced icon**
The 386 Enhanced dialog box opens, as shown in Figure G1.1.

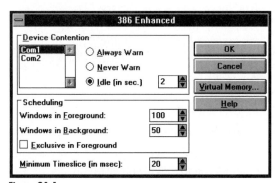

Figure G1.1

Pick: **Virtual Memory**
The Virtual Memory dialog box opens, as shown in Figure G1.2.

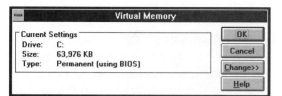

Figure G1.2

Pick: **Change>>**

The Virtual Memory dialog box expands and a section labeled New Settings appears at the bottom, as shown in Figure G1.3. Highlight the text box next to New Size.

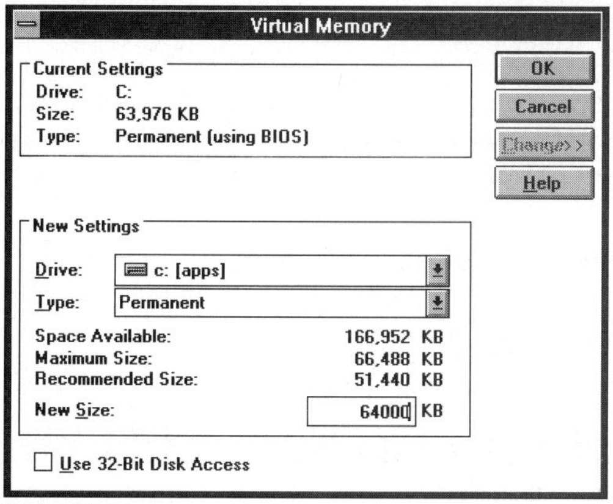

Figure G1.3

Type: **64000** ⏎

Pick: **OK** *(to exit the Virtual Memory dialog box)*

Pick: **OK** *(to exit the 386 Enhanced dialog box)*

Double-click: (the box in the top left corner of the Control Panel window to close it)

■ **Warning:** You may get an error message that the number you have specified is too large. If this happens, accept the largest file size Windows will allow and confirm that you want to make the changes to virtual memory. ■

VSHARE.386

Some Windows programs require the DOS driver *share.exe*, which may be called in your *autoexec.bat* file. AutoCAD is not compatible with *share.exe*. Microsoft has created a Windows version of the driver, *vshare.386*. It is shipped with some Microsoft applications and may already be on your system. It is also available on the Internet via FTP from the Microsoft Software Library at *ftp.microsoft.com*, on the World Wide Web at *www.microsoft.com*, or in the Cadence Forum on CompuServe. To retrieve it, connect to the library and search for the key words **vshare** or **ww1000**. Either of these key words will let you download *ww1000.exe*, which contains *vshare.386* and installation instructions.

Screen Resolution

The figures in this manual use a screen resolution of $800 \times 600 \times 256$ colors. This requires your computer to have 1 MB of video RAM.

If your computer has only 16 MB of memory, you may want to use a standard VGA display. The icons on your screen will be larger than they are in the manual and the screen menu will not be fully accessible, but you will be able to complete all the tutorials.

In Windows 3.1, you can change your screen resolution in the Windows Setup. Open the Main program group and then the Windows Setup program icon.

Pick: **Options, Change System Settings** *(from the pull-down menu)*

The Change System Settings dialog box appears on your screen, as shown in Figure G1.4a. Click on the down arrow to the right of the Display text box. Your choices for video display appear below the text box. Use the scroll bars to locate and pick a $800 \times 600 \times 256$

color display option *for your video card.* (Refer to your system documentation if you are unsure which display driver is appropriate for your display.)

Figure G1.4a Windows 3.1

Pick: **OK** *(to exit the Change System Settings dialog box)*

Pick: **OK** *(to exit Windows Setup)*

In Windows 95, you can change your screen resolution in the Control Panel. From Windows 95,

Pick: **Start, Control Panel** *(from the bottom left of your screen)*

Pick: **Display** *(from the Control Panel)*

Pick: **Settings** *(in the dialog box that appears)*

The Display Properties dialog box appears on your screen, as shown in Figure G1.4b. You can change the color selection and the pixel selection with the scroll bars. Set your display to 256 Color palette and 800×600 pixels. Windows 95 will look for and load the appropriate driver for your video display.(Refer to your system documentation if you are having trouble with your screen resolution settings.)

Figure G1.4b Windows 95

Pick: **OK** *(to exit the Display Properties dialog box)*

You will be prompted to restart Windows so that the change takes effect. Restart Windows now.

Separate AutoCAD Configurations

AutoCAD Release 13 for Windows allows you to save multiple configurations, making it possible to configure AutoCAD for different uses and peripherals. These different configurations are associated with separate AutoCAD program icons in Windows.

If you are working on a shared machine, you may want to create a special icon to use for these tutorials so the settings and defaults you set aren't changed between sessions.

If you are working on a networked computer, your professor or system administrator has probably configured the system already. Ask which icon you should use to launch AutoCAD.

Backing Up AutoCAD Defaults

Before you begin to configure and customize AutoCAD for the purposes of these tutorials, you will want to back up your configuration, initialization, and menu files to save the previous configuration. To do this, copy *acad.ini* and *acad.cfg* from the *c:\r13\win* directory and *acad.mnu* and *acad.mns* from the *c:\r13\win\support* directory to a floppy disk or to a safe location on your hard disk or a remote network drive where they will not be overwritten.

If you want to restore these AutoCAD defaults, you can copy them back to the *c:\r13\win* and *c:\r13\win\support* directories or your working directory, and overwrite the files there. (This manual assumes that you are working from the *c:* drive and that AutoCAD is installed on that drive. If you are not, substitute your drive name and path as needed in these instructions.)

■ **TIP** It's a good idea to create a backup copy of c:\av\autovis.pwd. This is the password file for AutoVision. AutoVision only allows one user at a time. If AutoVision quits unexpectedly, it may not release its license and it will think that a user is already logged in. If this happens, you can replace c:\av\autovis.pwd with your backup copy. ■

Creating a Working Directory

To store and use multiple configurations, you need to maintain and use different *acad.ini* and *acad.cfg* files for each configuration, and create separate program icons for each configuration. To do this, you will place copies of *acad.ini* and *acad.cfg* in your working directory, *c:\work*.

Use your Windows operating system now to create a new directory, *c:\work*. Copy *acad.ini* and *acad.cfg* from *c:\r13\win* to *c:\work*. If you have trouble creating directories or copying files, please refer to your Windows user's guide or Windows on-line help facility.

When you are finished, return to the Windows 3.1 Program Manager or Windows 95 main window.

Creating Multiple Icons

Open the AutoCAD R13 program group. If you have trouble locating or opening the AutoCAD R13 program group, refer to your Windows user's guide or Windows on-line help facility.

■ **TIP** In Windows 95, your program icons are available if you double-click the My Computer icon on your Windows 95 main window. You may need to open the folder or folders in which your application is located, for example, c:\drive, r13 and win directories. ■

Pick: (the AutoCAD R13 program icon and hold the pick button down)

Press: Ctrl *(and drag the AutoCAD R13 icon to copy it)*

This will create a new copy of your AutoCAD R13 icon. You will rename this copy of the icon *AutoCAD R13 Tutorials*. Windows 3.1 allows you to rename icons in the Properties dialog box. Windows 95 allows you to use the File, Rename option from the pull-down menu to rename icons. Rename the icon on your own now.

You will now set *c:\work* as your working directory.

Setting the Working Directory

The AutoCAD session will only use your new *acad.cfg* and *acad.ini* files and save your drawings to *c:\work* by default if it is set as your working directory.

Setting the Working Directory in Windows 3.1

To set *c:\work* as your working directory, you will use the Windows Program Manager.

Pick: (the new AutoCAD R13 Tutorials icon)

Pick: **File, Properties** *(from the pull-down menu)*

Pick: (the text box labeled Working Directory and highlight the existing text)

Type: **C:\WORK**

Your dialog box should look like Figure G1.5a. If you have created your *work* directory on another drive, substitute the correct drive letter.

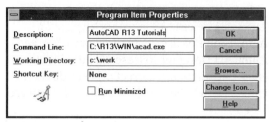

Figure G1.5a Windows 3.1

Pick: **OK** *(to exit the Properties dialog box)*

Setting the Working Directory in Windows 95

To set *c:\work* as your working directory, you will modify the shortcut for the AutoCAD R13 Tutorials icon.

Pick: (the new AutoCAD R13 Tutorials icon)

Pick: **File, Properties** *(from the pull-down menu)*

Pick: (the Shortcut tab at the top of the Properties dialog box)

Pick: (the text box labeled Start In and highlight the existing text)

Type: **C:\WORK**

Your dialog box should look like Figure G1.5b. If you have created your *work* directory on another drive, substitute the correct drive letter.

Figure G1.5b Windows 95

Pick: **OK** *(to exit the Properties dialog box)*

When you start AutoCAD from this icon or shortcut, the settings saved in the files in the directory *c:\work* will be used. When you save drawings to your default directory in AutoCAD, they will be saved to *c:\work*.

■ *TIP* To restore the AutoCAD settings from the original AutoCAD configuration, simply copy *acad.ini* and *acad.cfg* from *c:\r13\win* (or your backup disk or directory) to *c:\work* and choose to overwrite the previous files. ■

Installing Data Files for the Tutorials

The next step of your setup is to create a directory for and install the data files used in the tutorials. Throughout the tutorials, you will be instructed to work with files already prepared for you. You will not be able to fully complete the tutorials without these files. You can download these data files via FTP or with a web browser such as Mosaic. You will create a directory for them and then copy them into that directory.

To download via FTP, connect to *ftp.aw.com/cseng/authors/lockhart/r13* and look for the file called *r13data.exe*. If you have trouble, connect to *ftp.aw.com* and change to the *cseng/authors/lockhart/r13* directory. Use *anonymous* as your user name and your email address as your password when you log on.

To access the files from a web browser, use the URL *http://www.aw.com/cseng/authors/ lockhart/r13* and look for the file *r13data.exe*.

If you do not have access to the Internet, you may request a copy of the files from Addison-Wesley by calling (415) 594-4400, ext. 748, or by sending e-mail to aw.eng@aw.com.

After you have retrieved the files, use Windows to create a new directory, *c:\datafile*. Copy *r13data.exe* into the directory, then run the file to decompress it. To do so in Windows 3.1, highlight the file, then select File, Run from the File Manager pull-down menu. From Windows 95, you may double-click the file name in the Explorer, or use Start, Run and select or enter the file name in the Run dialog box. When the file is decompressed, you will see an assortment of files in the directory, most of which have the *.dwg* extension.

Now you are ready to customize your AutoCAD menus.

Adding AutoVision and Designer Menus

AutoCAD Release 13 is a menu-driven application. The standard AutoCAD menu does not include the AutoVision (AutoVis) and Designer options, because they are add-on applications. AutoCAD Release 13 allows you to configure menus "on the fly"; that is, you can add and delete menu items during any drawing session using the Options, Customize Toolbars selection from the pull-down menu. Menus customized on the fly will revert to the regular (uncustomized) AutoCAD menu the next time you load AutoCAD.

The tutorials in this manual use a menu that has the AutoVision and Designer menus integrated into the AutoCAD menu. To complete the tutorials as they are intended, you should compile your menu so that AutoVision and AutoCAD Designer are loaded whenever you load AutoCAD.

This chapter assumes that you have installed AutoCAD Release 13 into directory *c:\r13*, AutoVision Release 2 into directory *c:\av*, and AutoCAD Designer Release 1.2 into directory *c:\ad*. Substitute the correct drive or directory names in the following instructions if yours are different.

■ *TIP* If you are not sure what version of AutoCAD you are using, launch AutoCAD and pick About AutoCAD from the Help pull-down menu. The About AutoCAD dialog box appears on your screen with the current version of AutoCAD. If you are not sure what version of AutoVision you are using, pick AutoVis, Preferences from the pull-down menu. From the Render Preferences dialog box, pick Information. The dialog box that appears will tell you which version of AutoVision you are running. You may notice a number such as R13_c3 instead of just R13. This indicates the maintenance release in use. Maintenance releases are minor updates to the program. You can get the latest updates on-line in a variety of locations—from forums on America On line and CompuServe, from the Autodesk FTP site at *ftp.autodesk.com,* and on the World Wide Web at *http://www.autodesk.com.* ■

Preloading AutoCAD Designer

To preload AutoCAD Designer so that it is loaded every time you open AutoCAD, you need to add a command to the LISP program that sets up AutoCAD.

Open your text editor or word processor. Open the file *c:\r13\com\support\acadr13.lsp*. Move your cursor to the very end of the file. Make sure that you are at a new line.

Type: **(load "acdload")**

Your file should look like Figure G1.6.

Figure G1.6

Pick: **File, Save** *(from the pull-down menu)*

This will save the changes to *acadr13.lsp* so that it will pre-load AutoCAD Designer each time you start AutoCAD.

Pick: **File, Exit** *(or the equivalent command to exit your application)*

You will return to Windows.

> ■ **TIP** If you choose not to modify *acadr13.lsp* to preload AutoCAD Designer, you can load it during any AutoCAD session by using the Applications command from the Tools menu, or by typing *Appload* at the command prompt and selecting *c:\ad\adesign.arx* to load from the Application Load dialog box. ■

Menu Customization

The AutoVision install procedure automatically adds the AutoVision menu to your AutoCAD menu. You need to add the Designer menu yourself with the Popadd application provided with your AutoCAD Designer software.

■ *Warning:* If you are using a networked system, or a workstation in a classroom or lab setting, the menus may have already been customized for you. Please check with your technical support person before changing the menu configuration. ■

Compiling the Designer Menu

In Windows, select File, Run in the Windows 3.1 Program Manager or Start, Run in Windows 95. Substitute the drive letter and directory names to match your configuration in the instruction line below.

Type: **C:\AD\COM\POPADD C:\AD\WIN\AD_PRTL.MNU C:\R13\WIN\SUPPORT\ACAD.MNU**

Pick: **OK**

Your computer will run the Popadd application in a DOS window and return you to Windows when it is complete.

The file *acad.mnu* is a menu template. AutoCAD generates several menus from the *acad.mnu* file. When AutoCAD launches, it loads the menu from a compiled version of the menu file (*acad.mnc*), a file that contains the bitmaps for the menu (*acad.mnr*), and an ASCII file that is essentially the same as *acad.mnu* (*acad.mns*). If you modify *acad.mnu* you must also modify *acad.mns*, or the changes will not appear in your menu when you launch AutoCAD.

Use Windows to copy *c:\r13\win\support\ acad.mnu* to *c:\r13\win\support\acad.mns* and replace *acad.mns* now.

AutoCAD, AutoVision, and AutoCAD Designer should now be fully integrated for Windows.

AutoCAD Configuration for Tutorials

The tutorials in this manual assume that you are using AutoCAD's default settings and that AutoVision and AutoCAD Designer are properly installed. If your system has been customized for other uses, some of the settings may not match those assumed by the step-by-step instructions in the tutorials. You will now reconfigure your system so that you can work through the tutorials as instructed.

■ **Warning:** If you are working on a networked system or on a workstation in a classroom or lab, your system has already been configured for you. Please skip to Chapter 2, AutoCAD Basics. ■

To start AutoCAD from Windows 3.1, double-click the AutoCAD R13 Tutorials icon. To start AutoCAD from Windows 95, pick Start, Programs, AutoCAD R13, AutoCAD R13 Tutorials.

The AutoCAD graphics window appears on your screen, as shown in Figure G1.7, with the Draw, Modify, Designer Main, and Designer Viewing toolbars on and floating. The AutoVis menu option does not appear and the Designer icons appear as "smiley faces" until you add the correct support directories to the AutoCAD path. (You may want to move the Designer toolbars so that you can pick Options from the pull-down menu. You can move a toolbar by clicking and dragging on it.)

Designer toolbars

Figure G1.7

Pick: **Options, Configure**

An AutoCAD text window appears on your screen, instructing you to configure AutoCAD, as shown in Figure G1.8.

Figure G1.8

To follow the tutorials in this manual you should use the following AutoCAD configuration settings.

Available Video Display

The default video display driver is 'WHIP' – HEIDI (TM) Accelerated Display Driver – Autodesk, Inc. Accept the default settings for configuration and aspect ratio.

The WHIP display driver does not work with some systems. If your system does not appear to be working correctly with the WHIP driver, we recommend that you reconfigure AutoCAD to use the Accelerated Display Driver from Rasterex, or use an independently developed accelerated driver.

■ *Warning:* If you use the Accelerated Display Driver from Rasterex, there are two video display driver configuration options, Display List and GDI Bypass.

With the Display List option enabled, the driver is capable of maintaining the graphics window in a faster and more efficient manner; however, it requires more system memory.

The GDI Bypass option lets AutoCAD interact directly with the Windows display driver. This will increase the display speed when you are doing a redraw. However, since the GDI Bypass option lets AutoCAD communicate directly with the Windows display driver, an incompatibility may arise with some display drivers.

If your system is low on memory or you are experiencing problems running AutoCAD, AutoVision, or AutoCAD Designer, reconfigure your video display driver to disable Display List and GDI Bypass. This may be especially important when you are completing the AutoVision tutorials, Tutorials 9 and 17.

Display List must be enabled in order for you to use the AutoCAD Aerial Viewer in Tutorial 2. ■

Available Digitizer

The default digitizer is the Current System Pointing Device, or your mouse. AutoCAD can also support a tablet. To configure AutoCAD to use a tablet, refer to your AutoCAD documentation.

Available Plotter

The default plotter in AutoCAD is None. If you want to print your AutoCAD drawings, it is easiest to select the System Printer ADI 4.2 - by Autodesk, Inc. and accept the Control Panel default settings. This allows AutoCAD to use the printer you have already set up to work with Windows. If you want to choose one of the other plotter options, refer to your AutoCAD documentation.

If you are using a laser printer, the default orientation is probably portrait. In these tutorials you will be plotting to a landscape orientation. You may want to change the default orientation of your system printer to landscape so you do not have to rotate your plots in AutoCAD. In Windows 3.1 you can do this from the Windows Control Panel; choose the Printers, Setup options. In Windows 95 you can do this from the Start, Settings, Printers option; select your printer and choose the File, Properties, Paper options.

If you do not want to change your system printer to landscape, you can do a detailed plotter configuration in AutoCAD to rotate your plot 90° so that your drawings are plotted landscape. To do so, select Detailed Plotter Configuration. Accept all the defaults, except the Rotate Plot default. Rotate your plot 90°.

If you do not change your printer orientation or configuration, you will need to rotate your plot each time you print in a tutorial.

File-Locking

AutoCAD prevents users from changing the same file at the same time through an internal file-locking mechanism. The login name identifies the owner of a locked file. To use AutoCAD on your personal computer, it is recommended that you do not enable file-locking.

Available Spelling Dialects

AutoCAD Release 13 allows you to spell-check your drawing in American English or British English. The default spelling dialect is American English.

Accepting Configuration Changes

When you have made all your configuration changes, type *0* to exit to the graphics window. You will be asked to confirm that you want to keep the changes to your configuration. You must accept the default Yes response to this question, or you will lose your new configuration. The AutoCAD graphics window appears on your computer screen again.

Dialog Box Settings

There are three AutoCAD system variables that control whether dialog boxes appear when certain commands are issued. Filedia controls the display of dialog boxes for file operations, such as New, Open, and Save. Cmddia controls whether dialog boxes are displayed for Plot and external database commands. Attdia controls whether dialog boxes are displayed with the Insert command. The value for each of these variables (displayed in the angle brackets) should be 1.

Command: **FILEDIA** ⏎

New value for FILEDIA <1>: *(if the default value is 1, press* ⏎*; if the default value is 0, type 1* ⏎*)*

Command: **CMDDIA** ⏎

New value for CMDDIA <1>:*(if the default value is 1, press* ⏎*; if the default value is 0, type 1* ⏎*)*

Command: **ATTDIA** ⏎

New value for ATTDIA <0>: *(if the default value is 1, press* ⏎*; if the default value is 0, type 1* ⏎*)*

As you follow the instructions in the tutorials, you should see a dialog box on your screen every time one is used in the tutorials.

> ■ *TIP* If the Designer menu and toolbars (with "smiley faces") do not appear, make sure you copied *acad.mnu* to *acad.mns* and replaced the previous file. ■

AutoCAD Preferences

You will add the appropriate AutoVision and AutoCAD Designer directories to the AutoCAD path so that AutoCAD can find them when it is loaded in the future.

Pick: **Options, Preferences** *(from the pull-down menu)*

Pick: **Render** *(from the tabs at the top of the Preferences dialog box)*

Pick: (in the text box next to Map Files Path)

Type: **C:\AV\MAPS;C:\AV\TUTORIAL**

> ■ *TIP* If your drive or directory names are different, substitute those paths here and in the next step. ■

Your Preferences dialog box should look like Figure G1.9.

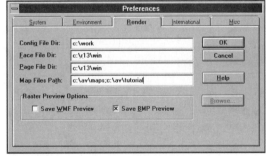

Figure G1.9

Pick: **Environment** *(from the tabs at the top of the Preferences dialog box)*

Position your cursor at the *end* of the text box next to Support by picking in the text box and pressing (END).

Type: **C:\AV\AVWIN;C:\AV\AVIS_SUP;C:\AD;C:\AD\WIN;C:\AD\COM\SUP**

Your Preferences dialog box should now look like Figure G1.10. Pick OK to exit the dialog box when you are done.

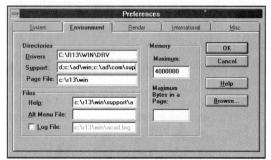

Figure G1.10

■ *TIP* You can check your AutoCAD path at the command prompt at any time. To do this, type *(getvar "acadprefix")* ↵. You will need to type the parentheses and quotation marks exactly. In response to this command your path should read: *C:\\R13\\COM\\SUPPORT\\;C:\\R13\\WIN\\SUPPORT\\;C:\\R13\\WIN\\TUTORIAL\\;C:\\R13\\COM\\FONTS\\;C:\\AV\\AVWIN\\;C:\\AV\\AVIS_SUP\\;C:\\AD\\;C:\\AD\\WIN\\;C:\\AD\\COM\\SUP\\.* ■

Checking Menus

When you launched AutoCAD, its menus were loaded from several files: *acad.mns, acad.mnc,* and *acad.mnr.* Because you had not yet added the correct AutoVision and AutoCAD Designer directories to your support paths, the pull-down menus at the top of your screen do not include the AutoVis and Designer selections. You need to use the New and Menu commands to open a new drawing with all the new

AutoCAD settings and recompile the menus to include AutoVision and the Designer icons using the new preferences you set.

Pick: **New icon**

The Create New Drawing dialog box appears on your screen.

Pick: **OK** *(to open an unnamed drawing)*

This will reload the AutoCAD defaults and force the menu on your screen to compile the new menu, with the AutoVis menu added. You will type the Menu command at the command prompt to force-load *acad.mnu* and make your menu recompile.

Command: **MENU** ↵

In this Select Menu File dialog box that appears, select the file *acad.mnu*. Pick OK to exit the dialog box. You will get a message that says that loading a *.mnu* file overwrites the *.mns* file, and asks if you want to continue loading the *.mnu* file. Pick OK in response to this warning. Once this operation is complete, the Designer icons should appear on the toolbar.

Your menu customization has been successful if the AutoVis and Designer menu options appear, and if the Designer icons have replaced the smiley faces, as shown in Figure G1.11. If the Designer menu and icons have not been incorporated into your new menu, review the Compiling the Designer Menu section and try again. If the AutoVis selection has not been incorporated into your new menu, you may need to manually compile it. Please refer to Manually Compiling the AutoVis Menu, on page G-17.

Figure G1.11

■ *Warning:* If you did a minimum installation of AutoCAD Designer, rather than installing all AutoCAD Designer files, you will not be able to use the Designer Help option. The Designer Help feature can be very useful while you are learning the application. You may want to reinstall all AutoCAD Designer files so that you are able to use Designer Help. ■

■ *TIP* If you select any AutoVision commands, the AutoVision window remains open, using up memory. To close the AutoVision window, pick its Windows Control box and choose Switch To from the pull-down menu. The Task List dialog box opens on your screen. Highlight AutoVision and pick End Task. AutoVision does automatically quit when you exit AutoCAD. ■

Once AutoCAD is configured and its menus compile correctly, you will exit and return to Windows.

Pick: *(the close box in the top left of each Designer toolbar)*

Pick: **File, Exit** *(from the pull-down menu)*

If you are prompted to save changes to the drawing, you can respond No. You have not created anything that you will need again. Your settings will not be changed if you choose not to save the drawing. Proceed to Chapter 2, AutoCAD Basics.

Manually Compiling the AutoVis Menu

If, after you install and configure AutoVision, add the correct AutoVision directories to the Preferences dialog box, and force *acad.mnu* to recompile by creating a new drawing, the AutoVis menu does not appear on your screen, you need to modify your *acad.mnu* file. (If AutoVis appears on your menu, proceed to Chapter 2, AutoCAD Basics.)

Open *c:\av\autovis.mnu* in your text editor or word processor, as shown in Figure G1.12. Copy all the text from *c:\av\autovis.mnu* to the Windows clipboard.

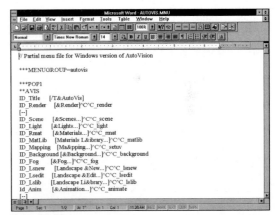

Figure G1.12

Open *c:\r13\win\support\acad.mnu*. Locate ***POP7 (the Help menu), as shown in Figure G1.13. Highlight ***POP7 so you can overwrite it with ***POP8.

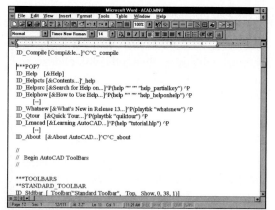

Figure G1.13

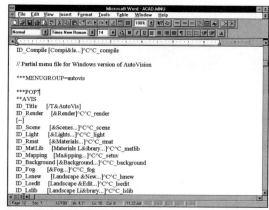

Figure G1.14

Type: ***POP8**

Position your cursor at the beginning of the new ***POP8 you just typed and paste the text from the Windows clipboard into *c:\r13\win\support\acad.mnu*. A section of text starting with *//Partial menu file...* is inserted into the document. Highlight ***POP1 from this new section.

Type: ***POP7**

When you have finished, this portion of your *c:\r13\win\support\acad.mnu* file should look like Figure G1.14.

Save *c:\r13\win\support\acad.mnu* and exit your text editor. Use Windows to copy *c:\r13\win\support\acad.mnu* to *c:\r13\win\support\acad.mns* again. Choose to overwrite the previous file again.

Return to the Checking Menus section above and follow the instructions to force-load your new menu.

You have completed the configuration and setup necessary to complete the tutorials in this manual. Proceed to Chapter 2 to start to learn AutoCAD.

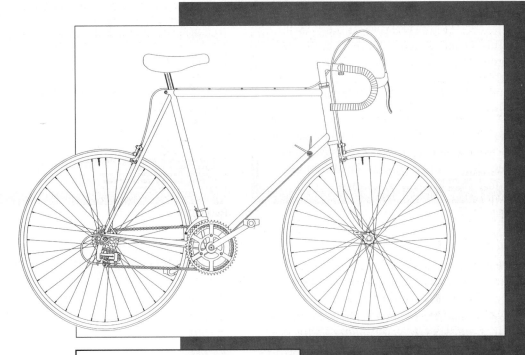

CHAPTER 2

AutoCAD Basics

Objectives

When you have completed this chapter, you will be able to

1. Minimize your AutoCAD drawing session and switch between AutoCAD and other Windows applications.

2. Start a new AutoCAD drawing.

3. Recognize the icons, menus, and commands used in AutoCAD Release 13 for Windows.

4. Use the mouse to pick commands, menu options, and objects.

5. Work with a dialog box.

6. Access AutoCAD's on-line help facility.

7. Save a drawing file.

8. Exit from the graphics window and return to the Windows operating system.

Introduction

In this chapter, you will learn the basics of AutoCAD Release 13's screen display, menus, and on-line help. The chapter also describes the type of instructions you'll encounter in the tutorials.

The tutorials in this manual assume that you are using AutoCAD Release 13 for Windows. AutoCAD should be installed and configured as described in Chapter 1, Preparing AutoCAD for the Tutorials. The tutorials assume you are using AutoCAD's standard configuration with the menus described in Chapter 1.

■ **Warning:** If you are using AutoCAD on a network, ask your technical support person about how the software is configured. Unique changes might have been made that differ from AutoCAD's standard configuration; for example, the program might be under a different directory name or require a special command or password to launch. ■

Loading AutoCAD Release 13 for Windows

Load AutoCAD Release 13 for Windows so the AutoCAD graphics window is displayed on your screen, as shown in Figures G2.1a and G2.1b. (Press ⏎ if necessary to get to the graphics window.) Figure G2.1a shows the AutoCAD screen in Windows 3.1; Figure G2.1b shows the AutoCAD screen in Windows 95. If you need help, refer to Chapter 1.

■ **Warning:** If AutoCAD does not appear, or you see a message stating that AutoCAD must be configured, make sure you have installed and configured AutoCAD properly for your system. ■

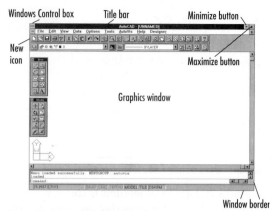

Figure G2.1a Windows 3.1

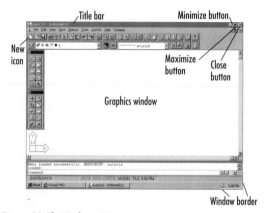

Figure G2.1b Windows 95

Microsoft Windows and AutoCAD

AutoCAD Release 13 for Windows uses many of the same conventions as other applications that run within Windows. This section will identify some of the techniques that you will be using to complete the tutorials in this manual.

The tutorials assume you will be using a mouse or pointing device to work with AutoCAD Release 13 for Windows. If you are unfamiliar with mouse operations, refer to Chapter 1 or your on-line Microsoft Windows tutorial for basic mouse techniques.

Navigating in Windows

Using a mouse is usually easier than using the keyboard, although a combination of mouse use and keyboard shortcuts is the most efficient way to navigate. For information on keyboard shortcuts in Windows, see your on-line Microsoft Windows tutorial.

Basic Elements of an AutoCAD Graphics Window

The AutoCAD *graphics window* is the main workspace. It has elements that are common to all applications written for the Windows environment. These elements are labeled in Figure G2.1 and described below.

- The *Windows Control box* is in the upper-left corner of each window in Windows 3.1. This box is sometimes referred to as the *close box* in the tutorials, because double-clicking it closes the window. Picking it once reveals a menu that you can use to close, move, or resize a window.

  In Windows 95, the close box is a button in the upper-right corner of each window. Picking this button closes the window.

- The *title bar* shows the name of the application (in this case, AutoCAD) followed by the document name (in this case, *unnamed*).

- The *window border* is the outside edge of a window. You can change the window size by moving the cursor over the border until it becomes a double-ended arrow. Holding the mouse button down while you move the mouse (dragging) resizes the window.

- The *maximize* and *minimize buttons* are in the upper-right corner of the window. Clicking the maximize button with the mouse enlarges the active window so that it fills the entire desktop; this is the default condition for AutoCAD. You will learn to use the minimize button in the next section.

Minimizing and Restoring AutoCAD for Windows

There may be occasions when you want to leave AutoCAD temporarily while you are in the middle of a work session, perhaps to access another application. Minimizing AutoCAD allows you to reduce the application to an icon and return to it more quickly than you could if you had to exit and start AutoCAD all over again.

Picking the minimize button reduces the window to an icon and makes the Windows desktop accessible. (You can also choose Maximize and Minimize from the Windows Control menu.)

Pick: (the minimize button)

AutoCAD is reduced to an icon at the bottom of your screen. (In a working AutoCAD session, you should always save your work before minimizing AutoCAD.)

AutoCAD is still running in the background, but other applications are accessible for your use. When you are ready to return to AutoCAD, double-click the minimized AutoCAD program icon, or pick the AutoCAD icon once and choose Restore from the menu that appears. (In Windows 95, only a single click is needed)

Double-click: (the minimized AutoCAD icon)

Microsoft Windows Multitasking Options

Microsoft Windows allows you to have several applications running and switch between them. This is called *multitasking*. You will use multitasking in some of the tutorials in this manual to access your spreadsheet or word processing application while you are still running AutoCAD. To switch between active applications, you press (ALT) (TAB). You will minimize AutoCAD and open another application to see how this works.

Pick: **(the minimize button)**

Pick: **(another Program group icon from the Windows Program Manager in Windows 3.1 or from Start in Windows 95)**

Pick: **(a Program Item icon in Windows 3.1 or an application name in Windows 95)**

Press: (ALT) (TAB) **(2 times)**

A small window appears on your screen with the name of an application in it each time you press (ALT) (TAB). (When you stop typing, your AutoCAD drawing session should be on your screen. If it is not, continue pressing (ALT) (TAB) until it appears.)

Press: (ALT) (TAB)

You should have returned to your other application. (If not, continue typing (ALT) (TAB) until you do.) Quit this application now. You will return to your AutoCAD drawing session automatically if it was the last application you were in. (If you do not see AutoCAD on your screen, press (ALT) (TAB) until you return to AutoCAD.)

■ *Warning:* If you do not have enough free conventional memory, you may get an *out of memory* message. Refer to your Windows documentation for the best multitasking settings and memory management techniques. ■

Opening a New Drawing

When you launch AutoCAD, it opens a new drawing for you, using default settings stored in a template, *acad.dwg*. You can start to draw immediately, then save the file when you have finished.

 Each time you start a new drawing with the New command, you can use this template or another of your choice.

Pick: **New icon**

The Create New Drawing dialog box appears on your screen, as shown in Figure G2.2. The name *acad.dwg* appears in the box to the right of the word Prototype. To begin your new drawing, AutoCAD will use the prototype drawing, or template, *acad.dwg*. If the check box in front of No Prototype is checked, your new drawing will lack even the basic settings that are included in *acad.dwg*. If the No Prototype box is already checked, click in the box again to unselect it. AutoCAD will use the default drawing or settings unless you specify otherwise.

■ *Warning:* If a dialog box does not appear when you pick the New icon, your Filedia system variable may be set to 0, instead of 1. Refer to Chapter 1 to set your dialog box system variables correctly. ■

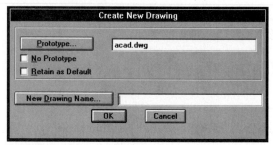

Figure G2.2

You are also prompted for the new drawing name. You can name the drawing now, or later when you save it. When you configured AutoCAD in Chapter 1, you created a working directory called *c:\work*. This is where AutoCAD will save your files. File names cannot be longer than eight characters and cannot contain any spaces. (AutoCAD will automatically add the drawing file suffix *.dwg* to your file name, so you need not type it. All file names in this manual will contain the appropriate suffix for clarity.) You will name this drawing later in this chapter.

Pick: **OK (to exit the Create New Drawing dialog box)**

Your drawing will still be called *unnamed*, and it will have the basic settings provided in *acad.dwg*.

Some tutorials in this manual instruct you to begin from prototype drawings provided with your AutoCAD software. Others use prototype drawings created for this manual. These drawings (and others) are in the *c:\datafile* directory that you created in Chapter 1.

Pointing Techniques in AutoCAD

A drawing is made up of separate elements, called *objects*, that consist of lines, arcs, circles, text, and other elements that you access and draw through AutoCAD's commands and menu options.

The pointing device (assumed in this text to be a mouse) is the most common means of picking commands and menu options, selecting objects, or locating points in AutoCAD.

Picking Commands and Menu Options

The tutorials in this manual assume that you will be using a mouse or pointing device to work with AutoCAD Release 13. The left mouse button is referred to as the *pick button*. The AutoCAD menus let you enter a command simply by pointing to the command and pressing the pick button to choose it. In this way you are instructed to *Pick:* specific commands.

The right mouse button is referred to as the *return button*; it duplicates the action of the ⏎ key when you are selecting objects of repeating commands and options from menus.

Entering Points

You can specify points in a drawing either by typing in the coordinates from the keyboard or by using your mouse to locate the desired points in the graphics window.

When you move the mouse around on the mouse pad or table surface, "crosshairs" (the intersecting vertical and horizontal lines on the screen) follow the motion of the mouse. These crosshairs form the AutoCAD cursor in the graphics window. When you are selecting points during the execution of some commands, the location of the intersection of the crosshairs will be the selected point when you press the pick button. This is how you pick a point. There are some modes in AutoCAD in which the crosshairs change to arrows or boxes with target areas during the execution of certain commands. You will learn about these modes in the tutorials. The cursor on the screen always echoes the motion of the mouse.

Dragging

Many AutoCAD commands permit dynamic specification, or *dragging*, of an image of the object on the screen. You can use your mouse to move an object, rotate it, or scale it graphically. You will learn about drag mode in Tutorials 1 and 2. Once you select an object in drag mode, AutoCAD draws tentative images as you move your pointing device. When you are satisfied with the appearance of the object, press the pick button to confirm it.

Object Selection

Many of AutoCAD's editing commands ask you to select one or more objects for processing. This collection of objects is called the *selection set*. You can use your mouse to add objects to, or remove objects from, the selection set. AutoCAD has various tools and commands you can work with when selecting objects; you use the cursor to point to objects in response to specific prompts. AutoCAD highlights the objects as they are selected. You will learn about the various ways to select objects in the tutorials.

AutoCAD Commands and Menu Options

AutoCAD gives you several options for entering commands. You can select them from the toolbars, pick them from the pull-down menus and from the screen menu, or type them at the command prompt. These elements are labeled in Figure G2.3 although you will not see the screen menu on your screen. (You will learn to display the screen menu in Tutorial 1.) If AutoVision and AutoCAD Designer are not installed, and if they have not been configured as instructed in Chapter 1, your menu will not match Figure G2.3.

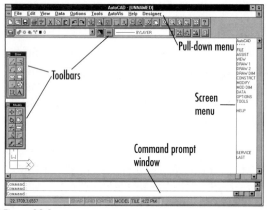

Figure G2.3

How you invoke a command may affect the order and wording of the prompts you see on the screen; these differences offer you options for more efficient use of AutoCAD. Most AutoCAD commands can be typed at the command prompt, but it is usually easier to pick them from a toolbar or a menu. For the purposes of the tutorials in this manual, you

should pick or type commands *exactly as instructed*. The command selection location you should use will be indicated after the command. When you are supposed to type a command it will appear in all capital letters; for example, **NEW**. You must hit the ⏎ key to activate the command. When you are supposed to invoke a command by picking its icon, the complete icon name, including the word *icon*, will appear in the instruction line; for example, **New icon**. When you are supposed to invoke a command by selecting it from the pull-down menu, you will see the list of menu names at the instruction line; for example, **File, New**.

As you progress through the tutorials, you will learn about the benefits of choosing commands in different ways in different situations. Remember, for the purposes of the tutorials in this manual, you should pick or type commands exactly as instructed.

Typing Commands

The command prompt window at the bottom of your screen is one means of interacting with AutoCAD. All commands that you select by any means are echoed there, and AutoCAD responds there with additional prompts that tell you what to do next. *Command:* in this window is a signal that AutoCAD is ready for a command. For commands with text output, you might need a larger command window display area. You can use the scroll arrows at the right of the command prompt window to review the lines that have passed. You can also enlarge the window by positioning the cursor at the border of the window and dragging to enlarge the window so that it shows more lines.

Position the cursor at the border of the command prompt window. Your cursor will change to a double-headed arrow. Hold down the cursor and drag the mouse up to enlarge the command prompt window to show two more lines. Your screen should look like Figure G2.4.

Figure G2.4

You can also review more lines of the command prompt by viewing an AutoCAD text window, as shown in Figure G2.5. You can use the F2 key to toggle to a text window. This will enlarge the command prompt window and move it in front of your AutoCAD graphics window. To return to your graphics window, use the F2 key again or close the text window with the Windows Control box or close box.

Figure G2.5

When you are supposed to type a command it will appear in all capital letters; for example, **LINE**. Watch the command prompt window as you type.

Command: **LINE** ⏎

The Line command is activated when you press ⏎, and the Line command prompt *From point:* is in the command prompt window. This prompt tells you to enter the first point from which a line will be drawn. You will learn about the Line command in Tutorial 1. For now,

Press: ESC

to cancel the command and return to the command prompt.

Using the Toolbars

Toolbars contain icons that represent commands. The Standard, Object Properties, Draw, and Modify toolbars are visible by default, as shown in Figure G2.6. The Standard toolbar contains frequently used commands, such as Redraw, Undo, and Zoom. The advantage of using icons to begin commands is that they are "heads-up"; that is, you do not need to look down to the keyboard to enter the command.

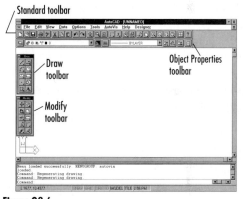

Figure G2.6

You can change the toolbar contents, resize them, and *dock* or *float* them. A docked toolbar, such as the Standard toolbar shown in Figure G2.6, attaches to any edge of the graphics window. A floating toolbar, such as Draw or Modify, can lie anywhere on the application screen, and can be resized. A docked toolbar cannot be resized and doesn't overlap the graphics window. You will learn to click and drag toolbars to reposition them in Tutorial 1.

When pointing to an icon, the cursor changes to an arrow that points up and to the left. You use this arrow to select command icons. When you move the pointing device over an icon, name of the icon appears below the cursor, as shown in Figure G2.7. This is called a *tool tip* and is an easy way to identify icons.

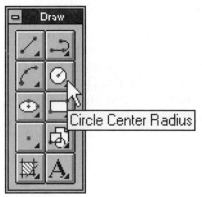

Figure G2.7

Icons with a small black triangle have *flyouts* that contain subcommands, such as the Circle flyout. Hold the pick button down with the cursor over the Circle icon on the Draw toolbar until the flyout appears, as shown in Figure G2.8.

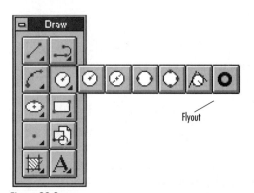

Figure G2.8

As new commands are introduced in the tutorials in this manual, their toolbar icons will appear at the beginning of the paragraph, as the Circle icon does here. Now pick the Circle Center Radius icon from the Circle flyout to see the options that

appear. The Circle Center Radius icon is on the far left of the flyout. It looks like a circle with a radius drawn. As you move your mouse over the icons (you may need to pause), the command names are confirmed by the tool tips that appear below your cursor, and the icon's function appears on the status bar at the bottom of the graphics window; in this case, *Creates a circle using a specified radius.* (Sometimes the tool tip command name is slightly different from the name echoed at the command prompt, or the command that you would type at the command prompt. If you are confused about a specific command name, check the AutoCAD Command Summary at the end of the manual.)

Pick: **Circle Center Radius icon**

You will see the following options appear at the command line:

Command: 3P/2P/TTR/<Center point>:

3P/2P/TTR are command options for the Circle command. Subcommands and command options work properly only when entered in response to the appropriate prompts on the command line. When a subcommand or option includes one or more uppercase letters in its name, it is a signal that you can type those letters at the prompt as a shortcut for the option name; for example, *X* for eXit. If a number appears in the option name, for example 2P for Circle 2 Points, which creates a circle using two endpoints of the diameter, you need to type both the number and the capitalized letter(s).

Backing Up and Backing Out of Commands

When you type a command name or any data in response to a prompt on the command line, the typed characters "wait" until you press the spacebar or ⏎ to instruct AutoCAD to perform the command or enter and act upon the entered data.

If you have not already pressed ⏎ or the spacebar, use the (BACKSPACE) key (generally located above and to the right of the ⏎ key on standard IBM PC keyboards and represented by a long back arrow) to delete one character at a time from the command/prompt line. Pressing (Ctrl)- *H* has the same effect as pressing (BACKSPACE).

Pressing (ESC) terminates the currently active command (if any) and reissues the *Command:* prompt. You can cancel a command at any time: while typing the command name, during command execution, or during any time-consuming process. A short delay may occur before the cancellation takes effect and the prompt *Cancel* confirms the cancellation.

Pressing (ESC) is also useful if you want to cancel a selection process. If you are in the middle of selecting an object, press (ESC) to cancel the selection process and discard the selection set. Any item that was highlighted because you selected it will return to normal.

If you complete a command and the result is not what you had expected or wanted, use the Undo command, or type *U*, to reverse the effect. You will learn more about Undo in Tutorial 1.

You will press (ESC) now to cancel the Circle command.

Command: 3P/2P/TTR/<Center point>: (ESC)

■ **TIP** The tutorials in this manual will instruct you to cancel operations with (ESC); however, if you have used AutoCAD Release 12 and are used to typing (Ctrl)- *C* to cancel a command, you can set your AutoCAD preferences to use AutoCAD Classic Keystrokes. To do this, select Options, Preferences from the pull-down menu to bring up the Preferences dialog box. In the System tab of the Preferences dialog box there is a Keystrokes section. Pick the AutoCAD Classic radio button. Pick OK to exit the Preferences dialog box and save this setting. ■

Repeating Commands

You can press the spacebar or ⏎ at the command prompt to repeat the previous command, regardless of the method you used to enter that command. You can also press the return button on your mouse to repeat the command. Some commands, especially those that prompt you for settings when first invoked, assume default settings when repeated in this manner. You will try this now, and you will use the (ESC) key to cancel the command again.

Press: (SPACEBAR)

The Circle command is invoked again and the prompt appears in the command window. You will cancel it with the (ESC) key.

Command: 3P/2P/TTR/<Center point>: (ESC)

Using Pull-Down Menus

The menu bar and its associated pull-down menus provide another means of executing AutoCAD commands. The menu bar in AutoCAD for Windows operates in the same way as menus in other applications for Windows. When in the menu area, the cursor changes to an arrow. You use this arrow to select a menu.

You open a menu in the menu bar by picking its name. To practice, you will open the Tools menu.

Pick: **Tools (from the menu bar)**

Picking a menu item executes the commands associated with it, opens a submenu of options, or opens a dialog box of options to be used to control the command. A triangle to the right of a menu item indicates that there is a submenu associated with it. You will pick Toolbars to see its submenu.

Pick: **Toolbars (from the Tools menu)**

The Toolbars submenu shown in Figure G2.9 appears next to the Tools menu.

Figure G2.9

A pulled-down menu remains displayed until you

- Pick an item from it
- Pick on the menu name again
- Pull down another menu by picking it from the menu bar
- Remove the menu by picking an unused area of the menu bar
- Pick a point on the graphics window
- Type a character on the keyboard

- Pick an item from a tablet or button menu
- Move your pointing device into the regular screen menu area and click

To remove the Tools and Toolbars menus from the screen,

Pick: **(a point in the graphics window)**

Using the Screen Menu

The screen menu is yet another way you can pick commands. You will learn to display the screen menu and go through an example of how to use it in Tutorial 1. In the figures in this manual, the screen menu will be on. (If your computer does not support a display of $800 \times 600 \times 256$, you may choose not to display the screen menu.)

Working with Dialog Boxes

An ellipsis (...) after a pull-down menu item indicates that it will open a dialog box. Several commands let you set AutoCAD modes or perform operations by checking boxes or filling in fields in a dialog box.

When a dialog box is displayed, the cursor changes to an arrow. You use this arrow to select items from the dialog box.

You will open the Drawing Aids dialog box by picking the Dimension Styles selection from the Data menu.

Pick: **Data, Dimension Styles**

The dialog box shown in Figure G2.10 appears on your screen.

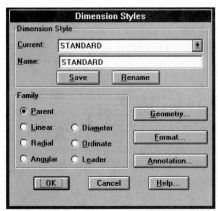

Figure G2.10

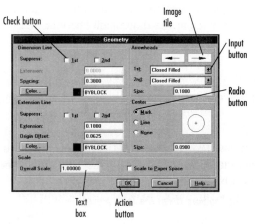

Figure G2.11

Dimension Styles is a typical dialog box. Most dialog boxes have an OK button that confirms the settings or options you have selected in the dialog box. Clicking it is analogous to pressing ⏎ to send a command to AutoCAD. Clicking the Cancel button disregards all changes made in the dialog box and closes the dialog box; it has the same effect as ⎋. You can use the Help button to get more information about a particular command.

■ *Warning:* The term "button" used in conjunction with a dialog box refers to these selection options and should not be confused with the buttons on your mouse or digitizer. ■

Some dialog boxes have sub-boxes that pop up in front of them. When this occurs, you must respond to the "top" dialog box and close it before the underlying one can continue. The Geometry, Format, and Annotation buttons at the right of the dialog box bring up such sub-boxes.

Pick: **Geometry**

The Geometry dialog box shown in Figure G2.11 appears on your screen. Any item in a dialog box that is "grayed out" cannot be picked, such as the Extension option in the Dimension Line section.

Most dialog boxes have several types of buttons that control values or commands in AutoCAD.

Check buttons: A check button is a small rectangle that is either blank or shows an "x." Check buttons control an on/off switch—for example turning Dimension Extension Line Suppression on or off—or control a choice from a set of alternatives—for example determining which modes are on. A blank check button is off.

Radio buttons: A radio button also turns an option on or off. The Dimension Center options are indicated by radio buttons labeled Mark Line and None in this dialog box. A filled-in radio button is on. You can select only one radio button at a time; picking one button automatically turns any other button off.

Action buttons: An action button doesn't control a value but causes an action. The OK action button causes the dialog box to close and all the selected options to go into effect. When an action button is highlighted (outlined with a heavy rule), you can press ⏎ to activate it.

Text boxes: A text box allows you to specify a value, such as the dimension line spacing in Figure G2.11. Picking a text box puts the cursor in it and lets you type values into it or alter values already there. If you enter an invalid value, the OK button has no effect; you must highlight the value, correct it, and select OK again.

Input buttons: An input button chooses among preset options, such as the arrowhead selection in Figure G2.11. Input buttons have a small arrow at the right end. Picking the arrow causes the value area to expand into a menu of options that you can use to select the value for the button.

Image tiles: An image tile displays choices as graphic images (image tiles) rather than words. Picking the arrowhead style image tile causes the next option in the image tile menu to appear.

You can also use the keyboard to move around in dialog boxes. Pressing (TAB) moves among the options in the dialog box. Try this by pressing (TAB) several times now. Once you have highlighted an input button, you can use the arrow keys to move the cursor in text boxes. The (SPACEBAR) toggles options on and off.

Another common feature of dialog boxes is the scroll bar. A dialog box may contain more entries than can be displayed at one time. You use the scroll bars to move (scroll) the items up or down. You will use a scroll bar in the section Accessing On-Line Help.

Most dialog boxes save changes to the current drawing only. Two exceptions to this rule are the Preferences dialog box and the Plot dialog box; these dialog boxes change the way AutoCAD operates and not the appearance of the current drawing itself.

The Cancel button ignores all the selections you have made while this dialog box was open and returns you to the most recent settings. To return to the graphics window without making any changes,

Pick: **Cancel (twice)**

You are returned to the graphics window.

> ■ *TIP* It is a feature of many Windows dialog boxes that you can double-click on the desired selection to select it and exit the dialog box in one action, without having to pick OK. ■

AutoCAD On-Line Help

AutoCAD has a *context-sensitive* on-line help facility. The tutorials in this manual explain the most efficient ways in which you can access a command. The AutoCAD Command Summary at the end of this manual also lists the various locations for a command, as well as the command names that you can enter at the command line. In addition, you can choose AutoCAD's on-line help to find other ways of selecting a command.

Accessing On-Line Help

 You can get help in a number of ways. You can pick the Help icon, type *HELP* or *?* at the command prompt, or press (F1) to bring up the Help window. You will use the Help icon, which looks like a question mark (?), from the Standard toolbar to get AutoCAD's on-line help.

Pick: **Help icon**

The AutoCAD Help window is shown in Figure G2.12.

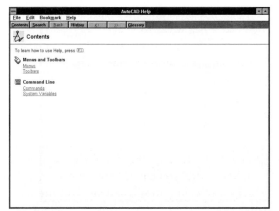

Figure G2.12

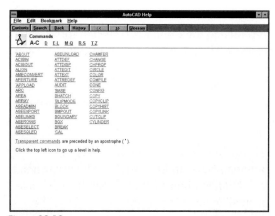

Figure G2.13

The window has a menu bar of its own that you can use to annotate and print on-line help. You are at the Contents screen of the Help window. Help is divided into help information on the menus and toolbars, and information about the command line, which addresses commands and system variables that you can or must type. You can select underlined topics by picking them with the mouse. To get help about a command, click on the underlined word Commands.

Pick: **Commands**

The commands that start with the letters A through C are shown on the screen, as shown in Figure G2.13. To access other commands, select the appropriate letters at the top of the screen to bring up more commands. Pick on the command to display the help for that command.

Pick: **E-L**

Pick: **Erase**

The help screen for the Erase command appears on your screen, as shown in Figure G2.14.

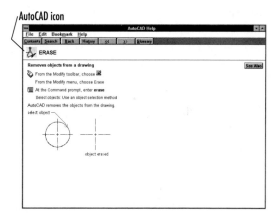

Figure G2.14

To return to the previous screen, you can select Back from the menu bar. To return to the Contents screen of the AutoCAD Help window, select Contents. To return to the menu, select the AutoCAD icon shown in Figure G2.14.

Pick: **Contents**

to return to the Contents screen of the Help window. This time you will pick on the Toolbars selection.

Pick: **Toolbars**

The list of toolbar names appears in the Help window.

Pick: **Draw Toolbar**

Help for the Draw toolbar now appears in the Help window, as shown in Figure G2.15. Picking once on icons that have the triangle symbol causes the additional buttons to fly out. Picking an icon that does not have a flyout symbol, or on a flyout, displays the help information for that item.

Figure G2.15

You can search for topics while you are in AutoCAD's on-line help facility. This is convenient if you do not know the name of the command, or the menu or toolbar on which the command appears. You will select Search from the menu bar to locate an item for which you need help.

Pick: **Search**

A dialog box with an alphabetical list of AutoCAD commands and a scroll bar appears, as shown in Figure G2.16.

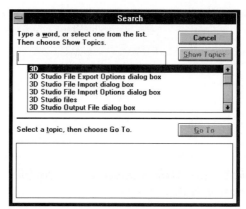

Figure G2.16

Using a Scroll Bar

A scroll bar allows you to access items on lists that are too long to be fully displayed in the box. Pick the up or down arrow to scroll in the desired direction. A slider box is located between the up and down arrows on the scroll bar. You can pick the slider box and drag it to move the entries up and down; with long lists this is often faster than using the arrows.

Use the up and down arrows now to move through the list. Then place the cursor on the slider box and click and drag it up and down to move more quickly through the list. To get information about the Circle command, you can pick it from the list or type the command in the box above the list of topics.

Type: **Circle**

Notice that as you type the word you are moved down the Help index. When you type c you are moved to the command *CAL*. Once you have typed *ci*, you are already at Circle in the Help index (you need not type the remainder of the word).

Pick: **Show Topics**

Your choices will show up in the bottom portion of the Search dialog box.

Pick: **Circle Command**

Pick: **Go To**

AutoCAD displays a text screen of information about the Circle command. If there is additional information that will not fit on the screen, use the scroll bar to move through it.

You will use the Windows Control box to close the Help window and return to the AutoCAD graphics window.

Double-click: **(the Windows Control box in the top left corner of the Help window)**

Transparent Help

You can easily get help while you are in the middle of any command using AutoCAD's transparent Help command. A transparent command is one that you can execute during another command. In AutoCAD you can pick transparent commands from an icon or button. If you type these commands, they must be preceded by an apostrophe ('). If you pick the Help icon from the Standard toolbar, type '? or 'HELP, or press (F1) during a command, you will enter the Help window, which shows help for the command you are currently using; this eliminates the need for searching a topic or navigating through the pages of the Help menu. You will begin the Line command, then type the transparent Help command to bring up the Help window for the Line command.

Command: **LINE** (↵)

From point: **'?** (↵)

The AutoCAD on-line Help window for the Line command opens. Double-click the Windows Control box to close the Help window.

Notice the message in the Command window: *Resuming LINE command.* The command is still active. Press the (ESC) key to cancel the Line command.

AutoCAD's on-line help is a great resource for learning this powerful software. You should use help whenever you want more information about commands and options as you complete the tutorials.

■ *TIP* For information on what's new in AutoCAD Release 13, you can select What's New in Release 13 from the Help menu. ■

Working with Documents

All the documents you will be instructed to use are located in the directories *c:\datafile* and *c:\work*, which you created in Chapter 1. The best procedure is to save all files on your hard drive, because it has more storage space and is faster than accessing a floppy disk drive. Saving directly from AutoCAD to your floppy disk can result in unrecoverable, corrupted drawing files.

AutoCAD Release 13 for Windows allows only one drawing file to be open at a time. If you have a drawing open when you select the Open or New commands, AutoCAD automatically closes the current drawing file before opening the new one. (If you have not saved your drawing, you will be prompted to save changes before you close the document.)

Using Dialog Boxes to Locate and Manage Files

AutoCAD offers many ways to manage files without exiting the program. You can use the dialog box that appears at the Open command to find a file if you cannot remember the directory or drive where it is located. In the Open dialog box, the Find File button will bring up the Browse/Search dialog box. You use this dialog box to select the drives and directories you want to search for your file name or file type. All

AutoCAD drawing files have the file suffix *.dwg*. You will use other file types in this manual; some of these are text files (*.txt*) and some are bitmap files (*.bmp*). You can use the wildcard characters asterisk (*) and question mark (?) to find and manage files as well. The asterisk wildcard character matches any string, including a null string. A search including *.dwg*, for example, would call up any drawing files in the directory being searched, as shown in Figure G2.17. (If you perform the search now, your computer will find no files because you have not created any drawings in your work directory yet.) The question mark, when used as a wildcard, matches any single character.

Figure G2.17

You can use the File Utilities dialog box to list the contents of a directory or delete, rename, or copy files. To access the File Utilities dialog box, select File, Management, Utilities.

Exiting AutoCAD

When you have finished an AutoCAD session, you will exit the program by choosing File, Exit. Before you exit this session, return your command prompt window to its original three-line display so that it will match the figures in the tutorials.

Pick: **File, Exit** *(from the pull-down menu)*

Because quitting AutoCAD unintentionally could cause the loss of a lengthy editing session, AutoCAD prompts you to save the changes if you want to. An AutoCAD dialog box appears that contains a *Save Changes to UNNAMED?* message and gives you three options: *Yes*, *No*, and *Cancel*.

Pick: **Yes**

The Save Drawing As dialog box appears, as shown in Figure G2.18. (Your dialog box will not list any files because you have not saved anything to your work directory yet.)

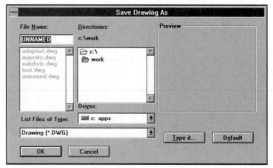

Figure G2.18

This dialog box is similar to other Windows dialog boxes for managing files. Along the right side you see a list of directories available on your system. By double-clicking on a directory, you can open it and select it as the place to store your file.

When you installed AutoCAD you created a working directory called *c:\work*, in which you will save your files. This directory should be open now. If it is not, refer to the Setting the Working Directory section of Chapter 1 to make *c:\work* your working directory. This change will eliminate the need to select the correct directory every time you create, open, or save a drawing file.

If you want to save your file to a disk, below the directory list you will see a text box labeled Drives; pressing the down arrow will display the drives available on your machine. Saving directly from AutoCAD to your floppy disk is a poor practice. However, if you have named your drawing so that it is on a floppy drive, such as drive *a:*, make sure to exit the AutoCAD program before you remove your disk from the drive. Always leave the same disk in the drive the entire time you are in AutoCAD. Otherwise, you can end up with part of your drawing not saved, which will result in an unrecoverable, corrupted drawing file. The best procedure is to create files and save on your hard drive.

Now you need to give your file a name. Your file name cannot be longer than eight characters and cannot contain any spaces. (AutoCAD will automatically add the drawing file suffix *.dwg* to your file name, so you need not type it. All file names in this manual will contain the appropriate suffix for clarity.) Since you have been practicing AutoCAD basics, name the file *basics.dwg*. The file name *unnamed* should be

highlighted in the File Name text box. (If it is not, click in the File Name text box and drag to highlight all the text.)

Type: **BASICS**

Pick: **OK**

Your file is saved, and AutoCAD automatically returns you to Windows.

> ■ *TIP* If you did not want to save your drawing, you would have picked No. If you have made some irreversible error and want to discard all changes made in an AutoCAD session, you can choose Exit from the File menu (or type *QUIT* at the command prompt) and select No so the changes are not saved to your file. ■

You are now ready to complete the tutorials in this manual.

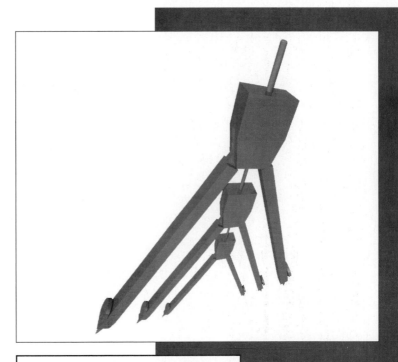

Tutorials

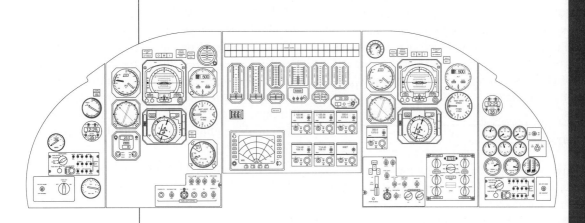

Introduction to AutoCAD

Objectives

When you have completed this tutorial, you will be able to

1. Select commands from the AutoCAD menus.

2. Use AutoCAD's toolbars.

3. Create a drawing file using AutoCAD.

4. Use AutoCAD's Help command.

5. Enter coordinates.

6. Draw lines, circles, and rectangles.

7. Erase drawing objects.

8. Select drawing objects, using implied Window and Crossing.

9. Add text to a drawing and edit it.

10. Save a drawing and transfer it from one drive to another.

Introduction

This tutorial introduces the fundamental operating procedures and drawing tools of AutoCAD. It explains how to create a new drawing, draw lines, circles, and rectangles, and name a new drawing and save it. You will learn how to erase items and select groups of objects. This tutorial also explains how to add text to a drawing.

Starting

Before you begin, launch Windows and AutoCAD Release 13 for Windows. If you need assistance in loading AutoCAD, refer to the Getting Started section of this manual.

The AutoCAD Screen

Your computer display screen should look similar to Figure 1.1, which shows AutoCAD's drawing editor. When you are first starting out, it is important to familiarize yourself with the major screen areas.

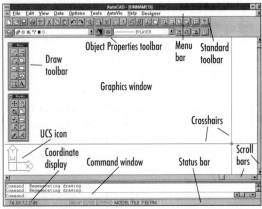

Figure 1.1

■ *Warning:* If your screen does not look similar to Figure 1.1, be sure that you have configured

AutoCAD properly for your computer system. Refer to the Getting Started portion of this manual. ■

The Graphics Window: The *graphics window* is the central part of the screen, which you use to create and display drawings.

The Graphics Cursor or Crosshairs: The *cursor*, or *crosshairs*, shows the location of your pointing device in the graphics window. You use the graphics cursor to draw, select, or pick menu items. The cursor can look different, depending on the command or option selected.

The Command Window: The *command window* is located at the bottom of the screen. It is a called a *floating* window because you can move it anywhere on the screen. You can also resize it to show more or fewer command lines. Notice that the command window also has scroll bars so that you can scroll to see commands that are not visible in the active area. Pay close attention to the command window because this is where AutoCAD prompts you when you need to enter information or make selections.

The User Coordinate System (UCS) Icon: The *UCS icon* helps you keep track of the current X, Y, Z coordinate system that you are using and the direction from which the coordinates are being viewed in 3D drawings.

The Toolbars: The *toolbars* are groups of *buttons* that let you select commands. The *Standard toolbar* described below is one of the many toolbars that are available. The toolbars allow you to pick commands quickly because you do not have to pick down through a menu structure and because they are *heads-up*; in other words, you are looking at the monitor the entire time you are using them. This usually increases the speed with which you can select commands. When the cursor is positioned over a button, its *tool tip* appears. A tool tip is text that describes the command invoked by the button. This is useful if you are unsure what the icon on the button represents. A help line for the button also appears at the bottom of the command window.

Toolbars can be floating or *docked* to the edge of the graphics window. You can float a toolbar around on your screen by picking on its title bar and holding the pick button down as you *drag* the toolbar to its new location. If you drag a toolbar near the edge of the window, the toolbar docks to the edge of the graphics window. The *title bar* at the top of a floating toolbar helps identify it. When a toolbar is docked, its title bar does not appear. You can also change the shape of a toolbar. To do this, position the arrow cursor near the edge of the toolbar until the cursor changes to a double-headed arrow. While the cursor has this appearance, pick on the edge of the toolbar and, keeping the pick button depressed, drag the edge of the toolbar to reshape it. You can customize the toolbars to show the commands you use most frequently and create new toolbars of your own. You will learn to do this in Tutorial 8.

Items on the toolbar can also be *flyouts*, where additional buttons are located. Figure 1.2 shows the floating Draw toolbar and the Polygon flyout. A flyout is identified by a small triangle in the button's lower right corner. To see the buttons that are located on a flyout, position the arrow cursor over the button that shows the flyout triangle and hold the pick button down. After you have held it down for a moment, you will see the additional buttons fly out. When a command is selected from a flyout, that selection becomes the top button shown on the flyout. At first you may find this confusing. Once you are familiar with the flyouts, it is a helpful feature because the most recently used command appears as the top button on the flyout, making it easy to select again.

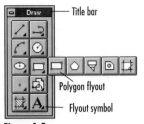

Figure 1.2

The Standard Toolbar: The top of the screen displays the Standard toolbar. The Standard toolbar shows buttons for frequently used commands. Move the graphics cursor up into the Standard toolbar area. Notice that the graphics cursor changes from the crosshairs to an arrow once it moves out of the graphics window. Position the arrow cursor over an item on the Standard toolbar, but do not press the pick button. After the arrow cursor remains over a button for a few seconds, the button's tool tip appears below it. Move the arrow cursor over each button in turn to familiarize yourself with the buttons on the Standard toolbar. If you are unsure what command is represented by an icon, use the help line at the bottom of the command window to help identify it. The Standard toolbar can float anywhere on the screen. Its default location is docked near the top of the screen.

The Object Properties Toolbar: The *Object Properties toolbar* contains tools to help manipulate the properties, such as color, linetype, and layer, of the graphical objects you create. On the Object Properties toolbar, from left to right, the icons are: Layer, Layer Control, Color Control, Linetype, Linetype Control, Object Creation, Multiline Style, Properties, and List. You can see the layer name 0 in the Layer Control display. The appearance of the Color Control button (it looks like several layers of paper) and the word BYLAYER in the Linetype Control display show that for newly created drawing objects these properties will be determined by the current layer. You will work more with these tools in Tutorial 2.

The Status Bar: The *status bar* at the bottom of the screen shows important settings and modes that may be in effect. The coordinate reference of the crosshairs location appears in a box at the left of the status bar as two numbers with the general form X.XXXX, Y.YYYY. The specific numbers displayed on your screen tell you the location of your crosshairs. The words SNAP, GRID, ORTHO, MODEL, and TILE that appear to the right of the coordinates are buttons that you can double-click to turn these special modes on and off. When the Snap, Grid, Ortho, or Tile modes are in effect, their buttons are *highlighted*; when they are turned off, their buttons are grayed out. The remaining button, MODEL, lets you quickly switch between model space and paper space, which you will learn to use in Tutorial 5. When you select paper space, the word PAPER will replace the word MODEL on this button. The current time set on your computer shows at the right of the status bar.

The Menu Bar: The row of words at the top of the screen is called the *menu bar*. The menu bar allows you to select commands. You pick commands by moving the cursor to the desired menu heading and pressing the pick button on your pointing device to pop down the available choices. To select an item from the selections that appear, position the cursor over the item and press the pick button. The menu bar shows the following major headings:

File	used to open existing files, create new files, save, import, export, recover, and perform other file operations
Edit	contains editing commands for cutting and pasting operations similar to other Windows applications
View	lets you select commands that manipulate the view of your drawing on the screen
Data	allows you to select commands to set color, linetype, text style, and other properties for new objects you create
Options	contains commands to set up various drawing parameters and system variables and control the appearance and configuration of AutoCAD
Tools	used to turn on and off and customize the toolbars, and to use the spell checker and the calculator, among other tools
Help	finds help for commands, toolbars, and variables

If you configured AutoCAD as directed in the Getting Started section of this manual, you will also see the items AutoVis and Designer appearing on your menu bar. These are specialized applications that run inside AutoCAD to provide additional capabilities. If you have not installed these applications, you will not see them on the menu bar. Refer to the Getting Started section of this manual for additional information about how to load these menu items.

Canceling Commands

You can easily cancel commands by pressing (ESC). If you make a selection by mistake, press (ESC) to cancel it. Sometimes you may have to press (ESC) twice to cancel a command, depending on where you are in the command sequence.

Picking Commands from the Menu Bar

Use your mouse to move the cursor to the menu bar.

Pick: **Tools**

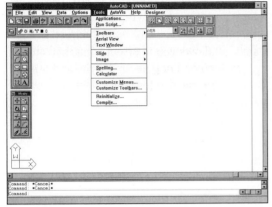

Figure 1.3

Figure 1.3 shows the Tools selection pulled down from the menu bar. Menu bar items with a triangle after the name activate another menu when picked. Items on the menu with three dots after them cause a dialog box to appear on the screen when picked. Notice that as you move the cursor along the selections of the menu, the status bar at the bottom of your screen shows a help line describing the menu selection that is currently highlighted.

■ *TIP* If you are unsure which selection from the menu bar contains the command you want to use, move the cursor to the menu bar area and hold down the pick button. Keeping the button down, move from menu to menu as you check for the location of the command. ■

Press: (the pick button and hold it down while moving down the Tools menu)

To remove the menu from the screen,

Pick: (in the graphics window)

Using the Preferences Dialog Box

A screen menu is also available. Before you can use it, you must turn it on. You will select Options from the menu bar and then Preferences from the list of items that appear, as shown in Figure 1.4.

Pick: **Options, Preferences**

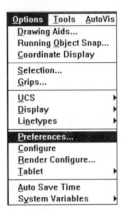

Figure 1.4

The Preferences dialog box appears on your screen, as shown in Figure 1.5.

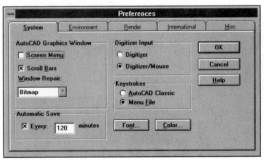

Figure 1.5

The Preferences dialog box uses the Windows interface to set up the appearance of AutoCAD's screen, environment variables, rendering environment variables, international appearance for drawings, and miscellaneous items such as how many lines are retained in the command window and which text editor is used. When you make changes to the Preferences dialog box and pick OK, AutoCAD starts using the changes in your current drawing session. These changes may also be saved in AutoCAD's Windows configuration file, *acad.ini*, so that the next time you start a drawing the changes will be retained. Make sure that you keep track of the changes that you are making so that you can change them back if you are not happy with the result.

■ *TIP* You can make backup copies of the files *acad.ini* and *acadenv.ini* before making changes so that you can easily restore the previous appearance of AutoCAD. You can also have multiple configurations and add different icons for them to the AutoCAD Windows group file so that you can easily start AutoCAD with several different appearances. You can read more about this in the Getting Started section of this manual. When you are starting out, you may not feel ready to customize AutoCAD, but after you have completed the tutorials in this book, you may want to work more with saving your custom configurations. ■

The Preferences dialog box looks like a stack of five index cards, labeled System, Environment, Render, International, and Misc. The System card appears to be at the top of the stack. Next you will look through the stack of cards to familiarize yourself with the selections you can make. To change from one card to the

next, you will pick on the index name at the top of the card. The card that you select will then appear at the top of the stack.

■ *TIP* You can also select from dialog boxes by typing the underscored letter appearing in the option name. For instance, you could type *E* in order to select Environment. When you are at a point in a dialog box where you could also enter text (say into an empty text box), you can press ⒶⓁⓉ and then the underlined letter to make a selection. ■

Pick: **Environment**

The Environment card appears at the top of the stack, as shown in Figure 1.6.

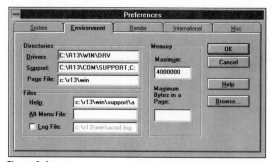

Figure 1.6

The Environment card lets you set the path that tells AutoCAD where to search for files. You can also use it to set the maximum amount of memory that AutoCAD will use in order to make it more compatible with other applications. If you are using a digitizer, the selection for Alt Menu File allows you to specify a menu file that can be swapped with the standard tablet menu. You will leave these settings as they are and look at the next card without making any changes at this time. (If you have made changes to the Environment card, pick Cancel to discard the changes and then select the Preferences dialog box again.)

Pick: **Render**

The Render preferences card is similar to the Environment card, but sets up the path for rendering. If you have installed your software as directed in Getting Started, the correct paths will be set for the AutoVision software to work properly. Again, don't change anything at this time.

Pick: **International**

The items under International let you select whether you want your initial settings to be metric or English units. Don't make any changes to this area for now.

Pick: **Misc**

You should now see the card for Misc preferences appearing at the top of the stack of cards. You can use it to select the text editor that you want to use with AutoCAD's Multiline text feature. You can also set up the appearance of the command window and the number of scroll lines it contains. Once again you will not make any changes.

Pick: **System**

Now the System card should be at the top of the pile once again. You will use it to turn on the screen menu.

Activating and Using the Screen Menu

Pick: **(the check box to the left of Screen Menu)**
Pick: **OK (to exit the Preferences dialog box)**

The screen menu appears as a column of words along the right-hand side of the screen, as shown in Figure 1.7.

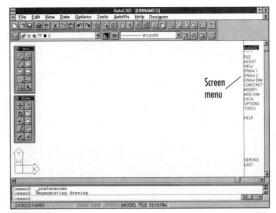

Figure 1.7

To select an item, position the cursor over the name of the item, just as you did when using the menu bar. The items that you see on the screen in capital letters are the names of menus. When you select a menu name, a new list of items replaces the previous list of words in the right-hand column. You will select Modify from the screen menu.

Pick: **Modify**

Notice the new items that appear on the screen menu, as shown in Figure 1.8. These are the names of commands that modify objects in your drawing. Items that have a colon after them are commands and not menus. Locate the Erase command on the screen menu. (Depending on the resolution of your screen, you may not be able to see Erase on the list of commands. If you want to change your screen display resolution, refer to the Getting Started section of this manual.)

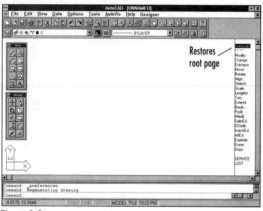

Figure 1.8

Next you will return to the *root page* of the menu. The word AutoCAD at the top of the screen menu has a special function. You can pick it to bring back the original screen menu selections (called the root page of the menu and shown in Figure 1.7). Picking AutoCAD will also cancel any current command. This function is very helpful when you are first learning AutoCAD. If you are lost or confused, picking AutoCAD will allow you to start with the root page of the menu again. The items SERVICE and LAST always appear at the very bottom of the screen menu. Picking SERVICE displays options you can use within commands.

Picking LAST returns the last screen menu page you were using. Next try picking AutoCAD from the screen menu to return to the root page.

Pick: **AutoCAD**

The screen menu is shown in these tutorials because it serves as a reference when you are learning AutoCAD. When you pick a command, its options appear on the screen menu, which helps you to learn them. The screen menu is not the fastest way to select commands, so you will use the toolbars and type commands during the tutorials.

Typing Commands

You can type commands directly at the *command prompt* in the command window. To do this you must type the exact command name. Remember that many of the words on the menu bar are menu names, not commands. Only the actual command name, not menu names, can be typed to activate a command. You will type the command *Line* in the command prompt area. You can type all capitals, capitals and lowercase, or all lowercase letters; AutoCAD is not case-sensitive.

Command: **LINE** ⏎
From point: (ESC)

Pressing (ESC) cancels the command. You will learn more about the Line command later in this tutorial.

> ■ **TIP** If you are used to typing (Ctrl)-C to cancel a command, as you did in AutoCAD Release 12, you can use the Preferences dialog box to turn on AutoCAD Classic keystrokes. ■

Typing a Command Option Letter

If, after you have selected a command, a number of *options* appear at the command prompt, you can type the letter or letters that are capitalized and then press (↵) to select the option you want. Options usually also appear on the screen menu, and you can highlight and pick them using your pointing device. On the screen menu, you can tell options from command names because options appear in capitals and lowercase and are not followed by a colon (:). Remember that you must first select a command before you can choose an option. Selecting an option from the screen menu does not automatically start the command with which it is associated.

Command Aliasing

You can type commands quickly through *command aliasing*. You can give any command a shorter name called an *alias*. Some sample commands already have aliases assigned to them to help you get started. These are stored in the file *acad.pgp*, which is part of the AutoCAD software, usually found in the directory *\r13\com\support*. You can create your own command aliases by editing the file *acad.pgp* with a text editor and adding lines that give the new, shorter names. (These lines take the form ALIAS, *COMMAND in the *acad.pgp* file.) After doing this, you can use the shortened name at

the AutoCAD command prompt. The following commands are some that already have aliases created for them. When you need one of the following commands, you can type its alias at the command prompt instead of the entire command name.

Alias	Command
A	Arc
C	Circle
CP	Copy
DV	Dview
E	Erase
L	Line
LA	Layer
LT	Linetype
M	Move
MS	Mspace
P	Pan
PS	Pspace
PL	Pline
R	Redraw
T	Mtext
Z	Zoom

The Command Summary found at the end of this book includes where to find the command, its icon, the actual command name you can type to start the command, and its alias, if there is one.

AutoCAD provides many different ways to select any command. After you have worked through the tutorials in this book, decide which methods work best for you. When you are working through the tutorials, however, it is important to select the commands from the locations specified, because the subsequent command prompts and options may differ, depending on how you selected the command.

Starting a New Drawing

When you start AutoCAD, it opens a new, unnamed drawing file. You can begin drawing in this file (which is the file you now have open), and save the drawing to a file name later. Or you can start a new file at any time. When you do, AutoCAD closes the current file and opens the one you specify. You will start a new file now and specify its file name. You will select the New command by picking File from the menu bar and then picking New from the list that pulls down.

Pick: **File, New**

You will see the message *Save Changes to UNNAMED?* as AutoCAD recognizes that changes have been made to the file that is open. Since you do not want to save this file,

Pick: **No**

The Create New Drawing dialog box appears on the screen as shown in Figure 1.9.

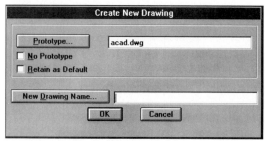

Figure 1.9

The Create New Drawing dialog box lets you start and name a new drawing. You can begin the new drawing from scratch or from a *prototype drawing*. You can use any drawing that has already been created as a prototype drawing. A prototype drawing can contain graphical *objects* or just basic settings. (You will learn more about creating and using prototype drawings in Tutorial 5.) A drawing called *acad.dwg* is the *default* prototype drawing provided with the AutoCAD software. A default

drawing or setting is the one that will be used unless you specify otherwise. The name *acad.dwg* appears in the box to the right of the word Prototype. For your new drawing, you will use the default prototype drawing *acad.dwg*. Make sure that the No Prototype box is turned off (i.e., the check box should be empty). If you start from no prototype, your drawing will lack even the basic settings that are included in *acad.dwg*. If the No Prototype box is already checked, click in the box again to unselect it. Make sure the prototype drawing name, *acad.dwg*, appears in the box to the right of Prototype.

Naming Drawing Files

The drawing name *shapes.dwg* will serve as a good name for this sample drawing. If you have not made any changes, the *typing cursor* will already be located in the empty text box for you to begin typing the file name. If not, move the cursor to the box to the right of New Drawing Name and press the pick button on your mouse so that the typing cursor appears in the empty box.

Type: **SHAPES**

AutoCAD follows the DOS rules for naming a file. Drawing file names can consist of 1 to 8 characters, followed by a *file extension*. The file extension is composed of a period, followed by 1 to 3 characters that help identify the file. All AutoCAD drawing files are automatically assigned the file extension *.dwg*. You don't usually need to type the *.dwg* drawing extension with the drawing name; AutoCAD automatically adds it when the drawing is opened or saved.

Drawing file names can contain letters, numbers, and certain other characters, such as underscores (_) and dashes (-). Upper- and lowercase letters are treated the same by AutoCAD. Most punctuation, such as commas (,), periods (.), and number signs (#), are not allowed in file names. No spaces are allowed in a file name. The tutorials in this book use drawing names that are descriptive of the drawing being created. For example, the drawing you will create for this first tutorial consists of basic shapes. Descriptive names make it easier to recognize completed drawings.

Drawing names can include a drive and directory specification. If no drive is specified, the drawing is saved on the computer's hard drive (usually drive *c:*) in the *default directory*. In the Getting Started section of this manual, you should have already created a default working directory for AutoCAD called *c:\work* where you will save your drawing files. If you have not done so, you may want to review the Getting Started section. The item New Drawing Name displays the dialog box shown in Figure 1.10, which you can use for naming your file and selecting the drive and directory. (It can also be used to confirm the current drive and directory.)

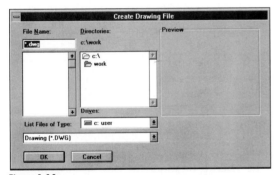

Figure 1.10

If a drawing is to be saved on a disk in drive *a:*, the drawing name *a:shapes.dwg* will automatically send the drawing file to the A drive. Saving directly from AutoCAD to your floppy disk is a poor practice. However, if you have named your drawing so that it is on a floppy drive, such as drive *a:*, make sure to exit the AutoCAD program before you remove your disk from the drive. Always leave the same

disk in the drive the entire time you are in AutoCAD. Otherwise, you can end up with part of your drawing not saved, which will result in an unrecoverable, corrupted drawing file.

> ■ *TIP* If you cannot open your drawing for use with the AutoCAD program because it is corrupted, you may be able to recover it with the selection File, Management, Recover from the menu bar. ■

Pick: **OK**

This returns you to the drawing editor, where you can begin work on your drawing called *shapes.dwg*. Unless you are using a different drive or working directory on your system, the entire default name of your drawing is *c:\work\shapes.dwg*. Notice that the name *shapes.dwg* now appears in the title bar area as the current file name.

■ *Warning:* It is suggested that you initially store all drawings on the hard drive (usually drive *c:*). As a drawing is being created, more and more data is being stored in the drawing file. If the file size exceeds the remaining capacity of a disk (say a floppy in drive *a:*), a **FATAL ERROR** may occur, resulting in the loss of the entire drawing file. To prevent problems with files that may become too large for a floppy, create all drawings on the hard drive and then transfer them to a floppy disk for storage after you exit AutoCAD. After completing the drawing for this tutorial, you will learn how to transfer drawings from the hard drive to a floppy disk. ■

Using Grid

It is often helpful to add a *grid* to the graphics window to act as a reference for your drawing. A grid is a background area in the graphics window covered with regularly spaced dots. The grid does not show up in your drawing when you print. You will add a grid background to the graphics window by typing the Grid command at the prompt.

Command: **GRID** ⏎

The grid spacing prompt displays the options available for the command. The default option, in this case <0.0000>, appears within *angle brackets*. Pressing ⏎ on your pointing device accepts the default. To set the size of a grid, you will type in the numerical value of the spacing and press ⏎.

Grid spacing(X) or ON/OFF/Snap/Aspect <0.0000>: **.5** ⏎

■ *TIP* If the chosen grid size is too small, so that it would effectively fill your graphics window with a solid pattern of grid dots, the error message *Grid too dense to display* appears. If this happens, restart the Grid command and change to a larger value. (You can simply press ⏎ to restart any command.) If the grid values are too large, nothing appears on the screen, because the grid is beyond the visual limits of the screen. ■

Your screen should look similar to Figure 1.11, which shows a background grid drawn with .5-unit spacing. Notice that GRID now shows clearly on the status bar to let you know that the grid is turned on.

Figure 1.11

Using Snap

The Snap command restricts the movement of the cursor to make it easier to select points with a specified spacing. When Snap is on, your cursor jumps to the specified intervals. Snap is a very helpful feature when you are locating points or distances on a drawing. If each jump on the snap interval is exactly .5 units (horizontally or vertically only, not diagonally), you could draw a line 2.00 units long by moving the cursor four snap intervals. You will activate Snap by typing the command.

Command: **SNAP** ⏎

It is useful to align the snap function with the grid size so the snap spacing is the same as the grid spacing, or at half the grid spacing, or some other even fraction. Since the grid spacing is .5, you will specify a numerical value of .5 at the prompt to align the snap function with the grid.

Snap spacing or ON/OFF/Aspect/Rotate/Style <1.0000>: **.5** ⏎

Notice that SNAP becomes highlighted on the status bar to let you know that Snap mode is in effect.

> ■ **TIP** If you set the snap spacing first, leaving the grid spacing at 0.0000, and then activate the grid, the grid spacing is automatically set to the same value as the snap spacing. ■

Move the crosshairs around on the screen. Notice that instead of the smooth movement that you saw previously, the crosshairs jump or "snap" from point to point on the snap spacing.

Grid and Snap Toggles

You can quickly turn the grid and snap on and off by double-clicking their respective buttons on the status bar. If their buttons on the status bar appear in black type, they are on; if in gray type, they are off. Each button on the status bar acts as a *toggle*: that is, when it is double-clicked once, the function is turned on; when it is double-clicked a second time, the function is turned off. You can toggle grid and snap on and off during other commands.

> ■ **Warning:** The buttons on the status bar will not function correctly when there is not enough free *conventional memory* on your system. If you have problems selecting Grid, Snap, and other toggles using the status bar, substitute using the function key or typing the command name instead. Grid can be toggled on and off by pressing F7 and Snap can be toggled on and off by pressing F9. ■

Double-click: SNAP button

Move the crosshairs around on the screen. Notice that they have been released from the snap constraint. To turn Snap on once again,

Double-click: SNAP button

The crosshairs jump from snap location to snap location again.

Double-click: GRID button

The grid disappears from your screen, but the snap constraint remains. To show the grid again,

Double-click: GRID button

Notice how the command window keeps a visual record of all the commands.

> ■ **TIP** Pressing Ctrl-G also toggles the grid on and off; Ctrl-B toggles the snap function. ■

Using Line

The Line command draws straight lines between endpoints that you specify. The Line command is located on the Draw toolbar. (AutoCAD arranges toolbars and menus according to general spoken English syntax. This is helpful when you are trying to remember where commands are located. For example, you might say, "I want to *draw* a *line*.")

For the next sequence of commands, you will use the Draw toolbar to pick the Line command. The Line flyout also contains the Ray and Construction Line commands, which you can use to draw lines that extend infinitely. When selecting the Line command, make sure you pick the icon that shows dots at both endpoints. This icon indicates that the command draws a line between the two points selected. You can continue drawing lines from point to point until you press ↵ to end the command.

Pick: **Line icon**

Your screen should look similar to Figure 1.12, which shows the Line command options on the screen menu.

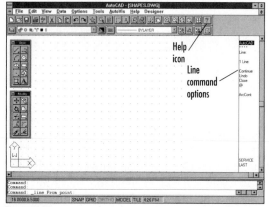

Figure 1.12

AutoCAD prompts you for the starting point of the line. You can answer prompts requesting the input of a point two different ways:

■ By moving the cursor into the graphics window and choosing a screen location by pressing the pick button (grid and snap help to locate specifically defined points)

■ By entering X, Y, *Z coordinate values* for the point

Getting Help

 You can get help during any command using AutoCAD's transparent Help command. A *transparent command* is one that you can execute during another command. In AutoCAD these commands are preceded by an apostrophe ('). If you type '? or '*Help* or press ⌈F1⌋ during a command, you will enter the Help dialog box, which shows help for the command you are currently using. While you are still at the *From point:* prompt, you will use the Standard toolbar to pick the transparent Help command. Its icon looks like a question mark (?).

From point: **(pick Help icon)**

The AutoCAD Help dialog box with help for the Line command is shown in Figure 1.13. AutoCAD's Help feature is *context sensitive*; it recognizes that you were using the Line command and therefore displays help for that command.

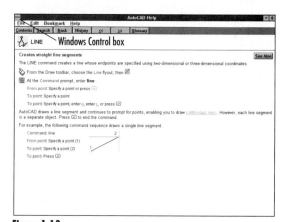

Figure 1.13

The AutoCAD Help window has a menu bar of its own. You will use the buttons at the top of the AutoCAD Help screen and pick to display the help contents. From the AutoCAD Help window,

Pick: **Contents button**

The contents are divided into help about items on the menus and toolbars, and help about the command line, which can be either commands or system variables. When you move the cursor over items displayed in a contrasting color, the cursor changes to a hand. These items can be selected.

Pick on the Toolbars selection.

Pick: **Toolbars**

The list of toolbar names appears in the Help window.

Pick: **Draw Toolbar**

Help for the Draw toolbar now appears in the Help window, as shown in Figure 1.14. Picking once on icons that have the flyout symbol causes the additional buttons to fly out. Picking on an icon that is already flown out or one that does not have a flyout symbol displays the command help for that item.

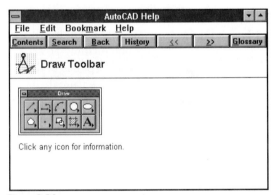

Figure 1.14

As you do with other Windows applications, you will use the Windows Control box to close the Help window.

Double-click: **(the Windows Control box in the top left corner of the AutoCAD Help window)**

Double-clicking the corner of the window should close it. Try again if the AutoCAD Help window did not disappear.

> ■ *TIP* Pressing Ⓕ1 also pops up the transparent Help window. ■

Entering Coordinates

AutoCAD stores your drawing geometry using *World Coordinates*, AutoCAD's default *Cartesian coordinate system,* where X, Y, Z coordinate values specify locations in your drawing. The UCS icon near the bottom left corner of your screen shows the positive X and positive Y directions on the screen. You can think of the default orientation of the Z axis as specifying the location in front of the monitor for positive values, inside for negative ones. AutoCAD functions according to the right-hand rule for coordinate systems. If you orient your right hand palm up and point your thumb in the positive X direction and index finger in the positive Y direction, the direction your other fingers curl will be the positive Z direction. Use the right hand rule and the UCS icon to figure out the directions of the coordinate system. The letter W appearing in the UCS icon indicates that the World Coordinate System is active.

You specify a point explicitly by entering the X, Y, Z coordinates, separated by a comma. You can leave the Z coordinate off when you are drawing in 2D, as you will be in this tutorial. If you do not specify the Z coordinate, it is assumed to be the current elevation in the drawing, for which the default value is zero. For now type only the X and Y values, and the default value of zero for Z will be assumed.

Using Absolute Coordinates

It is often necessary to type the exact location of a specific point. When you are creating a drawing, you will want to represent the geometry of the object you are creating exactly. You will often do this by typing the X, Y, Z coordinates to locate the point on the current coordinate system. These are called the *absolute coordinates*. Absolute coordinates specify a distance to move along the X, Y, Z axes from the origin (point 0,0,0 of the coordinate system).

> ■ *TIP* In AutoCAD the spacebar acts the same way as the ⏎ key, to end the command. When typing in coordinates, enter only commas between them. Do not enter a space between the comma and the following coordinate of a coordinate pair, because a space has the same effect as ⏎. Remember that you can press the spacebar to enter commands quickly. ■

Resuming the Line command, you will use absolute coordinates to specify the endpoints for the line you will draw. (Restart the Line command by picking the Line icon from the Draw toolbar if you do not see the *From point:* prompt.)

From point: **5.26,5.37** ⏎

As you move your cursor, you will see a line rubberbanding from the point you selected to the current location of your cursor. The next prompt asks for the endpoint of the line.

To point: **8.94,5.37** ⏎

Once you enter the second point of a line, the line appears on the screen. The rubberband line continues to stretch from the last endpoint you selected to the current location of the cur-

sor and the prompt asks for another point. This feature allows lines to be drawn end to end. Continue as follows:

To point: **8.94,8.62** ⏎
To point: **5.26,8.62** ⏎

> ■ *TIP* If you enter a wrong point while still in the Line command, you can back up one endpoint at a time by selecting the Undo option from the screen menu for the Line command or by typing the letter *U* and pressing ⏎. ■

■ *Warning:* Undo functions differently if picked as a command. If you select Undo when you are at the command prompt, you may undo entire command sequences. If necessary, you can use Redo to return something that has been undone; however, Redo only restores the last thing undone. ■

The Close option of the Line command joins the last point drawn to the first point drawn, thereby closing the lines and forming an area. To close the figure, type *c* in response to the prompt. (You can also pick Close from the screen menu.)

To point: **C** ⏎

Using Grid and Snap

Use the grid and snap settings to draw another rectangle to the right of the one you just drew, as shown in Figure 1.15. (Be sure Snap is highlighted on the toolbar. If it is not, double-click SNAP from the status bar or press ⏎ to turn it on.)

Pick: **Line icon**

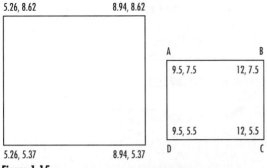

Figure 1.15

5.26, 8.62   8.94, 8.62

5.26, 5.37   8.94, 5.37

A   B

9.5, 7.5   12, 7.5

9.5, 5.5   12, 5.5

D   C

> ■ **TIP** You can restart the last command by pressing the ⏎ key, the spacebar, or the right mouse button. Since Line was the last command used, you could have pressed ⏎ or the spacebar to restart the Line command. Using these shortcuts will help you produce drawings in less time. Keep in mind, however, that some commands selected from the toolbar do not always function the same when restarted at the command prompt. Restarting a command has the same effect as typing the command name at the prompt. Some toolbar selections contain special programming to automatically select certain options for you. ■

On your own, choose the starting point on the grid by moving the cursor to point A, as shown in Figure 1.15, and pressing the pick button. Remember to use the coordinate display on the status bar to help you select the points.

Create the 2.5 unit long horizontal line by moving the cursor 5 snaps to the right and pressing the pick button to select point B.

Complete the rectangle by drawing a vertical line 4 spaces (2.00 units) long to point C, and another horizontal line 5 spaces (2.5 units) long to point D. Then use the Close command option to complete the rectangle and end the Line command on your own.

Using Last Point

AutoCAD always remembers the last point that was specified. Often you will find it necessary to specify a point that is exactly the same as the previous point. The @ symbol on your keyboard is AutoCAD's name for the last point entered. You will use last point entry with the Line command. This time you will restart the Line command by pressing ⏎ at the blank command prompt to restart the previous command.

> Command: ⏎
> From point: **2,5** ⏎
> To point: **4,5** ⏎
> To point: ⏎
> Press: ⏎ *(or the return button on your mouse to restart the Line command)*
> From point: **@** ⏎

> ■ **TIP** Once in the Line command, you can also press the spacebar, right mouse button, or ⏎ at the *From point* prompt to start your new line from the last point. ■

Your starting point is now the last point that you entered in the previous step (4,5). Notice the line rubberbanding from that point to the current location of your cursor.

> ■ **TIP** When using AutoCAD, it is often useful to move the cursor away from the point you have selected so that you can see the effect of the selection. You will not notice rubberband lines if the cursor is left over the previously selected point. ■

Using Relative X, Y Coordinates

Relative coordinates allow you to select a point at a known distance from the last point specified. To do this, you must precede your X, Y coordinate values with @. Continuing the Line command from above,

> To point: **@0,3** ⏎

Notice that your line is drawn 0 units in the X direction and 3 units (6 spaces on the grid) in the positive Y direction from the last point specified, which was 4,5. To complete the shape,

> To point: **@–2,0** ⏎

> To point: **@0,–3** ⏎

> To point: ⏎ *(or the return button on your mouse to end the command)*

Your drawing should look similar to Figure 1.16. You may need to scroll your screen to see all three rectangles.

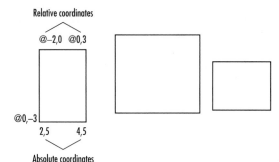

Figure 1.16

Using Polar Coordinates

When prompted to enter a point, you can specify the point using *polar coordinates*. Polar coordinate values use the input format @DISTANCE<ANGLE. When using the default decimal units, you don't need to specify any units when typing the angle and length. Later in this tutorial you will learn how to select different units of measure for lengths and angles. When you are using polar coordinates, each new input is calculated relative to the last point entered. The default system for measuring angles in AutoCAD defines a horizontal line to the right of the current point as 0°. As shown in Figure 1.17, angular values are positive in the counterclockwise direction. Both distance and angle values may be negative.

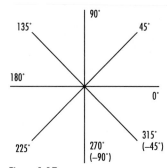

Figure 1.17

Figure 1.18 shows an example of lines drawn using relative polar coordinates. Using relative coordinate entry, a line is drawn from the starting point to the distance and at the angle specified. In the figure, the line labeled 1 goes from the starting point a distance of 3 units at an angle of 30°. Remember that the default is that the angle is measured from a horizontal line to the right of the starting point. Line 2 starts from the last point of line 1 and is drawn a distance of 2 units at an angle of 135°, again measured from a horizontal line to the right of that line's starting point. Polar coordinates are very important for creating drawing geometry.

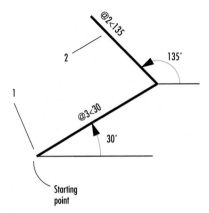

Figure 1.18

You will now draw a rectangle, using polar coordinate values. The starting point for this rectangle is specified in Figure 1.19. Use the toolbar to select the Line command. You will draw a horizontal line 3.5 units to the right of the starting point by responding to the prompts.

Pick: **Line icon**

From point: *(pick point A, coordinates 1.0,4.0)*

To point: **@3.5<0** ⏎

Complete the rectangle, using polar coordinate values.

To point: **@2<–90** ⏎

To point: **@3.5<180** ⏎

To point: **C** ⏎

Your drawing should look like Figure 1.19.

■ *TIP* Double-clicking on the coordinates displayed on the status toggles their display, i.e., you can toggle the coordinate display off and on. When you toggle the coordinate display off, the coordinates do not change when you move the crosshairs and they appear grayed out on the status bar. When you toggle the coordinates back on, they once again display the X, Y location of the crosshairs. During commands there is a third toggle for the coordinate display: in a command like Line, double-clicking the coordinate display displays the length and angle from the last point picked. This can be very useful in helping to determine approximate distances and angles for polar coordinate entry. You can also toggle the coordinates by pressing F6 or Ctrl-D. ■

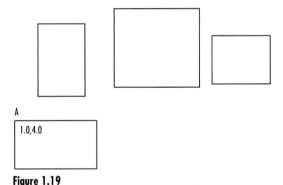

Figure 1.19

Using Save

The Save command lets you save your work to the file name you have previously specified, in this case *shapes.dwg.* It is good practice to save your work frequently. If the power went off or your computer crashed, you would lose all of the work that you had not previously saved. If you save every ten minutes or so, you will never lose more than that amount of work in case of a problem. Save your work at this time. Use the Standard toolbar to pick the Save icon, which looks like a floppy disk.

Pick: **Save icon**

Your work will be saved to *shapes.dwg.* If you had not previously named your drawing, the Create Drawing File dialog box would appear, allowing you to specify a name for the drawing. Save your work periodically as you are working through these tutorials.

Using Erase

 The Erase command removes objects from a drawing. The Erase command is located on the Modify toolbar.

■ *TIP* Typing *E* ⏎ (the command alias) will also start the Erase command. ■

Pick: **Erase icon**

You are prompted to select objects. The select cursor, a small rectangle, replaces the crosshairs. You identify objects to be erased by placing the cursor over them and pressing the pick button. Each line that you have drawn is a separate drawing object.

You will erase the objects labeled A and B in Figure 1.20. On your own, turn Snap off if it is still on to make selecting easier.

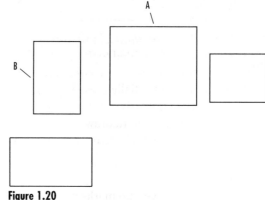

Figure 1.20

Select objects: *(pick on line A)*
Select objects: *(pick on line B)*

The lines you have selected become highlighted but are not erased. When using AutoCAD, you can select more than one item to create a *selection set.* You will learn more ways to do this later in this tutorial. When you are done selecting, you will press the return button on your pointing device or ⏎ to tell AutoCAD that you are done selecting. Pressing ⏎ or the return button completes the selection set and erases the selected objects.

Press: ⏎ *(or the return button)*

■ *TIP* Don't position the selection cursor at a point where two objects cross, because you cannot be certain which will be selected. The entire object is selected when you point to any part of it, so select the object at a point that is not ambiguous. ■

The selected lines are erased from the screen. Some marks, called *blipmarks,* may remain on the screen. Blipmarks indicate where you have selected a point from the screen. You can remove them by using the Redraw View command.

Using Redraw View

 The Redraw command removes the excess marks added to the drawing screen when you are drawing objects and restores objects partially erased while you are editing other objects. Redraw is located on the Standard toolbar. The Redraw command has no prompts; it is simply activated when you pick it.

Pick: **Redraw View icon**

Your screen is redrawn; the marks are eliminated.

> ■ *TIP* You can also just type *R* ⏎ at the command prompt. R is the command alias for Redraw. ■

If an object is erased by mistake, activate the Oops command from the Erase screen menu. Oops is also available from the Miscellaneous toolbar. You will type the command name, as this toolbar is not turned on at present.

Command: **OOPS** ⏎

The lines are restored on your screen. Oops restores only the most recently erased objects. It will not restore objects beyond the last Erase command. However, it will work to restore erased objects even if other commands have been used in the meantime.

Erasing with Window or Crossing

You need to clear your drawing editor to make room for new shapes. You will erase all the rectangles on your screen. You can pick multiple objects by using the Window and Crossing selection modes.

To use AutoCAD's *implied Windowing mode*, at the *Select objects:* prompt, pick a point on the screen that is not on an object and move your pointing device from left to right. A window-type box will start to rubberband from the point you picked. In the implied Windowing mode (a box drawn from left to right), the window formed selects everything that is *entirely enclosed* in it. However, if you draw the box from right to left, the *implied Crossing mode* is used. Everything that either crosses the box drawn or is enclosed in the box is selected. You can use implied Windowing and implied Crossing to select objects during any command in which you are prompted to select objects.

You can turn implied Windowing on and off. Before using it, you will check to see that implied Windowing is turned on.

Pick: **Options, Selection**

The dialog box you see in Figure 1.21 appears on your screen. For implied Windowing to work, it must be selected (an X must appear in the box to the left of Implied Windowing). On your own, select this mode if it is not currently selected, then pick OK to exit the dialog box. If it is already turned on, pick Cancel to exit without making any changes.

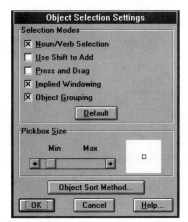

Figure 1.21

Pick: **Erase icon**

Select objects: *(pick a point above and left of your upper three rectangles, identified as A in Figure 1.22)*

Other corner: *(pick a point below and right of your upper three rectangles, identified as B)*

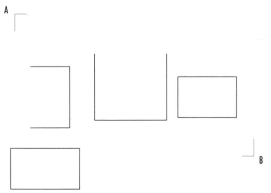

Figure 1.22

You will notice that a "window" box formed around the area specified by the upper left and lower right corners. Your screen should look like Figure 1.23.

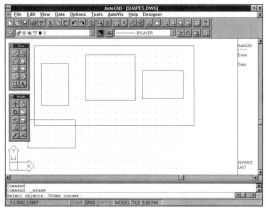

Figure 1.23

When you picked the second corner of the window (point B), the objects that were entirely enclosed in the window became highlighted. You will end the object selection by pressing ⏎ and erasing the selected objects:

Select objects: ⏎

Only objects that were entirely within the window were erased. The fourth rectangle was not completely enclosed and therefore was not erased.

Docking the Toolbars

The toolbars can float, as you have seen, or they can be docked to the edge of the screen. Next you will dock the Draw toolbar to the right edge of the screen and the Modify toolbar to the left edge of the screen so that it is easier to select the remaining rectangle. You will move each toolbar by picking on its title bar and, keeping the pick button depressed, moving it to a location near the edge of the screen. When the toolbar is near the correct location, its outline will change to the docked shape. Release the pick button to leave the toolbar in its new docked location.

Pick: (on the title bar of the Draw toolbar and drag it to the right edge of the screen)

After you have docked the Draw toolbar to the right edge of the screen,

Pick: (on the title bar of the Modify toolbar and drag it to the left edge of the screen)

When you have finished docking the toolbars, your screen should look similar to Figure 1.24.

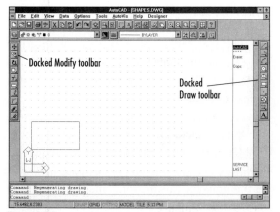

Figure 1.24

Next, you will use the Erase command and implied Crossing to erase the fourth rectangle.

Pick: **Erase icon**

Select objects: *(pick a point below and right of the remaining rectangle)*

Other corner: *(pick a point above and left of the rectangle)*

Select objects: ⏎

The remaining rectangle is erased from the drawing.

Using Undo

 The Undo command removes the effect of previous commands. If you make a mistake, pick the Undo command from the Standard toolbar, and the effect of the previous command disappears.

You will try this now to see the last rectangle you erased reappear.

Pick: **Undo icon**

■ *TIP* If you undo the wrong object, there is a Redo command, which lets you restore the last item you undid. It is the next icon to the right of Undo on the Standard toolbar. ■

Line and other commands also include Undo as a command option that allows you to undo the last action within the command. This Undo option appears on the screen menus for these commands. Notice the difference between these uses.

Pick: **Line icon**

From point: *(pick any point)*

To point: *(pick any point)*

To point: *(pick any point)*

To point: *(pick any point)*

To point: **U** ⏎

To point: ⏎

One line segment disappeared while you remained within the Line command. Now you will undo the entire Line command with the Undo command. You will pick Undo from the Standard toolbar.

Pick: **Undo icon** ⏎

All the line segments drawn with the last instance of the command disappear.

Typing *Undo* at the command prompt offers more options for undoing your work. You will draw some lines to have some objects to undo.

On your own, use Erase from the toolbar and erase any remaining objects in your graphics window. Then use Redraw to clean up any blip-marks. Use Line on your own to draw six parallel lines anywhere on your drawing screen. Use Snap and Grid as needed.

In the next step you will type the Undo command to remove the last three lines that you drew.

Type: **UNDO** ⏎

Auto/Control/BEgin/End/Mark/Back/<number>: **3** ⏎

The last three lines drawn should disappear, corresponding to the last three *instances* of the Line command. You could have selected any number, depending on the number of command steps you wanted to undo.

> ■ **TIP** Often it is easier to back up one step at a time by using Undo from the Standard toolbar (the U command). Some commands do several steps in one sequence and you may not be sure of the exact number you want to back up. ■

Using Redo

If the last command undid something you really did not want to undo, you can use the Redo command to reverse the effects. The Redo icon appears on the Standard toolbar to the right of Undo.

Pick: **Redo icon**

The three lines should reappear. Redo reverses only the last Undo; it must follow immediately after the Undo command.

Using Back

The Back option of the Undo command takes the drawing back to a mark that you set with the Mark option or to the very beginning of the drawing session if you haven't set any marks. Be careful selecting these options or you may undo too much.

Type: **UNDO** ⏎
Auto/Control/BEgin/End/Mark/Back/<number>: **B** ⏎
This will undo everything. OK?<Y>: ⏎

All operations up to the beginning of the drawing session (or mark, if you had previously set one using the Mark option of the Undo command) will be undone. (You can use Help to find out what the remaining options, Auto, Control, Begin, and End, are used for.)

Pick: **Redo icon**

The previous appearance of the drawing should return. On your own, erase all objects in this drawing session from the screen. Use the Redraw View selection from the Standard toolbar to clean up the screen.

Next you will create a new drawing showing a plot plan. A plot plan is a drawing showing a plan view of a lot boundary and the location of facilities. You will set the units for this drawing, instead of using the defaults. To start a new file, you will pick the New command from the Standard toolbar. It is the left-most icon.

Pick: **New icon**

Since you have been working in a drawing already, AutoCAD will ask you whether you want to save your changes. You do not need to save *shapes.dwg*.

Use the Create New Drawing dialog box on your own to name your new drawing *plotplan.dwg* in the empty text box to the right of New Drawing. Use the default prototype drawing *acad.dwg*. Check to see that your new drawing is being created in the *c:\work* directory. To leave the dialog box when you are finished,

Pick: **OK**

Setting the Units

AutoCAD allows you to work in the type of *units* that are appropriate for your drawing. The default units are decimal. You can change the type of units for lengths to *architectural units* that appear in feet and fractional inches, *engineering units* that appear in feet and decimal inches, *scientific units* that appear in exponential format, or *fractional units* that are whole numbers and fractions. When you use architectural or engineering units, one drawing unit is equal to one inch; to specify feet you must type the feet mark after the numerical value (examples are 50.5' or 20'2" or 35'-4"). You can think of the other types of units as representing any type of real-world measurement you want: decimal miles, furlongs, inches, millimeters, microns, or anything else. When the time comes to plot the drawing, you determine the final relationship between your drawing database, in which you create the object the actual size it is in the real world, and the paper plot.

Angular measurements can be given in decimal degrees; degrees, minutes, and seconds; gradians; radians; or surveyors units, such as the *bearing* N 45d0'0"E.

You will use the Data selection from the menu bar to access the Units Control dialog box.

Pick: **Data, Units**

The Units Control dialog box appears on your screen, as shown in Figure 1.25.

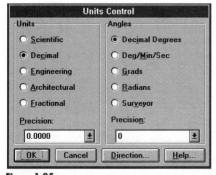

Figure 1.25

In the Units area, the radio buttons to the left let you select the type of unit for the drawing. Pick the button to the left of Architectural. The center of the button becomes filled in to tell you it is selected. You can select only one button at a time. Notice that the units displayed under the heading Precision change to architectural units. Pick the button for Decimal once again. Notice that the units in the Precision

area change back to decimal units. The dimensions for the plot plan you will create will be in decimal feet, so you will leave the units set to Decimal. Next pick on the number 0.0000 displayed in the box below Precision. The choices for length unit precision pull down below where you picked, as shown in Figure 1.26.

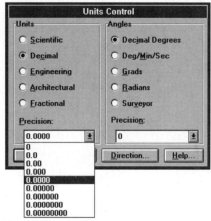

Figure 1.26

Pick on the selection 0.00 to set the display to show two decimal places. When specifying coordinates and lengths, you can still type a value from the keyboard with more precision and AutoCAD's drawing database will keep track of your drawing with this accuracy. However, because you have selected this precision, only two decimal places will be displayed. Units precision determines the display of the units on your screen and in the prompts, not the accuracy internal to the drawing. Though AutoCAD keeps track internally to at least 14 decimal places, only eight decimal places of accuracy will ever appear on the screen. You can change the type of unit and precision at any point during the drawing process.

The right side of the dialog box controls the type and precision of angular measurements. Remember that the default is that angles are measured in a counterclockwise direction, starting from a horizontal line to the right as 0°. You can change this by picking on the Direction button and making a new selection. Leave the direction set at the default of 0° towards East.

Select the button for Surveyor angles. Notice that the display in this Precision area changes to list the angle as a bearing. When this mode is active, you can type in surveyor's angles. AutoCAD will measure the angle from the specified direction, North or South, toward East or West, as specified. The default direction, North, is straight toward the top of the screen. If you want to see more precision for the angles, pick on the box containing the precision, N 0d E. The list of the available precisions pops down, as shown in Figure 1.27, allowing you to select to display degrees, minutes, and seconds. On your own, pick the selection N0d00'00"E to display degrees, minutes, and seconds.

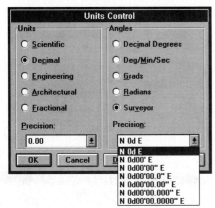

Figure 1.27

Pick: **OK** *(to exit the dialog box)*

Look at the status bar. You will see that the coordinate display has changed to show two instead of four places after the decimal.

Sizing Your Drawing

In AutoCAD you always create your drawing geometry in *real-world units*. This means that if the object is 10' long, you make it exactly 10' long in the drawing database. If it is a few millimeters long, you create the drawing to those lengths. You can think of the decimal units as representing whatever decimal measurement system you are using. For this drawing, the units will represent decimal feet, so ten units will equal 10 feet in the drawing. After you have created the drawing geometry, you can decide on the scale at which you want to plot your final hardcopy drawing on the sheet of paper. This is one of the powerful features of CAD. You can create very exact drawings from which you can make accurate measurements and calculations using the computer. Also, you can plot the final drawing to any scale, saving a great deal of time because you don't have to remake drawings just to have different scaling, as you would with paper drawings.

For your plot plan drawing, you need to create a larger graphics window to accommodate the site plan shown in Figure 1.28.

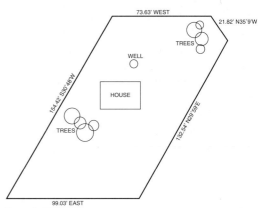

Figure 1.28

Using Limits

The Limits command sets the size of your drawing. Limits can also be turned off if you do not want to preset the size of the drawing. Use the Drawing Limits selection from the Data menu to change the overall size of the drawing to reflect the metric drawing units.

Pick: **Data, Drawing Limits**

Reset model space limits:

ON/OFF/<Lower left corner> <0.00,0.00>: ⏎

   (to accept default of 0,0)

Upper right corner <12.00,9.00>: **300,225** ⏎

Notice the message that read *Reset model space limits.* The default space where you create your drawing geometry or model is called model space. This is where you accurately create real-world-size models. In Tutorial 5, you will learn to use paper space, where you lay out views, as you do on a sheet of paper. You can set different sizes for model space and paper space with the Limits command.

Using Zoom

The Zoom command enlarges or reduces areas of the drawing on your screen. It is different from the Scale command, which actually *makes* the selected items larger or smaller *in your drawing database.* You can use the Zoom command when you want to enlarge something on the screen so that it is easier to see the details. You can also zoom out so that objects appear smaller on your display. The Zoom command options are located on the Standard toolbar and on the View menu.

Using Zoom All

 The Zoom All option displays the drawing limits, or shows all of the drawing objects on the screen,

depending on which is larger. Zoom All is located on the Zoom flyout on the Standard toolbar. You will select View, Zoom All from the menu bar to show this larger area on your display.

Pick: **View, Zoom, All**

Move the crosshairs to the upper right-hand corner of the screen. You will see that the status bar displaying the coordinate location of the crosshairs indicates that the size of the graphics window has changed to reflect the limits of the drawing.

> ■ *TIP* If the coordinate display does not show a larger size, check to see that you set the limits correctly and that you picked Zoom All. Make sure that the coordinates are turned on by double-clicking on the coordinates located on the status bar until you see them change as you move the mouse. ■

Now you will set the grid spacing to a larger value so that you can see the grid on your screen.

Command: **GRID** ⏎

Grid spacing(x) or ON/OFF/Snap/Aspect <0.00>: **10** ⏎

To draw the site boundary you will use absolute and polar coordinates as appropriate. You will be using the surveyor's angles to specify the directions for the lines. AutoCAD's surveyor's angle selection, which you selected when setting the type of units, uses bearings. Bearings give the angle to turn from the first direction stated toward the second direction stated. For example, a bearing of N29°59'E means to start out at the direction North and then turn an angle of 29°59' toward the East.

Now you are ready to start drawing the lines of the site boundary. Using the Draw toolbar,

Pick: **Line icon**

From point: **50,30** ⏎

To point: **@99.03<E** ⏎

To point: **@132.54<N29d59'E** ⏎

To point: **@21.82<N35d9'W** ⏎

To point: **@73.63<W** ⏎

To point: **c** ⏎

Your screen should look like Figure 1.29.

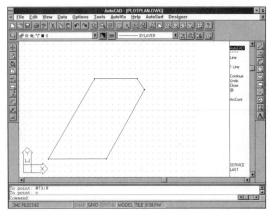

Figure 1.29

Using Rectangle

You are going to draw a rectangle to represent a house on the plot plan. Using coordinate values, you will place the rectangle roughly in the center of your plot plan, as shown in Figure 1.30. You will pick Rectangle from the Draw toolbar.

Pick: **Rectangle icon**

AutoCAD will prompt for the first corner of the rectangle. You will type in the coordinates for the corners of the rectangle.

First corner: **120,95** ⏎

The first point acts as an anchor for the first corner of the rectangle. As you move the crosshairs away from the point, a box stretches from the first point to the current location of the crosshairs.

Other corner: **150,115** ⏎

Your drawing should look similar to Figure 1.30.

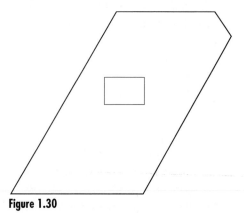

Figure 1.30

Saving Your Drawing

Having completed the site boundary, you should now save the drawing on your disk. It is a good practice to save any time you have completed a major step or every 10 to 15 minutes. This way, if your computer crashes, you will not have lost more than a few minutes of work. Also, if you make a mistake that you don't know how to correct, you can pick File, Open, then discard your drawing changes and open the saved version.

Pick: **Save icon**

AutoCAD saves your file to the name you assigned when you began the new drawing, *plotplan.dwg.*

Drawing Circles

You will draw a circle to represent the location of the well on the plot plan. You draw circles using the Circle command located on the Draw toolbar. The Circle button is a flyout that contains several different circle icons depicting different methods of specifying a circle. Figure 1.31 shows the Circle flyout.

■ *Warning:* When you are first learning AutoCAD, the flyouts may get confusing. This is because when you select a command, it appears as the top button on the flyout the next time. For instance, the Circle flyout contains options for Circle Center Radius, Circle Center Diameter, and so on. Once an icon, such as Circle Center Diameter is selected, it appears as the top button on the flyout after that. If you have trouble locating an icon, use Help to get information about the location of the icon. ■

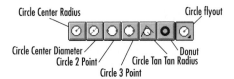

Circle Center Radius   Circle flyout
Circle Center Diameter   Donut
Circle 2 Point   Circle Tan Tan Radius
Circle 3 Point

Figure 1.31

Pick: **Circle Center Radius icon**

3P/2P/TTR/<Center point>: **145,128** ⏎

As you move the cursor away from the center point, a circle continually reforms, using the distance from the center you selected to the cursor location as the radius value. You will type the value to specify the exact size for the circle's radius.

Diameter/<Radius>: **3** ⏎

Now draw some circles to represent trees or shrubs in the plot plan. You will specify the locations for the trees by picking from the screen.

Pick: **Circle Center Radius icon**

3P/2P/TTR/<Center point>: *(pick a point for the center of circle 1, as shown on Figure 1.32)* ⏎

Diameter/<Radius>: *(move the crosshairs away from the point you picked until the circle appears similar to Figure 1.32 and press the pick button)* ⏎

On your own, draw the remaining circles representing trees using this method.

Your drawing should look similar to Figure 1.32.

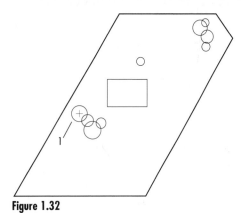

Figure 1.32

Adding Text

The Dtext, or Dynamic Text, command adds single lines of text, words, or numbers to a drawing. Dtext is located on the Text flyout on the Draw toolbar.

Pick: **Dtext icon**

Unless you specify otherwise, text is added to the right of a designated starting point. You can also center text about a point or add it to the left of a point if you use the Justify option. To specify the starting point for the text, you will pick from the screen. (You can also type the coordinates of the location for the starting point.) You will label the lengths and bearings of the lot lines shown in the plot plan. Start with the bottom lot line by picking a point below the lot line.

Command: Dtext Justify/Style/<Start Point>: *(select a point at about coordinates 90,25)*

The .20 value shown in the next prompt is the default text height. Since the current drawing must be scaled to fit on a piece of paper, this height is much too small. A height of 3 will create text that is in proportion with the drawing.

■ *TIP* To determine the actual text height necessary, you must know the size of paper and scale you will use to plot the final drawing. In Tutorial 5 you will learn to use paper space to set up scaling and text for plotting. If you are unsure what text height to use, move the crosshairs on the screen and read the coordinate display on the toolbar. Use the lengths shown to get an idea of an appropriate size for your text. ■

Height<0.20>: **3** ⏎

The default value of E in the *Rotation angle <E>* prompt generates horizontal (East) text. You can specify any angle. Text on technical drawings is usually drawn horizontally and is called *unidirectional* text. To accept the default rotation angle,

Rotation angle <E>: ⏎

AutoCAD is now ready to accept the typed-in text.

Text: **99.03' EAST** ⏎

Text: ⏎

■ *TIP* Notice that the flyouts keep track of which button you selected most recently and bring it to the top of the flyout. Since you picked the Dtext command, it is now on the top of the Text flyout. The next time you want to use the Dtext button, you can just pick once without having to look through the buttons on the flyout. ■

You can type the text just as you would with a word processing program. Dtext will not wrap text, so you must designate the end of each line by pressing ⏎. If you make a mistake before exiting Dtext, backspace to erase the text. After exiting the Dtext command, you can use Edit Text to make corrections to lines of text. The *Text:* prompt always appears after every line of text. Press ⏎ without entering any text (a null reply to the prompt) to end the Dtext command. (You *must* do this from the keyboard, not with the return button on your pointing device.) When you exit Dtext, AutoCAD recreates the text as a permanent object. Your drawing should look like Figure 1.33.

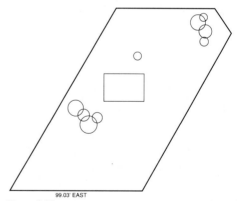

99.03' EAST

Figure 1.33

Using Special Text Characters

You will add the text labeling the length and angle for the right-hand lot line. When you are using the standard AutoCAD fonts, the special text item %%d creates the degree symbol. Type the line of text with %%d in place of the degree symbol. When you are finished typing in the text and press ⏎, AutoCAD will replace %%d with a degree symbol. (You will use more special text characters in Tutorial 7.) You will also specify a rotation angle for the text so that it is aligned with the lot line.

Pick: **Dtext icon**

Justify/Style/<Start Point>: *(pick a point slightly to the right of the right-hand lot line)*

Height <3.00>: ⏎

Rotation angle <E>: **N29d59'E** ⏎

> ■ *TIP* You can also specify the rotation by picking when the line that rubberbands from the point you picked for the text location is oriented at the rotation angle you desire. If you do not want to type in bearings for the text angle, you can change the units at any time. ■

Text: **132.54' N29%%d59'E** ⏎

Text: ⏎

The text should appear in your drawing aligned with the lot line.

On your own, use the techniques that you have learned to add the bearings and distances for the remaining lot lines. To have the text on the left lot line be parallel to the line, but face the same direction as the text on the right lot line, reverse the direction of the bearings (for example, if it was SW, use NE), and keep the same degree and minute values. You can also specify the angle by picking two points that align with the lot line along which you are creating text.

Add more text identifying the house and the well locations. Your drawing should look like Figure 1.34.

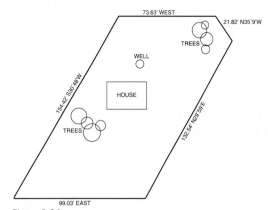

Figure 1.34

Editing Text

 The Edit Text command allows you to edit text using a dialog box. You can select it from the Special edit flyout on the Modify toolbar. The flyout is shown in Figure 1.35.

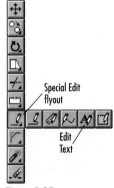

Figure 1.35

Pick: **Edit Text icon**

<Select a TEXT or ATTDEF object>/Undo: *(pick the top line of text)*

After you activate the command, you are prompted to select a text object (line of text). The selected text appears in a dialog box like the one in Figure 1.36. You can add or change text without erasing the entire line. Use the arrow keys on your keyboard to move within the line of text without deleting. Use the (DEL) key to delete the letter to the right of the cursor. Use (BACKSPACE) to delete the letter to the left of the cursor. You can also retype the entire line. Make sure that the text in the dialog box reads: *73.63′ WEST*. To exit the dialog box,

> Pick: **OK**
>
> <Select a TEXT or ATTDEF object>/Undo: ⏎ *(to exit the command)*

Figure 1.36

Setting the Text Style

The Style command allows you to create a style specifying the *font* and other characteristics for the shapes of the text in your drawing. A font is a character set comprising letters, numbers, punctuation marks, and symbols of a distinctive style and design. AutoCAD supplies several *shape-compiled fonts* that you can use to create text styles. In addition to the AutoCAD shape-compiled fonts, you can also use True Type and Type 1 Postscript fonts. Some sample True Type fonts are provided with AutoCAD in the directory *c:\r13\com\fonts*. You can also use your own True Type and Type 1 Postscript fonts. AutoCAD also supports Unicode fonts for many languages that use large numbers of characters.

The Style command allows you to have different appearances of the same font. Using the Style command, you can control the appearance of the font so that it is slanted (oblique), backwards, vertical, or upside-down. You can also control the height and proportional width of the letters.

If you don't use the default style, you must create the style you want to use prior to adding text to your drawing, (or later change the properties of the text to that style).

To create a style, you will select Data, Text Style from the menu bar. A text style is composed of several characteristics that are presented after you select a new font from the list of available fonts. Once you create (or select) a style, that style will remain current for all text until you set a new style as the current one. An AutoCAD shape font that works well for engineering drawings is called Roman Simplex.

> Pick: **Data, Text Style**

At the prompt for text style name, you will type in the name of the style you want to create. Use a descriptive name for the style that will remind you of its purpose later.

> Text style name (or ?)<Standard>: **NOTES** ⏎

The dialog box shown in Figure 1.37 appears on your screen.

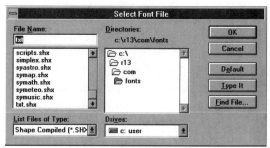

Figure 1.37

At the left side of the dialog box are the names of the AutoCAD shape fonts. If you want to use True Type or Postscript fonts you will change the selection under the heading List Files of Type near the lower left corner of the dialog box. Refer to Figure 1.38 while you are making the next selections.

Pick: **Shape Compiled (*.SHX)**

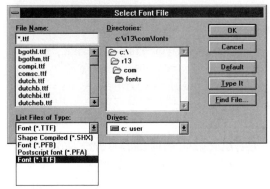

Figure 1.38

The list showing the types of font files available pops down. To show the list of True Type fonts provided by AutoCAD,

Pick: **Font (*.TTF)**

After you make this selection, the list of True Type fonts replaces the shape-compiled fonts at the left of the dialog box. (You can also use the dialog box to change to a different directory to see True Type fonts stored in other places on your system.) Next you will set the selection back to AutoCAD shape fonts and select the *romans* font as the shape of the letters for your style, NOTES.

Pick: **Font (*.TTF)**

From the list of font types that pulls down, once again,

Pick: **Shape Compiled (*.SHX)**

Now use the slider bar to scroll down the list of shape fonts until you see the selection romans.shx. To make it the current font,

Pick: **romans.shx**

Pick: **OK**

You will return to the command prompt for the remaining selections. You will accept the default text height of 0.00. Setting the text height in the Style command causes you not to be prompted for the height of the text when you use the Dtext command.

New style. Height <0.00>: ⏎
Width factor <1.00>: ⏎
Obliquing angle <0d0'0">: ⏎
Backwards? <N>: ⏎
Upside-down? <N>: ⏎
Vertical? <N>: ⏎
NOTES is now the current text style.

Any new text added to your drawing will use the current text style until you set a different style as current. You will now add a title to the bottom of the drawing.

Pick: **Dtext icon**

Dtext Justify/Style/<Start Point>: J ⏎
Align/Fit/Center/Middle/Right/TL/TC/TR/ML/MC/MR/BL/
  BC/BR: C ⏎
Center point: *(select a point near the middle bottom of the drawing)*
Height<3.00>: ⏎
Rotation angle <E>: ⏎ *(make sure your rotation angle is set to E)*
Text: **PLOT PLAN** ⏎
Text: ⏎

Using Mtext

 Next you will use the Multiline Text feature to add a block of notes to your drawing. Unlike Dtext, the Mtext command automatically adjusts text within the width you specify. You can create text with your own text editor and then import it with the Mtext command, or you can create text

using the Mtext command. Like Dtext, Mtext can be edited easily using the Edit Text command after it has already been placed in the drawing. You can select Mtext by picking the Text button from the Draw toolbar.

Pick: **Text icon**

Attach\Rotation\Style\Height\Direction\<Insertion point>: *(pick a point to the right of the plot plan drawing where you want to locate the lower corner of the notes; a window will form)*

Attach/Rotation/Style/Height/Direction/Width/2Points/ <Other corner>: *(pick the second corner of the window to size the area for the notes)*

If you want to only enter the width of the text you will type in, type *W* at the prompt shown above instead of specifying the corners of the box, and then type the width you want to specify.

Because shape fonts like *romans.shx* cannot be displayed in the Mtext text editor, the Select Font dialog box shown in Figure 1.39 appears, allowing you to temporarily choose a different type of font to represent the romans.shx font until you are done creating the text. When you exit the Mtext command, the letters will return to the romans.shx font. If you do not want to select a different font, you can pick Cancel and AutoCAD will use a default.

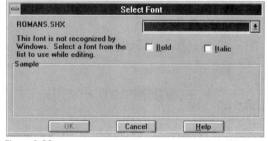

Figure 1.39

Pick: **Cancel**

The Edit Mtext dialog box appears on your screen, as shown in Figure 1.40. The light-colored area of the Edit Mtext dialog box is

shaped like the area you selected at the first set of prompts. You will use it to type in the notes for the plot plan.

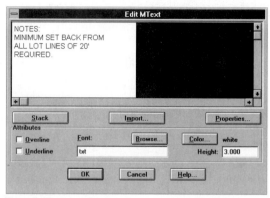

Figure 1.40

Type: **NOTES:** ⏎ **MINIMUM SET BACK FROM ALL LOT LINES OF 20' REQUIRED.**

You can use the standard Windows control keys to edit text in the Edit Mtext dialog box. They are listed below:

Ctrl-C	Copy selection to the Clipboard
Ctrl-V	Paste Clipboard contents over selection
Ctrl-X	Cut selection to the Clipboard
Ctrl-Z	Undo and Redo
Ctrl-SHIFT-SPACEBAR	Insert a nonbreaking space
⏎	End the current paragraph and start a new line

You can create stacked fractions or stack text using the Mtext command. To do this, separate the text to be stacked with either / or ^. Using / draws a line between the numerator and denominator of fractions. Using ^ stacks the text with no line.

Picking on the Import button allows you to select a text file that you have already created with a text editor. Picking Properties will bring up the Mtext Properties dialog box, as shown in Figure 1.41. You can use it to select from the different text styles you have already created, set the text height and direction, and control where Mtext objects are placed in the previously selected area, as well as the width and rotation of the area.

Pick: **Properties**

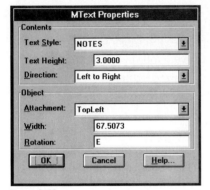

Figure 1.41

Pick: **OK**

(If the Select Font dialog appears again, pick Cancel)

Pick: **OK *(to exit the Edit Mtext dialog box)***

The notes are placed in your drawing, which should appear similar to Figure 1.42.

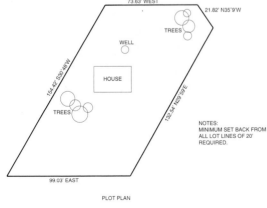

Figure 1.42

Spellchecking Your Drawing

You can use AutoCAD's spell checker to detect spelling errors in your drawing. To use the spell checker,

Pick: **Tools, Spelling**

Select objects: *(pick on the notes you added)* ⏎

If the text contains spelling errors, the Check Spelling dialog box will appear, as shown in Figure 1.43. If your drawing does not contain any misspellings, you will see the message *Spelling complete*.

Pick: **OK**

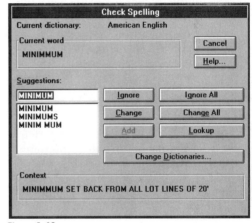

Figure 1.43

Saving Your Drawing

To save your drawing,

Pick: **Save icon**

The default file name is always the current drawing name. When you started the new drawing, you named it *plotplan.dwg*. The drawing is saved on the hard drive, since you did not specify a drive letter.

Transferring Files

■ *Warning:* To complete this section you will need a blank formatted floppy disk. ■

You can transfer files from one drive to another. To transfer a file, place a blank formatted disk in drive *a:* (or substitute your drive letter). To transfer the file named *plotplan.dwg* from the hard drive to a floppy disk in the *a:* drive, you will use the Windows File Manager.

Pick: **(the minimize button in the upper right corner of the AutoCAD screen)**

AutoCAD is minimized and you return to the Windows Program Manager. Double-click on the Main program group. Double-click on the Windows File Manager icon. The Windows File Manager appears on your screen. Use it to copy your saved AutoCAD files from the hard drive to your floppy drive. To change the directory selection, position the arrow cursor over the name of the directory you want to select and press the pick button. If you need to step back out of a subdirectory to a directory one level closer to the root or main directory, click on the *c:* selection. If you need to select a different drive, click on its name, such as *a:*, *b:*, or *c:*. Once you have selected the drive and directory where your file is stored, pick on the file you want to copy. Then, holding the pick button down, drag the file to the floppy disk icon.

A Confirm Mouse Operation dialog box with the message *Are you sure you want to copy the selected files or directories to A:\?* appears.

Pick: **Yes**

Close the Windows File Manager. Once you have done this, double-click on the minimized icon that represents your AutoCAD *session*. If you do not see it near the bottom of the screen, press (ALT)-(ESC) to bring it to the front, so that it is not hidden behind some other window.

You are returned to your AutoCAD drawing session.

Exiting AutoCAD

You can use the Exit selection from the File menu to return to Windows, or you can type *QUIT* at the command prompt. If you have not previously saved your drawing, the File, Exit selection prompts you to save changes, discard changes, or cancel. If you have already saved your drawing, File, Exit exits immediately. If you have not saved your changes and would like to, pick on Save Changes in the dialog box. This will save your drawing to the file name *plotplan.dwg* that you selected when you began the new drawing.

■ *TIP* You can also double-click the *Windows Control box* above the File selection on the menu bar to exit the program. ■

The system returns to the Windows Program Manager. Now you have completed Tutorial 1.

■ *TIP* The End command in AutoCAD saves the file to the current default file name and directly exits the program. However, the Save and Exit series eliminates unused and erased items from the drawing database, resulting in smaller file sizes. The Save, Exit series is thus a better practice. ■

absolute coordinates
alias
angle brackets
architectural units
bearing
blipmarks
buttons
Cartesian coordinate system
command aliasing
command prompt
command window
context sensitive
coordinate values
cursor (crosshairs)
default
default directory
docked
drag
engineering units

file extension
floating
flyouts
font
fractional units
graphics window
grid
heads-up
highlight
implied Crossing mode
implied Windowing mode
instance
menu bar
Object Properties toolbar
objects
options
polar coordinates
prototype drawing
real-world units
relative coordinates

romans
root page
scientific units
selection set
session
shape-compiled font
Standard toolbar
status bar
submenu
title bar
toggle
tool tip
toolbars
transparent command
typing cursor
UCS icon
unidirectional
units
Windows Control box
World Coordinates

Circle Center Radius
Dynamic Text
Edit Text
End
Erase
Grid
Help
Limits

Line
New Drawing
Oops
Multiline Text
Quit
Rectangle
Redo
Redraw View

Save
Snap
Style
Undo
Units
Zoom All

Redraw the following shapes. If dimensions are provided, use the dimensions and create your drawing to show the exact geometry of the part shown. The letter M after an exercise number means that the given dimensions are in millimeters (metric units). If no letter follows the exercise number, the dimensions are in inches. The Ø symbol indicates that the following dimension is a diameter. Do not include dimensions on the drawings.

 1.1 Baseplate

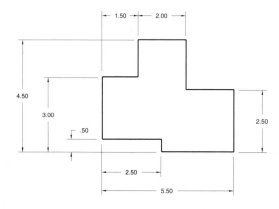

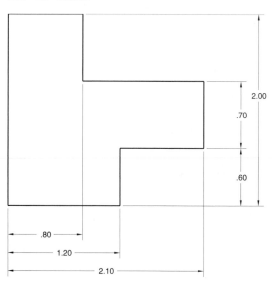 **1.2 Bracket**

1.3 Site Boundary

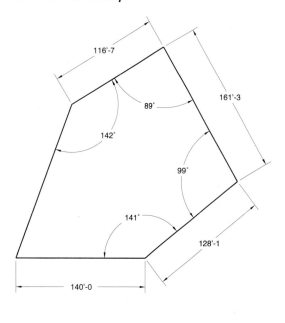

1.4 Filter Plate

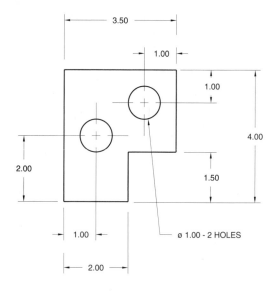

1.5M Gasket

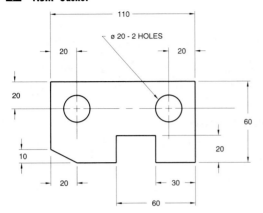

1.6 Spacer

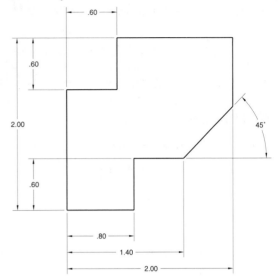

1.7M Guide Plate

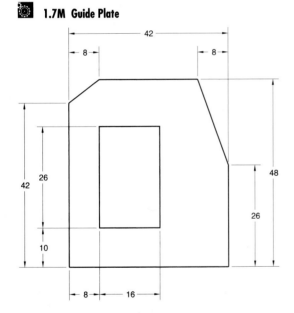

PLATE - 1020 STEEL
2 REQUIRED - FULL SIZE

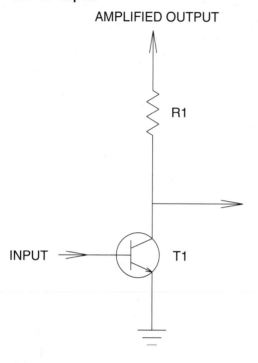

1.8 Amplifier

AMPLIFIED OUTPUT

R1

INPUT

T1

1.9 Dorm Room

Reproduce this drawing using Snap and Grid set to 0.25. The numerical values are for reference only. Use Dtext to label the items.

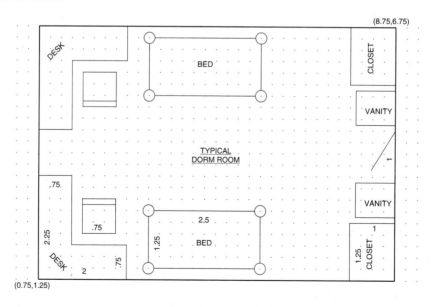

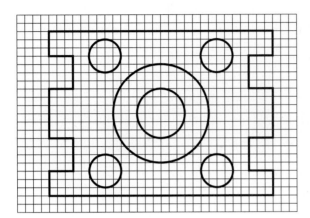

1.10 Template

Draw the shape using one grid square equal to 1/4" or 10mm.

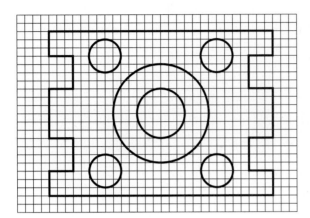

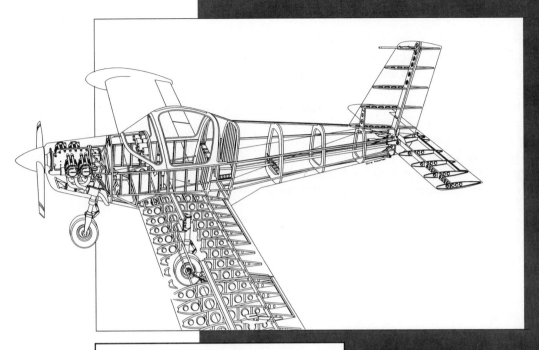

Basic Construction Techniques

Objectives

When you have completed this tutorial, you will be able to

1. Open existing drawings.

2. Work with existing layers.

3. Draw, using the Arc and Circle commands.

4. Set and use running object snaps.

5. Change the display, using Zoom and Pan.

6. Use the Aerial Viewer.

Introduction

You usually create technical drawings by combining and modifying several different basic shapes called *primitives*, such as lines, circles, and arcs, to create more complex shapes. This tutorial will teach you how to use AutoCAD to draw more of the most common basic shapes. As you are working through the tutorial, keep in mind that one of the advantages of using AutoCAD over drawing on paper is that you are creating an accurate model of the drawing geometry. In the next tutorial, you will learn to list information from the drawing database. Information extracted from the drawing is only accurate if the drawing is created accurately in the first place.

Starting

Before you begin, launch Windows and AutoCAD Release 13 for Windows. If you need assistance in loading AutoCAD, refer to the Getting Started section in this manual.

■ *Warning:* If the main AutoCAD screen, including the graphics window, Standard toolbar, Object Properties toolbar, status bar, command window, and menu bar, is not on your display screen, check to be sure that you have configured AutoCAD properly for your computer system. If you are still having difficulty, ask your technical support person for help. ■

Opening an Existing Drawing

In this tutorial you will add arcs and circles to the subdivision drawing provided with the data files that came with this manual. In Tutorial 3 you will finish the subdivision drawing so that the final drawing looks similar to Figure 2.1.

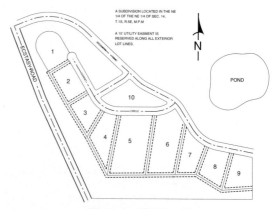

Wannabe Heights Estates

Figure 2.1

 To open an existing drawing you can use the File, Open selection on the menu bar, or pick the icon that looks like an open folder from the Standard toolbar.

Pick: **Open icon**

The Select File dialog box appears on your screen. Use the center portions, which show the default directory and drive, to select the location where your data files have been stored. In the Getting Started portion of this manual, you should have already created a directory called *c:\datafile*, and copied all the AutoCAD data files into it. If you have not done so, you may want to review the Getting Started section of this book. If the correct directory is not showing, double-click on *c:* to move to the root directory. Use the scroll bars if necessary to scroll down the list of directories until you see the appropriate one. The files are shown on the left side of the dialog box. Use the scroll bars to scroll down the list of files until you see the file named *subdivis.dwg*. When you select a file, a preview of the file appears in the box to the right. (AutoCAD Release 13 automatically creates the preview and saves it inside the drawing file when the drawing is saved. You can use the Makepreview command to create a preview for drawings created in older releases.) Figure 2.2 shows the Select File dialog box and a preview of drawing *subdivis.dwg*.

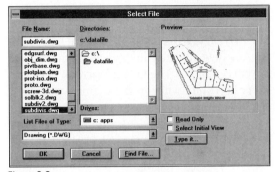

Figure 2.2

If you want to search through existing drawing files, you can use the Find File button near the bottom of the Select File dialog box.

Pick: **Find File**

The Browse/Search dialog box appears on your screen, as shown in Figure 2.3.

Figure 2.3

Use the directory tree at the right to select the directory you want to view the files in. The figure shows AutoCAD's sample files directory. When you have picked on the directory name, all of the drawing files in the directory that can be previewed appear in the box to the left. You can double-click on the preview picture of the file you want to open.

Pick: **(the c:\datafile directory from the directory list at the right)**

Double-click: **(on the picture of file subdivis.dwg)**

When you have opened the file, it appears on your screen, as shown in Figure 2.4. Note that it opens with its own defaults for Grid, Snap, etc. These settings are saved in the drawing file. When you open a drawing, its own settings are used.

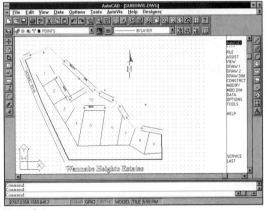

Figure 2.4

Saving as a New File

The Save As command allows you to save your drawing to a new file name. You can select this command from the File selection on the menu bar. Don't use the Save command, because that will save your changes into the original data file.

Pick: **File, Save As**

The Save Drawing As dialog box appears on your screen, similar to Figure 2.5. On your own, use it to select the drive and directory *c:\work* as shown and specify the name for your drawing *subdivis.dwg*. The new file name is the same as the previous file, but the directory is different. This will save a new copy of drawing *subdivis.dwg* in the directory *c:\work*.

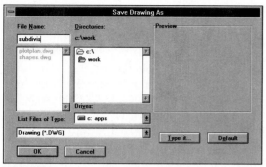

Figure 2.5

The original file *c: \datafile\subdivis.dwg* will remain unchanged on your drive. When you use the SaveAs command and specify a new file name, AutoCAD sets the newly saved file as current.

Notice the message at the command prompt stating that the current drawing name is set to *c:\work\subdivis.dwg*.

Using Layers

You can organize drawing information on different *layers*. You can think of a layer as a transparent drawing sheet that you place over the drawing and that you can remove at will. The coordinate system remains the same from one layer to another, so graphical objects on separate layers remain aligned. You can create an unlimited number of layers within the same drawing. The Layer command controls the color and linetype associated with a given layer, and allows you to control which layers are visible at any given time. Using layers allows you to overlay a base drawing with several different levels of detail (such as wiring or plumbing schematics over the base plan for a building). By using layers, you can also control which portions of a drawing are plotted, or remove dimensions or text from a drawing to make it easier to add or change objects. You can also

lock layers so they are inaccessible but still visible on the screen; you can't change anything on a locked layer until it is unlocked.

Current Layer

The *current layer* is the layer you are working on. Any new objects you draw are added to the current layer. The default current layer is Layer 0. If you do not create and use other layers, your drawing will be created on Layer 0. You used this layer when drawing the plot plan in Tutorial 1. Layer 0 is a special layer that is provided in AutoCAD. You cannot rename it or delete it from the list of layers. Layer 0 has special properties when used with the Block and Insert commands, which you will learn in Tutorial 8.

Layer POINTS is the current layer in *subdivis.dwg*. There can be only one current layer at a time. The name of the current layer appears on the Object Properties toolbar.

Controlling Layers

The Layer Control feature on the Object Properties toolbar is an easy way to control the visibility of existing layers in your drawing. You will learn more about creating and using layers in Tutorial 4. In this tutorial you will use layers that have already been created for you.

To select the Layer Control pull-down feature,

> Pick: **(on the layer name POINTS, which appears on the Object Properties toolbar near the top of the screen)**

The list of available layers pulls down, as shown in Figure 2.6. Notice the special layer 0 displayed near the top of the Layer Control list.

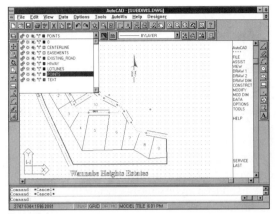

Figure 2.6

Pick: **(on the layer name CENTERLINE from the Layer Control list)**

It becomes the current layer listed on the Object Properties toolbar. Any new objects will be created on this layer until you select a different current layer. Your screen should look similar to Figure 2.7. Notice the layer name CENTERLINE shown at the upper left.

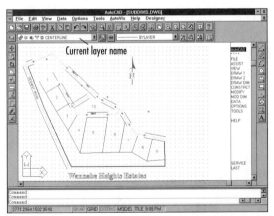

Figure 2.7

Use the Line command you learned in Tutorial 1 to draw a line off to the side of the subdivision drawing. You will notice that it is green and has a centerline linetype (long dash, short dash, long dash). The line you drew is on Layer CENTERLINE. Now erase or undo the line

on your own; you do not want to add it to your drawing.

Color in Layers

Each layer can have a color associated with it. Using different colors for different layers helps you visually distinguish differing types of lines in the drawing. An object's color also controls which pens are used when plotting. You can use different-colored layers to make it easy to set pen colors and widths during plotting.

 AutoCAD provides two different ways of selecting the color for objects on your screen. The best way is to set the layer color and draw the objects on the appropriate layer. This keeps your drawing organized. The other method is to set the object color using Color Control from the Object Properties toolbar. When you do this, all new objects drawn will have this color, regardless of what layer is current, until you change the color again.

Pick: **Color Control icon**

The Select Color dialog box shown in Figure 2.8 appears on your screen.

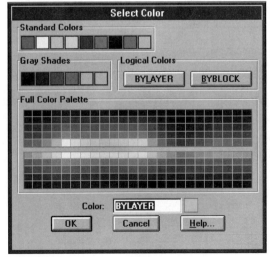

Figure 2.8

The default option for the Color (and also for the Linetype) command is BYLAYER. This is the best selection, because when you draw a line, the color and linetype will be those of the current layer. Otherwise the color in your drawing can become very confusing. You will pick Cancel to leave the Select Color dialog box without making any changes. You will continue to have the color for your new objects determined by the layer they are created on.

Pick: **Cancel**

Linetype in Layers

Layers can have a linetype associated with them as well as a color, as Layer CENTERLINE does. For example, you could create a layer that not only drew all lines in red, but also drew only hidden (short dashed) lines. You will learn how to load the linetypes and create new layers that use them in Tutorial 4.

Layer Visibility

One of the advantages of using layers in the drawing is that you can choose not to display selected layers. This way, if you want to create projection lines, or even notes about the drawing, you can draw them on a layer that you will later turn off so that it is not displayed or plotted. Or you may want to create a complex drawing with many layers, such as a building plan that contains the electrical plan on one layer and the mechanical on another, as well as separate layers for the walls, windows, etc. You can store all of the information in a single drawing, and then you can plot different combinations of layers to create the electrical layout, first-floor plan, and any other combination you want. Next you will use Layer Control to turn *off* Layer POINTS, *freeze* Layer TEXT, and *lock* Layer LOTLINES.

Pick: **(on Layer CENTERLINE, shown on the Object Properties toolbar)**

The list of layers pulls down. Refer to Figure 2.9 as you make the following selections.

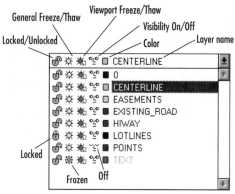

Figure 2.9

Pick: **(on the On/Off icon, which looks like eyes, to the left of Layer POINTS)**

Pick anywhere away from the layer list. Notice that now the points have been turned off so that they no longer appear. Invisible (off) layers are not printed or plotted, but objects on these layers are still part of the drawing. Next you will freeze Layer TEXT.

Freezing Layers

Freezing a layer is very similar to turning it off. You use the Freeze option not only to make the layer disappear from the display, but also to cause it to be skipped when the drawing is regenerated. This feature can noticeably improve the speed with which AutoCAD regenerates a large drawing. You cannot freeze the current layer because that would create a situation where you would be drawing objects you could not see on the screen. The icon for freezing and thawing layers looks like a snowflake when frozen and a shining sun when thawed.

Pick: **(the General Freeze/Thaw icon to the left of Layer TEXT)**

Pick anywhere in the graphics window for the selection to take effect. Layer TEXT is still on, but frozen and therefore invisible. A layer can both be turned off and frozen, but the effect is the same as freezing the layer.

Locking Layers

A locked layer can be seen on the screen, but no changes can be made to it. This is useful when you need the layer for reference, but do not want to change it. An example is when you are trying to move a number of items so that they line up with something on the locked layer, but do not want things on the locked layer to move. You will lock layer LOTLINES. This way you cannot accidentally change the lines that are already on the layer. You can still add new lines to a locked layer.

> Pick: (the Lock/Unlock icon to the left of Layer
> LOTLINES)

Your screen should now look similar to Figure 2.10. Layer CENTERLINE is the current layer. Layer TEXT is frozen and so does not appear. Layer POINTS is turned off and does not appear. Layer LOTLINES is locked so that it can be seen and added to, but not changed. On your own, try erasing one of the lot lines. You will see a message saying that the object is on a locked layer. The object will not be erased.

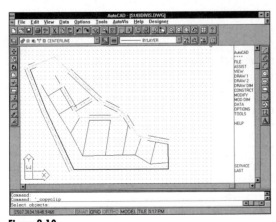

Figure 2.10

Next you will create the curved sections of the road centerline for the subdivision. The straight-line sections that the curves are tangent to have been drawn to get you started. First you will turn on the running object snap for Node.

Using Object Snap

You use the object snap feature in AutoCAD to accurately select locations based on existing objects in your drawing. When you just pick points from the screen, without using object snaps, the resolution of your screen makes it impossible to select points with the accuracy with which AutoCAD stores the drawing geometry in the database. You have seen how it is possible to pick an accurate point on the screen by snapping to a grid point. Object snap makes it possible to accurately pick points on your drawing geometry by snapping to an object's center point, endpoint, midpoint, and so on. Whenever you are prompted to select a point, you can use an object snap to help make an accurate selection. Without this command, it is virtually impossible to locate two objects with respect to each other in a way that gives you correct and useful geometry. Object snap is one of the most important tools that AutoCAD provides. The many specialized object snaps are listed on the Object Snap flyout on the Standard toolbar, shown in Figure 2.11. The tool tip for the default top tool on the Object Snap flyout says *Snap From*.

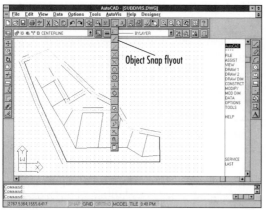

Figure 2.11

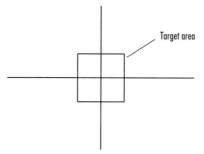

Target area

Figure 2.12

Object snap can operate in two different ways. The first is called *override mode*. With this method, you select the object snap *during* a command. The object snap acts as a modifier within the command string to target the next point you select. When you activate an object snap from within a command in this manner, it is active for one pick only. Remember, you can only use this method during a command that is prompting you to select points or objects. You activate object snaps from within other commands by picking the appropriate icon from the Standard toolbar. The object locations they select are indicated by red dots on the icons.

> ■ *TIP* You can also activate an override by typing the three-letter name any time you are prompted to enter points or select objects. Refer to the Command Summary for the three letter codes. ■

When an object snap is active, AutoCAD has a special cursor, the *aperture*, shown in Figure 2.12.

The square area around the horizontal and vertical lines is called the *target area*. When you are selecting a location using an object snap, the object needs to be within the target area.

The second method for using object snap is called *running mode*. With this method, you turn on the object snap and leave it on before you select any commands. When a running mode object snap is on, during any future command that requests a point location, object selection, or other pick, the aperture box appears on the crosshairs and whatever object snap you have selected is used.

 Next you will turn on the running mode object snap. Use the Standard toolbar to select Running Object Snap from the Object Snap flyout.

Pick: **Running Object Snap icon**

The Running Object Snap dialog box appears on your screen. On your own, use it to select Node by picking the check box to the left of the word Node. An X appears in the box when it is selected, as shown in Figure 2.13. Node snaps to objects drawn with AutoCAD's Point command. If there is more than one point within the aperture box, it finds the one closest to the center of the box, where the crosshairs are located. The bottom portion of the dialog box is used to resize the aperture box. It can be useful to make the aperture box smaller when working in a crowded drawing, making it easier to select only the object you want. When the drawing is not crowded, a larger aperture box makes selecting faster, because it doesn't require delicate hand-eye coordination to position. Try using the slider to make the aperture box smaller and then larger. When you have finished, pick OK.

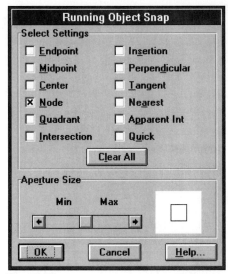

Figure 2.13

> ■ *TIP* To eliminate any running mode object snaps you may have turned on, use the Running Object Snap dialog box and select Clear All. ■

Now you are ready to start creating arcs at accurate locations in the drawing. When you are prompted to select, look for the aperture box on the crosshairs. This lets you know that Snap to Node is being used.

On your own, use the Layer Control button located on the Object Properties toolbar to turn Layer POINTS back on. Do this now before continuing. Make sure that CENTERLINE is the current drawing layer.

Using Arc

You can use the Arc command in eleven different ways to create arcs in your drawing. The Arc command is located on the Draw toolbar. You can also type *ARC* at the command line. The Arc flyout is shown in Figure 2.14.

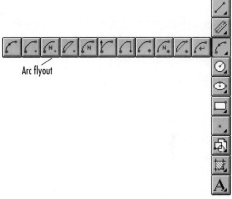

Arc flyout

Figure 2.14

Each of the Arc command options requires you to input point locations. You can define point locations by defining the coordinate values of the points, or by locating the points with the cursor and pressing the pick button on your pointing device. For the exercises presented in this tutorial, follow the directions carefully so that your drawing will turn out correctly. Keep in mind that if you were designing the subdivision, you might not necessarily use all of these command options; this tutorial will demonstrate many different ways to draw circles and arcs. When you are using AutoCAD later for design, select the command options that are appropriate for the geometry in your drawing.

Pick: (on the Arc flyout)

The Arc command selections fly out on your screen.

Arc 3 Points

 The 3 Points option of the Arc command draws an arc through three points that you specify. Remember, to specify locations you can pick them or type in absolute, relative, or polar coordinates. You will draw an arc using the 3 Points option; refer to Figure 2.15 as you are selecting. The Snap to Node running object snap will help you pick the points drawn in the data file. Be sure to have only one point in the aperture box when you press the pick button.

Pick: **Arc 3 Points icon**

Center/<Start point>: *(select point 1)*

Center/End/<Second point>: *(select point 2)*

AutoCAD enters drag mode. When you are in drag mode, you can see the arc move on the screen as you move the cursor, as shown in Figure 2.15. Many AutoCAD commands permit dynamic specification, or dragging, of the image on the screen.

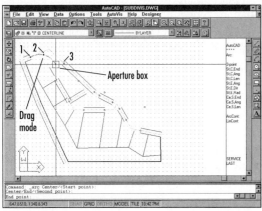

Figure 2.15

Move the cursor around the screen and see how it affects the way the arc would be drawn. You should remember this feature from drawing your circles in Tutorial 1.

End point: *(select point 3)*

The third point defines the endpoint of the arc. The radius of the arc is calculated based on the locations of the three points. Your drawing should now show the completed arc, as in Figure 2.16.

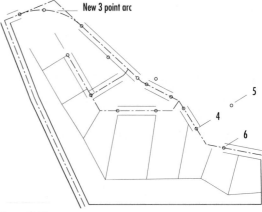

Figure 2.16

Arc Start Center End

 Next, you will draw an arc by specifying Start Center End.

Pick: **Arc Start Center End icon**

Figure 2.16 shows the points that you will select to create this arc. At the prompt:

Center/<Start point>: *(select point 4)*

AutoCAD prompts you for the center point.

Center: *(select point 5)*

The distance between the start point and the center point will be used as the radius for the arc. You must specify the endpoint location to define the end of the arc. You are in drag mode.

Angle/Length of chord/<End point>: *(select point 6)*

An arc is always drawn in a counterclockwise direction from the start point. It is therefore important to correctly define the start point and the center point. Figure 2.16 shows the point locations needed to draw a concave arc.

If the start point was located where the endpoint is, a convex arc outside the centerlines would have been drawn. When you have added the arc correctly, your result should look like Figure 2.17.

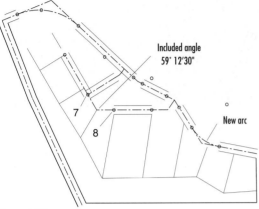

Figure 2.17

On your own, try drawing another arc with the Arc Start Center End option, this time picking point 6 first, then the center, and then point 4. You can see that it is drawn counterclockwise, resulting in a convex arc. Undo this backwards arc by typing *U* ⏎ at the command prompt.

Arc Start End Angle

 Arc Start End Angle draws an arc through the selected start and endpoints using the *included angle* you specify. Figure 2.17 notes the points that you will select. To draw an arc with the Start End Angle selection,

> *Pick:* **Arc Start End Angle icon**
>
> Center/<Start point>: *(select point 7)*

Next, specify the endpoint for the arc.

> End point: *(select point 8)*

The arc is defined by the included angular value (often called the *delta angle* in survey drawings) from the start point to the endpoint. Angular values are positive in the counterclockwise direction. You can use negative values to get an arc drawn in the clockwise direction. (Type *d* for ° when typing in surveyor's angles. Use the single quote and double quote for minutes and seconds.)

> Included angle: **59d12'30"** ⏎

When you have drawn the arc, your screen should look like Figure 2.18.

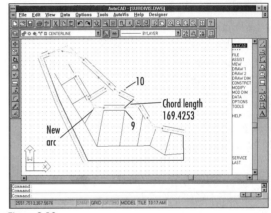

Figure 2.18

Arc Start Center Length

Arc Start Center Length draws an arc specified by the start point and center point of the arc, and the *chord length*. A chord length is the straight-line distance from the start point of the arc to the endpoint of the arc. You can enter negative values for the chord length to draw an arc in the opposite direction. Figure 2.19 shows the information that is input to create this type of arc.

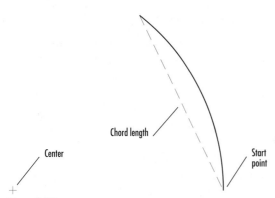

Center

Chord length

Start
point

Figure 2.19

You will draw the next arc using Arc Start
Center Length. Refer back to Figure 2.18 for
the points to select.

Pick: **Arc Start Center Length icon**

Center/<Start point>: *(select point 9)*

Center: *(select point 10)*

Length of chord: **169.4253** (↵)

Your new arc should look like the one shown in
Figure 2.20.

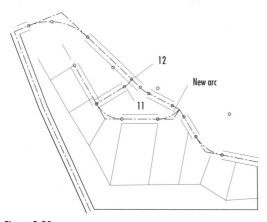

12

New arc

11

Figure 2.20

Arc Start End Radius

 Draw the next arc using Arc Start End
Radius. The locations of the start and
endpoints are shown in Figure 2.20.

Pick: **Arc Start End Radius icon**

Center/<Start point>: *(select point 11)*

End point: *(select point 12)*

Radius: **154.87** (↵)

> ■ **TIP** You can use negative radius values
> to create a *major arc.* A major arc is an arc
> which comprises more than 180 degrees of
> a circle ■

Your drawing should look similar to Figure 2.21.

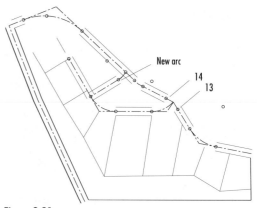

New arc

14

13

Figure 2.21

Continuing an Arc

Arc Continue allows you to join an arc
to a previously drawn arc or line. You
will give this a try off to the side of the
subdivision drawing and then erase or undo it,
as it is not a part of the drawing. To draw an
arc that is the continuation of an existing line,

Pick: **Line icon**

and draw a line anywhere on your screen.
Once you have a line drawn,

Pick: **Arc Continue icon**

The last point drawn for the line becomes the first point for the arc. AutoCAD is in drag mode and an arc appears from the end of the line. AutoCAD prompts you for an endpoint.

End point: *(select an endpoint)*

Next you will use the Arc flyout to select Arc Start Center End and draw another arc. Then you will continue an arc from the endpoint of that arc.

Pick: **Arc Start Center End icon**

Center/<Start point>: *(select a start point)*

Center: *(select a center point)*

Angle/Length of chord/<End point>: *(select an endpoint)*

Pick: **Arc Continue icon**

End point: *(select a location for the endpoint)*

> ■ *TIP* After you draw an arc, restarting the command and then pressing the return button or ⏎ also engages the Continue option. You can use this to continue a line from the previously drawn arc or line. ■

Your drawings should look similar to Figure 2.22.

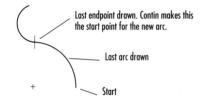

Last endpoint drawn. Contin makes this the start point for the new arc.

Last arc drawn

Start

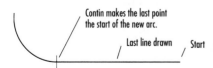

Contin makes the last point the start of the new arc.

Last line drawn   Start

Figure 2.22

On your own, erase or undo the extra arcs and lines you have created.

Quick Selections for Arc

To select the Arc command, you can also type the command alias, *A*, at the command prompt. When you start the Arc command this way, the default for drawing the arc is the 3 Points option you learned earlier in this tutorial. If you want to use another option, you can type the command option letter at the prompt. You will start the Arc command by typing its alias. You will then specify the start, end, and angle of the arc by typing the command option letters at the prompt. Refer to Figure 2.21 for the points to select.

> ■ *TIP* Remember that keyboard entry in AutoCAD is not case-sensitive. You can type uppercase, lowercase, or mixed-case letters. In this book your keyboard entry is shown as uppercase to make it easy to identify what you are supposed to type, but AutoCAD does not require uppercase letters. ■

Command: **A** ⏎

The command Arc is echoed at the command prompt, followed by the prompt,

Center/<Start point>: *(pick point 13)*

Center/End/<Second point>: **E** ⏎

End point: *(pick point 14)*

Angle/Direction/Radius/<Center point>: **A** ⏎

Included angle: **30d54'04"** ⏎

The arc is added to your drawing. However, it may be very hard to see because of the size of the arc, relative to the size of the screen and drawing. AutoCAD provides Zoom commands to change the size of the image on your display.

Using Zoom

The Zoom flyout is located on the Standard toolbar shown in Figure 2.23.

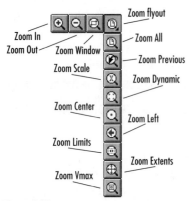

Figure 2.23

 To zoom in on an area of the drawing, you can select the Zoom In icon.

Pick: **Zoom In icon**

Your drawing is enlarged to twice its previous size on the screen so that you can see more detail. Your screen should appear similar to Figure 2.24.

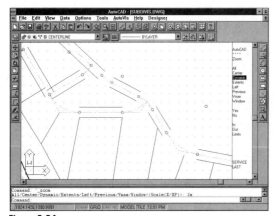

Figure 2.24

 Now you will zoom back out using the Zoom Out selection from the Standard toolbar.

Pick: **Zoom Out icon**

Your drawing should return to its original size on the screen. The Zoom In and Zoom Out options use the Scale feature of the Zoom command to zoom to a scale of 2X (twice the previous size) and .5X (half the previous size).

Zooming Using Scale Factors

 You can also zoom using *scale factors* when you pick the Scale option of the Zoom command. Scale factor 1.00 shows the drawing limits. Scale factor .5 shows the drawing limits half-size on the screen. Typing *X* after the scale factor makes the zoom scale relative to the previous view. For example, entering *2X* causes the new view to be shown twice as big as the view established previously, as you saw when using Zoom In. A scale factor of .5X reduces the view to half the previous size, as you saw when using Zoom Out. Zoom Scale uses the current left corner or 0,0 coordinates as the base location for the zoom. Typing *XP* after the scale factor makes the new zoom scale relative to *paper space*. A scale factor of .5XP means that the object will be shown half-size when you are laying out your sheet of paper. You will learn more about paper space in Tutorial 5.

This time, you will select the Zoom command by typing its alias at the command prompt.

Type: **Z** ↵

All/Center/Dynamic/Extents/Left/Previous/Vmax/Window/
<Scale(X/XP)>: **2X** ↵

The objects are enlarged on the screen to twice the size of the previous view. Repeat the command.

Command: ↵

All/Center/Dynamic/Extents/Left/Previous/Vmax/Window/
<Scale(X/XP)>: **.5** ↵

The drawing limits appear on the screen at half their original size. The area shown on the screen is twice as big as the drawing limits. To restore the original view,

Command: ↵

All/Center/Dynamic/Extents/Left/Previous/Vmax/Window/
<Scale(X/XP)>: **1** ↵

Zoom Left and Zoom Center let you specify a fixed point, the center point or the lower left corner respectively, before you enter the zoom scale factor. You will learn more about these options in Tutorial 9.

 You can use Zoom Window and create a window around the area that you want to enlarge to fill the screen. This lets you quickly enlarge the exact portion of the drawing that you are interested in seeing in detail. You will select this command using the icon on the Standard toolbar.

Pick: **Zoom Window icon**

The Zoom command is echoed at the command prompt. Notice that the Zoom command has an apostrophe in front of it in the command prompt area. This means that it is a transparent command and can be selected during another command. You will zoom in on the area shown in Figure 2.25. To create the window, you will first select a point at the top left corner of the area to be zoomed, and then select a point on the diagonal in the lower right corner, as shown in Figure 2.25.

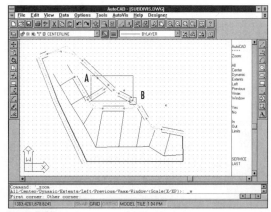

Figure 2.25

First corner: *(select point A)*

Other corner: *(select point B)*

The defined area is enlarged to the full screen size, as shown in Figure 2.26.

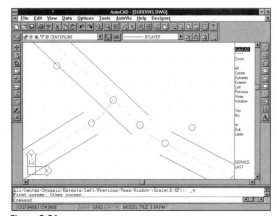

Figure 2.26

 To return an area to its previous size, you will pick Zoom Previous from the Standard toolbar.

Pick: **Zoom Previous icon**

Your drawing is returned to its original size. Areas can be repeatedly zoomed, that is, you can zoom in on a zoomed area; in fact you can continue to zoom until the portion shown on the display is ten trillion times the size of the original.

Using the Aerial Viewer

 To allow you to quickly zoom in and move around in a large drawing, AutoCAD provides the Aerial Viewer. To select the Aerial Viewer, you will pick the icon that looks like an airplane from the Standard toolbar.

Pick: **Aerial View icon**

The Aerial Viewer appears on your screen, as shown in Figure 2.27. If the Aerial Viewer window is not as large as you would like, you can pick on its corners and drag to increase its size. You can move it around on your screen by picking on its title bar and moving it to a new location while holding down the pick button.

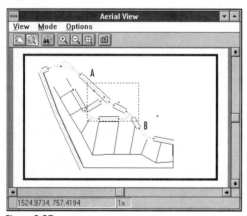

Figure 2.27

■ *Warning:* You cannot use the Aerial Viewer unless you have AutoCAD configured to use a Windows accelerated display list driver. If you do not have enough system memory to use the display list driver, skip the next section on

using the Aerial Viewer. If you have trouble configuring the Aerial Viewer, refer to the Getting Started section of this manual. The Aerial Viewer does not work when you are in paper space or when you are using dynamic 3D viewing. ■

You will use the crosshairs and form a window in the Aerial Viewer around the area of the drawing you would like to enlarge on the screen.

Pick: **(point A in Figure 2.27)**

Pick: **(point B)**

Once you have selected point B, the area enclosed in the window in the Aerial Viewer is enlarged to fill the graphics window, as shown in Figure 2.28. The thick border around the area in the Aerial Viewer shows which portion is your current view. The Aerial Viewer remains on your screen, available for continued use.

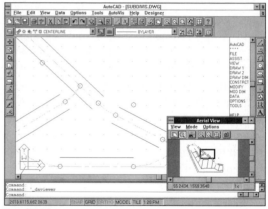

Figure 2.28

 Next, you will select the Pan option from the toolbar in the Aerial Viewer. Be sure you pick the Pan button from the Aerial Viewer toolbar and not the Pan command from the Standard toolbar, as this command is also located there. The two commands have the same effect of moving the view while maintaining the same zoom factor, but the two versions are implemented differently. Using the Aerial Viewer, you can easily position the area of the screen you want to view. From the Aerial Viewer toolbar,

Pick: **Pan icon**

The Pan icon on the Aerial Viewer toolbar becomes highlighted. Notice that now, instead of crosshairs that move around the Aerial Viewer screen to let you select, you see a box. Position the box over the area shown in Figure 2.29 to move the view at the same enlargement to the new area. Once you have the box positioned,

Press: **(the pick button)**

The view on your screen moves so that it is similar to Figure 2.29.

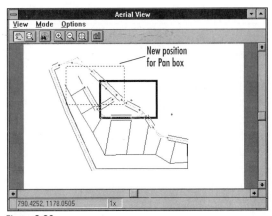

Figure 2.29

 Next, try the Locator button on the Aerial Viewer toolbar. To use the Locator, you will pick its icon and drag onto your drawing in the graphics window. While you are *holding down* the pick button, the Locator displays a target box on the screen. Keeping the pick button depressed, position this target box over the area of the drawing in the graphics window that you want to examine.

The area of the drawing over which the Locator is positioned is shown enlarged in the Aerial Viewer. (You use the Options, Locator Magnification selection from the Aerial Viewer menu bar to change the magnification for the Locator.)

When you release the pick button, the area over which the locator is positioned is enlarged in the graphics window. This is useful for finding small details in a large drawing.

Pick: **Locator icon**

Keeping the pick button depressed, move the Locator target around in the graphics window. Notice that the same area is shown in the Aerial Viewer. You can use this technique to locate details in the drawing and quickly enlarge them on the screen. When you are satisfied with the area shown in the Aerial Viewer, release the pick button. The area of the drawing where the Locator was positioned is enlarged on the screen. To return to the original view you will pick Zoom Previous from the Standard toolbar.

Pick: **Zoom Previous icon**

To remove the Aerial Viewer from your screen, double-click the Windows Control box in the upper left corner of the Aerial Viewer.

You will return to the original view by using Zoom All.

Zoom All

 Zoom All returns the drawing to its original size by displaying the drawing limits, or displaying the drawing extents (all of the drawing objects), whichever is larger. You can select the Zoom All icon from the Standard toolbar.

Pick: **Zoom All icon**

> ■ *TIP* If you open a drawing that has been saved with the view zoomed in, you can use Zoom All to return to the original drawing limits. ■

The drawing should return to its original size, that is, as it was before you began the Zoom command. Experiment on your own with the other options of the Zoom command and read about them using the Help screen. The Vmax option lets you zoom out as far as possible without causing AutoCAD to *regenerate* the drawing. AutoCAD uses the *virtual screen*, a file from which the screen display is created. If you try to zoom out to areas that are not calculated in the current virtual screen file, AutoCAD will have to regenerate the drawing to calculate the display for the new area. Regenerating can take quite a while on a complex drawing, so it is useful to be able to zoom out without causing a regeneration. Zoom Dynamic works in a similar way to the Aerial Viewer. Using Zoom Dynamic, you position and size a window on the area of the drawing you want to zoom. When you have the window positioned and sized the way you want, you press the return button or ⏎ to select that area to zoom in on.

Using Pan

 As you saw using the Aerial Viewer, the Pan command lets you position the drawing view on the screen without changing the zoom factor. Unlike the Move command, which moves the objects in your drawing to different locations on the coordinate system, the Pan command does not change the location of the objects on the coordinate system. Rather, your view of the coordinate system and the objects changes to a different location on the screen. The Pan command prompts you to select two points. The first point you select is a base point; the second point is the new location you want that point to have on the screen.

Pick: **Pan Point icon**

Displacement: **(pick a point near the center of the screen)**

Second point: **(pick a point about 1 inch to the right of the first point)**

The drawing should shift over to the right on the screen by about 1 inch. Notice the apostrophe that precedes the command, indicating that the Pan command can be used transparently during another command.

Using Circle 2 Point, Circle 3 Point, and Circle Tangent Tangent Radius

You have already seen how you can use the Circle command by specifying a center point and a radius value in Tutorial 1. You can also use the Circle command to draw circles by specifying any two points (Circle 2 Point), any three points (Circle 3 Point), or two tangent references and a radius (Circle Tan Tan Radius).

Before adding the next two circles, you will turn on the Endpoint running mode object snap so that the circles you create will line up exactly with the ends of the existing lot lines. You will turn off the Node running object snap that you have been using. (You can use more than one object snap at the same time, but to make sure that you select the endpoints and not point objects in the drawing, you will turn Node off.)

Pick: **Running Object Snap icon**

Pick: **(to turn on Endpoint)**

Pick: **(to turn off Node)**

Pick: **OK**

The Endpoint running mode object snap is turned on and Node is turned off. When you see the aperture box appear on the crosshairs, AutoCAD will find the nearest endpoint of the line or arc object that you select.

On your own, pick the Layer Control from the Object Properties toolbar. When the list of layers pulls down, pick on the layer name LOTLINES to set it as the current layer. Do this before continuing, so that the next circle you create is on layer LOTLINES. It is still locked so you cannot change any of the objects on it, but new objects will be created on layer LOTLINES. If you need to make corrections you will have to unlock the layer first.

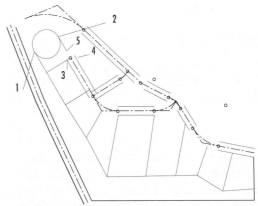

Figure 2.30

Circle 2 Point

 First you will draw a circle using Circle 2 Point. The two points that you select are the endpoints of the circle's diameter. Refer to Figure 2.30.

Pick: **Circle 2 Point icon**

First point on diameter: *(select point 1)*

You are in drag mode and you can see the circle move around on your screen with your cursor.

Second point on diameter: *(select point 2)*

The circle is drawn, using the endpoints of the lot lines you selected as the endpoints for the diameter of the circle, as shown in Figure 2.30.

Circle 3 Point

 To draw a circle using Circle 3 Point, specify any three points on the circle's circumference. Refer to Figure 2.30 for the points to select.

Pick: **Circle 3 Point icon**

First point: *(select point 3)*

Second point: *(select point 4)*

You are in drag mode again; you can see the circle being created on your screen as you move the cursor.

Third point: *(select point 5)*

The circle has been defined by three points on its circumference. Your screen should appear similar to Figure 2.31.

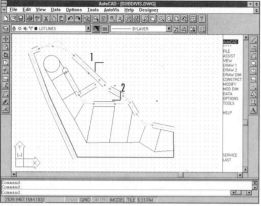

Figure 2.31

You will use Circle Tangent Tangent Radius to draw a circle tangent to two angled centerlines. First, zoom up the area shown in Figure 2.31.

Pick: **Zoom Window icon**

First corner: *(pick point 1)*

Other corner: *(pick point 2)*

The area should be enlarged on your screen, as shown in Figure 2.32.

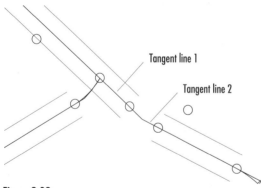

Tangent line 1

Tangent line 2

Figure 2.32

Turn off the Endpoint running object snap. To quickly turn off all running mode snaps, you can pick Clear All from the Running Object Snap dialog box.

Pick: **Running Object Snap icon**

Pick: **Clear All**

Pick: **OK**

Any running mode object snaps are now turned off. It is a good idea to do this, because sometimes object snaps can interfere with the selection of points and with the operation of certain commands. Since the Circle Tangent Tangent Radius option for drawing circles uses the Tangent object snap, it is better to make sure other object snaps are turned off.

The next circle you draw should be created on Layer CENTERLINE, so set that layer current at this time. On your own, use the Object Properties toolbar to select the Layer Control list. Pick on the layer name CENTERLINE to make it current.

Circle Tangent Tangent Radius

Circle Tangent Tangent Radius requires that you specify two objects that the resulting circle is tangent to, and the radius of the resulting circle. This method is frequently used in laying out road centerlines, by selecting the two straight sections of the road centerline to which the curve is tangent and then specifying the radius.

Pick: **Circle Tan Tan Radius icon**

You will notice that your cursor now shows the aperture box, with the horizontal and vertical lines crossing through it. Refer to Figure 2.32 for the lines to select. You can select either line first.

Enter Tangent spec: *(pick line 1)*

Enter second Tangent spec: *(pick line 2)*

Radius <34.0000>: **267.3098** (↵)

A circle with a radius of 267.3098 is drawn that is tangent to both the original lines, as shown in Figure 2.33.

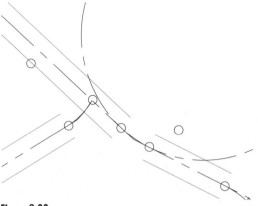

Figure 2.33

You can also use Circle Tangent Tangent Radius to draw a circle tangent to two circles.

Draw two circles off to the side of the subdivision drawing on your own, using Circle 2 Point for the first one and Circle 3 Point for the second. Then you will draw a circle tangent to them.

Pick: **Circle Tan Tan Radius icon**

Enter Tangent spec: *(select one of the circles)*

Enter second Tangent spec: *(select the other circle)*

Radius<267.3098>: **150** ⏎

A circle with a radius of 150 is drawn that is tangent to both circles.

> ■ *TIP* If you get the message, *Circle does not exist*, the radius you specified may be too small or too large to be tangent to both lines. ■

Erase the extra circles on your own. Use Zoom All to return to the original view. Save your drawing at this time. When you have successfully saved your drawing,

Pick: **File, Exit**

You are returned to the Windows Program Manager. You will finish the subdivision drawing in the next tutorial.

turn all layers on

aperture	layer	regenerate
chord length	lock	running mode
current layer	major arc	scale factors
delta angle	override mode	target area
freeze	paper space	virtual screen
included angle	primitives	

Aerial Viewer	Circle 3 Point	Snap to Endpoint
Arc 3 Points	Circle Tan Tan Radius	Snap to Node
Arc Continue	Color Control	Save As
Arc Start Center End	Layer Control	Zoom In
Arc Start Center Length	Locator	Zoom Out
Arc Start End Angle	Open Drawing	Zoom Previous
Arc Start End Radius	Pan	Zoom Scale
Circle 2 Point	Running Object Snap	Zoom Window

Redraw the following shapes. If dimensions are given, create your drawing geometry exactly to the specified dimensions. The letter M after an exercise number means that the given dimensions are in millimeters (metric units). If no letter follows the exercise number, the dimensions are in inches. Do not include dimensions on the drawings. The Ø symbol means diameter; R indicates a radius.

2.1M Bracket

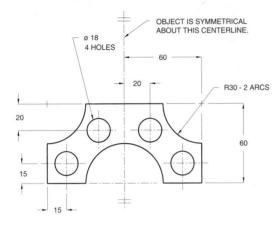

2.2 Gasket

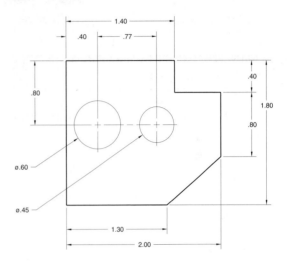

2.3M Flange

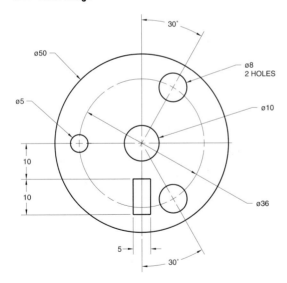

2.4M Puzzle

Draw the figure shown according to the dimensions provided. From your drawing determine what the missing distances must be (hint: use DIST).

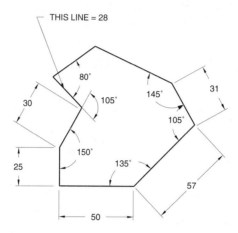

THIS LINE = 28

2.5 Plot Plan

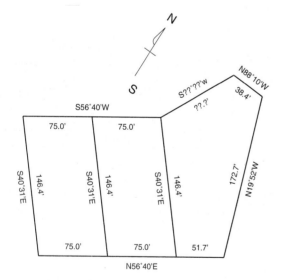

2.6 Power Supply

Draw this circuit using the techniques you have learned.

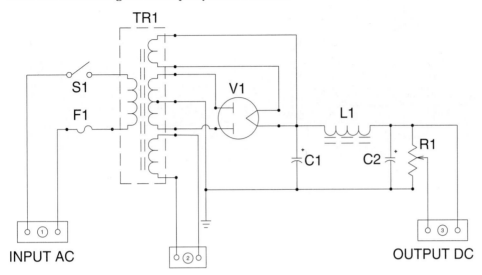

POWER SUPPLY

2.7 Link

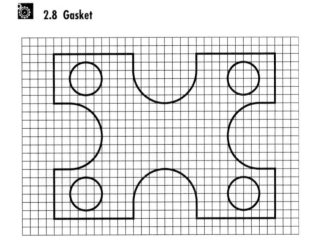

ø0.76
X TWO HOLES

.825 R BOTH ENDS

3.17

2.8 Gasket

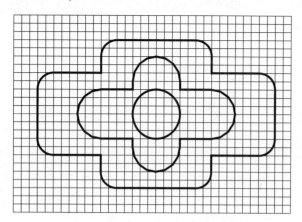

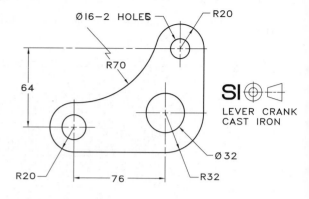

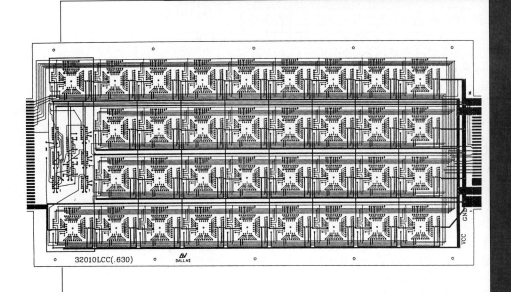

32010LCC(.630)

DALLAS

Basic Editing and Plotting Techniques

Objectives

When you have completed this tutorial, you will be able to

1. Modify your drawing, using the Fillet, Chamfer, Offset, and Trim commands.

2. Create and edit polylines and splines.

3. List graphical objects, locate points, and find areas from your drawing database.

4. Change properties of drawing objects.

5. Create multilines and multiline styles.

6. Print or plot your drawing.

Introduction

This tutorial will teach you how to modify some of the basic shapes you have created to create a wider variety of shapes required for technical drawings. As you are working through the tutorial, keep in mind that one of the advantages of using AutoCAD over drawing on paper is that you are creating an accurate model of the drawing geometry. You will see how to use this accurate drawing database to find areas, lengths, and other information. In later tutorials, you will also learn to use basic two-dimensional shapes as a basis for three-dimensional solid models.

In this tutorial you will finish the subdivision drawing that you started in Tutorial 2. A drawing has been provided with the data files that accompany this manual. You will start from it, in case the drawing that you created in Tutorial 2 has some settings that are different from the data file. You will trim lines and arcs, and add fillets, multilines, and text to finish the subdivision drawing. When the drawing is completed, you will plot your final result.

Starting

Before you begin, launch Windows and AutoCAD Release 13 for Windows. If you need assistance in loading AutoCAD, refer to the Getting Started section of this manual.

Starting from a Prototype Drawing

To continue the subdivision drawing, you will start a new file, using the drawing file *subdiv2.dwg* as a prototype drawing. Any drawing can be used as the prototype to start a new file. When you start a drawing from a

prototype, a copy of the prototype drawing is named with the new file name and this newly copied drawing is the one that appears in the AutoCAD drawing editor. Use the New selection from the Standard toolbar to start a new file.

Pick: **New icon**

The Create New Drawing dialog box shown in Figure 3.1 appears on your screen. You will use it to select drawing *subdiv2.dwg* from the data files to act as the prototype.

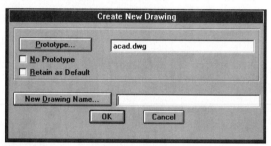

Figure 3.1

■ *Warning:* AutoCAD has three system variables that control the use of dialog boxes. They are Filedia, Cmddia, and Attdia. Filedia controls whether a dialog box is used for file selection. Cmddia controls whether dialog boxes are used with other commands, for example, the Plot command. Attdia controls whether a dialog box is used with the Insert command. Setting these variables to 1 uses the dialog box in each instance. Setting the variables to 0 suppresses the use of the dialog box, so that the command line is used instead. If you have trouble getting the Create New Drawing dialog box to appear, type *FILEDIA* at the command prompt and make sure its value is set to 1. ■

Pick: **Prototype**

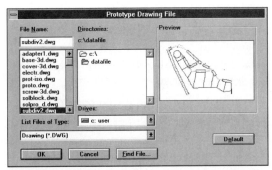

Figure 3.2

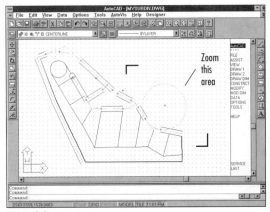

Figure 3.3

Figure 3.2 shows the Prototype Drawing File dialog box with the file *c:\datafile\subdiv2.dwg* selected. Notice that you can see its picture in the Preview area. On your own, use the center area of the dialog box to select the correct drive and directory, *c:\datafile*. Select the file *subdiv2.dwg* from the list of files at the left. When you have the correct drawing shown,

>*Pick:* **OK**

You are returned to the Create New Drawing dialog box. Pick in the box to the right of New Drawing Name so that the typing cursor appears in the box. For the new drawing name,

>*Type:* **MYSUBDIV**

>*Pick:* **OK**

The new drawing is created in the default directory, *c:\work*, that you created in the Getting Started portion of this book. You can use the New Drawing Name button to select a different drive and directory if necessary.

The drawing on your screen should appear similar to Figure 3.3.

On your own, use Zoom Window to enlarge the view of the area indicated in Figure 3.3. When you are finished zooming, your screen will look similar to Figure 3.4.

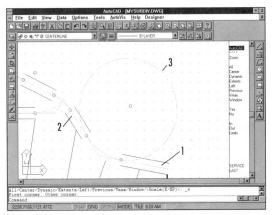

Figure 3.4

Using Trim

 The Trim command removes part of an object. It is located on the Trim and Extend flyout on the Modify toolbar. Trim has two steps. First, you are prompted to select the objects that you will use as *cutting edges*. The cutting edges are drawing objects that you use to cut off the portions that you want to trim. The cutting edge objects you select must cross the objects that you want to trim at the point at which you want to trim. You press ⏎ to indicate that you are done selecting cutting edges and want to begin the second step. After you press ⏎, you are prompted to select the portions of the objects that you want to trim. Pick on the portions you want removed by the Trim command.

The Project option of the Trim command gives you three choices for the projection method used by the command:

View	trims objects where they intersect as viewed from the current viewing direction.
None	trims objects only where they intersect in 3D space.
UCS	trims objects where they intersect in the current User Coordinate System.

The Edge option lets you select whether objects are trimmed only where they intersect in 3D space, or where they would intersect if the edge were extended. The Undo option lets you undo the last trim without exiting the command, similar to the Undo option you used with the Line command. The Project and Edge options are very useful when you are working with 3D models as you will be in Tutorial 9.

You will pick the Trim command from the Modify toolbar and use it to remove the excess portion of the circle. Refer to Figure 3.4 as you make your selections.

Pick: **Trim icon**

Select cutting edges: (Projmode = UCS, Edgemode = No extend)

Select objects: *(select lines 1 and 2)*

The cutting edges selected will be highlighted. Note that the command line reported the projection mode set to UCS and the edge mode set to no extension.

Select objects: ⏎

Now you are done selecting the cutting edge. Next, select the portion of the circle to be removed.

<Select object to trim>/Project/Edge/Undo: *(pick on the circle near point 3)*

You will press ⏎ to end the command,

<Select object to trim>/Project/Edge/Undo: ⏎

When you are done, your figure should look similar to Figure 3.5.

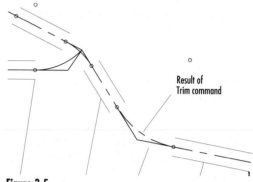

Result of Trim command

Figure 3.5

Refer to Figure 3.6 to understand how the cutting edge works in relation to the object to be trimmed.

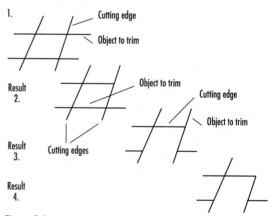

1. Cutting edge
   Object to trim

Result 2. Object to trim
   Cutting edge
   Object to trim

Result 3. Cutting edges

Result 4.

Figure 3.6

When lines come together, as at a corner, you can select one or both of the lines as cutting lines, as shown in Figure 3.6, parts 1 and 3. Again, the cursor location determines which portion of the line is removed.

On your own, restore the previous zoom factor by selecting Zoom Previous from the Standard toolbar. Next, unlock layer LOTLINES so that you can make changes to the objects on that layer. Currently, the layer is locked, so you cannot trim the lines and circles on layer LOTLINES. Use the Layer Control list, which pulls down from the Object Properties toolbar, to unlock layer LOTLINES on your own. Leave layer CENTERLINE as the current layer. When you have finished these steps, your screen should look like Figure 3.7.

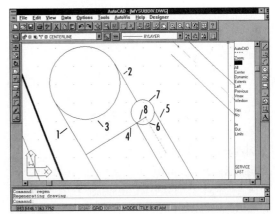

Figure 3.8

■ *TIP* When you zoom in on an area, circles may not be shown correctly. On the screen, AutoCAD draws circles by approximating them with a number of straight line segments. The number of segments used may look fine when the circle appears small on the screen, but when you zoom in on it, it may appear jagged. To eliminate this poor appearance, you can type *REGEN* at the command prompt to cause the display to be recalculated from the drawing database and redisplayed on the screen correctly. ■

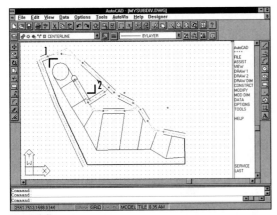

Figure 3.7

Next, use Zoom Window to enlarge the area shown on your own. Pick near point 1 for the first corner of the window and near point 2 for the other corner. When you have finished this step your screen should look like Figure 3.8.

Next, you will trim the excess portions of the circles that you created using the 2 Point and 3 Point options. The arcs that remain when you are finished trimming will form the lot line and the cul-de-sac for the road.

Pick: **Trim icon**

Select cutting edges: (Projmode = USC, Edgemode = No extend)

Select objects: *(select lines 1 and 2)*

Select objects: ⏎

<Select object to trim>/Project/Edge/Undo: *(select on the circle near point 3)*

<Select object to trim>/Project/Edge/Undo: ⏎

Command: ↵ *(to restart the Trim command)*

Select objects: *(select lines 4 and 5)*

Select objects: ↵

<Select object to trim>/Project/Edge/Undo: *(pick on the circle near point 6)*

<Select object to trim>/Project/Edge/Undo: ↵

Command: ↵ *(to restart the Trim command)*

Select objects: *(select circle 7)*

Select objects: ↵

<Select object to trim>/Project/Edge/Undo: *(pick on the line near point 8)*

<Select object to trim>/Project/Edge/Undo: ↵

When you have finished trimming the circles, your screen should look like Figure 3.9. Redraw your screen if necessary on your own.

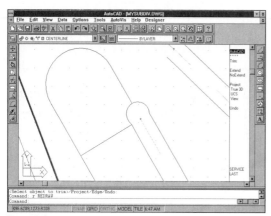

Figure 3.9

On your own, use Zoom Previous to restore the previous zoom factor.

Using Offset

 The Offset command creates a new object parallel to a given object. The Offset command is located on the Modify toolbar, under the Copy flyout; see

Figure 3.10. You will use Offset to create parallel curves 30 units from either side of the curved centerlines that you drew in Tutorial 2.

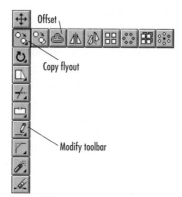

Figure 3.10

Pick: **Offset icon**

To offset an object, you need to determine the *offset distance* (the distance away from the original object) or the *through point* (the point through which the offset object is to be drawn). To specify the offset distance indicated in Figure 3.11,

Offset distance or Through <Through>: **30** ↵

Select object to offset: *(select curve 1)*

Side to offset? *(pick a point below the curve, like point A)*

Select object to offset: *(select curve 1)*

Side to offset? *(pick a point above the curve, like point B)*

Once you have defined the offset distance, AutoCAD continues to repeat the prompt *Select object to offset:*, allowing you to create additional parallel lines that have the same spacing.

Repeat the steps above on your own to create lines offset 30 units from either side of the remaining curved centerlines, so that your drawing looks like Figure 3.11. If necessary, use the Zoom commands to enlarge the smaller curves so you can see them better on your screen to make selections. Use Zoom

Previous to return to the previous view. Remember that the Zoom commands (with the exception of Zoom All) can be transparent, so you can select them during another command without canceling that command.

If you select the wrong item to offset, either erase the incorrect lines when you are finished or press (ESC) to cancel the command and then start again. When you have finished using the Offset command, press the return button or (↵) to end the command. Notice that the new objects created by the Offset command are on the same layer as the object that you selected to offset. It does not matter which layer is current; the newly offset object will always be on the same layer as the original object selected.

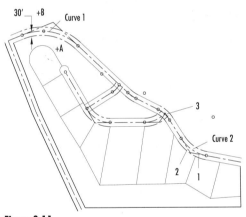

Figure 3.11

Trimming the Remaining Lines

Referring to Figure 3.11, on your own, select curve 2 as the cutting edge. Be sure to press (↵) or the return button when you are finished selecting cutting edges. Trim off the portions of the centerline by picking the points labeled 1 and 2, which extend past the cutting edge. Repeat this process and trim all of the remaining centerlines where they extend beyond the centerline arcs on your own. Note that line 3 does not extend past any cutting edges so it cannot be trimmed. Use the Erase command on your own to remove it.

Next, zoom in on the center area of the drawing on your own so that you can trim the excess lines where the roadways join. The center portion of the drawing will be enlarged on your screen.

This time, instead of selecting one cutting edge at a time, you will use implied Crossing to select all the lines in that area as cutting edges. Refer to Figure 3.12 to make your selections.

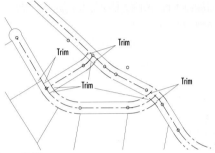

Figure 3.12

Pick: **Trim icon**

Select cutting edges: (Projmode = UCS, Edgemode = No extend)

Select objects: *(pick the first corner of a window near the bottom right of the screen. Be sure not to select any lines or objects.)*

Other corner: *(pick the second corner of the window, at the upper left of the screen)*

All of the objects that crossed the selection box become highlighted, indicating that they are selected as cutting edges.

Select objects: (↵)

<Select object to trim>/Project/Edge/Undo: *(select the extreme ends of the lines you want to remove and continue doing so until all the excess lines are trimmed)*

<Select object to trim>/Project/Edge/Undo: (↵)

When you have finished trimming, on your own use Zoom All to show the entire drawing on the screen. If you have some short line segments left which no longer cross the cutting edges, use Erase to remove them on your own. Your drawing should look like Figure 3.13.

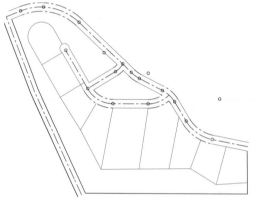

Figure 3.13

Using the Change Properties Dialog Box

 Next you will change the properties of the centerlines that you offset so that they are on layer LOTLINES, using the Change Properties dialog box. The new offset lines are the edges of the road and should be on the same layer as the other lot lines. The Change Properties dialog box lets you change the linetype, layer, and color of the objects you select, as well as their linetype scale, which you will learn more about in Tutorial 4, and their thickness. You can select the Change Properties dialog box by picking the Properties icon near the right side of the Object Properties toolbar.

Pick: **Properties icon**

Select objects: (select all of the roadway edges that appear as centerlines)

Select objects: ⏎

The Change Properties dialog box appears on your screen, as shown in Figure 3.14. You will use it to change the objects you selected from layer CENTERLINE to the layer LOTLINES.

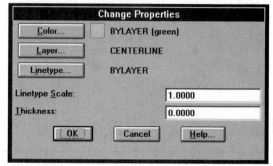

Figure 3.14

Pick: (the button with the word Layer)

A dialog box showing the existing drawing layers appears, as shown in Figure 3.15. From the list, select layer LOTLINES.

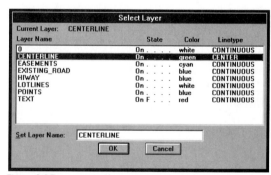

Figure 3.15

Pick: **OK** *(to exit the Select layer dialog box)*

Pick: **OK** *(to exit the Change Properties dialog box)*

You will see the lines you selected change to black (or white, depending on your configuration) and linetype CONTINUOUS, as set by layer LOTLINES's properties.

Next you will use the Fillet command to add small rounded corners to some of the lots. Before doing this, change your current layer to LOTLINES on your own. Pick on the current layer name listed on the Object Properties toolbar and use the list of layer names that appears to select layer LOTLINES. Its name should appear on the toolbar as the current layer. Use Zoom Window to enlarge the center area of the drawing as shown in Figure 3.16.

> ■ *TIP* It is a good idea to save your drawing whenever you have finished a major step that you are satisfied with. Take a minute to use the Save icon from the Standard toolbar and save your drawing on your own. ■

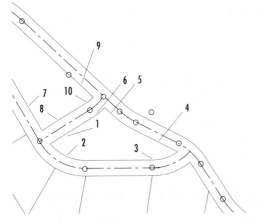

Figure 3.16

Using Fillet

 The Fillet command connects lines, arcs, or circles with a smoothly fitted arc, or *fillet*. The Fillet command is located on the Fillet/Chamfer flyout of the Modify toolbar.

Pick: **Fillet icon**

Polyline/Radius/Trim/<Select first object>: **R** ⏎

Typing *R* tells AutoCAD that you want to enter a radius. You will enter a radius value of 10.

Enter fillet radius <0.0000>: **10** ⏎

Command: ⏎ *(to restart the Fillet command)*

AutoCAD prompts you to select the two objects. Refer to Figure 3.16 to select the lines.

Command: Fillet Polyline/Radius/Trim/<Select first object>: *(select line 1)*

Select second object: *(select line 2)*

A fillet should appear between the two lines, as shown in Figure 3.17.

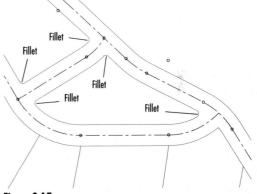

Figure 3.17

Command: ⏎

AutoCAD repeats the previous command when you press ⏎ or the return button at the blank command prompt. In this case, it returns to the original Fillet prompt, which allows you to repeat the selection process, drawing additional fillets of the same radius. Create the next fillet between the lines shown in Figure 3.16.

Polyline/Radius/Trim/<Select first object>: **(select line 3)**

Select second object: **(select line 4)**

Continue to fillet lines 5 through 10 on your own.

You can use the Fillet command to fit a smooth arc between any combination of lines, arcs, or circles. Once you have defined the radius value, the direction of the fillet is determined by the cursor location used to identify the two objects. Figure 3.18 shows some examples of how you can use Fillet to create different-shaped fillets by choosing different point locations.

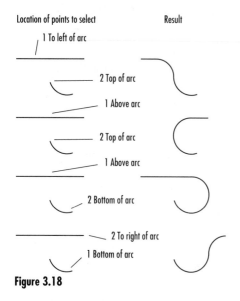

Figure 3.18

Each example starts with a line located directly above an arc, as shown in the left column. Picking the objects where indicated yields the results in the right column.

You can also set the Fillet command so that it does not automatically trim the lines to join neatly with the fillet, which is the default. You do this by selecting the command option Trim by typing the letter *T* at the command prompt, and then choosing the option No trim by typing *N*. Experiment with this on your own.

■ *TIP* If you use a radius value of 0 with the Fillet command, you can use the command to make lines intersect that do not meet neatly at a corner. This also works to make a neat intersection from lines that extend past a corner. You can also do this using the Chamfer command (which you will learn next) by setting both chamfer distances to 0. ■

Using Chamfer

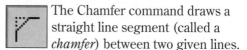

The Chamfer command draws a straight line segment (called a *chamfer*) between two given lines. Chamfer is the name for the machining process of flattening off a sharp corner of an object. Use the AutoCAD Help window to read about the options for the Chamfer command on your own. The Chamfer command is located on same flyout as the Fillet command on the Modify toolbar.

On your own, use the Line command to draw a rectangle, off to the side of the subdivision drawing, as shown in Figure 3.19.

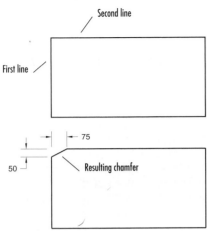

Figure 3.19

Pick: **Chamfer icon**

Polyline/Distances/Angle/Trim/Method/<Select first line>:
D ⏎

Typing *D* tells AutoCAD you want to enter distance values.

Enter first chamfer distance <0.0000>: **50** ⏎

Enter second chamfer distance <0.0000>: **75** ⏎

Command: ⏎

AutoCAD returns to the Chamfer command and prompts for selection of the first and second lines to add the chamfer between. Refer to Figure 3.19 to select the lines.

Polyline/Distances/ Angle/Trim/Method/<Select first line>:
(select the first line)

Select second line: *(select the second line)*

A chamfer should appear on your screen that looks similar to the one shown in Figure 3.19. The size of a chamfer is defined by the distance each end of the chamfer is from the corner. As with the Fillet command, once you have entered these distances, you can draw additional chamfers of the same size by pressing ⏎ after each chamfer is completed to restart the command.

On your own, use the commands you have learned to erase the chamfered rectangle from your drawing. Use Zoom All to restore the original appearance and then save the drawing.

Using Polyline

 The Polyline command draws a series of connected lines or arcs that AutoCAD treats as a single graphic object, called a *polyline*. The Polyline command is used on technical drawings to draw irregular curves, and lines that have a width. An irregular curve is any curve that does not have a constant radius (a circle and an arc have a constant radius). The Polyline command is located on the Polyline flyout of the Draw toolbar. The Polyline icon appears similar to an arc and line connected.

You will use the Polyline command to create the shape for a pond to add to the subdivision drawing. On your own, make layer LOTLINES current before continuing.

Pick: **Polyline icon**

From point: *(select any point to the right of the subdivision where you want to locate the pond)*

Current line-width is 0.0000

Arc/Close/Halfwidth/Length/Undo/Width/<Endpoint of line>:
(select 9 more points)

You will use the Close option as you did in Tutorial 1 when drawing closed figures with the Line command.

Arc/Close/Halfwidth/Length/Undo/Width/<Endpoint of line>:
C ⏎

Your screen should show a line made of multiple segments, similar to the one in Figure 3.20.

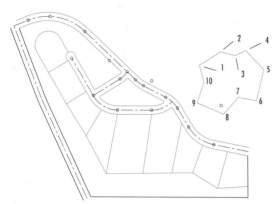

Figure 3.20

A shape drawn using Polyline is different from a shape created by using the Line command, in that AutoCAD treats the polyline as one line. You cannot erase one of the polyline segments. If you try, the entire polyline is erased. Other Polyline options allow you to draw an arc as a polyline segment (Arc), specify the starting and ending width (or half-width) of a given segment (Width, Half-width), specify the length of a segment (Length), and remove segments already drawn (Undo).

Edit Polyline

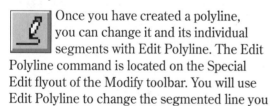

Once you have created a polyline, you can change it and its individual segments with Edit Polyline. The Edit Polyline command is located on the Special Edit flyout of the Modify toolbar. You will use Edit Polyline to change the segmented line you just created to a smooth curve.

The Edit Polyline command provides two different methods of fitting curves. The Fit option joins every point you select on the polyline with an arc. The Spline option produces a smoother curve by using either a cubic or a quadratic B-spline approximation (depending on how the system variable Splinetype is set). You can think of the spline operation as working like a string that is stretched between the

first and last points of your polyline. The vertices on your polyline act to pull the string in their direction, but the resulting spline does not necessarily reach the point.

Pick: **Edit Polyline icon**

Select polyline: **(select any part of the polyline)**

Open/Join/Width/Edit vertex/Fit/Spline/Decurve/
Ltype gen/Undo/eXit <X>: **F** ⏎

■ *TIP* You can use Edit Polyline to convert objects into polylines. At the *Select polyline:* prompt, pick the line or arc you want to convert to a polyline. You will see the message *Entity selected is not a polyline. Do you want to turn it into one? <Y>:*. Press ⏎ to accept the yes response to convert the selected object into a polyline and continue with the prompts for the Edit Polyline command. ■

The straight line segments change to a curved line, as shown in Figure 3.21.

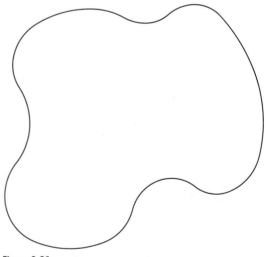

Figure 3.21

Fit curve connects all the vertices of a 2D poly-line by joining each pair of vertices with an arc. You will use the Undo option of the command to return the polyline to its original shape.

Close/Join/Width/Edit vertex/Fit/Spline/Decurve/
Ltype gen/Undo/eXit <X>: **U** ⏎

The original polyline returns on your screen. Now try the Spline option:

Close/Join/Width/Edit vertex/Fit/Spline/Decurve/
Ltype gen/Undo/eXit <X>: **S** ⏎

You should see a somewhat flatter curve, similar to the one shown in Figure 3.22, replace the original polyline. Among other things, splined polylines are useful for creating contour lines on maps.

> ■ **TIP** If you have a polyline that has already been curved with Fit or Spline, Decurve returns the original polyline. ■

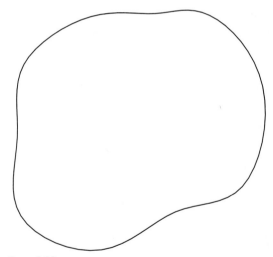

Figure 3.22

Now try the Width option:

Close/Join/Width/Edit vertex/Fit/Spline/Decurve/
Ltype gen/Undo/eXit <X>: **W** ⏎
Enter new width for all segments: **5** ⏎

The splined polyline is replaced with one that is 5 units across. You will accept the default by pressing ⏎ to exit the Edit Polyline command.

Close/Join/Width/Edit vertex/Fit/Spline/Decurve/
Ltype gen/Undo/eXit <X>: ⏎

Your drawing should appear similar to Figure 3.22.

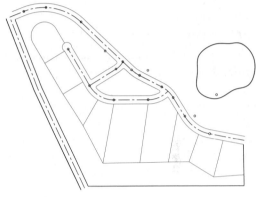

Figure 3.23

Using Spline

AutoCAD Release 13's Spline command lets you create quadratic or cubic spline (or NURBS) curves. NURBS stands for non-uniform rational B-spline, which is the name of the method used to draw the curves. These curves can be more accurate than spline-fitted polylines because you control the tolerance to which the spline curve is fit. Splines also take less file space in your drawing than do spline-fitted polylines.

You can also use the Object option of the Spline command to convert spline-fitted polylines to splines. (The variable Delobj controls whether the original polyline is deleted or not after it is converted.) The Spline command is located on the Polyline flyout of the Draw toolbar. You will draw a spline off to the side of the subdivision and then later erase it.

Pick: **Spline icon**

Object/<Enter first point>: *(pick any point)*

Enter point: *(pick any point)*

Close/Fit Tolerance/<Enter point>: *(pick any point)*

Close/Fit Tolerance/<Enter point>: *(pick any point)*

Close/Fit Tolerance/<Enter point>: *(pick any point)*

Close/Fit Tolerance/<Enter point>: ⏎

Enter start tangent: ⏎

Enter end tangent: ⏎

Pressing ⏎ at the tangency prompt causes AutoCAD to calculate the default tangencies. You can make the spline tangent to other objects by picking the appropriate points at these prompts. In addition, the Spline command has a Close option similar to the Line and Polyline commands you have used.

The spline appears on your screen. Experiment by drawing splines on your own. Splines can be edited using the Splinedit command located on the Special Edit flyout. When you have finished experimenting with the Spline command, erase all of the splines you created.

Now use the commands you have learned to thaw layer TEXT so that the text returns to your screen. Set TEXT as the current layer and use the Dtext command to add the label POND to the pond. Set the height for the text to 30 units. When you have finished these steps on your own, your screen should look like Figure 3.24.

Figure 3.24

Getting Information about Your Drawing

Because your AutoCAD drawing database contains accurate geometry and has been created as a model of real-world objects, you can find information about distances and areas, and locate coordinates. You can also find volumes and mass properties of solid models, which you will learn to do in Tutorial 9. Next you will use AutoCAD's ability to list objects and make calculations to determine information about the subdivision. You will use the Inquiry flyout on the Object Property toolbar shown in Figure 3.25.

Figure 3.25

Using List

The List command shows information that AutoCAD contains in the drawing database about the object selected. Different information may be shown, depending on the type of object. The layer, coordinate location for the object, color and linetype if different from the layer, and other information are listed. Refer to Figure 3.24 to make your selections.

Pick: **List icon**

Select objects: *(pick line 1)*

Select objects: ⏎

The AutoCAD text window appears on your screen. If you want, you can resize it by picking on its corners and dragging it to a new size. You can also pick on the title bar and drag it to a new location while you hold the return button down. Refer to Figure 3.26 for the information listed about the line you selected. You can use this information to make calculations or to label the lot lines of this subdivision.

Figure 3.26

Close the text window by double-clicking its Windows Control box.

Pick: **List icon**

Select objects: *(pick on the polyline representing the pond)*

Select objects: ⏎

The text window appears on your screen again. It contains all of the information about the vertices of the polyline. Keep pressing ⏎ until all of the information has scrolled by. At the bottom of the information you will see that AutoCAD lists the area and perimeter of the

polyline, as shown in Figure 3.27. This can be very useful for calculating the area of irregular shapes like the pond.

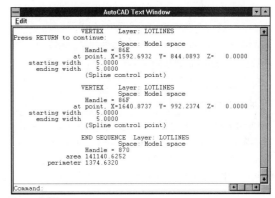

Figure 3.27

Locating Points

Now turn on the Endpoint running object snap on your own. You will use it with Locate Point to select accurate point locations for the next steps. You can select Locate Point from the Inquiry flyout on the Object Properties toolbar. Locate Point shows the coordinates of the location that you select. When you use this command in conjunction with object snaps, you can find the exact coordinates of an object's endpoint, center, midpoint, etc.

Pick: **Locate Point icon**

Select point: *(pick on the left end of line 2 in Figure 3.24)*

The command window displays the coordinates of the point, X=912.5026, Y=334.8196, Z=0.0000

Using Area

The Area command finds areas that you define by picking points to define a boundary, picking a closed object, or adding and subtracting boundaries. The options for the Area command are First point,

Object, Add, and Subtract. To use the default method, First point, you begin by selecting points to define straight-line boundaries for the area you want to measure. When you press ⏎ to indicate that you are finished selecting, AutoCAD automatically closes the boundary for the area from the last point selected back to the first point. (You can also close the area by picking the point again if you want.) The area inside the boundary you specify is reported. By selecting the Object option, you can pick a closed object, such as a polyline, circle, ellipse, spline, region, or a solid object, and its area is reported. You could use this method to find the area of the pond instead of using List as you did above. The Add and Subtract options let you define more complex boundaries by adding and subtracting from the first boundary selected. Next you will use the Area command to find the area of lot 5. Remember that you still have the Endpoint object snap selected, so the nearest endpoint is found when you select a line. This way the area calculation will be accurate.

> *Pick:* **Area icon**
>
> <First point>/Object/Add/Subtract: *(pick point a on Figure 3.24)*
>
> Next point: *(pick b)*
>
> Next point: *(pick c)*
>
> Next point: *(pick d)*
>
> Next point: ⏎

The area and perimeter values of lot 5 are listed in the command window.

On your own, set the current layer to EASEMENTS before continuing.

Using Multilines

Next you will use Multiline to easily add multiple lines to the drawing. This feature is especially useful in architectural drawings for creating walls; and in civil engineering for highways, among other things, where you can draw the centerline of the road, and automatically add the edge of pavement and edges of right-of-way lines a defined distance away. Multiline also allows you to fill the area between two lines with a color. Using the command for Multiline Style, you can define up to 16 lines called *elements*, which you will use when drawing multilines. You can save multiline styles with different names so they can easily be recalled. You can also set the multiline justification so that you can create lines by specifying the top, middle, or bottom of the style. You do this by setting the variable Cmljust to 0, 1, or 2 respectively.

Creating a Multiline Style

 You can select the Multiline Style command from the Object Properties toolbar. Multiline Style creates the spacing, color, linetype, and name of the style, as well as the type of endcaps that will be used to finish off the lines.

> *Pick:* **Multiline Style icon**

The Multiline Styles dialog box appears on your screen, as shown in Figure 3.28. The default style name is STANDARD. Notice that it is the current style. To create a new style, you will use the Name portion of the dialog box and type in the new name *EASEMENT*. Pick in the box to the right of Name if the typing cursor is not already shown there. Then you will set the properties of the elements to form this new style.

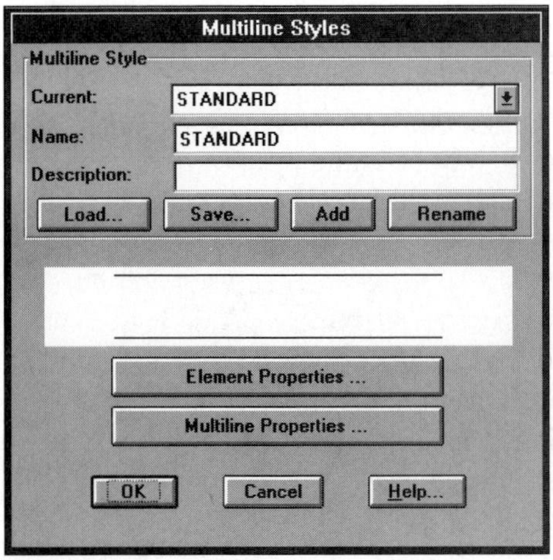

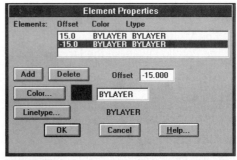

Figure 3.28

Figure 3.29

Type: **EASEMENT**

Pick: **Element Properties**

The Element Properties dialog box appears. You will use it to set the spacing of the elements, as shown in Figure 3.29. To change the spacing for an element, pick the element shown in the list at the top of the dialog box. When it is highlighted, use the box to the right of Offset to set its offset spacing. On your own, type *15* in for the offset spacing. Now set the offset spacing for the second element to *-15*. To see the element spacing change for a highlighted item, pick once again on its name in the box at the top of the dialog box. When you have set the element spacing, your dialog box should look like Figure 3.29. You will leave the color and linetype set to BYLAYER, as layer EASEMENTS already has these properties set to cyan and HIDDEN linetype. If you wanted to have different-colored lines or different linetypes, you could use the appropriately named portions of the dialog box to set them. When you have finished,

Pick: **OK**

You return to the Multiline Styles dialog box. Selecting Multiline Properties will cause the Multiline Properties dialog box to appear on your screen. You can use it to select different types of endcaps, including angled lines or arcs, to finish off the multilines you create. You can also use this area to turn on fill if you want the area between multiline elements to be shaded. Experiment with this area on your own and return to the Multiline Style dialog box when you are finished.

The Add button in the Multiline Style dialog box sets a style in the Name box as current. You will set Style EASEMENT as current.

Pick: **Add**

The Save button saves a style to the library of multiline styles. When you pick the Save button, the Save Multiline Style dialog box appears on your screen. AutoCAD saves multiline styles to *c:\r13\com\support\acad.mln* by default. You can create your own file for your multiline styles if you want, by typing in a different name. The Load button loads saved multiline styles from a library of saved styles. Rename lets you rename the various styles you have created. You cannot rename the STANDARD style. When you have finished examining the dialog box,

Pick: **OK**

Drawing Multilines

The Multiline command is located on the Polyline flyout of the Draw toolbar. You will pick it to add some multilines for easements along the lot lines of the subdivision drawing. You will use the current multiline style, EASEMENT, that you created.

Pick: **Multiline icon**

Now you are almost ready to draw some multilines. Before you do, decide on the justification you will need to use. You can choose to have the multilines you draw align so that the points you select align with the top element, with the zero offset, or with the bottom element. You will set the justification so that the points you pick are the zero offset of the multiline, in this case the middle of the multiline.

Justification/Scale/STyle/<From point>: **J** ⏎

Top/Zero/Bottom: **Z** ⏎

You will see the message *Justification = Zero, Scale = 1.00, Style = EASEMENT.* Now you are ready to start drawing the easement lines. Refer to Figure 3.30 for the points to select. Remember that the Endpoint running object snap should still be on.

Justification/Scale/STyle/<From point>: **(pick point 1)**

<To point>: **(pick 2)**

Undo/<To point>: **(pick 3)**

Close/Undo/<To point>: **(pick 4)**

Close/Undo/<To point>: **(pick 5)**

Close/Undo/<To point>: **(pick 6)**

Close/Undo/<To point>: **(pick 7)**

Close/Undo/<To point>: **(pick 8)**

Close/Undo/<To point>: ⏎

Your drawing should look like Figure 3.30.

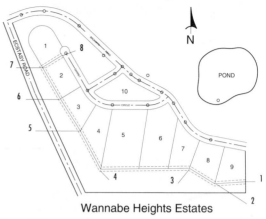

Wannabe Heights Estates

Figure 3.30

> ■ *TIP* If you want to trim a multiline, you must first use the Explode command to change it back into individual objects before you can trim it. Use AutoCAD's Help command for more information about using Explode.

On your own, use the Mtext command to add text describing the location of the subdivision and the easements, as you see in Figure 3.31. Save your drawing now.

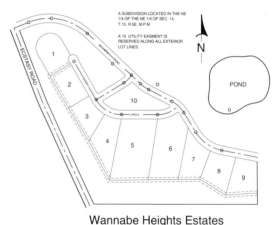

Wannabe Heights Estates

Figure 3.31

■ **Warning:** You should always save your drawing before you plot. If for some reason the plotter is not connected or there is another problem, AutoCAD may not be able to continue. To go on, you would have to reboot the computer and would lose any changes you had made to your drawing since the last time you saved. ■

The Plot Configuration Dialog Box

 Depending on the types of printers or plotters you have configured with your computer system, the Plot command causes your drawing to be either printed on your printer or plotted on the device that you configured to use with AutoCAD. The same command is used for printing and for plotting. Selecting Print causes the Plot Configuration dialog box to appear on your screen, as shown in Figure 3.32.

■ **Warning:** The Plot Configuration dialog box will not appear unless the system variable Cmddia is set to 1. If the Plot Configuration dialog box does not appear when you pick the Print icon, type *CMDDIA* at the command prompt, and make sure its value is set to 1. ■

Pick: **Print icon**

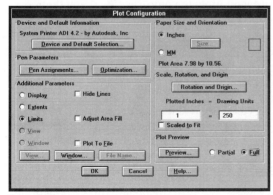

Figure 3.32

Among the things that the Plot Configuration dialog box lets you choose are:

What portion of your drawing to plot or print

The plotter or printer where you will send the drawing

The pen or color selections for the plotter or printer

The scale at which the finished drawing will be plotted

Whether to use inches or millimeters as the plotter units

Where your plot starts on the sheet

Whether the plot is rotated or moved on the origin

The scale relating the number or plotted units to the number of drawing units

Because there are so many factors, take care when you are starting out not to change too many things about the plotter at once. That way you can see the effect that each option has on your plot.

Device and Default Selection

Click on the Device and Default Selection box. A dialog box appears that lets you select from the plotters you have configured. If there are no printers or plotters listed, then you must use AutoCAD's Configure command to configure your output devices. Refer to the Getting Started section of this manual for details about configuring your output device. To select a listed device, highlight the name of the device and pick OK. The options you see for pen selection and other items depend on your particular output device. If an item in the Plot Configuration dialog box appears grayed instead of black, it is not available for selection. You may not be able to choose certain items, depending on the limitations of your printer or plotter.

Additional Parameters

This area of the Plot Configuration dialog box lets you specify what area of the drawing will be printed or plotted. Display selects the area that appears on your display as the area to plot. Extents plots any drawing objects you have in your drawing. Limits plots the predefined area set up in the drawing with the Limits command. View plots a named view you have created with the View command. (Notice that if you have not made any views, this area is grayed.) Window lets you go back to the drawing display and create a window around the area of the drawing you want to plot. The radio button for Window is grayed out until you use the Window button at the bottom of the Additional Parameters area to define a window. You can do this by picking from your drawing or by typing in the coordinates.

Select Limits by picking the button to its left on your own, because the exact limits of the area you want to plot were already set in the datafile. If you have not set up the size with Limits, then Extents is often useful.

Hide Lines hides the back lines in a 3D drawing. You will not need this for your 2D plot of the subdivision drawing. When Adjust Area Fill is turned on in a drawing that contains areas solidly filled with color, they are adjusted for the width of the plotter pen so that it does not draw over the boundary edge for the area. Plot To File sends your plot to a file rather than directly to the plotter or printer. If you want to do this, check the box to the left of the words and the File Name item will turn black, indicating that it can now be used to display a dialog box where you can enter the file name. The plot file will be created in the format of whatever output device you have selected in the Device and Default Selection area.

Paper Size and Orientation

Select either inches or millimeters (MM) by picking the appropriate button to the left of the measurement you want to use for your paper. For the drawing you have just set up, pick Inches. A solid circle fills the center of the button to indicate that it is selected. Pick Size to see the dialog box where you can set the paper size. The paper sizes that you can select depend on your printer. Standard 8.5 × 11 paper is called size A. 11 × 17 paper is size B. (Smaller values indicate the area your printer can image on the sheet.) Max is the maximum size of paper your printer will handle. Notice the message *Orientation is portrait* (or landscape) near the bottom of the Paper Size dialog box. This refers to the default orientation for the paper. Portrait orientation is a vertical sheet layout. Landscape is horizontal. (If you have a laser printer and specified that the drawing be rotated when you configured the printer, the paper sizes shown may reflect this.) Certain options may not be available with your model of printer. If so they will be shown grayed out.

Scale, Rotation, and Origin

You set the scale for the drawing by entering the number of plotted inches for the number of drawing units in your drawing. If you do not want the drawing plotted to a particular scale, you can fit it to the sheet size by checking the box to the left of Scaled to Fit. Most of the time it is useful to have your engineering drawing plotted to a known scale.

If Scaled to Fit is on (there is a check mark in the box), turn it off. On your own, type in *1* for Plotted Inches and *250* for Drawing Units if they do not appear in the boxes already. This will give you a drawing scale of 1"=250' for your drawing, because the default units in the drawing represented decimal feet.

Depending on whether you have a printer or plotter, you may need to rotate your drawing to fit on the sheet correctly. Most printers need to have a horizontal-shape drawing, such as the subdivision drawing, rotated 90° to print it out correctly. Most plotters will not need to have the drawing rotated. You may have to experiment to find out what will work for your particular printer or plotter.

Use the Rotation and Origin button to place the 0,0 coordinates or bottom-most left point in the drawing at the location you specify on the paper. You can move the drawing to the right by using a positive value for the X origin on the paper; move the drawing up on the paper by specifying a positive value for the Y origin. (Keep in mind that choosing to rotate your plot will affect the directions for X and Y origins.) Also, you should be aware that moving the origin for the paper may cause the top and right lines of the drawing not to print if they are outside the printer limits.

Plot Preview

You may choose to have a partial preview, where you see only the overall border of your graphics window as it will appear on the paper, or a full preview, where your entire drawing as it will fit on the paper is shown on the screen. If you have a very detailed drawing, it may take some time to do a full preview. The subdivision drawing is not too large to preview in a reasonable amount of time, so pick the radio button next to Full so that you can see approximately how the drawing will appear on the sheet. Click on Preview to show the drawing on the screen. Your screen will look similar to Figure 3.33. Plot rotation is set to 90° in Figure 3.33. You may not need to do this for your printer.

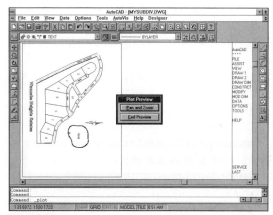

Figure 3.33

> ■ **TIP** You can move the Plot Preview message box shown in Figure 3.32 around on your screen by picking on its title bar and dragging it to a new location with the pick button held down.■

If the drawing appears to fit the sheet correctly, pick End Preview to return to the Plot Configuration dialog box and then pick OK to plot your drawing.

You will see a prompt similar to the following, depending on your hardware configuration:

Effective plotting area: 10.50 wide by 8.00 high

Position paper in plotter.

Press RETURN to continue or S to stop for hardware setup.

Press ⏎ if you are ready to print or plot.

If the drawing does not fit on the sheet correctly, review this section and determine which setting in the Plot Configuration dialog box needs to be changed. You should have a print or plot of your drawing that is exactly scaled so that 1" on the paper is equal to 250' in the drawing.

When you have a successful plot of your drawing, you have completed this tutorial. Exit AUTOCAD by choosing,

Pick: **File, Exit**

KEY TERMS

chamfer	fillet	through point
cutting edges	offset distance	
elements	polyline	

KEY COMMANDS

Area	List	Plot/Print
Chamfer	Locate Point	Polyline
Change Properties	Multiline	Spline
Edit Polyline	Multiline Style	Trim
Fillet	Offset	

Redraw the following shapes. If dimensions are given, create your drawing geometry exactly to the specified dimensions. The letter M after an exercise number means that the given dimensions are in millimeters (metric units). If no letter follows the exercise number, the dimensions are in inches. Do not include dimensions on the drawings.

3.1 Clearance Plate

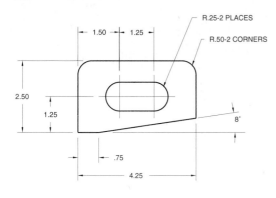

3.2M Bracket

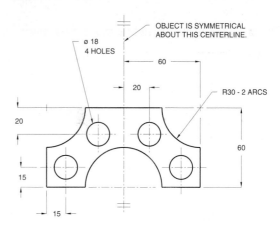

3.3 Roller Arm

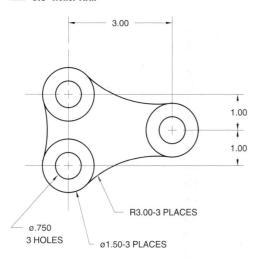

3.4 Roller Support

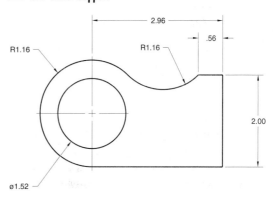

3.5 Slotted Ellipse

Draw the figure shown (note the symmetry). Do not show dimensions.

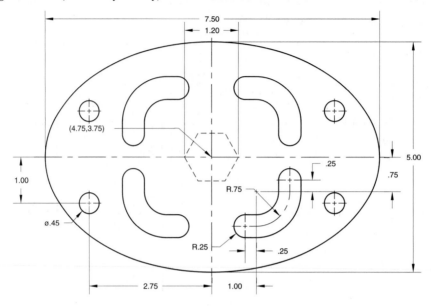

3.6 Roadway

Shown is a center line for a two-lane road (total width is 20 feet). At each intersection is a specified turning radius for the edge of pavement. Construct the center line and edge of pavement. Do not show any of the text.

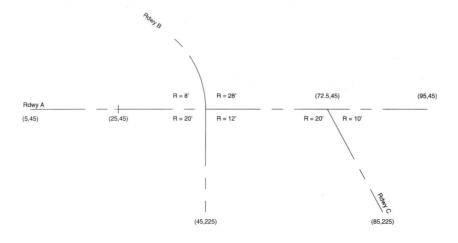

3.7 Circuit Board

Draw circuit board below, using Polylines with the width option to create wide paths.

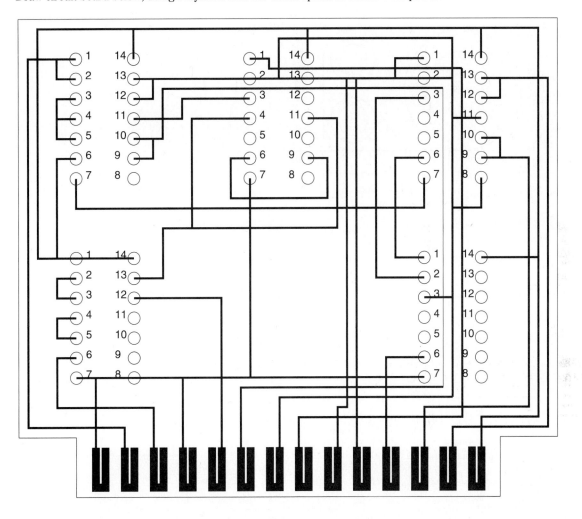

3.8M Shaft Support

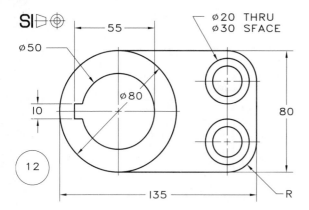

SI

∅50

55

∅20 THRU
∅30 SFACE

∅80

10

80

12

135

R

3.9M Puller Base

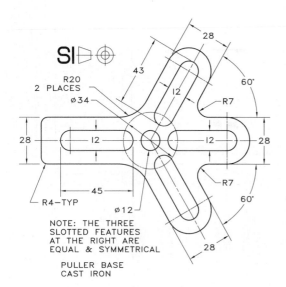

SI

R20
2 PLACES

∅34

28

43

28

12

12

12

R7

60°

28

45

R4–TYP

R7

∅12

60°

28

NOTE: THE THREE
SLOTTED FEATURES
AT THE RIGHT ARE
EQUAL & SYMMETRICAL

PULLER BASE
CAST IRON

3.10 Concrete Walkway

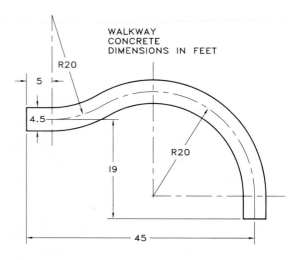

WALKWAY
CONCRETE
DIMENSIONS IN FEET

R20

5

4.5

19

R20

45

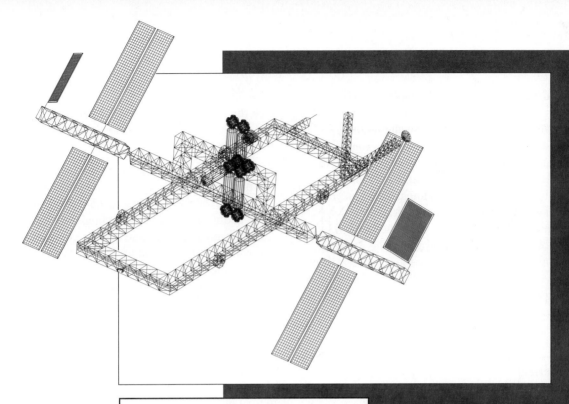

TUTORIAL 4

Geometric Constructions

Objectives

When you have completed this tutorial, you will be able to

1. Draw polygons and rays.

2. Use object snaps to pick geometric locations.

3. Load linetypes and set their scaling factors.

4. Use the Copy, Extend, Rotate, Move, Mirror, Array, and Break editing commands.

5. Build selection sets.

6. Use hot grips to modify your drawing.

Introduction

This tutorial expands your skills in creating and editing drawing geometry by introducing several of the AutoCAD drawing techniques used for geometric constructions. You will learn more about how to use object snaps to select locations, such as intersections, end-points, and midpoints of lines, based on your existing drawing geometry. This tutorial synthesizes the techniques you have learned by showing you how to coordinate the drawing commands to create shapes for technical drawings. In this tutorial you will create three drawings: a wrench, a coupler, and a geneva cam. You will learn how to apply editing commands to quickly create drawing geometry.

Starting

Launch Windows and AutoCAD Release 13 for Windows.

Your computer display shows AutoCAD's drawing editor. Notice that at the top of the screen *unnamed* appears as the current file name. This tells you that you are starting out in a new AutoCAD drawing, and that you have not yet assigned a name for the drawing. Later, when you save your drawing, you will be prompted for a file name. If you use the Save icon without specifying a file name first, the Save Drawing As dialog box appears on your screen, prompting you for a file name.

Begin the Wrench Drawing

You will create a drawing named *wrench.dwg,* and draw the basic shapes required for a wrench. You will do many of the steps on your own, using the commands you have learned in

the previous tutorials. Pay careful attention to the directions, making sure that you complete each step before going on.

Pick the Save icon from the Standard toolbar and save your drawing as *wrench.dwg* on your own.

You will create the wrench drawing using decimal inch measurements. For example, 4.5 units in your drawing will represent 4.5 inches on the real object. AutoCAD's decimal units can stand for any measurement system you want. You will have them stand for inches. When you are creating drawing geometry, make the objects in your drawing the actual size they would be in the real world. Do not scale them down, as you would when drawing on paper. Keep in mind that one advantage of the accurate AutoCAD drawing database is that you can use it to directly control machine tools to create parts. You would not want your actual part to turn out half-size because you scaled your drawing to half-size. When you plot the final drawing, you specify the ratio of plotted inches or millimeters to drawing units to produce scaled plots.

First you will draw construction lines in the drawing; you will later change these to center-lines. Refer to Figure 4.1.

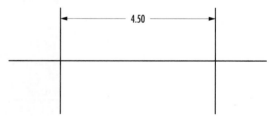

Figure 4.1

On your own, set Grid and Snap to .5-inch spacing and make sure they are turned on. If necessary, pick Zoom All so that the grid area fills the graphics window. (Depending on the aspect ratio of your screen, the grid may not fill it in all directions.)

Now use Line to draw a horizontal line 7.5 units long near the middle of your graphics window. The line does not have to be at the exact center, since it is easy to move your drawing objects later if they do not appear centered. You will learn how to move objects in this tutorial.

Next, draw two vertical lines that are 3.00 units long and 4.5 units apart, as shown in Figure 4.1. (Hint: Draw one line and then use the Offset command you learned in Tutorial 3, with an offset distance of 4.5, to create the other line.) The exact location of the vertical lines on the horizontal construction line is not critical, as long as they are 4.5 units apart.

Remember, you can never pick a point from the screen exactly unless you use Snap or an object snap.

 Now use Circle Center Diameter on your own to draw a circle with a 2-inch diameter. The center point of the circle should be the intersection of the horizontal line and the right vertical line and should be a point you can use Snap to pick.

Using Copy

 Copy Object copies an object or group of objects within the same drawing. The original objects remain in place and during the command you can move the copies to a new location. You can pick Copy Object quickly from the Copy flyout on the Modify toolbar, as shown in Figure 4.2. You will pick Copy Object to create a second circle in your drawing.

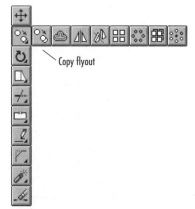

Figure 4.2

Pick: **Copy Object icon**

> ■ *TIP* It might be difficult to get the cursor to select the circle with Snap on. Using implied Windowing or Crossing would allow you to easily select the circle. You might also find that turning Snap off makes it easier to select the circle. Be sure to turn it back on again before selecting the base point or displacement, or you may not pick the exact center point. ■

Select objects: *(select the circle)*
Select objects: ⏎

Pressing ⏎ tells AutoCAD you are done selecting objects and want to continue with the command. Be sure Snap is on again so you can accurately select the center point of the circle, which you created on the snap increment.

<Base point or displacement>/Multiple: *(select the center point of the circle)*

AutoCAD switches to drag mode so you can see the object move about the screen and prompts you for the second point of displacement. You can define the new location by typing new absolute or relative coordinate values or by picking a location from the screen. Since this object was created on the snap increment, use the snap and pick the second point of displacement.

The second circle must be centered at the intersection of the horizontal and left vertical lines, which should be a snap location.

Second point of displacement: *(pick the left intersection)*

Your drawing should now look like Figure 4.3.

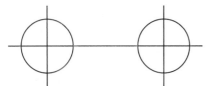

Figure 4.3

Next, use the Offset command on your own to offset the construction line you created a distance of .5 units to either side of the center construction line to create the body of the wrench. Your resulting drawing should look like Figure 4.4.

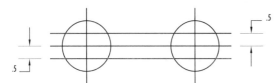

Figure 4.4

Now use the Trim command on your own to trim the lines you created with the Offset command so that they intersect neatly with the circles. When you have finished, your drawing should appear similar to Figure 4.5.

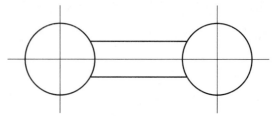

Figure 4.5

Loading Linetypes

Once you have finished removing the excess lines, you will change the middle construction lines in your drawing to centerlines. Before you can use various linetypes in AutoCAD, you must load them into the drawing. You will load the linetype CENTER.

> **TIP** You can load linetypes when you create and assign linetypes to layers. You will learn to create layers in Tutorial 5. ■

The Linetype icon is on the Object Properties toolbar.

Pick: **Linetype icon**

The Select Linetype dialog box appears on your screen, as shown in Figure 4.6. You will use it to load the linetype CENTER into your drawing so that it is available for use. Pick the Load button at the lower right of the dialog box.

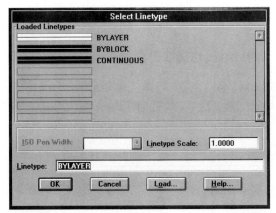

Figure 4.6

Pick: **Load**

The Load or Reload Linetypes dialog box shown in Figure 4.7 appears on your screen. Use the dialog box to highlight the CENTER linetype.

Figure 4.7

Pick: **OK** *(to exit the Load or Reload Linetypes dialog box)*

Pick: **OK** *(to exit the Select Linetype dialog box)*

The linetype is loaded into your drawing and ready for use in other commands.

Changing Properties

You will use the Properties selection from the Object Properties toolbar to change the lines you have drawn to centerlines. Changing objects in this way is not always a good practice. In Tutorial 5 you will learn to create your own layers and set their linetypes and colors, as well as other properties. Changing the color or linetype of a layer does not affect objects on that layer that have had their color or linetype set with the Change Properties dialog box, or with the Color, Linetype, or Change commands. In general, it is better to use layers to set the color and linetype. For this reason, the default method for setting the color of objects is the selection BYLAYER, which means that the color set for the layer will be used as the color for the objects.

Pick: **Properties icon**

Select objects: *(select the middle horizontal line and the two vertical lines)* (↵)

The Change Properties dialog box, which you used in Tutorial 3, appears on your screen. It lets you change a number of properties of the objects that you have drawn. You can change an object's color, layer, linetype, and thickness (this gives the object thickness in the Z direction to create a 3D wireframe).

Pick: **Linetype**

The Select Linetype dialog box appears on the screen, as shown in Figure 4.8.

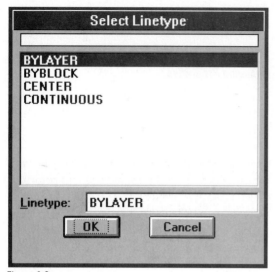

Figure 4.8

Pick: **CENTER** *(by highlighting it and pressing the pick button)*

Pick: **OK** *(to exit the Select Linetype dialog box)*

Pick: **OK** *(to exit the Change Properties dialog box)*

Your drawing should now look like Figure 4.9.

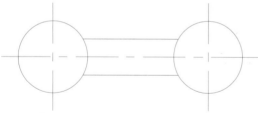

Figure 4.9

Setting the Global Linetype Scaling Factor

The *global linetype scaling factor* is a value that you set that adjusts the lengths of the lines and dashes used to make up various linetypes. Each linetype pattern is stored in a file with the *.lin* file extension which is external to your drawing. The lengths of the lines and dashes are set in that external file. But when you are working on large-scale drawings where the distances are in hundreds of units, a linetype with the dash length of 1/8 unit will not appear correctly. In order to adjust the linetypes, you specify a factor by which the length of the dashes in the *.lin* file should be multiplied. The global linetype scaling factor adjusts the scaling of all of the linetypes in your drawing.

When you are working on a large drawing such as the subdivision that you created in Tutorials 2 and 3, you must scale the dashed lines of the linetype up in order to make them visible. When you are working on a drawing that you will plot full scale, generally you should set the global linetype scaling factor to 1.00, which is the default. Setting the global linetype scaling factor to a decimal less than 1 (like .75) results in a drawing with smaller dashes comprising the linetypes. In general, the linetype scaling factor you set in your drawing should be the reciprocal of the scale at which you will plot your drawing. When you learn about paper space in Tutorial 5, you will see that you can set the linetype scaling factor both in model space, where you are working now, and in paper space. You will type the command to set the global linetype scale at the command prompt.

Type: **LTSCALE** ⏎

New scale factor <1.0000>: **.75** ⏎

The centerlines should now cross in the center of each circle, as shown in Figure 4.10.

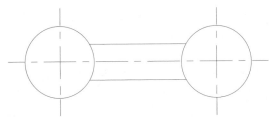

Figure 4.10

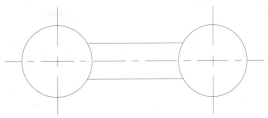

Figure 4.11

Changing on Object's Linetype Scale

You can also use the Linetype Scale area of the Change Properties dialog box to set the linetype scale for any particular object independent of the rest of the drawing. Next you will set the linetype scale for the two vertical lines to 1.5, giving them a different linetype scale than the horizontal line. You will use the Object Properties toolbar and pick Properties to use the Change Properties dialog box.

Pick: **Properties icon**

Select objects: *(pick the two vertical lines)*

Select objects: (⏎)

The Change Properties dialog box appears on your screen. Use the box to the right of Linetype Scale to set the linetype scaling factor for the objects you previously selected. You will set the linetype scale for the two lines to 1.5. Highlight the text in the box to the right of Linetype Scale and overtype the new value.

Type: **1.5**

Pick: **OK**

Now your drawing should look like Figure 4.11.

> ■ *TIP* The Dimcenter command creates correctly shown centerlines for circles. You will learn to use it in Tutorial 6. ■

Save your drawing before continuing.

Now you are ready to add the fillets to your drawing, as you see in Figure 4.12. On your own, use the Fillet command, with a radius of .5, to add the fillets. (Refer to Tutorial 3 if you need to review Fillet.)

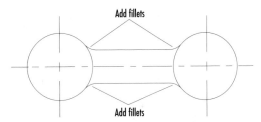

Figure 4.12

Using Polygon

The Polygon command draws regular *polygons* with 3 to 1024 sides. AutoCAD creates polygons as polyline objects, which you used in Tutorial 3. Polygons act as a single object because they are created as a connected polyline. A regular polygon is one in which the lengths of all sides are equal. The size of a polygon is usually expressed in terms of a related circle. This derives from the classic straight edge/compass construction techniques most students learn in their first geometry course. Polygons are either *inscribed* in or *circumscribed* about a circle. A pentagon is a five-sided polygon. Figure 4.13 shows a pentagon inscribed within a circle and a pentagon circumscribed about a circle.

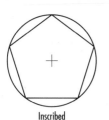
Inscribed

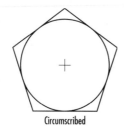
Circumscribed

Figure 4.13

Polygon is located on the Polygon flyout on the Draw toolbar, as shown in Figure 4.14. You will add a pentagon to the right side of the wrench. You are going to circumscribe the pentagon about a circle.

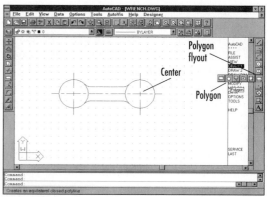

Figure 4.14

Pick: **Polygon icon**

Number of sides <4>: **5** ⏎

Edge/<Center of polygon>: *(select the center point of the right circle)*

Inscribed in circle/Circumscribed about circle (I/C)<I>: **C** ⏎

You are now in drag mode; observe the pentagon changing size on the screen as you move your cursor. You must specify the radius of the circle, either by picking with the pointing device or by typing in the coordinates. For this object you will specify a radius of one-half inch.

Radius of circle: **.5** ⏎

A five-sided regular polygon (a pentagon) is drawn on your screen. Your drawing should look similar to Figure 4.15.

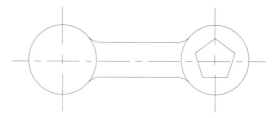

Figure 4.15

Drawing Hexagons

Hexagons are a very common shape in technical drawings. Hexagons are six-sided polygons. The heads of bolts, screws, and nuts often have a hexagonal shape. The size of a hexagon is sometimes referred to by its *distance across the flats*. This is because the sizes of screws, bolts, and nuts are defined by the distance across their flat sides. For example, a 16mm hexhead screw would measure 16 mm across its head's flats and would fit a 16mm wrench. The distance across the flats of a hexagon is not the same as the length of an edge of the hexagon. Figure 4.16 illustrates the difference between the two distances. If a hexagon is circumscribed about a circle, the diameter of the circle equals the distance across the hexagon's flats. If a hexagon is inscribed in a circle, the diameter of the circle equals the distance across the corners of the hexagon.

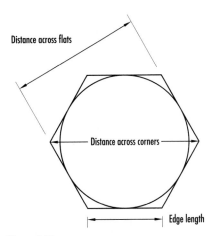

Figure 4.16

You can use the Polygon command's Edge option to draw regular polygons by specifying the length of the edge. This function is helpful when you are creating side-by-side hexagonal patterns (honeycomb patterns). Note that AutoCAD draws polygons in a counterclockwise direction. If you use the Edge option, the sequence in which you select points will affect the position of the polygon. See Figure 4.17.

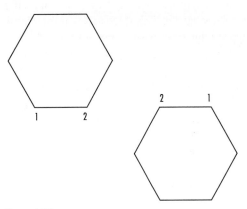

Figure 4.17

Next, you will add the hexagon to the other side of the wrench. You will use the circumscribed option again.

> Command: ⏎ *(or press the return button)*
> Number of sides <5>: **6** ⏎
> Edge /<Center of polygon>: *(pick the center point of the left circle)* ⏎
> Inscribed in circle/Circumscribed about circle (I/C)<C>: ⏎
> Radius of circle: **.5** ⏎

A hexagon measuring 1.00 across the flats appears in your wrench. The final drawing should look similar to Figure 4.18.

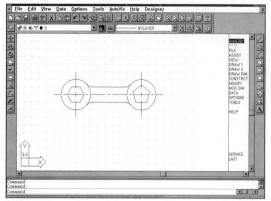

Figure 4.18

You have now completed your wrench. You will save this drawing and begin a new drawing to continue this tutorial.

> *Pick:* **Save icon**

In this next section, you will practice with additional drawing commands and with object snaps. You will start a new drawing and name it *coupler.dwg*.

> *Pick:* **New icon**

Type *coupler.dwg* in as the name of the new drawing. Make sure that the default drawing, *acad.dwg*, is the prototype.

> *Pick:* **OK**

You are returned to the drawing editor. Notice that *coupler.dwg* appears as the drawing name in the title bar.

Using Object Snaps

As you saw in Tutorial 2, you use object snaps to accurately select locations in relation to other objects in your drawing. You learned that object snaps can operate in two different ways, the override mode and the running mode. You will practice more with object snaps in this tutorial when you create the coupler drawing.

Remember, when an object snap is active, AutoCAD has a special cursor, the aperture box, which appears around the crosshairs. When you are selecting a point, the point can be anywhere within the target area of the aperture box.

Object Snap Overrides

You will draw the coupler shown in Figure 4.19, using the object snap overrides. You will use the object snap overrides to position lines and circles relative to this figure. (You may see other ways that you could use editing commands to create parts of this figure, but in this example object snaps will be used as much as possible. When you are working on your own drawings, you will use a combination of the methods you have learned.)

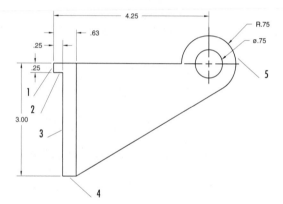

Figure 4.19

You will use the Line and Circle commands to create lines 1–4 and circle 5. On your own, set Grid and Snap to .25 and make sure that they are turned on. Use Zoom All if necessary so that the grid area fills the graphics window.

> *Pick:* **Line icon**
> From point: **3.75,6.5** ⏎
> To point: **@.25<270** ⏎
> To point: **@.25<0** ⏎
> To point: **@2.75<270** ⏎

To point: **@.375<0** ⏎

To point: ⏎

Pick: **Circle Center Radius icon**

3P/2P/TTR/<Center point>: **8,6.5** ⏎

Diameter/<Radius>: **.75** ⏎

When you have finished drawing these objects,
your drawing should look like Figure 4.20.

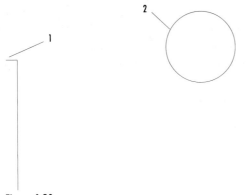

Figure 4.20

Turn Snap off. You will use the object snaps to
accurately locate points. Leaving Snap turned
on may interfere with selection.

Showing a Floating Toolbar

You will show the floating Object Snap toolbar
to make it easy to select object snaps. As you
recall, the object snaps are also located on the
Standard toolbar as a flyout selection. You
can show their floating toolbar by using the
Toolbars selection on the Tools pull-down
menu.

Pick: **Tools, Toolbars, Object Snap**

The Object Snap toolbar appears on your
screen, as shown in Figure 4.21. You can position it anywhere on the screen or dock it to the
edge of the graphics window. Refer to Tutorial 1
if you need to review floating windows and how
to dock toolbars.

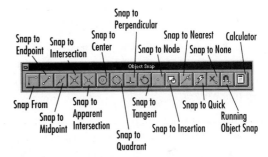

Figure 4.21

Snap to Endpoint

 The Endpoint object snap snaps to the
closest endpoint of an arc, line, or polyline vertex. You will draw a line from
the endpoint of line 1, shown in Figure 4.20, to
touch the circle at the *quadrant point.* The
quadrant points are at 0°, 90°, 180°, and 270°
of a circle or an arc. You will use the Endpoint
and Quadrant object snaps to locate the exact
points. Use the floating Object Snap toolbar to
select the object snaps.

> ■ *TIP* You can also type a three-letter code
> for the object snap at the *Select objects*
> prompt instead of picking it from the menu.
> The command-line equivalent for the
> Endpoint object snap is END. ■

Pick: **Line icon**

From point: *(pick Snap to Endpoint icon)*

Notice that the cursor changes to show the
aperture box and the command line reflects
the object snap you chose.

endp of *(place the aperture box on the upper end of
line 1 and press the pick button)*

Notice that the cursor has jumped to the exact
endpoint of the line. The endpoint you want to
select does not have to be inside the aperture
box; the box merely has to be positioned on
the object nearer to that endpoint than the

other one. Positioning the aperture box slightly away from the point you want to pick is useful because then you can see when the cursor "jumps" to the object snap location. This way you are able to determine that the object snap has worked.

> ■ *TIP* When you have an object snap running mode turned on, you can still use a different object snap during a command. The object snap override takes precedence over the object snap running mode for that pick. You can use the Snap to None override so that no object snap is used during a selection. ■

Snap to Quadrant

The Quadrant object snap attaches to the quadrant point on a circle nearest to the position of the crosshairs. The command-line equivalent for Snap to Quadrant is QUA. The quadrant points are the four points on the circle that are tangent to a square that encloses it. They are also the four points where the centerlines intersect the circle. Next you will finish the line, using the Quadrant object snap to pick the quadrant point of the circle as the second endpoint for your line. You should see the *To point:* prompt for the Line command.

To point: *(pick Snap to Quadrant icon)*

qua of *(pick on the circle near point 2 in Figure 4.20)*

To point: ⏎

Your drawing should look similar to Figure 4.22.

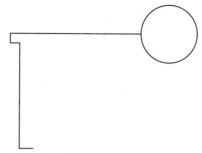

Figure 4.22

Snap to Center

The Center object snap finds the center of a circle or an arc. You will use it to create concentric circles by selecting the center of the circle you have drawn as the center for the new circle you will add. The command-line equivalent for Snap to Center is CEN.

Pick: **Circle Center Diameter icon**

3P/2P/TTR/<Center point>: *(pick Snap to Center icon)*

cen of: *(pick on the edge of the circle)*

> ■ *TIP* A point on the circle must be within the target area of the aperture box. Do not position the target area over the center point. ■

AutoCAD finds the exact center of the circle and a circle rubberbands from the center to the location of the crosshairs. Notice that the aperture box disappears once the circle is picked. Object snap overrides stay active for only one pick. AutoCAD now prompts you to specify the diameter.

Diameter<1.5000>: *.75* ⏎

The circle should be drawn concentric to the original circle in the drawing, as shown in Figure 4.23.

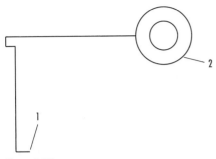

Figure 4.23

To position the next line from the endpoint of line 1 in Figure 4.23 tangent to the outer circle, you will use the Endpoint object snap and then the Tangent object snap.

Pick: **Line icon**

From point: ***(pick Snap to Endpoint icon)***

endp of: ***(pick near the right endpoint of line 1)***

A line rubberbands from the endpoint of line 1 to the location of the crosshairs. Next, you will use the Tangent object snap to locate the second point of a line tangent to the circle.

Snap to Tangent

 The Tangent object snap attaches to a point on a circle or an arc; a line drawn from the last point to the referenced object is drawn tangent to the referenced object. The command-line equivalent is TAN.

To point: ***(pick Snap to Tangent icon)***

tan to: ***(pick the lower right side of the circle)***

To point: ⏎

Your screen should look similar to Figure 4.24.

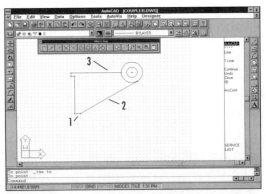

Figure 4.24

> ■ *TIP* To draw a line tangent to two circles, pick Line. At the *From point* prompt, pick Snap to Tangent and pick on one of the circles in the general vicinity of the tangent line you want drawn. You will not see a line rubberband. At the *To point* prompt, pick Snap to Tangent again and pick the other circle. When you select the second object, the tangent is defined and it will be drawn on your screen. ■

Next you will draw a line from the intersection of the short horizontal line and the angled line, perpendicular to the upper line of the object. You will use two more object snaps to draw this line.

Snap to Intersection

 Snap to Intersection finds the intersection of two graphical objects. The command-line equivalent for Snap to Intersection is INT.

When you are using 3D solids modeling in Tutorial 9, remember that when you are looking for the intersection of two objects when selecting Snap to Intersection, the objects must *actually* intersect in 3D space, not just *apparently* intersect on the screen (one line may be behind the other in the 3D model). To select lines that do not intersect in space, use the

Apparent Intersection object snap. It finds the intersection of lines that would intersect if extended and also the apparent intersection of two lines (where they appear to intersect in a view), when one line is really behind the other in 3D space.

Refer to Figure 4.24 when selecting your points.

Pick: **Line icon**

From point: *(pick Snap to Intersection icon)*

int of *(pick so that the intersection of lines 1 and 2 is anywhere inside the target area)*

A line rubberbands from the intersection to the current position of the crosshairs. To select the second endpoint of the line, you will use Snap to Perpendicular to draw the line perpendicular to the top horizontal line.

Snap to Perpendicular

The Perpendicular object snap attaches to a point on an arc, a circle, or a line; a line drawn from the last point to the referenced object forms a right angle with that object.

You will invoke the Perpendicular object snap by typing its command-line equivalent, PER, at the command line.

To point: **PER** ⏎

to *(pick on line 3)*

To point: ⏎

The line is drawn at a 90° angle, perpendicular to the line, regardless of where on the line you picked. Your screen should look similar to Figure 4.25. Although this example draws a perpendicular line that touches the target line, it need not touch it. If you choose to draw perpendicular to a line that does not intersect at an angle of 90°, the perpendicular line is drawn to a point that would be perpendicular if the target line were extended.

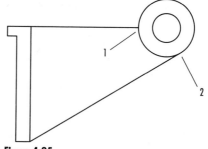

Figure 4.25

■ *TIP* Remember that in order to draw a perpendicular line, you must define two points. When you are defining a perpendicular from a line, no line is drawn until you define the second point, because there are an infinite set of perpendicular lines to a given line, but there is only one line perpendicular to a line through a given point. Your perpendicular line will be drawn after you have selected the second point. ■

Use the techniques you have learned to save your drawing now.

Practice on your own with each of the object snaps. Try selecting the rest of the object snaps from the toolbar.

Practicing with Running Mode Object Snaps

You can also use object snap in running mode, as you did in Tutorial 2 with Snap to Node. When you use object snaps this way, you turn the mode on and leave it on. Any time a command calls for the input of a point or selection, the current object snap is used when you pick. The running mode object snaps are very useful, as they reduce the number of times you must pick from the menu to achieve the desired drawing results. You can use any of the object snaps in either running mode or in override mode.

- ***Warning:*** When using running mode object snaps, make sure to turn them off when you are done. If you forget, you may have trouble with certain other commands. For example, if you turn on the running object snap for Perpendicular and leave it on and later try to erase something, you may have trouble selecting objects because AutoCAD will try to find a point perpendicular to every object you pick. ■

Pick: **Running Object Snap icon**

The Running Object Snap dialog box appears on the screen, as shown in Figure 4.26.

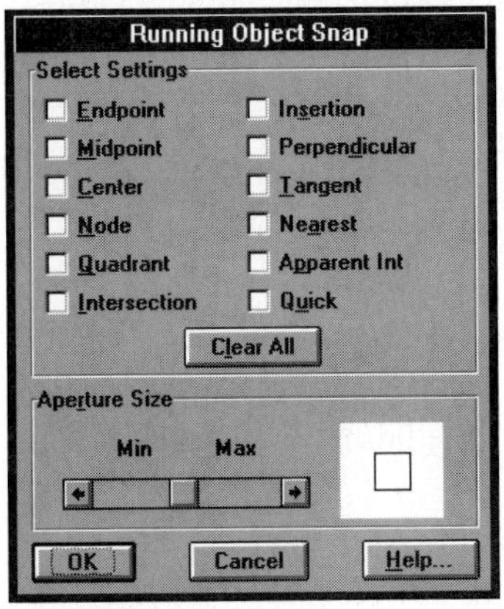

Figure 4.26

Pick the empty box to the left of Intersection on your own. An X appears in the box to indicate that Intersection has been selected. Pick OK to exit the Running Object Snap dialog box.

Now Intersection is turned on. Any time you are prompted to select, you will see the aperture box on the crosshairs. If the intersection of two objects is inside the aperture area when you press the pick button, the exact intersection is selected. If you do not have an intersection inside the box, the closest point to the

center of the crosshairs is selected. Unlike override mode, running mode does not display a message to tell you no intersection was found.

- ***Warning:*** Because the closest position to the center of the crosshairs is returned when there is no intersection in the box, you must be careful when using running mode object snap. If your drawing geometry does not intersect properly, you will not be made aware of this fact. For a beginner, the safest method is to use the overrides so you will be prompted if the object snap condition is not met. ■

You will use the Break command to break the circle between intersection 1 and intersection 2, shown in Figure 4.25. You will use the running mode object snap Intersection while selecting points during the Break command. (Notice that you could also accomplish this by using Trim.) As you break the lines, notice that the aperture appears each time you are prompted to select a point—you do not have to select the object snap each time.

Using Break

The Break command erases part of a line (or an arc or circle). The Break flyout is located on the Modify toolbar. The Break command options, selected from the Modify toolbar, are shown in Figure 4.27.

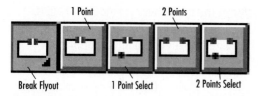

1 Point   2 Points

Break Flyout   1 Point Select   2 Points Select

Figure 4.27

When using the Break command, you can specify a single point at which to break the object, you can select the object and then specify a single point at which to break it, you can specify two points on the object and the Break command will automatically remove the portion between the points selected, or you can select the object and then specify the two points at which to break it. You will use the 2 Points Select option. Refer to Figure 4.25 to make your selections.

Pick: **2 Points Select icon**

Select object: *(pick the large circle)*

Enter first point: *(pick intersection 1)*

Enter second point: *(pick intersection 2)*

Notice that when you are prompted to select points, the aperture box automatically appears on the crosshairs. The Intersection running object snap is in effect. After you select the second break point, the portion of the circle between the two selected points is removed. Your drawing should look similar to Figure 4.28.

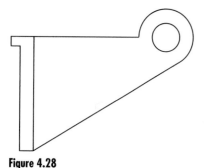

Figure 4.28

The other object snaps are described below. Try them on your own until you are familiar with how they work. When you are finished, return to the Running Object Snap dialog box and turn off any object snaps you may have left on.

Snap to Apparent Intersection

The Apparent Intersection object snap finds two different kinds of intersections that would not be found with the regular Intersection object snap. One type of intersection it will find is where two objects would intersect if they were extended. The other type of intersection it will find is where two 3D lines intersect on the screen, when they do not in fact intersect in space. The command-line equivalents are APPINT and APP.

Snap to Nearest

The Nearest object snap attaches to the point on an arc, circle, or line, or to the point object, that is closest to the middle of the target area of your cursor.

A line drawn using the Nearest function may look similar to a line that could have been drawn simply using the Line command, but there is a subtle difference. Many of AutoCAD's operations, such as hatching, require an enclosed area: all lines that define the area must intersect (touch). This is particularly true when you are working with three-dimensional functions. When you draw lines by picking two points on the screen, sometimes the lines are not really touching. They appear to touch on the screen, but when you zoom them sufficiently, you will see that they don't touch. They may only be a hundredth of an inch apart, but they don't touch. This means that you do not have an enclosed area and many of AutoCAD's functions will not recognize the apparent area as an area. When creating a drawing with AutoCAD, you should always strive to create the drawing geometry accurately. The Nearest function ensures that the nearest object is selected. The command-line typing equivalent is NEA.

Snap to Node

The Node object snap finds the exact location of a point object in your drawing. You used it in Tutorial 2 to locate exact points that had already been placed in the drawing with the Point command. The command-line typing equivalent is NOD.

Snap to Quick

You can use the Quick object snap in conjunction with the other object snaps you have learned to limit the search through the database, making selection using object snaps faster. For example, if you have Intersection and Quick turned on, the search through the drawing database stops as soon as one intersection is found, even if there are several within the aperture box. When using Quick, it is best to have just one of the type of object you are trying to locate within the aperture box. The command-line typing equivalent is QUI.

Snap to Insertion

The Insertion object snap finds the insertion point of text or of a block (you will learn about blocks in Tutorial 8). This is useful when you want to determine the exact point where existing text or blocks are located in your drawing. The command-line typing equivalent is INS.

■ **TIP** The Locate Point command that you learned in Tutorial 3 lists the coordinates of a selected point. To find the exact insertion point of text or a block, pick Locate Point, and then use the Insert object snap and pick on the text or block. Similarly, you can find the coordinates of an endpoint or intersection by combining those object snaps with the Locate Point command. ■

Snap From

The From object snap is a special object snap tool. It establishes a temporary reference point from which you can specify the point to be selected. Usually it is used in combination with other object snap tools. For example, you would use it if you were going to draw a new line that you want to start a certain distance away from an existing intersection. To do this, pick the Line command; when AutoCAD prompts you for the starting point of your line, pick the Snap From icon. You will then be prompted for a base point. This is the location of the reference point from which you would like to locate your next input. At the base point prompt, pick Snap to Intersection. You will be prompted to select the intersection of two lines. Pick on the intersection. Then you will see the additional prompt *of <Offset>:*. This is the result of your having picked Snap From. At this prompt, use relative coordinates to enter the distance you want the next point to be from the base point you selected.

On your own, turn any running mode object snaps off and save your drawing before continuing.

Using Extend

The Extend command extends the lengths of existing lines and arcs to end at a selected boundary edge. Its function is the opposite of Trim's. Like Trim, Extend has two parts: first you select the object to act as the boundary, then you select the objects you want to have extended. Extend is located on the Modify toolbar on the same flyout as Trim.

Similar to the Trim command, the command-line options for the Extend command are Project, Edge, and Undo. The default option is just to select the objects to extend. You can use the Edge option to extend objects to the point at which they would meet the boundary edge if it were longer. This is useful when the boundary edge is short and the lines to be extended would not intersect it. The Project

option allows you to specify the plane of projection used when extending. This is useful when working in 3D drawings, because it allows you to extend objects to a boundary selected from the current viewing direction or User Coordinate System, even though the 3D objects may not actually intersect the boundary edge but just appear to in the view. The Undo option lets you undo the last object extended, while remaining in the Extend command.

On your own, use the commands you have learned to erase the lines labeled 1, 2, and 3. Draw a new line .5 units (two snap increments) from the previous left-most line. Refer to Figure 4.29 when making your selections.

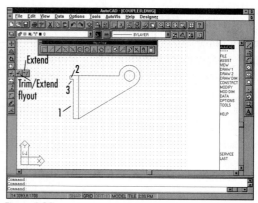

Figure 4.29

When you have completed this step, your drawing should appear similar to Figure 4.30.

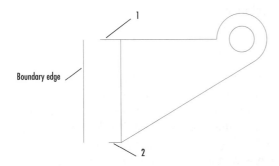

Figure 4.30

Boundary edge

1

2

Now you will use the Extend command to extend the existing lines to meet the new boundary.

Pick: **Extend icon**

Select objects: *(select boundary edge)*

Select objects: (⏎)

You are finished selecting boundary edges. To extend the horizontal lines, select them by picking points on them near the end closer to the boundary edge. (If the points are picked closer to the other end, you will get the message *No edge in that direction,* and the lines will not be extended.)

<Select object to extend>/Project/Edge/Undo: *(select lines 1 and 2)* (⏎)

The screen will look like Figure 4.31 once the lines have been extended.

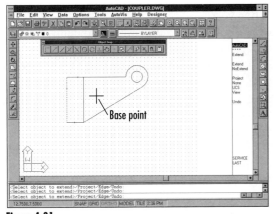

Figure 4.31

Using Rotate

The Rotate command allows you to rotate a drawing object or group of objects to a new orientation in the drawing. Rotate is located on the Modify toolbar. In this section, you will use Rotate to rotate objects in your drawing.

Pick: **Rotate icon**

Select objects: *(select the entire coupler using a Crossing window)*

7 found

Select objects: (⏎)

AutoCAD prompts you for the base point. You will select the middle of the object as the base point, as shown in Figure 4.31.

■ *TIP* The base point need not be part of the object chosen for rotation. You can use any point on the screen; the object rotates about the point you select. It is often useful to select the base point for rotation in the center of the object you want to rotate, so that as the object rotates it stays in a position on the screen similar to the original one. ■

Base point: (*pick in the middle of the coupler*)

<Rotation angle>/Reference: **45** ⏎

Your coupler should be rotated 45°, as shown in Figure 4.32. Angles are positive in the counterclockwise direction. A horizontal line to the right of the base point is defined as 0°. You can also enter negative values.

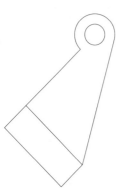

Figure 4.32

On your own, pick the Undo icon from the Standard toolbar to undo the Rotate command so that the object is restored to its original appearance before continuing.

Using Move

 The Move command moves existing objects from one location on the coordinate system in the drawing to another location. Do not confuse the Move command and the Pan command. Pan moves your view and leaves the objects where they were. Move actually moves the objects on the coordinate system. Move is located on the Modify toolbar.

Pick: **Move icon**

Select objects: (*use Crossing or Window to select the entire coupler*)

Select objects: ⏎

Base point or displacement: (*pick the Snap to Endpoint icon*)

endp of (*select the upper corner of the coupler*)

AutoCAD switches to drag mode so you can see the object move about the screen. It should appear similar to Figure 4.33.

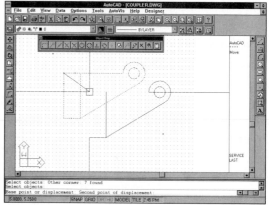

Figure 4.33

Second point of displacement: (*select a point so the coupler is to the lower right of its old location*)

Methods of Selecting Objects

One advantage of AutoCAD is the many ways that you can select graphical objects for use with a command. You have already seen how you can use Window and Crossing. You can save lots of time in creating and editing your drawing by clever use of the selection methods. Generally, whenever you are asked to select objects, you can continue selecting until you have all of the objects that you want highlighted in one *selection set*. You can combine the various selection modes to select the objects you want. The command then operates on the selection set you have built.

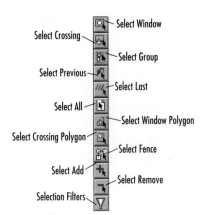

Figure 4.34

The Standard toolbar has a flyout for the available selection modes. It is shown in Figure 4.34.

During any command that prompts *Select objects*, you can type the option letters for the method you want to use or pick the selection icon from the toolbar shown in Figure 4.34. You can also pick the Select Objects option from the screen menu to display the list of object selection options there. AutoCAD allows you to continue selecting, using any of the methods in the table below, until you indicate that you are done building the selection set by pressing ⏎ or the return button. Then the command will take effect on the objects that you have selected.

Name	Option Letter(s)	Method
Picking	none	picks objects by placing selection cursor and pressing pick button
Select Window	W	specifies diagonal corners of a box that only selects objects that are entirely enclosed
Select Crossing	C	specifies diagonal corners of a box that selects all objects that cross or are enclosed in the box
Select Group	G	selects all objects within a named group, prompts to enter group name
Select Previous	P	reselects the previous selection set
Select Last	L	selects the last object created
Select All	ALL	selects all the objects in your drawing unless they are on a frozen layer
Select Window Polygon	WP	similar to Window except you draw an irregular polygon instead of a box around the items to select
Select Crossing Polygon	WC	similar to Window Polygon except that all objects that cross or are enclosed in the polygon are selected

Name	Option Letter(s)	Method
Select Fence	F	similar to WC except that you draw line segments through all the objects you want to select
Select Add	A	use after Remove to add more objects to the selection set; can continue with any of the other selection modes once Add has been chosen
Select Remove	R	selects objects to remove from the current selection set; removal continues until Add or ⏎ is used; you can use Window, Crossing, etc. while selecting items to remove
Undo	U	during object selection, unselects the last item or group of items you selected

Selection filters are a special method of selecting objects. You can use them to select types of objects, such as all of the arcs in the drawing or objects that were created by setting the color or linetype independently of the layer they are on.

Pick: **Selection Filters icon**

The Object Selection Filters dialog box shown in Figure 4.35 appears on your screen. You can use this dialog box to filter the types of objects to select. You can save named filter groups for reuse.

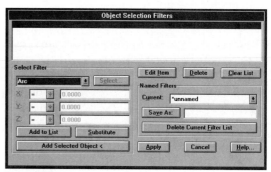

Figure 4.35

Pick: (on Arc below the heading Select From)

The list of types of drawing objects pops up. These are drawing object types that you could use to filter the objects in your selection.

Pick: **Arc** *(so that it is highlighted)*

Pick: **Add to List**

Pick: **Apply**

You return to the command prompt to select the objects in your drawing. Notice the message *Applying filter to selection*.

Select objects: *(use Crossing to select all of the objects)*

You should see the message *7 found, 6 were filtered out*. Only the arc object matched the filter list. The lines of the drawing were filtered out.

Select objects: ␛ESC *(to cancel)*

Additional methods of selecting are Auto, Box, Single, and Multiple. Consult AutoCAD's on-line help about the Select command for definitions of their use.

On your own, create a new drawing called *geneva.dwg*. Discard the changes to *coupler.dwg*.

Creating the Geneva Cam

You will create the geneva cam you see in Figure 4.36, using many of the editing commands you have learned in this tutorial. You will learn to use the Array command to create rectangular and radial patterns.

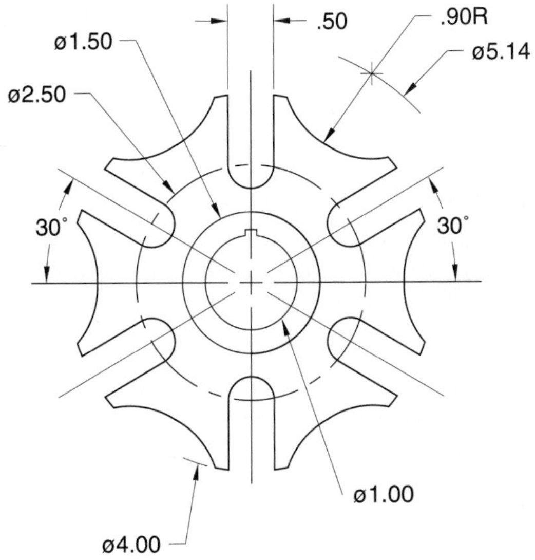

On your own, set Snap to .25. Make sure that Snap and Grid are turned on by checking to see if the words Snap and Grid are displayed on the status bar. If not, turn them on. Use Zoom All if needed so that the grid area fills the graphics window.

Next, you will use the Circle command to create the innermost circle of diameter 1.00. Locate the center of the circle at coordinates 5.5,4.5.

Pick: **Circle Center Diameter icon**

3P/2P/TTR/<Center point>: 5.5,4.5 ⏎

diameter: 1 ⏎

On your own, use Offset or the Circle command to draw the 1.5-diameter circle and the outer 4-diameter circle concentric to the circle you drew in the previous step and add the centermarks. (If you use Offset, remember to calculate the offset distance. You will need to subtract the diameter of the inner circle from the diameter of the outer circle and then divide the result by two to find the value to use for the offset distance.)

Your drawing should look similar to Figure 4.37.

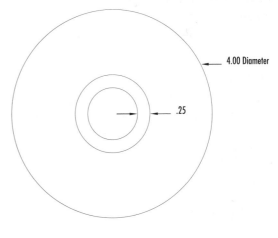

Figure 4.37

Figure 4.36

Now continue adding the other features, according to the dimensions given in Figure 4.36. On your own, draw a vertical and a horizontal construction line through the center of the circles. Next you will use the Offset command to create lines parallel to the vertical centerline.

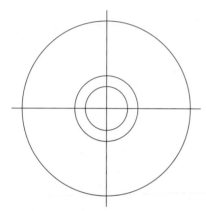

Figure 4.38

Pick: **Offset icon**

Offset distance or Through <Through>: **.25** ⏎

Select object to offset: *(pick the vertical centerline)*

Side to offset? *(pick on the right side of the line)*

Select object to offset: *(pick the same vertical centerline)*

Side to offset? *(pick on the left side of the line)*

Select object to offset: ⏎

Next you will use the Ray command to draw a line angled at 60° and of infinite length, as shown in Figure 4.39.

Using Ray

The Ray command and the Construction Line command located on the Line flyout of the Draw toolbar create lines that extend infinitely and are very useful in creating construction geometry. Ray creates an infinite line (through a point that you select) that extends in only one direction

from the first point selected. Construction lines extend infinitely in both directions.

Pick: **Ray icon**

From point: **5.5,4.5** *(or pick the center point of the circles, using Snap)*

Through point: **@3<60** ⏎

Through point: ⏎

The ray appears in your drawing, as shown in Figure 4.39. The ray will extend as far as you zoom in the direction in which it extends. When rays are trimmed, they become lines.

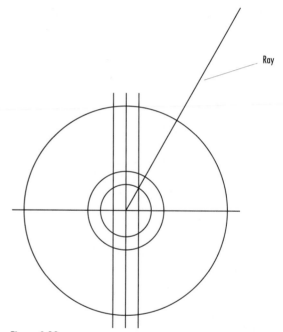

Figure 4.39

Use the Circle command to add the construction circles to your drawing.

Pick: **Circle Center Diameter icon**

3P/2P/TTR/<Center point>: **5.5,4.5** ⏎ *(or use a snap increment to select the same center point as before)*

diameter: **5.14** ⏎

Command: ⏎ *(or press the return button)*

3P/2P/TTR/<Center point>: **5.5,4.5** ⏎ *(or use a snap increment to select the same center point as before)*

Diameter/<Radius> <2.5700>: **d** ⏎

diameter <5.14000>: **2.50** ⏎

Command: ⏎

3P/2P/TTR/<Center point>: *(pick the point where the vertical centerline crosses the 2.5-diameter construction circle)*

Diameter/<Radius> <1.2500>: **d** ⏎

diameter (2.5000): **.50** ⏎

The small circle is added to your drawing between the two lines you offset. Your drawing should look like Figure 4.40.

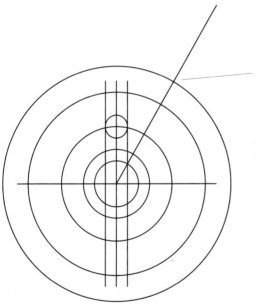

Figure 4.40

Draw the final construction circle by specifying the center coordinates, where the 60° ray and the circle intersect.

Command: ⏎

3P/2P/TTR/<Center point>: *(pick Snap to Intersection icon)*

of *(pick point 1 in Figure 4.40)*

Diameter/<Radius> <.2500>: **.90** ⏎

Your screen should look like Figure 4.41.

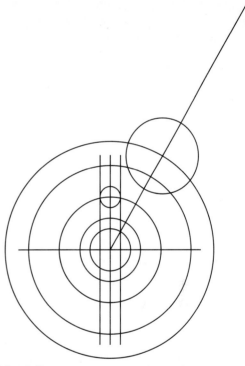

Figure 4.41

The next step is to remove the portion of the lines and circles that you will not need in the Array command.

On your own, use the Trim and Erase commands to remove the unwanted portions of the figure until your drawing looks like Figure 4.42.

> ■ *TIP* If you use the implied Crossing method to select all of the objects in the drawing as the cutting edges and then press ⏎, you can quickly trim the object to the shape shown by picking on the extreme ends of the objects you want to have removed. Use Erase to remove any of the objects that no longer cross a cutting edge. ■

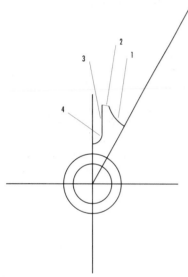

Figure 4.42

Using Mirror

 The Mirror command makes a mirror-image copy of the objects you select around a *mirror line* that you specify. The Mirror command is located on the Copy flyout on the Modify toolbar. Its icon is the mirror image of a shape. The Mirror command also allows you to delete the old objects or to keep them.

To mirror an object, you must specify a mirror line. A mirror line defines both the angle and the distance at which the mirror image is to be drawn. The mirrored image is drawn perpendicular to the mirror line. You can use slanted mirror lines. The mirrored object is the same distance from the mirror line as the original object is, but on the other side of the mirror line. The mirror line does not have to exist in your drawing before you use the Mirror command; when asked to specify the mirror line, you can pick two points from the screen that define it.

The vertical centerline will serve as the mirror line as you mirror the lines and arcs of the geneva cam. Refer to Figure 4.42 as you pick points. Be sure that Snap is turned on or use object snaps before picking the mirror line.

Pick: **Mirror icon**

Select objects: *(pick objects 1, 2, 3, and 4)*

> ■ *TIP* If you select objects that you do not want to mirror, you can use the Remove option of the object selection modes to remove them from the selection set. To do this, type *R* ⏎ or pick Remove from the Select Objects flyout on the Standard toolbar while still at the *Select objects:* prompt. Pick the object to be removed from the selection. If you want to add more objects after using Remove, pick Add or type *A* ⏎ while still at the *Select objects* prompt. Press ⏎ when you have selected the objects you want. This tells AutoCAD that you are done selecting and want to continue with the command. ■

Select objects: ⏎

First point of mirror line: *(select the center of the concentric circles)*

Your cursor is now dragging a mirrored copy of the objects.

Second point: *(select a point straight down from the center point)*

Delete old objects? <N> ⏎

Your screen should look like Figure 4.43.

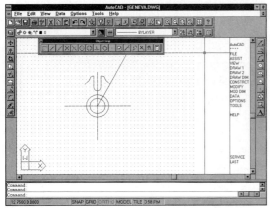

Figure 4.43

■ *TIP* Figure 4.44 shows an object mirrored about one of its own edge lines. If you want, you can add mirror lines to the drawing and then erase them after the construction is complete; or you can pick any two points from the screen that would define the line you want to specify. ■

Mirror line

Mirrored object

Object

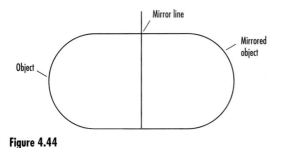

Figure 4.44

Using Array

The Array command copies an object multiple times to form a regularly spaced rectangular or circular pattern. Any time that you have a regularly spaced pattern of objects or objects in your drawing, you can use the Array command to quickly create the pattern. Examples are creating a circular pattern of holes in a circular hub, creating rows of desks in laying out a classroom, or creating the teeth on a gear.

The Array command is located on the Modify toolbar on the Copy flyout. In the Array command, you are prompted to select the items you want to array. The array can be either rectangular or polar. A *rectangular array* is composed of a specified number of rows and columns of the items with a specified distance between them. In a *polar array*, the items are copied in a circular pattern around a center point you specify. The copies can fill 360° of the circle or some portion of 360°. You can choose to have the items rotated as they are copied with the Array command so that they have the proper orientation. Icons for Polar Array and Rectanglar Array and their 3D counterpart are all on the Copy flyout.

 You are now ready to use the Polar Array command to copy the lines and arcs around in a radial pattern to finish creating the geneva cam.

Pick: **Polar Array icon**

Select objects: *(pick the items 1–4 that you mirrored and the mirrored copies of them)*

Select objects: ⏎

Center point of array: *(pick the center of the concentric circles)*

Number of items: **6** ⏎

Angle to fill (+=ccw, -=cw) <360>: ⏎

Rotate objects as they are copied? <Y>: ⏎

On your own, use Erase to erase the ray from the drawing. Your drawing should look like Figure 4.45.

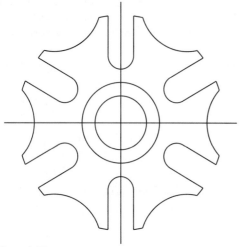

Figure 4.45

Your drawing is now complete; save it on your own.

Making Changes with Hot Grips

A quick way to make changes to your drawing is by using *hot grips*. Hot grips let you grab onto an object that is already drawn on your screen and use certain editing commands directly from the mouse or pointing device without having to select the commands from the menu or keyboard. To use the hot grips you must have a two-button mouse as your minimum hardware configuration. You will check to see that hot grips are enabled, using the Grips dialog box.

Pick: **Options, Grips**

The Grips dialog box appears on your screen, as shown in Figure 4.46. The box to the left of Enable Grips should be selected.

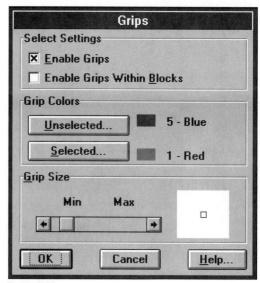

Figure 4.46

AutoCAD allows you to turn grips on or off in your drawing and to change their color and size. To do this on your own, move the selection cursor on the arrow at the right-hand side of the slider bar toward the bottom of the dialog box. Press the pick button several times to enlarge the size of the grips box. When you are finished, restore it to its original size. You can also change the color of the grips, and the base or activated grip, and turn grips off completely if you do not want to use them.

Exit the dialog box by picking OK. You are returned to the drawing editor.

Activating an Object's Grips

Move the crosshairs over the upper vertical centerline of the geneva cam and press the pick button to select the line. The line becomes dashed and small boxes appear at the endpoints, as shown in Figure 4.47.

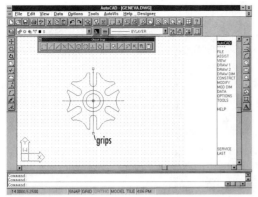

Figure 4.47

These are called the hot grips. You can use them to stretch, move, rotate, scale, and mirror the object.

Using Stretch with Hot Grips

Move the crosshairs to the grip at the upper end of the line and press the pick button to select it. You will see it change to the highlighting color and become filled in solid. This is now the *base grip*. It will act as the base point for the command you will select by pressing the buttons on your pointing device.

In the command line area, you should see the prompt for the Stretch command.

STRETCH

<Stretch to point>/Base point/Copy/Undo/eXit:

The upper end of the line, where you picked the base grip, rubberbands from the position of the crosshairs. Move the crosshairs to a point above and to the right of the old location of the corner and press the pick button. The point you selected should move to the new stretched location.

> ■ *TIP* If you want to cancel using the hot grips, press Ⓔˢᶜ twice to eliminate the hot grips. ■

Using Move with Hot Grips

Now make a crossing box that crosses the entire geneva cam. The hot grips appear as small boxes on all the selected objects. Select the center of the circles as the base grip by moving the crosshairs over the box and picking. You will see the prompt for the Stretch command in the command line area. Press the return button on your pointing device. You will see the command prompt cycle to the Move command.

> ■ *TIP* You can also press ↵ to cycle through the command choices that are available for use with hot grips. ■

The command prompt has the following options for the Move command:

MOVE

<Move to point>/Base point/Copy/Undo/eXit: *(Move the crosshairs down and to the left. You will see the faint outline of the object attached to the crosshairs. Move it to a new location down and to the left and press the pick button when you have it positioned where you want it.)*

Using Move with the Copy Option

The hot grips should still be highlighted. Pick the grip at the center of the circles as the base grip. Press the return button on your pointing device once to cycle to the Move command. You will type *C* ↵ to pick the Copy option when you see the prompt.

MOVE

<Move to point>/Base point/Copy/Undo/eXit: **C** ↵

A new prompt appears, similar to the previous one:

MOVE (multiple)

<Move to point>/Base point/Copy/Undo/eXit: *(Move the crosshairs to a location where you would like to make a copy of the object and press the pick button; repeat this procedure to make several copies. When you are finished, press ⏎ or the return button to end the command.)*

On your own, use the Erase command to erase some of the copies if your screen is too crowded.

Using Rotate with Hot Grips

Now you will use implied Windowing to activate the hot grips for these lines.

Pick: *(above and to the left of one of the copies of the geneva cam)*

You will see a window box form as you move the crosshairs away from the point you selected. Move downwards and to the right to enclose all of the lines in the drawing in the forming window. When you have the window sized correctly,

Pick: *(a point below and to the right of the geneva cam)*

You will see the hot grips appear on the objects in the drawing.

Select the grip at the center of the circles as the base grip by positioning the crosshairs over it and pressing the pick button. You will see it turn the highlighting color and be filled in solid. The Stretch command appears at the command prompt. Press the return button or ⏎ once to cycle to the Move command. Press it again to cycle to the Rotate command. If Snap is turned on, turn it off so you can see the effect of the rotation command clearly.

You will see the prompt:

ROTATE

<Rotation angle>/Base point/Copy/Undo/Reference/eXit: *(You will see the faint object rotating as you move the crosshairs around on the screen. You can press the pick button when the object is at the desired rotation or type in a numeric value for the rotation. Keep in mind that 0° is to the right and positive values are measured counterclockwise.)*

Type: **45** ⏎

Notice that the object rotated so that it is at an angle of 45°.

Using Scale with Hot Grips

The Scale command changes the size of the object in your drawing database. Make sure that you use Scale only when you want to make the actual object larger. Use Zoom Window when you just want it to appear larger on the screen so that you can see more detail.

On your own, use hot grips to scale the object. Activate the hot grips, using implied Windowing again. This time, however, use the Crossing option. In order to use Crossing instead of Window, start your window box in the lower right of the drawing and select the first point. Move the crosshairs with the window up and to the left. This specifies a crossing box. Unlike Window, which selects only the objects that are entirely enclosed in the box, Crossing selects anything that crosses the box or is enclosed. When you have done this successfully, you will see the hot grips appear at the corners and midpoints of the lines.

Pick: *(the grip at the center of the circles as the base grip)*

It becomes solid, filled with the highlighting color. You will see the Stretch command in the command prompt area.

Press: *(the return button or ⏎ three times to cycle through the commands until the Scale command appears)*

You will see the prompt:

SCALE

<Scale factor>/Base point/Copy/Undo/Reference/eXit: *(As you move the crosshairs away from the base point, the faint image of the object becomes larger; as you get closer to the base point it appears smaller. When you are happy with the new size of the object, press the pick button.)*

■ *TIP* To change the scale to known proportions, you can type in a scale factor. A value of 2 makes the object twice as large; a value of .5 makes it half its present size. ■

Using Mirror with Hot Grips

On your own, once again activate the hot grips with implied Windowing. When you see the grips appear on the object, select the center grip as the base grip. Press the return button or ⏎ to cycle through the commands Stretch, Move, Rotate, and Scale until the command Mirror appears in the command prompt area.

The Mirror command uses a mirror line and forms a symmetrical image of the selected objects on the other side of the line. You can think of this line as rubberbanding from the base hot grip to the current location of the crosshairs. Notice that as you move the crosshairs to different positions on the screen, the faint mirror image of the object appears on the other side of the mirror line. You will see the prompt:

MIRROR

<Second point>/Base point/Copy/Undo/eXit: **B** ⏎

The Base point option lets you specify some other point besides the hot grip you picked as the first point of the mirror line for the object.

Base point: *(pick a point to the left of the figure)*

Move the crosshairs on the screen and notice the object being mirrored around the line that would form between the base point and the location of the crosshairs. When you are happy with the location of the mirrored object, press the pick button to select it. The old object disappears from the screen and the new mirrored object remains. End the command by pressing ⏎ or the return button.

Noun/Verb Selection

You can also use hot grips with other commands for *noun/verb selection*. Noun/verb selection is a method of selecting the objects that will be affected by a command first, instead of first selecting the command and then the group of objects. Think of the drawing objects as nouns, or things, and the commands as verbs, or actions. You will use this with the Erase command to clear your screen.

Pick: *(a point below and to the right of all your drawing objects)*

Other corner: *(pick a point above and to the left of the drawing objects)*

You will see the hot grips for the objects that were crossed by the implied Crossing box appear in your drawing.

Pick: **Erase icon**

Notice that you do not have to select the items. They were preselected with the hot grips and as soon as you pressed ⏎, the items you selected were erased.

Pick the close box at the upper left corner of the Object Snaps toolbar to hide it. On your own, hide any other toolbars you may have left on your screen, except for the Draw and Modify toolbars. Exit AutoCAD. You have completed Tutorial 4.

base grip

circumscribed

distance across the flats

global linetype scaling factor

hot grips

inscribed

mirror line

noun/verb selection

polar array

polygon

quadrant point

rectangular array

selection filters

selection set

Break

Copy

Extend

Grips

Linetype

Mirror

Move

Polar Array

Polygon

Ray

Rotate

Scale

Selection Filters

Snap From

Snap to Apparent
 Intersection

Snap to Center

Snap to Insertion

Snap to Intersection

Snap to Nearest

Snap to Perpendicular

Snap to Quadrant

Snap to Quick

Snap to Tangent

Stretch

The letter M after an exercise number means that the given dimensions are in millimeters (metric units). If no letter follows the exercise number, the dimensions are in inches. Do not include dimensions on the drawings.

 4.1M Gasket

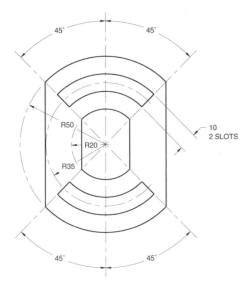

 4.2 Starboard Rear Rib

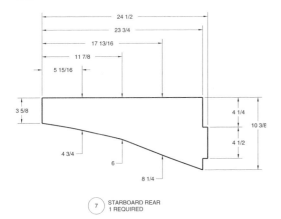

4.3 The Cycle

Starting with the parallelogram shown, use Object snap to draw the lines, arcs, and circles indicated in the drawing. Object snap commands needed are Int, End, Cen, Mid, Per, and Tan.

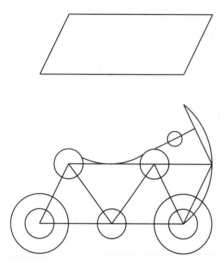

4.4 Park Plan

Create a park plan similar to the one shown here. Use Polyline and Polyline Edit, Spline to create a curving path. Note the symmetry. Add labels with Dtext.

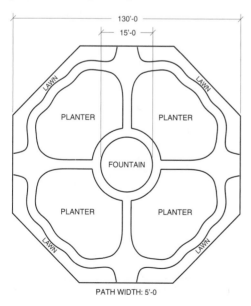

4.5 Floor Plan

Draw the floor plan according to the dimensions shown. Add text, border, and title block.

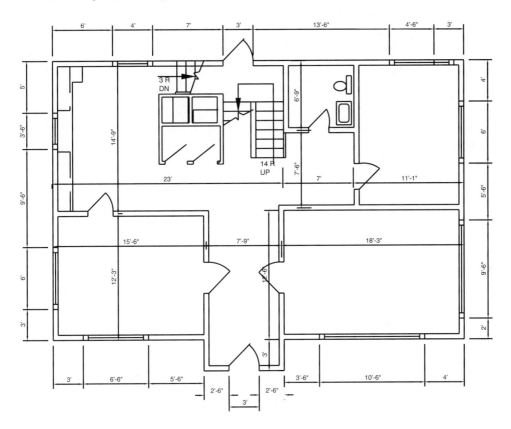

4.6 Interchange

Draw this intersection, then mirror the circular interchange for all lanes.

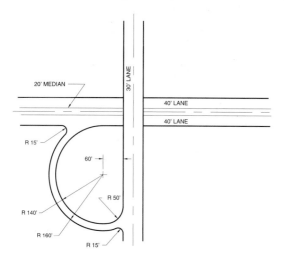

20' MEDIAN

30' LANE

40' LANE

40' LANE

R 15'

60'

R 50'

R 140'

R 160'

R 15'

4.7M Plastic Knob

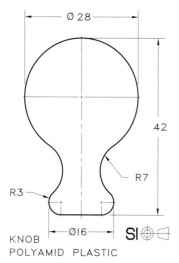

Ø 28

42

R7

R3

Ø16   SI

KNOB
POLYAMID PLASTIC

4.8M Foundry Hook

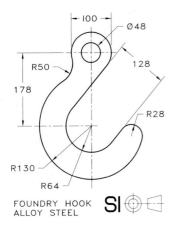

FOUNDRY HOOK
ALLOY STEEL   SI ⬤⊲

4.9M Support

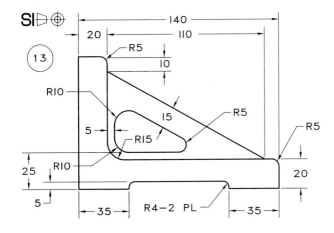

4.10 Hanger

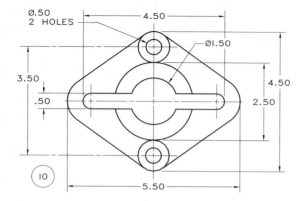

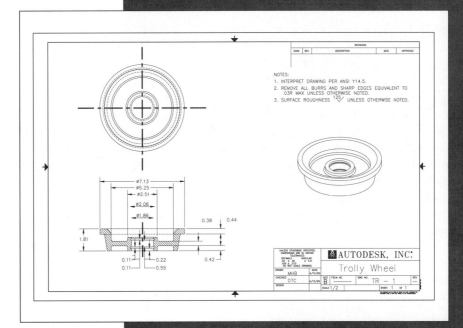

Prototype Drawings

Objectives

When you have completed this tutorial, you will be able to

1. Create and save a prototype drawing for later use.

2. Create a system of basic layers for mechanical drawings.

3. Use the Drawing Aids dialog box.

4. Preset Viewres, Limits, and other defaults in a prototype drawing.

5. Set up paper space and model space in a prototype drawing.

6. Insert one drawing into another as a block.

7. Set the style for drawing points.

8. Use the Divide command.

9. Use the Purge command.

10. Plot drawings using paper space.

Introduction

One of the advantages of using AutoCAD is that you can easily rescale, change, copy, and reuse drawings. Up to this point, most of the drawings you have created began from a drawing called *acad.dwg*, which is part of the AutoCAD program. In *acad.dwg*, many variables are preset to help you begin drawing. A drawing in which specific default settings have been selected and saved for later use is called a prototype drawing. You can use any existing drawing as a prototype from which to start a new drawing. You may need to establish more than one prototype drawing to use for different scales and types of drawings. In this tutorial you will make a drawing containing default settings of your own from which you will start drawings in succeeding tutorials.

Setting up prototype drawings can eliminate repetitive steps and make your work with AutoCAD more efficient. The amount of time you spend creating one prototype is roughly the amount of time you will save on each subsequent drawing that you start from the prototype drawing. Your prototype drawing will contain your custom defaults for layers, limits, grid, snap, and text size and font for use in future drawings.

You will also use the layer commands to create your own layers for use in your drawings. In Tutorial 2 you used Layer Control to control layers that had already been created in the subdivision drawing. In this tutorial you will learn to create your own layers and set their linetype and color. You will also use the Drawing Aids dialog box to set up Grid, Snap, and other drawing tools.

You will enable paper space and set up a *viewport* where you can view model space. Using paper space allows you to easily control where your drawing views are placed on the

actual sheet of paper when you plot your drawing. You will also insert *c:\work\mysubdiv.dwg* into your current drawing as a block and use the settings you have created there to plot it.

Starting

Before you begin, launch AutoCAD. You should be in the AutoCAD drawing editor. Make sure that the Draw and Modify toolbars, as well as the Standard and Object Properties toolbars, are turned on.

> *Pick:* **New icon**

to begin a new drawing. Check to make sure that the No Prototype box is *not* selected (no X should appear in the box to its left). Make sure that *acad.dwg* is the prototype drawing from which your new drawing will be started. Type the name *myproto.dwg* in the text box to the right of New Drawing Name.

> *Pick:* **OK *(to exit the dialog box)***

The main AutoCAD drawing editor appears on your display screen, with *myproto.dwg* in the title bar.

Making a Prototype Drawing

You will set the limits, grid, snap, layers, text size, and text style in this drawing, as well as setting up the paper space limits, viewport, and linetype scale. This will save time, make plotting and printing easier, and provide a system of layers to keep future drawings neatly organized.

Effective Use of Layers

As you recall from Tutorial 2, you can think of layers as clear overlay sheets in your drawing. However, unlike a stack of overlays, the coordinate system always aligns exactly from one layer to the next. Layers are similar to the system of *pin registry* drafting, which is often used in manually drafting maps. In pin registry drafting, a series of transparent sheets are punched with a special hole pattern along one edge, allowing the sheets to be fitted onto a metal pin bar. The metal pin bar keeps the drawings aligned from one sheet to the next. Each sheet is used to show different map information. For example, one sheet may have the streets, another the lot lines, and another the political boundaries; yet another may show the rivers and other geography. When a client wants a map prepared, only the sheets that contain the information needed are attached to the pin bar and printed using a vacuum frame printer. You can use layers for similar purposes, to organize the information in your drawing and make it possible to turn off or freeze information you do not want to show. Unlike pin registry drafting, where the sheets of punched transparent paper sometimes slip and cause a misalignment, layers always stay aligned unless *you* change them.

Using layers helps you organize the information in your drawing. To use layers effectively, choose layer names that make sense and separate the objects you draw into logical groups.

Object color in the drawing controls the pen selection when printing from AutoCAD. This is also true for the line thickness, on printers that are capable of printing different line weights. As discussed in Tutorial 2, the default in AutoCAD is to set color by layer. Using different layers helps standardize the colors of types of objects in the drawing, which in turn helps standardize plotting the drawing. You must use different colors for different types of objects in order to plot more than one color or line thickness effectively.

Using a prototype drawing helps to maintain a consistent standard for layer names. Using consistent and descriptive layer names makes it possible for more than one person to work on the same drawing without puzzling over the

purpose of various layers. On networked computer systems, many different people can use or work on a single drawing. A prototype drawing is an easy way to standardize layer names and other basic settings, such as linetype.

Using Layer

 The Layer command controls the color and linetype associated with a given layer. You can also use Layer to control which layers are visible or plotted at any given time, and to set the current layer. Remember, only one layer at a time can be current. New objects are created on the current layer. Use the Layers icon on the Object Properties toolbar to create new layers and set their properties.

Pick: **Layers icon**

The Layer Control dialog box appears on the screen, as shown in Figure 5.1.

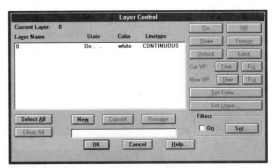

Figure 5.1

There should be one layer name listed in the Layer Name column, 0. Layer 0 is a special layer that is provided in AutoCAD. You cannot rename it or delete it from the list of layers. Layer 0 has special properties when used with the Block and Insert commands. You will use Insert later in this tutorial. In Tutorial 8 you will use the Block command and work more with Insert. Layer 0 is the current layer.

You will now create a layer named HIDDEN_LINES for drawing hidden lines, and set its color and linetype. Then you will make it the current layer.

The typing cursor appears in the empty text box (above the word OK), ready for you to type in the new layer name. Layer names can be up to 31 characters long. Layer names cannot have spaces in them, nor can they have any illegal DOS characters, such as a period (.), comma (,) or pound sign (#). Letters, numbers, and the characters dollar sign ($), underscore (_), and hyphen (-) are valid.

Type: **HIDDEN_LINES**

Pick: **(on the New button just above where you typed HIDDEN_LINES)**

The layer name HIDDEN_LINES appears on the list near the top of the screen, below layer 0.

Next, set the color for layer HIDDEN_LINES. On the right-hand side of the dialog box is the Set Color button. On your own, pick to highlight layer name HIDDEN_LINES and then

Pick: **Set Color**

A dialog box containing color choices pops up on the screen. Figure 5.2 shows the Select Color dialog box.

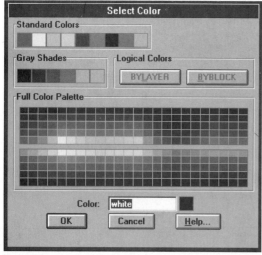

Figure 5.2

Color in Layers

This dialog box allows you to specify the color for objects drawn on a layer. You will select the color blue for the hidden line layer that you are creating. The color helps you visually distinguish linetypes and layers in drawings. You also use color to select the pen and pen width for your printer or plotter.

You use the Select Color dialog box to select color in other dialog boxes, as well as in the Layer Control dialog box. It has the choices BYLAYER and BYBLOCK on the right-hand side. Since you are specifying the color for layer HIDDEN_LINES only, you cannot select these choices, so they are shown grayed. Move the arrow cursor into the Standard Colors box, where blue is the fifth color from the left.

Pick: **(the blue Standard Colors box at the top of the dialog box)**

The name of the color you have selected appears in the box next to the word Color at the bottom of the screen. (If you select one of the standard colors, the name appears in the box; if you select one of the other 255 colors under Full Color Palette, the color number appears.)

Pick: **OK**

Now the color for layer HIDDEN_LINES is set to blue. Check the listing of layer names and colors to see that blue has replaced white (the default color) in the Color column to the right of the layer name HIDDEN_LINES.

> ■ *TIP* If you selected dark lines on a light background (the default) when configuring your video display, white lines will be black on your monitor and color boxes indicating white will be black. ■

Linetype in Layers

The Set Ltype button allows you to set the linetype drawn for the layer. You will select the linetype HIDDEN for your layer named HIDDEN_LINES.

Pick: **Set Ltype**

The Select Linetype dialog box shown in Figure 5.3 appears on your screen.

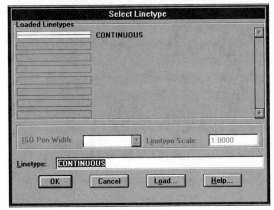

Figure 5.3

You will notice there is only one choice of linetype available, CONTINUOUS. Before you can select a linetype in the Layer Control dialog box, it must be loaded into AutoCAD. You only need to do this one time in the drawing. You do not always need all of the linetypes loaded. To keep your prototype drawing size smaller and see a shorter list of linetypes when you use the Set Ltype option in the Layer Control dialog box, load only the linetypes that you use frequently into your drawing. You can always load other linetypes as needed during the drawing process. You will select Load, the second button from the right at the bottom of the dialog box, to load the linetypes you want to use into your drawing.

Pick: **Load**

Loading the Linetypes

The Load or Reload Linetypes dialog box appears, as shown in Figure 5.4. To the right of the selection File is the name of the default file, *acad.lin*, where the predefined linetypes are stored. You can also create your own linetypes, using the Linetype command. You can store your custom linetypes in the *acad.lin* file or in another file that ends with the extension *.lin*. Below the file name you see the list of available linetypes and a picture of each. Use the scroll bar at the right of the list to move down the items until you see the selection HIDDEN. On your own, pick on the name HIDDEN so that it becomes highlighted, then pick OK to exit the dialog box.

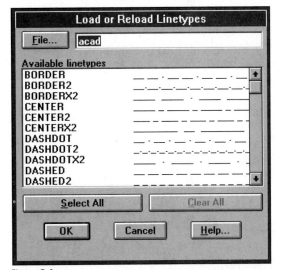

Figure 5.4

You will now see the linetype HIDDEN listed below the linetype CONTINUOUS in the Select Linetype dialog box. You will use the dialog box to select HIDDEN as the linetype for layer HIDDEN_LINES.

> *Pick:* **HIDDEN**
>
> *Pick:* **OK**

You return to the Layer Control dialog box. The list of layer names should now show layer HIDDEN_LINES having color blue and linetype HIDDEN. To make HIDDEN_LINES the current layer,

> *Pick: (the name HIDDEN_LINES in the layer name list; it becomes highlighted)*

While the name is highlighted,

> *Pick: (the Current box below the layer names)*
>
> *Pick:* **OK**

to exit the dialog box.

HIDDEN_LINES is now indicated as the current layer on the Object Properties toolbar and the color blue appears in the square to the left of the layer name. Now use the Line command to draw a few random lines on the screen on your own. These lines are on the HIDDEN_LINES layer; they will appear blue and have linetype HIDDEN (dashed lines) on your color monitor.

Remember that using layers to control the color and linetype of new objects that you create will only work if BYLAYER is selected as the method for establishing object color and object linetype. It is the default, so you should not have to change anything. (To check, pick Color Control from the Object Properties toolbar. It should be set to BYLAYER. *Be careful—* Color Control uses the same dialog box as the Select Color option of Layer, but the effects of using Color Control and Layer Control are very different! Be sure that you leave Color Control set to BYLAYER.) To check that the object linetype is set to BYLAYER, pick the Linetype icon from the Object Properties toolbar. In the dialog box that appears the linetype should be set to BYLAYER.

Using Layer at the Command Prompt

You can also use the Layer command from the command prompt. You will type *LA*, the alias for the Layer command, at the prompt to create another layer for text. You will set its color to red and leave its linetype CONTINUOUS.

> Command: **LA** ⏎
>
> ?/Make/Set/New/ON/OFF/Color/Ltype/Freeze/Thaw/LOck/ Unlock: **N** ⏎
>
> New layer name: **TEXT** ⏎
>
> ?/Make/Set/New/ON/OFF/Color/Ltype/Freeze/Thaw/LOck/ Unlock: **C** ⏎
>
> Color: **RED** ⏎
>
> Layer name(s) for color 1 (red) <HIDDEN_LINES>: **TEXT** ⏎
>
> ?/Make/Set/New/ON/OFF/Color/Ltype/Freeze/Thaw/LOck/ Unlock: ⏎

Pick: (on the current layer name on the Object Properties toolbar)

Use the list of layers that appears to set the current layer to TEXT. On your own, draw some lines on your screen. They should be drawn with color red and linetype CONTINUOUS.

Erase your screen on your own before you continue.

Defining the Layers

Next you will use the Layer Control dialog box to create the remaining layers for your prototype drawing.

Pick: **Layers icon**

The Layer Control dialog box appears on your screen. The typing cursor is positioned in the empty box directly below the list of layer names. From the keyboard, type in the new layer names listed below one at a time. Pick the New box after each name.

> ■ *TIP* Many Windows programs use the convention of showing typing shortcuts by an underlined letter. Notice the underlined letter w in the word New on the button. This indicates that you can quickly select this button by typing the key combination (ALT)-W. Look for the quick key combinations in other dialog boxes. Try it when you create the layers listed below. ■

Type: **VISIBLE**
Pick: **New**
Type: **THIN**
Pick: **New**
Type: **CENTERLINE**
Pick: **New**
Type: **PROJECTION**
Pick: **New**

Type: **HATCH**
Pick: **New**
Type: **DIM**
Pick: **New**
Type: **BORDER**
Pick: **New**
Type: **VPORT**
Pick: **New**

Each time you pick New, the new name pops up onto the list of layer names. If the list is long, sometimes you may need to scroll up and down it, using the boxes that are located near the right-hand side of the dialog box.

Setting the Color and Linetype

Now you will set the colors and linetypes for the layers you have created. On the right of the dialog box are buttons for On, Off, Thaw, Freeze, Unlock, Lock, Cur VP:Thw and Frz (Current Viewport Thaw and Freeze), New VP: Thw and Frz (New Viewport Thaw and Freeze), Set Color, and Set Ltype. Notice that the layers you have created are turned on. The color for all the layers is white.

Move the arrow cursor to the name THIN and press the pick button to highlight that layer name. Use the same method to select layer HATCH and layer TEXT from the list of layers. You should now have three layers selected.

Pick: **Set Color**

The Select Color dialog box, which you used earlier in this tutorial, pops up on the screen. Use the standard colors from the top row. You will pick red as the color for the layers you selected.

Pick: (the red box in the very top row of colored boxes, under the heading Standard Colors)

The word red appears in the Color box, indicating that it is the color choice for the layers that you selected.

Pick: **OK**

Pick: **Clear All** *(to unselect the layers)*

Next you will set the linetype for layer CENTERLINE. First you will select the layer CENTERLINE from the list of layers, and then you will pick to set the linetype.

Pick: **CENTERLINE** *(to highlight the layer name)*

Pick: **Set Ltype**

The Select Linetype dialog box appears on your screen. Do the following steps on your own. Pick the Load button and select the linetype named CENTER to load. Pick OK to return to the Select Linetype dialog box. You will see the linetypes CENTER, CONTINUOUS, and HIDDEN listed. To select linetype CENTER, pick on the centerline pattern to the left of the name CENTER. Notice that the word CENTER appears next to Linetype at the bottom of the dialog box. Pick OK to close the Select Linetype dialog box. This sets the linetype for layer CENTERLINE to linetype CENTER.

To unselect layer CENTERLINE before you continue, pick once on the layer name so that it is no longer highlighted.

Repeat these steps to set the colors and line-types in the table below for the other layers you have created. When you are setting the colors for the layers listed below, keep in mind that cyan is the aqua color on the top row of stand-ard colors, magenta is a purplish pink color, and that the color white appears black if you have chosen to draw on a light background.

> ■ *TIP* When you are selecting the color and linetype, make sure you have high-lighted only the names of the layers you want to set. If other layer names are already highlighted, pick on their names again to turn the highlighting off. ■

Layer	Color	Linetype
0	WHITE	CONTINUOUS
HIDDEN_LINES	BLUE	HIDDEN
TEXT	RED	CONTINUOUS
VISIBLE	WHITE	CONTINUOUS
THIN	RED	CONTINUOUS
CENTERLINE	GREEN	CENTER
PROJECTION	MAGENTA	CONTINUOUS
HATCH	RED	CONTINUOUS
DIM	CYAN	CONTINUOUS
BORDER	WHITE	CONTINUOUS
VPORT	MAGENTA	CONTINUOUS

Before you exit the Layer Control dialog box, set the current layer to VPORT. Highlight layer name VPORT on your own and

Pick: **Current**

When you are finished creating the layers and setting the colors and linetypes, the dialog box on your screen should appear similar to Figure 5.5.

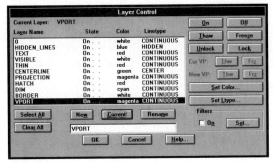

Figure 5.5

Pick: **OK**

You have now created a basic set of layers that you can use in future drawings. When you return to the Layer Control dialog box in the future, you will notice that the list of layer names has become alphabetized. (The variable Maxsort controls the number of layers that will be sorted in the Layer Control dialog box. It is set at 200 as the default. If you do not want your layer list to become sorted, you can type *MAXSORT* at the command prompt and then follow the prompts to set its value to 0.)

■ TIP When using the Layer command at the command prompt, you can quickly control several layers at once by using wildcards in the layer names. The standard DOS wildcard conventions work. For example, to freeze all of the layers that start with the characters CE and have any combination of letters following, at the command prompt type *LA* ⏎ (to start the Layer command). At the prompt *?/Make/Set/New/ON/OFF/Color/Ltype/ Freeze/Thaw/LOck/Unlock,* type *F* ⏎ (to select the freeze option). At the prompt *Layer name(s) to Freeze,* type *CE** ⏎ (to select all of the layer names that start with CE). You can also do this when you want to control just a single layer with a long name. Instead of typing the entire name, just type enough so that only the layer you want will match and then type * for the remainder. You can also use layer filters within the Layer Control dialog box, by picking the Set button in the Filters area near the lower right of the dialog box. From there, you can use wildcards to select layers by their name, color, or linetype. Using the filter causes only the layers that match the filter condition you set to appear on the layer list. This makes it easy to control a logical group of layers quickly. Keep these features in mind when you are creating layer names and assigning colors and linetypes, especially in large drawings. ■

Save your drawing before continuing.

Setting Drawing Aids

Next you will select the Drawing Aids dialog box from the Options menu bar and use it to set Snap and Grid.

Pick: **Options, Drawing Aids**

The Drawing Aids dialog box appears on your screen, as shown in Figure 5.6.

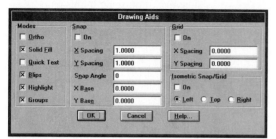

Figure 5.6

You will set the X spacing for the snap to .25.

Pick: **(highlight the text in the input box next to X Spacing under Snap by clicking and dragging)**

Type: **.25** ⏎

The Y Spacing box automatically changes to match the X Spacing box when you press ⏎. If you want unequal spacing for X and Y, you can change the Y spacing also.

At the left of the dialog box are six boxes for Ortho, Solid Fill, Quick Text, Blips, Highlight, and Groups. An X appearing in the box to the left of any name indicates that the setting is active.

■ **TIP** Blips are the little crosses that appear on the screen when you select a point. Sometimes these are useful. If you find them annoying, you can turn them off here. Otherwise, Redraw View removes them from the screen. ■

Move to the On buttons below Snap and Grid and turn them on. Leave the values for Grid set to 0.0000. This way the grid will automatically be set to whatever value you set for the snap.

Pick: **OK (to exit from the dialog box)**

Using Zoom All

You should see a grid of dots on your screen. You will use Zoom All if the grid does not fill the entire screen area. Do not be concerned if the grid does not fill entirely to the right edge of the screen. Your drawing limits are currently set to 12 × 9; if your graphics window on the screen does not have the same *aspect ratio* (the ratio of the height to the width), only one dimension will be filled completely. If you want, you can figure out the aspect ratio of your graphics window and always set your drawing limits so that the grid fills the entire screen, but this is unnecessary. When you start a drawing from your prototype, you will set the drawing limits at that time to a value that is large enough for the particular part you will draw. Then you can use the Zoom command to view the new drawing limits. For now,

Pick: **Zoom All icon**

Selecting the Default Text Font

AutoCAD offers a number of fonts for different uses. The best font for lettering engineering drawings is called romans, for Roman Simplex. Select text font *romans* as your default font.

Pick: **Data, Text Style**

Text style name (or ?) <Standard>: **MYTEXT** ⏎

The Select Font dialog box, which you used in Tutorial 1, pops up, showing AutoCAD's choices of text fonts. Select the correct directory, *c:\r13\common\fonts*. Move the arrow cursor to the name *romans.shx* and pick it to select it as the current font. Pick OK or press ⏎. Next you will accept the defaults by pressing ⏎ at the prompts. Remember, accepting the defaults now means you will be prompted for these values when you create the text; this offers you greater flexibility within your drawings. Do not become confused between style names and font names. A font is a set of characters with a particular shape. When you create a *style*, you can assign it any name you want, but you must specify the name of a font that already exists for the style to use.

Height <0.0000>: ⏎

Width factor <1.0000>: ⏎

Obliquing angle <0>: ⏎

Backwards? <N>: ⏎

Upside down? <N>: ⏎

Vertical? <N>: ⏎

MYTEXT is now the current text style.

Setting the Viewres Default

Viewres controls how many line segments are used to draw a circle on your monitor. This does not affect the way that circles are plotted, just how they appear on the screen. Have you noticed that when you use Zoom Window to enlarge a portion of the drawing, circles may appear as octagons? This is because Viewres is set to a low number. The default setting is low to save time when circles are drawn on the screen. With faster processors and high-resolution graphics, you can use a larger value. You will type *VIEWRES* at the command prompt.

Command: **VIEWRES** ⏎

Do you want fast zooms? <Y>: ⏎

Enter circle zoom percent (1-20000) <100>: **5000** ⏎

■ *TIP* If you are using a slower computer system, you may notice that performance on your computer slows down. It may be because of this setting. If you need to, you can reset Viewres to a lower number. Type *REGEN* at the command prompt to regenerate circles that do not appear round. ■

Save your drawing now. Before you continue creating the prototype drawing, and make a paper space viewport to be used for plotting,

you may want to complete the next section to determine the limits of your output device. In order to center your drawing exactly on the sheet of paper when plotting it, you must know the limits of your printer or plotter. If you already know the limits of your output devices, skip to the topic "Creating a Paper Space Viewport."

Determining the Limits of Your Output Device

Output devices, such as printers and plotters, cannot plot or print all the way to the edge of the sheet of paper. Here is a simple test you can perform to determine the limitations of your output device.

Make sure that you have saved your drawing *myproto.dwg*. Begin a new drawing and call it *test.dwg*. You will set the drawing limits to 11×8.5 by typing the Limits command.

Command: **LIMITS** ⏎

ON/OFF/ <Lower left corner><0.0000,0.0000>: **0,0** ⏎

Upper right corner <12.0000,9.0000>: **11,8.5** ⏎

Next you will draw a horizontal line from point 0,0 to point 11,0 and then a vertical line from 0,0 to 0,8.5. These two lines show the width and height of an 8.5" × 11" sheet of paper.

Pick: **Line icon**

From point: **0,0** ⏎

To point: **11,0** ⏎

To point: ⏎

Command: ⏎ *(to restart the Line command)*

From point: **0,0** ⏎

To point: **0,8.5** ⏎

To point: ⏎

Use the command Plot and plot or print the drawing limits on 8.5" × 11" paper (size A). In the Plot Configuration dialog box, you will see the defaults you selected when you configured your output device. Select Limits, under

Additional Parameters, as the graphics window to plot. Be sure that the drawing origin is set to 0,0, and the scale of plotted inches to drawing units is 1=1, not Scaled to Fit. If either is not set correctly, make corrections in the dialog box, as you learned to do in Tutorial 3. When the settings are correct, pick OK to plot your drawing.

Your output shows as much of the two lines you have drawn as will fit on the paper at full scale. Measure the actual length of the lines that were plotted to determine the limits of your output device. Where the two lines intersect at point 0,0 in your drawing is the origin for the paper; in other words, it is the spot closest to the lower left corner of the paper that the printer can reach. Knowing this location will help you in figuring out how to correctly center drawings on the sheet of paper for your printer. (If you have difficulty determining the limits and origin of your output device, ask your technical support person for help.)

If you rotate your plot, your printer driver may not locate the drawing origin correctly. If one or both lines do not appear on your plot, draw more lines and symbols to help you determine where the driver is locating the origin. You will use this information later in the tutorial.

> ■ *TIP* Setting your system printer to a default landscape orientation from the Windows Control Panel eliminates the need to rotate your plot and will allow both lines to plot correctly at the paper's origin. ■

For the purposes of the tutorials, we will use the values 10.00" × 7.75" as the limits of the output device. You should substitute the correct limits for your output device. If your output device uses more than one paper size, determine the limits for each paper size.

■ *Warning:* If you created *test.dwg*, make sure to open your saved drawing, *myproto.dwg*, before continuing or you will have difficulty with the remainder of the tutorial. ■

Creating a Paper Space Viewport

AutoCAD, Release 13 allows you to set up your drawings using a method similar to hand drafting to create plots to any exact scale on your paper. Up to this point you have been working in *model space*, where you create your drawing geometry. When you are ready to plot, you can use *paper space*, where you lay out the views of your drawing on the "sheet of paper." Paper space is basically two-dimensional; model space is three-dimensional. Your drawing geometry is created in model space, whereas paper space contains things like borders, title blocks, and viewports.

This is an example of the differences between model space and paper space. Imagine that you are watching the Rose Bowl on television. The actual Rose Bowl game is being played in Pasadena at the stadium. On your television set is a picture of the game being taken by the cameraperson, who is actually there at the game. If the cameraperson zooms in, the objects in the view of the Rose Bowl become larger on your TV screen. In your AutoCAD drawing, model space is like being at the actual game; paper space is like the picture on your television. The actual game is three-dimensional; the picture on your TV screen is two-dimensional. If you have one of the picture-within-a-picture TV sets, you can even show more than one picture in separate "windows" on your TV. These windows on the TV are like "viewports" in paper space. You can create more than one viewport in paper space if you want; these are called *floating viewports*. These viewports can be overlapping or separated from each other, as you wish. You can create any number of these viewports to lay out your drawing on the paper sheet in the way you want. Each viewport contains a view of the model space drawing at the zoom factor and line of sight that you specify. Paper space viewports are very useful for plotting the drawing, adding drawing details, showing an enlarged view of the object, or showing multiple views of the object.

Keep this television example in mind as you are doing the next steps.

In three-dimensional model space, it is also possible to create multiple viewports, called *tiled viewport*s, on your screen. The number you can create depends on your hardware configuration. These viewports cannot overlap each other. They must meet exactly at the edges. They are called tiled viewports because they are like floor tiles that you would lay edge to edge. Tiled viewports are created when you highlight the Tile button or set Tilemode to 1. You will not be using tiled viewports during these tutorials. You will use paper space viewports.

Enabling Paper Space

Before you can use paper space, you must turn Tilemode off. Tilemode is a toggle. Double-clicking the Tile button on the status bar turns it off if it is on, and on if it is off. The default is on, signified by the highlighted word TILE on the status bar. (If you have trouble using the status bar, type *TILEMODE* at the command prompt and set its value to 0 for off.)

Double-click: TILE button *(so that it becomes grayed out)*

Now that Tilemode is turned off, leave it off. You will use the PAPER and MODEL buttons on the status bar, or the aliases, PS and MS, to switch between paper space and model space from now on. Turning Tilemode on disables paper space.

Don't be alarmed when your grid disappears from the screen. This is normal. Entering paper space is analogous to changing from being *at the Rose Bowl game* to looking at the blank TV screen. In order to see model space (or the game in the Rose Bowl example), you must create a viewport (i.e., turn on the TV).

Notice the change in the UCS icon. It now shows the paper space icon, which looks like a triangle. Your screen should look similar to Figure 5.7; notice that PAPER appears where MODEL used to be on the status bar.

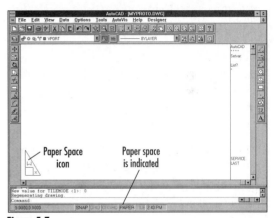

Figure 5.7

Next you will create a viewport, so that you can see model space.

Sizing the Viewport

You will size the viewport so that it will fit inside the limits for your output device when paper space is plotted full size. The values used in this tutorial are general and may work for your printer or plotter. You can substitute the limit values that you have determined for your specific output device.

Figure 5.8 shows an example of how to determine the drawing coordinates for the viewport you will create.

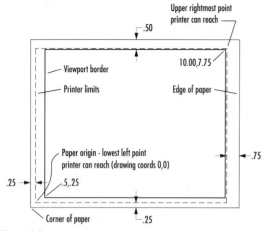

Figure 5.8

The outer line in Figure 5.8 represents the edge of the paper. The distance from the edge that a printer can reach is different for each printer. It is also not necessarily the same distance from the left edge as it is from the right edge or the top and bottom edges. The limits of the printer may also be different for each sheet size on which you are printing.

The dashed line represents the limits of the printer. (When you preview a plot in the Plot Configuration dialog box, the printer limits are shown by the outer line in the preview.) The thick line represents the viewport border you will create. In this example, the printer places the lowest corner of the drawing at a location .25" above the bottom of the paper and .25" in from the left edge of the paper when you choose to plot the origin at 0,0 (in the Plot Configuration dialog box). The printer in this example can only reach to .5" from the top border and .75" from the right edge of the sheet. In order to create a viewport that is centered on the paper, you would specify .5,.25 for the lower left corner of the viewport (this will move the left edge of the viewport an additional .5 so that it is .75" from the left edge of the paper and move the bottom of the viewport an additional .25 so that it is .5" from the

bottom). Specify 10.00,7.75 for the upper right corner of the viewport (that is as far as the printer will reach and is .75" from the right edge and .5" from the top of the paper). This will produce a viewport that is centered on your sheet when you print the drawing from paper space, so that the area from 0,0 to 10.00,7.75 is full-size on a sheet of 8.5" × 11" paper.

When you are doing the steps listed below, use your printer limits in place of the suggested values. Decide what values you will need to use in order to get your drawing centered on the paper. Remember, each style of printer will have its own limits. Take the time to set your prototype drawing up correctly, so that when you print your drawings they will look their best.

> ■ **TIP** If you have access to several different printers or work with several different sheet sizes, it is very useful to have several different prototype drawings. ■

Your current layer should now be VPORT (see the toolbar). If it is not, set the layer to VPORT on your own before continuing.

Pick: **View, Floating Viewports, MV Setup**

Align/Create/Scale viewports/Options/Title block/Undo:
   C ⏎

Delete objects/Undo/<Create viewports>: ⏎

Pressing ⏎ takes you to a text screen where you will see the choices for common arrangements of viewports. You will select 1 for a single viewport.

Redisplay/<Number of entry to load>: **1** ⏎

You will create a viewport to correspond to the output device limits. The values .5,.25 and 10.00,7.75 may work for you, but if you have determined the exact size that your output device can print, substitute those values when specifying the viewport boundary. You are returned to the AutoCAD drawing editor screen for the remaining selections.

Bounding area for viewports. First point: **.5,.25** ⏎
Other point: **10.00,7.75** ⏎
Align/Create/Scale viewports/Options/Title block/Undo: ⏎

You will see the magenta lines of the viewport boundary drawn on the screen with the grid contained inside them, similar to Figure 5.9. The grid inside the viewport is in model space; you are seeing it in your paper space viewport, as though it were the Rose Bowl on your television!

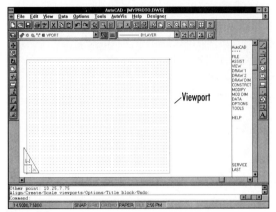
Viewport

Figure 5.9

Using Limits

The Limits command lets you predefine a boundary in your drawing. When the limits are turned on, you cannot draw outside the area that is specified. You can turn the limits off at any time if you want to draw outside this area. The Limits command can be useful in producing plots located exactly where you want them on the sheet. You can define separate limits in the drawing when you are in paper space and when you are in model space. Next you will set the limits for paper space. You will set them to the size of your printer's limits, since you cannot print outside that area. Returning to the Rose Bowl analogy, the paper space limits are like the size of the screen on your TV. Whatever size it is, the only way to make it

larger is to buy a larger TV. Set the Limits command in paper space to reflect the limits of your printer.

Setting the Paper Space Limits

The Limits command is under Data on the menu bar. Use limits in your prototype drawing to represent the edge of the printer limits in paper space. (The viewport represents the area inside the printer limits where you want to center your drawing.) Setting the upper right corner of the limits slightly beyond the viewport is useful because then when you use Zoom All in paper space, the viewport border will be slightly in from the edge of the AutoCAD drawing screen, which can make it easier to select.

Pick: **Data, Drawing Limits**
Reset Paper Space limits.
ON/OFF/<Lower left corner> <0.0000,0.0000>: **0,0** ⏎
Upper right corner <12.0000,9.0000>: **10.5,8** ⏎

You will use ⏎ to restart the Limits command and turn the limits on. Then you will use Zoom All to show the entire limits area on the screen.

Command: ⏎
ON/OFF/<Lower left corner> <0.0000,0.0000>: **ON** ⏎

Pick: **Zoom All icon**

Next, you will switch to model space by picking the PAPER button from the status bar. (If you have trouble with the status bar, type the alias, MS ⏎ at the command prompt to switch to model space.)

Double-click: **PAPER button**

to return to model space where you can create your drawing geometry. Notice that the button now says MODEL and the UCS icon has returned to your screen. When you see the UCS icon, you are in model space. In the Rose Bowl analogy, you are now at the Rose Bowl. Changes you make in model space are in the "real" 3D world. Changes in paper space are on the 2D TV screen. Both model space and paper space can have settings for Grid, Snap, and Limits. The settings can be different for model space and for paper space.

> ■ *TIP* It can be confusing when the grid is turned on in both model space and paper space. This produces two patterns of grid dots that do not necessarily align. Generally, you will want to turn off the paper space grid before you return to model space. ■

Now you have a viewport border in your prototype drawing. It is on a separate layer so that you can freeze it when you do not want it to print. Your prototype drawing at this point should look like Figure 5.10.

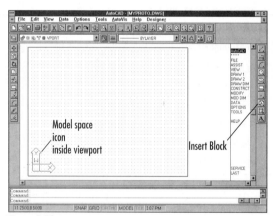

Figure 5.10

On your own, save drawing *myproto.dwg* before continuing.

Inserting an Existing Drawing

 You can insert any drawing into any other AutoCAD drawing. You will use Insert Block to insert the subdivision drawing that you finished in Tutorial 3 into the current prototype drawing. This way you will

be able to see the effects of using paper space viewports, which you will create in the next steps. The Insert Block button on the Draw toolbar is shown in Figure 5.10.

Pick: **Insert Block icon**

The Insert dialog box appears on your screen, as shown in Figure 5.11. You will pick the File button to insert an existing drawing into the current drawing. Only the portion of the drawing that is created in model space will be inserted.

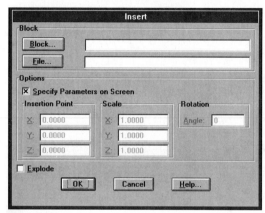

Figure 5.11

Pick: **File**

Use the Select Drawing File dialog box that appears on your screen to select the file *mysubdiv.dwg* that you created in Tutorial 3, and pick OK to return to the Insert dialog box.

Notice the X in the Options box to the left of Specify Parameters on Screen. This selection returns you to your drawing to select the insertion point, scale, and rotation for the drawing you are inserting. If it is not selected, the grayed-out areas of the dialog box for insertion point, scale, and rotation are selectable. You could use them to type in the values you want to use. If there is no X in the Specify Parameters on Screen box, pick the box now.

Pick: **OK** *(to return to the drawing)*

Insertion point: **0,0** ⏎

X scale factor <1>/Corner/XYZ: ⏎

Y scale factor <default=X>: ⏎

Rotation angle <0>: ⏎

The image of the subdivision does not show in the viewport, because the size of the subdivision is much too large for the viewport at the current zoom factor. In the Rose Bowl analogy, this is as if the cameraperson zoomed into an area which did not show anything, for example, a blank wall. In order to see the Rose Bowl again (or the subdivision) you must zoom out. To make the subdivision fit inside the window,

Pick: **Zoom All icon**

Now all of the subdivision drawing should appear inside the viewport, as shown in Figure 5.12.

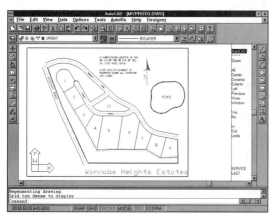

Figure 5.12

However, as it is shown in Figure 5.12, if you were to plot the drawing you would not be sure at what scale it would be plotted. In order to have the drawing at a particular scale, you must establish a relationship

between the number of units in model space and the number of units in paper space. This is very similar to the way you would determine the plot scale that you would note in the title block of a drawing. In the Rose Bowl analogy, you have to go to Pasadena (model space) and tell the camera person exactly how much to zoom in or out. This would mean that when you were back in your living room watching TV, you could hold up a ruler on the TV screen and there would be an exact scale between the real size of the stadium and the size shown on your set.

Using Zoom XP

The next step will be to establish a relationship between the number of units in model space and the number of units in paper space. You will set up the drawing so that one unit in paper space equals 250 units in the subdivision in model space (a scale of 1" =250'). To establish this relationship, you will use the Zoom command to specify the XP (times paper space) scale factor. The XP scale factor is a ratio. It is the number of units from the object in paper space divided by the number of units in model space. For example, if you want the model space object to appear twice its size on the paper, specify 2XP as the scale factor (2 paper space units/1 model space unit). To specify a scale of 1"=250' for the subdivision, You will set the XP scaling factor to .004 (1 unit in paper space/250 units in model space). You will do this by typing the alias for the Zoom command.

Command: **Z** ⏎

All/Center/Dynamic/Extents/Left/Previous/VMax/Window/
Scale (X/XP): **.004XP** ⏎

Your screen should appear similar to Figure 5.13. Notice that the image is now smaller.

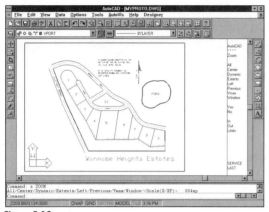

Figure 5.13

You have set up the Zoom XP scale factor so that if you plot paper space at a scale of 1=1, the model space object is shown at a scale of 1"=250' on the paper. Remember that in the original subdivision drawing, each unit represented one foot. In order to preserve the relationship between model space and paper space, be sure that if you use Zoom Window to enlarge your drawing, you use Zoom Previous, not Zoom All, to return to this XP size before you plot. Additionally, you can always use Zoom and specify a different XP scale factor at a later time if you want.

■ *TIP* You can use the LISP program MVSetup to calculate the Zoom scale factor for you by selecting the number of model space units for the number of paper space units. To do this, pick View, Floating Viewports, MV Setup from the pull-down menus (or type *MVSETUP* at the command prompt). You will see the options Align, Create, Scale viewports, Options, Title block, and Undo. Type *S* ⏎ to scale the viewports. You will be prompted to select the viewports to scale. Pick on the viewport border and press ⏎ (the magenta line in your drawing). Next you are prompted to enter the ratio of paper space units to model space units. Start by entering the number of paper space units and pressing ⏎. Then enter the number of model space units. The MVSETUP program converts this to a ratio and determines the correct value, which it will then use in the Zoom XP command. ■

Using Pan

You can use the Pan command (on the Standard toolbar) to drag the model space drawing around in the paper space viewport without changing the scale or the location of the drawing on the model space coordinate system.

Pick: **Pan Point icon**

Displacement: *(pick a point near the center of the subdivision)*

Second point: *(pick the new location at the center of the viewport for the point you selected)*

The drawing should be centered in the viewport.

Creating a Second Floating Viewport

You can use floating viewports to create enlarged details, location drawings, or additional views of the same object. (You will learn more about this when you create solid models in Tutorial 9.) You will pick the Mview command to create another viewport. The Mview command also has options to turn viewports on and off. You cannot see any objects in viewports that are turned off, regardless of whether the objects' layers are frozen or thawed. The option Hideplot allows you to select viewports in which you want hidden lines removed from the view when plotting in paper space. The Fit options fit a viewport to the available paper space graphics window. The options for 2, 3, and 4 viewports create patterns of multiple viewports in the drawing. Restore converts tiled viewports to individual viewports of similar configurations.

At the prompt for first point, you will pick a location for one corner of the viewport. Refer to Figure 5.14 for the placement of the viewport you will create.

Pick: **View, Floating Viewports, 1 Viewport**

Switching to paper space

ON/OFF/Hideplot/Fit/2/3/4/Restore/<First point>:
    (pick near point 1 for the lower left corner of the viewport)

Other corner: **(pick near point 2 for the upper right corner to locate the viewport)**

A second viewport is added to the drawing. It shows the entire subdivision drawing, zoomed so that all of the drawing limits fit inside the viewport. Notice also that the second viewport you created has its own set of crosshairs and its own UCS icon. Only one viewport can be active at a time. To make a viewport active, move the arrow cursor into that viewport and press the pick button. When a viewport is active, its border becomes highlighted and the crosshairs appear completely inside of it.

■ **TIP** If you have trouble picking a viewport to make it active, press Ctrl-R to select it. ■

Next you will use the Zoom Window command to enlarge an area inside of the smaller viewport you created.

Pick: **(inside the smaller viewport to make it active)**

Pick: **Zoom Window icon**

First corner: **(pick the first corner so that you enlarge the rounded end of lot 1)**

Other corner: **(pick to create a window around lot 1)**

The area inside the window you selected will be enlarged inside the viewport, as shown in Figure 5.14.

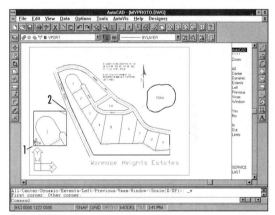

Figure 5.14

The viewport border is on layer VPORT. If you do not want to have the boundary of the viewport plotted in your drawing, you can freeze this layer.

Adding a Title Block and Text to Paper Space

You can easily add a title block to your drawing, using lines and text in paper space. The
advantage of adding the title block, notes, and
border to paper space is that the measurements there are the same as you would make
on a regular sheet of paper. For instance, the
standard size for text on 8.5" × 11" drawings is
1/8". To add text of this height to paper space,
you would set the text height to 1/8" and create the text. However, if you wanted to add text
to model space so that it would appear 1/8" on
the final plot, you would have to take into consideration that model space is going to be plotted 1/250th of its actual size. So you would
need to create the text 31.25 units tall in model
space in order for it to be 1/8" tall on the final
plot. Also, in model space you can only draw in

one viewport or another once they are enabled.
Using paper space, you can draw across the
viewports.

Return to paper space on your own by double-
clicking the MODEL button on the status bar,
or by typing the alias, PS and pressing ⏎,
before adding the lines and text to make up
the title block.

You will see the paper space icon return to
your screen and the button you picked now
says PAPER.

Next use Layer Control on the Object
Properties toolbar and set BORDER as the current layer on your own. Once you have
selected BORDER as the current layer, turn
layer VPORT off by picking the selection
across from VPORT that looks like eyes, so
that they look as though they're asleep. The
viewport borders no longer appear because
they are on a layer that is turned off. Before
continuing, check the toolbar to see that BOR-
DER is the current layer. If you need help,
refer to Tutorial 2.

You will use the commands you've learned to
add borders and a title block on your own.
First, draw the lines for the border of the
drawing you see in Figure 5.15.

Next, set Snap to .125; this will be useful for positioning lettering and the lines for the title block. Use the Offset command to offset a line across the bottom of the viewport .375 units (three snap increments) up from the bottom line of the viewport. Then use the Circle command to draw a circle around the area from the second viewport, which is going to serve as a detail of lot 1. Make sure to add the border lines to paper space. Lines added to model space will show up in every viewport (unless you control the visibility with viewport layer visibility control). Keep in mind that paper space represents the sheet of paper on which you are laying out the drawing. Things like borders belong on the sheet of paper and not in model space.

When you are finished, your drawing should look similar to Figure 5.15.

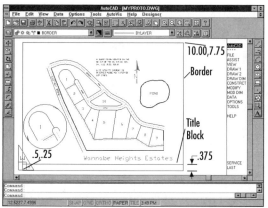

Figure 5.15

Using Divide

The Divide command places points along the object you select, dividing it into the number of segments you specify. You can also choose to have a block of grouped objects placed in the drawing, instead of points. You will use the Divide command to

place points along the line you just drew, dividing it into three equal segments. Divide is on the Point flyout on the Draw toolbar. Refer to Figure 5.15.

Pick: **Divide icon**

Select object to divide: **(pick the line you offset to form a title strip)**

<Number of segments>/Block: **3** ⏎

Because the Point Style is set at just a dot, you will probably not be able to see the points that mark the equal segment lengths. You will use the Point Style dialog box to change the display of points in the drawing to a larger style so you can see them easily.

Pick: **Options, Display, Point Style**

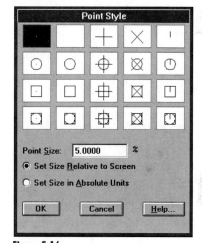

Figure 5.16

The Point Style dialog box shown in Figure 5.16 is on your screen. On your own, select one of the point styles that has a circle or target around the point so it is easier to see. To exit the dialog box,

Pick: **OK**

You will need to tell AutoCAD to recalculate the display file for your drawing in order to see this change. This is done with the Regen command. You will type the command at the prompt.

Command: **REGEN** ⏎

The points should appear larger on the screen now. Your drawing should look similar to Figure 5.17.

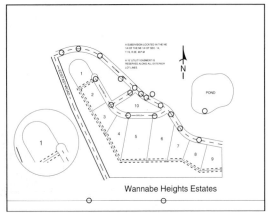

Figure 5.17

You can use the Node object snap to find point objects in your drawing. To draw lines dividing the title area exactly into thirds,

Pick: **Line icon**

From point: *(pick Snap to Node)*

nod of: *(target one of the points)*

Draw a line straight down from the point by using the Perpendicular object snap.

To point: *(pick Snap to Perpendicular icon)*

perpendicular to: *(pick the bottom line of the viewport)*

To point: ⏎

Now repeat this process to draw another line at the other point. Use the Point Style dialog box to change the point style back on your own and then use Regen to regenerate the points on the screen.

■ *TIP* The Measure command is similar to the Divide command, except that instead of specifying the number of segments into which you want to have an object divided, with Measure you specify the length of the segment you would like to have. Like Divide, the Measure command puts points or groups of objects called blocks (which you will learn about in Tutorial 8) specified distances along the line. ■

On your own, use the Object Properties toolbar to set TEXT as the current layer. You will use the Dtext command from the Text flyout on the Draw toolbar to add titles to the drawing.

Pick: **Dtext icon**

Justify/Style/<Start point>: **C** ⏎

Center point: **5.25, .375** ⏎

Height <0.2000>: **.125** ⏎

Rotation angle <0>: ⏎

Text: **SUBDIVISION** ⏎

Text: ⏎

The word SUBDIVISION appears, centered around the point you selected. The centering is only horizontal; otherwise the letters appear above the point selected for the center. If you want both horizontal and vertical centering, pick the Justify option, then Middle.

Now repeat this process, using the Justify, Left option of the Dtext command to position the words DRAWN BY: YOUR NAME in the left-hand area of the title block. The default Start point option prompts you for the bottom left starting point for the text you will enter. Use the Right justified option to right justify the words Scale: 1"=250' in the right-hand area. The Justify R option prompts you for the lowest, right-most point for the text you will enter.

On your own, add text identifying DETAIL A, as shown in Figure 5.18.

Now you have completed a simple title block for your drawing and are ready to plot. Your drawing should look like Figure 5.18. Save it before you go on.

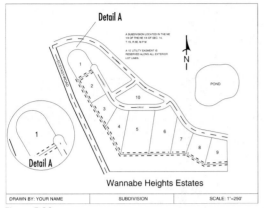

Figure 5.18

Plotting from Paper Space

You have correctly set up your drawing in paper space and have used the Zoom command with the XP scaling factor to establish the number of units in model space that you want to equal one unit in paper space; you are now ready to plot. Make sure you are still in paper space for plotting your drawing. If you are not, your plotted drawing may not fit on the sheet correctly and the title block will not be shown. Since you used paper space, you will type 1=1 for the scaling in the Plot command.

Pick: **Print icon**

Use the Plot Configuration dialog box that appears on your screen to select the limits of your drawing as what to plot. Select size A paper and plot the drawing at a scale of 1=1 on the paper.

Now that you understand paper space, you are ready to erase the subdivision and the detail viewport from the drawing and save the drawing to use as a prototype for starting future drawings.

If you have the grid on in paper space, turn it off on your own by pressing (F7) or by double-clicking the word GRID on the status bar while in paper space. Turn layer VPORT on and use the Erase command to pick on the edge of the viewport border containing Detail A. Viewports are much like any other drawing object when you are in paper space; you can scale, stretch, move, and erase them as desired. The viewport and its contents are erased. Also erase the text referring to Detail A.

Then continue on your own and return to model space by typing *MS* (↵) at the command prompt or by double-clicking on the PAPER button on the status bar. Now use the Erase command on your own to erase the entire subdivision. Since you inserted it into the drawing, it should act as one object when you select it to erase.

Using Purge

The Purge command eliminates unused layers, styles, blocks, and other named objects. A named object is just that, any type of AutoCAD object which can have a name, such as layers, linetypes, views, blocks, and others. Notice that when you inserted the subdivision into your drawing, it automatically brought all of its layers that contained drawing objects and other settings along. To eliminate these unwanted layers from your prototype drawing, you will use the Purge command. Purge is also very useful for keeping your drawing database as small as possible. This is very important when you are working with 3D solids, which can result in very large files. Be careful not to purge any of the layers which you created for the prototype.

Command: **PURGE** (↵)

Purge unused

Blocks/Dimstyles/LAyers/LTypes/SHapes/STyles/APpids/
  Mlinestyles/All: **A** (↵)

At the Purge layer prompt, Respond Y to each of the subdivision layers and text styles on your own. Press ⏎ to accept No as the default for the prototype drawing layers and styles. (Refer to pages 142 and 144 for the layers and styles you created.)

Set the current layer to VISIBLE. You will save this drawing with VISIBLE set as the current layer. This way, when you begin a new drawing from this prototype, you will be ready to start drawing on the layer for VISIBLE lines. Now save your drawing to the file name *myproto.dwg*.

   Pick: **Save icon**

Now you have completed *myproto.dwg*. Be sure that you keep a copy of the drawing on your own floppy disk. You also should keep a second copy of your drawings on a separate floppy as a backup disk, in case the first disk becomes damaged.

It is easy to use the Edit Text command to make changes to the standard information you provide in the title block. Refer to Tutorial 1 if you need to review the Edit Text command.

Beginning a New Drawing from a Prototype Drawing

You can use any AutoCAD drawing as a starting point for a new drawing. The settings that you have made in drawing *myproto.dwg* will be used to start future drawings. An identical prototype drawing, called *proto.dwg*, is in your data files. If you want to use the prototype you just created, substitute *myproto.dwg* whenever you are asked to use *proto.dwg*. Next you will start a new drawing from the prototype provided with the data files.

   Pick: **New**

   Pick: **Prototype**

Use the dialog box shown in Figure 5.19 to select the correct drive and directory and pick *proto.dwg*.

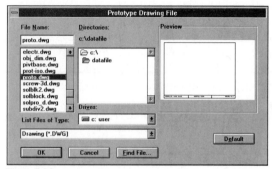

Figure 5.19

When you are finished making this selection, pick OK to return to the Create New Drawing dialog box, where you will provide a name for the new drawing that you will start from the prototype.

> ■ **TIP** You can select the box Retain as Default in the Create New Drawing dialog box if you want your prototype drawing name to always appear in the box instead of the default drawing *acad.dwg*. ■

   Type: **TRY1** *(next to New Drawing Name)*

   Pick: **OK**

This starts a new drawing (called *try1.dwg*) from a copy of the drawing called *proto.dwg*. The drawing *proto.dwg* remains unchanged, so that you can use it to start other drawings. Your current drawing name is now *try1.dwg*. For future drawings in this book, use *proto.dwg* or your drawing *myproto.dwg* as a prototype drawing unless you are directed otherwise.

Changing the Title Block Text

The border, text, and lines of the title block were created in paper space. To make a change to them, you must first return to paper space. Then you will use the Edit Text selection (Ddedit command) to change the title block text so that it is correct for the new drawing you are starting. You will pick the Edit Text command from the Modify toolbar under the Special Edit flyout.

Command: **PS** ⏎

Command: **Edit Text icon**

<Select a text or ATTDEF object>/Undo: *(select the text DRAWING TITLE)*

The Edit Text box appears on your screen, containing the text you selected. Use the ⏎, (BACKSPACE), and/or (DEL) keys to remove the word DRAWING TITLE. Change the entry to TRY1. When you are finished editing the text,

Pick: **OK** *(to exit the dialog box)*

<Select a text or ATTDEF object>/Undo: ⏎ *(to end the command)*

■ *TIP* You can change the style of text that has already been added to your drawing by picking the Properties icon from the Object Properties toolbar. At the prompt, select a text object and press ⏎. The Modify Text dialog box will appear on your screen. Use the Style area near the lower right of the dialog box, to pull down the list of available styles and make a new selection. Remember that styles must be created using the Style command before they can be used. You can also use Modify Text to change the text entry, its height, width factor, rotation, obliquing angle, and location. ■

Remember to return to model space before continuing to draw when using your prototype.

Exit AutoCAD and discard the changes to drawing *try1.dwg*.

KEY TERMS

aspect ratio   pin registry   tiled viewports
floating viewports   style   viewports
model space

KEY COMMANDS

Divide   Mview   Tile
Insert Block   Purge   Viewres
Layer   Regen
Measure

Draw the following objects. The letter M after an exercise number means that the given dimensions are in millimeters (metric units).

5.1 Amplifier Circuit

Draw the amplifier circuit. Use the grid at the top to determine the sizes of the components. Each square=0.0625.

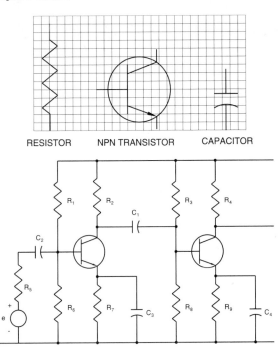

RESISTOR   NPN TRANSISTOR   CAPACITOR

5.2 Support

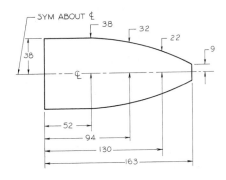

5.3 Vee Block

FILLETS & ROUNDS R.25

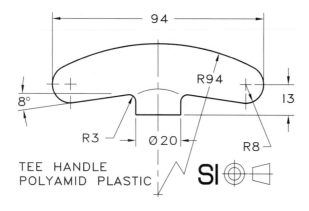

5.4M Tee Handle

94

R94

8°

13

R3

Ø20

R8

TEE HANDLE
POLYAMID PLASTIC

SI

5.5M Grab Link

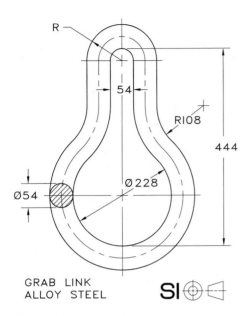

GRAB LINK
ALLOY STEEL

5.6 Idler Pulley Bracket

Create the front view for the object shown.

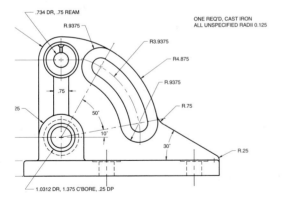

ONE REQ'D, CAST IRON
ALL UNSPECIFIED RADII 0.125

Draw the object shown below. Set Limits to –4,–4 and 8,5. Set Snap to .25 and Grid to .5. The origin (0,0) is to be the center of the left circle. Do not include the dimensions.

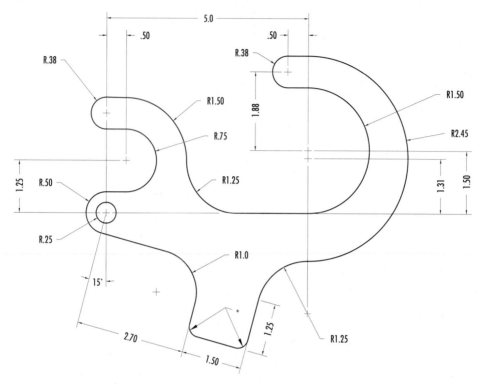

*Corners have R.25

Redraw the window schedule shown below. Experiment with different fonts.

WINDOW SCHEDULE

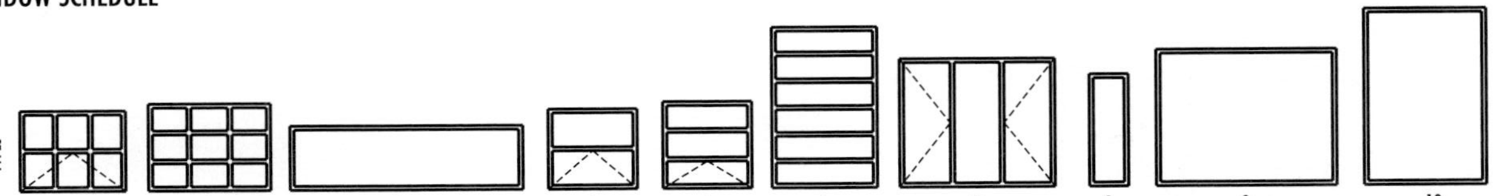

TYPES

SYM	TYPE	SIZE H	SIZE W	MATL FIN	FRAME	FIN SCRN AREA	AREA	VENT	VENT	GLAZING MAT'L	DETAILS H	DETAILS J	DETAILS S	REMARKS
Ⓐ	1	5'-0"	6'-4"	MTL	ST	MTL	ST	Y	32.5 SF	65 SF				
Ⓑ	2	5'-5"	7'-4"	MTL	ST	MTL	ST	N		40.5 SF				FIXED GLAZING
Ⓒ	3	4'-0"	14'-8"	MTL	ST	MTL	ST	N		58.8 SF				FIXED GLAZING
Ⓓ	4	7'-0"	5'-0"	MTL	ST	MTL	ST	N		70 SF				FIXED GLAZING
Ⓔ	5	5'-5"	4'-6"	MTL	ST	MTL	ST	N		48.6 SF				FIXED GLAZING
Ⓕ	6	5'-0"	5'-4"	MTL	ST	MTL	ST	Y	13.3 SF	26.5 SF				
Ⓖ	7	5'-5"	5'-4"	MTL	ST	MTL	ST	Y	9.9 SF	29.7 SF				
Ⓗ	8	8'-0"	6'-4"	MTL	ST	MTL	ST	N		100.8 SF				FIXED GLAZING
Ⓘ	9	5'-6"	4'-4"	MTL	ST	MTL	ST	N		126.1 SF				FIXED GLAZING
Ⓙ	10	8'-0"	9'-4"	MTL	ST	MTL	ST	Y	24.8 SF	74.4 SF				
Ⓚ	11	5'-6"	4'-8"	MTL	ST	MTL	ST	N		25.3 SF				FIXED GLAZING
Ⓛ	12	6'-0"	3'-0"	WD	ST	WD	ST	N		36 SF				FIXED GLAZING
Ⓜ	13	7'-6"	11'-6"	WD	ST	WD	ST	N		86.3 SF				FIXED GLAZING WITH CURVED GLASS
Ⓝ	14	10'-0"	8'-0"	MTL	ST	MTL	ST	N		160 SF				FIXED GLAZING
Ⓞ	15	10'-0"	14'-0"	MTL	ST	MTL	ST	N		140 SF				FIXED GLAZING
Ⓟ	16	6'-0"	6'-6"	WD	ST	WD	ST	N		32.5 SF				FIXED GLAZING

5.9 Saw Blade

Draw the saw blade shown below. Use Array and Polyline. Do not include dimensions. Use the Arc option of the Polyline command to create the thick arc. Add the arrow to the arc by making a Polyline with a beginning width of 0 and a thicker ending width that shows the blade's rotation.

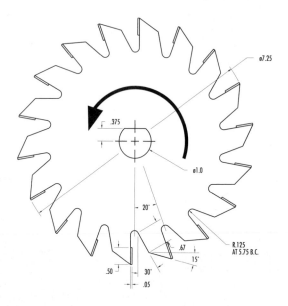

5.10M Five Lobe Knob

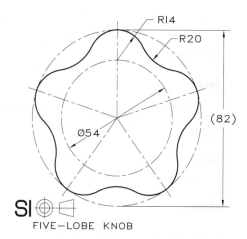

FIVE-LOBE KNOB

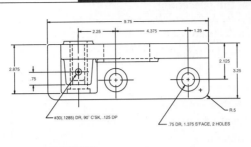

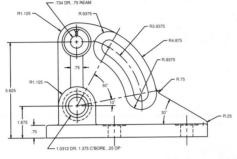

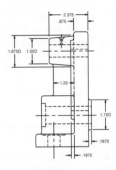

2D Orthographic Drawings

Objectives

When you have completed this tutorial, you will be able to

1. Use AutoCAD to create 2D orthographic views.

2. Use Ortho.

3. Create construction lines.

4. Draw hidden, projection, center, and miter lines.

5. Set the global linetype scaling factor.

6. Draw ellipses.

7. Add correctly drawn centermarks to circular shapes.

In this tutorial you will apply many of the commands that you have learned in the preceding tutorials to create orthographic views. *Orthographic views* are two-dimensional drawings that you use to accurately depict the shape of three-dimensional objects. You will learn to look at a three-dimensional object and draw a set of two-dimensional drawings that define it. In Tutorial 9 you will learn to create a three-dimensional solid model of an object.

The Front, Top, and Right-Side Orthographic Views

Technical drawings usually require front, top, and right-side orthographic views to completely define the shape of an object. Some objects require fewer views and others require more. All the objects in this tutorial require three views. Each orthographic view is a two-dimensional drawing that shows only two of the three dimensions (height, width, and depth). This means that no individual view contains sufficient information to completely define the shape of the object. You must look at all three views together to get a complete understanding of the object's shape. For this reason it is important that the views are shown in the correct relationship to one another.

Figure 6.1 shows a part and Figure 6.2 shows the front, top, and right-side orthographic views of the part.

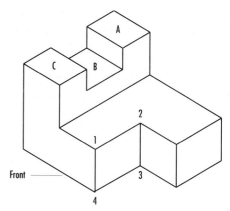

Figure 6.1

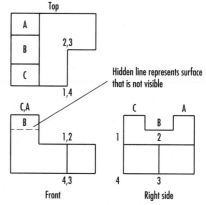

Figure 6.2

There are three rectangular surfaces, labeled A, B, and C, located on the left side of the top view. Which surface is the highest? The top view alone is not sufficient to answer this question. The three surfaces must be located on the other views for you to get a complete understanding of the relationships among them. The right-side view (often just referred to as the side or profile view) shows that surfaces A and C are at the same height, and that surface B is lower.

The side view shows the relative locations of surfaces A, B, and C, but the surfaces appear as straight lines; therefore you need the top view to see the overall shape of the surfaces. You need both the top view and the side view to define the size, shape, and location of the surfaces.

Look at surface 1-2-3-4 in the side view of Figure 6.2. You see its shape in the right-side view, but it appears as a straight line in the front and top views, because it is perpendicular to the views—like a sheet of paper viewed looking onto the edge. A plane surface appears as a straight line when viewed from a direction where the surface is perpendicular to the viewing plane. Surfaces that are perpendicular to two of the three principal orthographic views are called *normal surfaces* (normal meaning 90°). Normal surfaces show the true size of the surface in one of the principal views. All surfaces in the object shown in Figure 6.1 are normal surfaces.

Surface B is shown in the front view of Figure 6.2, using a *hidden line*. All surfaces must be drawn in all views. Hidden lines represent surfaces that are not directly visible; that is, that are hidden from view by some other surface on the object.

View Location

The locations of the front, top, and side views on a drawing are critical. The top view must be located directly above the front view. The side view must be located directly to the right of the front view. An alternative position for the side view is to rotate it 90° and show it aligned with the top view. By aligning the views precisely with each other, you can interpret them together to understand the three-dimensional object they represent. Because views are shown in alignment, you can *project* information from one view to another. Refer to Figures 6.3 and 6.4.

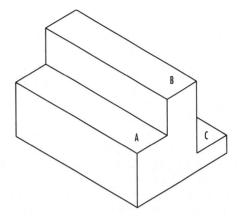

Figure 6.3

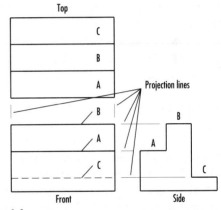

Figure 6.4

Surfaces A, B, and C in Figure 6.3 appear as three straight lines in the front view. Which line represents which surface? Because the front and side views are aligned, you can draw horizontal lines from the vertices of the surfaces in the front view to locate them in the side view. These lines are called *projection lines* and each surface is located between its projection lines in both views. You can locate surfaces between the front and top views by using vertical projection lines. Without exact view alignment, it would be impossible to accurately relate the lines and surfaces of one view to those of another view, making it difficult or impossible to understand the drawing.

Starting

Before you begin, launch AutoCAD. You will begin this tutorial by drawing an orthographic view for the adapter in Figure 6.5. The file *proto.dwg* has been provided with the data files so that you can be sure you are using the same settings as are used in the tutorial.

To start a drawing from a prototype as you did at the end of Tutorial 5,

Pick: **New icon**

Pick: **Prototype**

Use the dialog box that appears on your screen to select the file *proto.dwg* from the data files that accompany this text. When you have finished selecting it, pick OK to return to the Create Drawing File dialog box and type the new file name in the empty box near the bottom of the dialog box.

Type: **ADAPTORT**

Pick: **OK**

> ■ **TIP** If you want to create the new file in a directory other than the default working directory, *c:\work*, pick on the button that says New File Name when you specify the file name, and then you can select the directory name in which you want to save the file. You will see the directory name appear near the top of the dialog box when you have selected it successfully. ■

The main AutoCAD drawing editor should reappear on your display screen. On your own, make sure that Grid and Snap are set to .25 spacing and turned on; the grid should be on your screen and the Snap button should be darkened on the status bar.

Figure 6.5 shows the adapter you will create. All dimensions are in decimal inches.

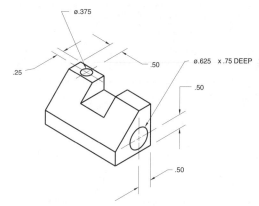

Figure 6.5

Deciding the Model Space Limits

Remember from Tutorial 5 that you can use the Limits command in paper space to set the limits of the paper area that you want to draw on. You can also use the Limits command in model space to set the limits of the area that you will use to create your real-world model of the object. The limits set in paper space and in model space do not have to be the same. Thinking back to the Rose Bowl analogy, your TV screen may be 27", but the actual football stadium is much larger than that. In the prototype drawing you set the limits for paper space to the printer limits. Set the limits in model space to whatever value is necessary to fit the views of the object you are creating. You can change the limits at any time during a drawing session, using the Limits command.

Examine the size of the part and the amount of space the views will require. Then set the model space limits to allow a big enough area to create the views. The adapter is 3 inches wide, 1.5 inches high, and 1.5 inches deep. The slot in the top of the adapter is .5 inches deep. The default limits for model space are 12 × 9. The adapter that you will draw does not require much space; therefore, you will leave the limits set to 12 × 9.

Viewing the Model Space Limits

Use the command Zoom All in model space to show the drawing limits or extents (of the drawing objects), whichever is larger, in the viewport.

Pick: **Zoom All icon**

Move the cursor around inside the viewport. You should see that the coordinates and grid match the limits of the drawing.

Using Ortho

You can use the Ortho command to restrict Line and other commands to operate only horizontally and vertically. This feature is very handy when you are drawing orthographic views and when you are projecting information between the views. You toggle the Ortho command on and off by double-clicking on the Ortho button on the status bar, or pressing the (F8) function key, so it is easy to activate when you are in a different command.

Double-click: **Ortho button**

> ■ *TIP* If you have difficulty getting the Ortho button on the status bar to work, your system may not have enough free RAM available. If you have trouble using the buttons on the status bar, use the (F8) function key to turn Ortho on. ■

The name on the Ortho button becomes dark to indicate that it is turned on. You can also use the Drawing Aids dialog box to toggle the Ortho command.

Pick: **Options, Drawing Aids**

Notice that an X now appears in the box to the left of Ortho. If it is not selected, pick it now.

Pick: **OK (to return to the drawing editor)**

The Ortho button should be darkened on the status bar of your screen.

Next, you will draw the horizontal and vertical construction lines, as shown in Figure 6.6. These lines will represent the left-most and bottom-most margins of your orthogonal views. The coordinates are given for the lines you will create to ensure that your results will be the same as in the tutorial. In general, you would make the construction lines at any location and then move the views if necessary. Don't worry if views are not perfectly centered when you begin your drawing. You will use AutoCAD to center the views after they are drawn and you see how much room is needed.

Drawing Construction Lines

 Construction lines extend infinitely. The default method for drawing a construction line is to specify two points through which it passes. When you use the Horizontal option, the line appears parallel to the X axis through the point you select. The Vertical option is the same, but creates a line parallel to the Y axis. By selecting the Angle option, you can specify the construction line by entering the angle and a point through which the line passes. The Bisect option lets you define an angle by three points and create a construction line that bisects it. Finally, the Offset option allows you to specify the offset distance or through point, as when you use the Offset command, to create an infinite construction line.

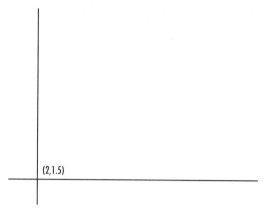

Figure 6.6

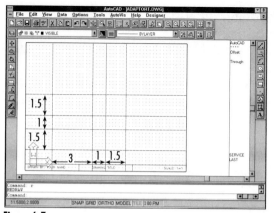

Figure 6.7

Pick: **Construction Line icon**

Hor/Ver/Ang/Bisect/Offset/<From point>: **2,1.5** ⏎

Through point: *(pick to the right to define a horizontal line)*

Through point: *(pick above to define a vertical line)*

Through point: ⏎

Two infinite construction lines appear in your drawing, one vertical through point 2,1.5, the other horizontal through the same point.

Next, you will use the Offset command to create a series of parallel horizontal and vertical lines to define the overall dimensions of each view, as shown in Figure 6.7. Then you will trim the lines to remove the excess portions. If you need to, review the Offset command in Tutorial 3.

For the first horizontal line,

Pick: **Offset icon**

Offset distance or Through<Through>: **1.5** ⏎

Select object to offset: *(pick the horizontal line)*

Side to offset? *(pick any point above the horizontal line)*

A new line is created, parallel to the bottom line and exactly 1.5 units away. You will end the command with the ⏎ key because the next line will be a different distance away.

Select object to offset: ⏎

You will restart the Offset command by pressing ⏎ so that you are prompted again for the offset distance.

Command: ⏎

Offset distance or Through<1.5000>: **1** ⏎

Select object to offset: *(pick the newly created line)*

Side to offset? *(pick any point above the line)*

A line appears 1.00 unit away from the line you selected.

Select object to offset: ⏎

Now repeat this process on your own until you have created all of the horizontal and vertical construction lines according to the dimensions shown in Figure 6.7. The lines are parallel to the horizontal line at distances of 1.5 (the given height of the object), 2.5 (the 1.5-inch height and an arbitrary 1-inch spacing between the front and top views), and 4 (the 1.5-inch height plus 1 plus the 1.5-inch depth of the object). Your screen should look similar to Figure 6.7.

You will define the areas for the front, top, and side views by using the Trim command to remove excess lines. You will pick Trim from the Trim/Extend flyout on the Modify toolbar. You will use implied crossing with the Trim command to select all of the lines as cutting edges. When construction lines have one end trimmed, they become rays. When trimmed again, they become lines. Use Figure 6.8 to determine which construction lines to trim.

Pick: **Trim icon**

Select cutting edges(s). . .

Select objects: *(start your selection at corner A, shown in Figure 6.8, then pick corner B)*

Select objects: (⏎)

<Select object to trim>/Undo: *(select segments 1–24 in the order in which they are numbered in Figure 6.8)* (⏎)

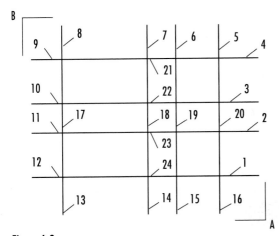

Figure 6.8

> ■ *TIP* Turn Snap off while trimming to make it easier to select. ■

Next, use the Erase command to remove the unwanted lines, as shown in Figure 6.9.

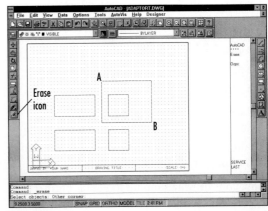

Figure 6.9

Pick: **Erase icon**

Select objects: *(pick corner A, as shown in Figure 6.9)*

Other corner: *(pick corner B)*

Select objects: (⏎)

Your drawing should look similar to Figure 6.10.

Figure 6.10

The overall dimensions of the views are established and aligned correctly. Before continuing, redraw the screen on your own.

Next, draw the slot in the front view. Use the points listed below, or create the lines on your own by looking at the dimensions specified on the object in Figure 6.5.

From point: **3,3** ↵
To point: **3,2.5** ↵
To point: **4,2.5** ↵
To point: **4,3** ↵
To point: ↵

On your own, use the Trim command from the Modify toolbar and remove the center portion of the top horizontal line.

Now your drawing should look like Figure 6.11.

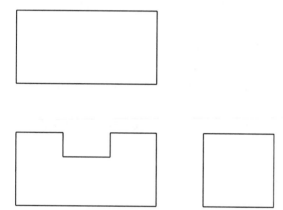

Figure 6.11

Hidden Lines

You will use hidden lines to represent the lines that are not visible on the side view. Remember, each view is a view of the entire object drawn from that line of sight. All surfaces are shown in every view. A hidden line in the drawing represents one of three things:

1. An *intersection* of two surfaces that is behind another surface and therefore not visible.

2. The *edge view* of a hidden surface.

3. The outer edge of a curved surface that is hidden. This is also called the *limiting element* of a contour.

There are a few general practices to use when drawing hidden lines that help prevent confusion and to make the drawing easier to read.

Clearly show intersections, using intersecting line segments.

Clearly show corners, using intersecting line segments.

Leave a noticeable gap (about 1.16″) between aligned continuous lines and hidden lines.

See Figure 6.12.

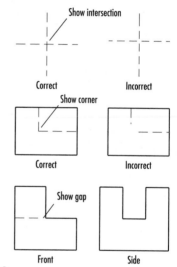

Figure 6.12

These hidden line practices are sometimes difficult to implement. If you change one hidden line so that it looks better, using the global linetype scale command, all the other hidden lines take on the same characteristics and may be adversely affected. In general, hidden line practices are not followed as strictly as they once were, partly because with CAD drawings plotted on a good-quality plotter, the thick visible lines can easily be distinguished from the thinner hidden lines. The results of a reasonable attempt to conform to the standard are considered acceptable in most drawing practices.

Hidden lines are usually drawn using a different color than the one used for the continuous object lines on the AutoCAD drawing screen. This helps to create a visual difference between the different types of lines and makes them easier to interpret. Also, you control printers and plotters by using different colors in the drawing to represent different thicknesses of lines on the plot. You can use any color, but be consistent. Make all hidden lines the same color. It is best to set the color and linetype by layer, and draw the hidden lines on that separate layer with the correct properties. This is why BYLAYER is the default choice for color and linetype in AutoCAD. There is already a separate layer for hidden lines in the prototype drawing from which you started *adaptort.dwg*.

Drawing Hidden Lines

Now you will create a hidden line in the side view of your drawing to represent the bottom surface of the slot.

On your own, set layer HIDDEN_LINES as the current layer.

You will draw a horizontal line from the bottom edge of the slot in the front view into the side view. This line will be used to project the depth of the slot into the side view. Check to see that the Ortho button on the status bar is highlighted.

Show the Object Snap toolbar by selecting Tools, Toolbars, Object Snap on your own. Position it on the screen where it will be handy to pick the object snaps during commands.

Pick: **Line icon**

From point: **(pick Snap to Intersection icon)**

int of **(pick the lower right corner of the slot in the front view)**

To point: **(pick any point to the right of the side view)**

To point: ⏎

Your drawing should look like Figure 6.13.

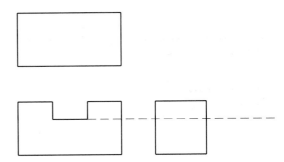

Figure 6.13

On your own, trim the projection line so that only the portion within the side view remains.

Your drawing should look like Figure 6.14.

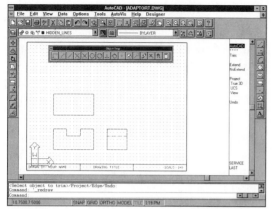

Figure 6.14

Next, you will use the Line command to project the width of the slot from the front view into the top view, using vertical lines.

Do the following steps on your own. First, set the current layer back to VISIBLE for the next lines you will create. To help make the projection lines straight, be sure that Ortho is on. Since you will be drawing a number of lines from intersections, pick the running object snap from the Object Snap toolbar to turn on the Intersection running mode object snap. Now you are ready to draw the lines for the slot in layer VISIBLE.

> *Pick:* **Line icon**
>
> From point: *(target the upper left corner of the slot in the front view)*
>
> To point: *(pick Snap to Perpendicular icon)*
>
> per to: *(pick a point on the upper line of the top view)*
>
> To point: (⏎)
>
> Command: (⏎)
>
> From point: *(target the upper right corner of the slot in the front view)*
>
> To point: *(pick Snap to Perpendicular icon)*
>
> per to: *(pick a point on the upper line of the top view)*
>
> To point: (⏎)

Use the Trim command to remove the excess lines. Your drawing should look similar to Figure 6.15.

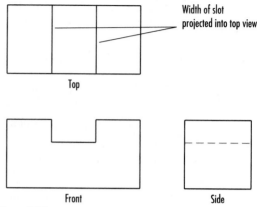

Figure 6.15

Save *adaptort.dwg* before you continue.

Line Precedence

Different types of lines often line up with each other within the same view, as illustrated in Figure 6.16.

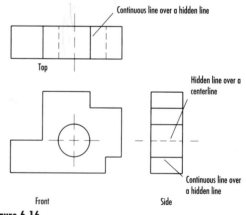

Figure 6.16

The question arises, which type of line takes *precedence*; that is, which type do you draw? The rule is that continuous lines take precedence over hidden lines, and hidden lines take precedence over centerlines. Note that in the side view of Figure 6.16, the short end segments of the covered-up centerline show beyond the edge of the object. This practice is sometimes used to show the centerline underlying the hidden line. If you show the short end segments where a centerline would extend, be sure to leave a gap so that the centerline does not touch the other line, as this makes it difficult to interpret the lines.

AutoCAD does not determine line precedence for you. You must decide which lines to show in your 2D orthographic views. If you draw a line over the top of another line in AutoCAD, both lines will be in your drawing. If you are using a plotter, both will plot, making a darker or thicker line than should be shown. On your screen you may not notice that there are two lines, because one line will be exactly over the top of the other.

Slanted Surfaces

Orthographic views can only distinguish *inclined surfaces* (ones that are tipped away from one of the viewing planes) from *normal surfaces* (ones that are parallel to the viewing plane) if the surfaces are shown in profile. Inclined surfaces are perpendicular to one of the principal viewing planes, and tipped away, or foreshortened, in the other views. As illustrated in Figure 6.17, there is no way to tell by looking at the top and side views which lines are inclined and which are normal.

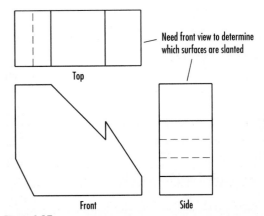

Figure 6.17

The front view is required, along with the other two views, to completely define the object's size and shape. In the next steps you will add a slanted surface to the adapter so that the object looks like that shown in Figure 6.18.

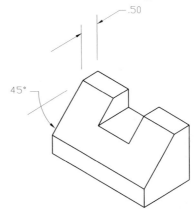

Figure 6.18

You will now use the Line command to add the slanted surface to the side view, using relative coordinates. You will locate a point .5 inches to the left of the top right corner of the right-side view and draw and draw a slanted line from this point for the 45° surface.

Pick: **Line icon**

From point: **7,3** ⏎

To point: **@3<–135** ⏎

To point: ⏎

The distance 3 was chosen because the exact distance is not known and 3 is obviously longer than needed, since the entire object is only 1.5 inches high.

Your screen should look like Figure 6.19.

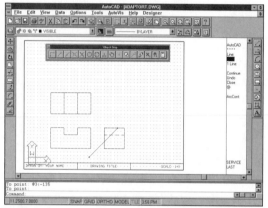

Figure 6.19

Before continuing, use the Trim command on your own to trim this line and any lines above the slanted surface, and then set the current drawing layer to layer PROJECTION. When you are finished, your drawing should look like Figure 6.20.

■ *TIP* Use Zoom Window to zoom the side view to help you locate the points to trim. Return to your drawing with Zoom Previous when all the excess lines are removed. You may want to turn Snap and Ortho off to help you trim the lines. ■

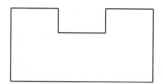

Figure 6.20

Top-View to Side-View Projection

You can project information from the top view to the side view and vice versa by using a 45° *miter line*. You can draw the miter line anywhere above the side view and to the right of the top view, but it is often drawn from the top right corner of the front view, as shown in Figure 6.21.

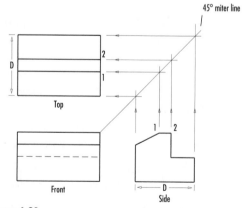

Figure 6.21

To project information from the side view to the top view, you would draw vertical projection lines from the points in the side view so that they intersect the miter line. In the example in Figure 6.21, points 1 and 2 are projected. Then horizontal lines would be projected from the intersection of the vertical lines and the miter line across the top view.

Drawing the Miter Line

You will draw a 45° ray, starting where the front edge of the top view and the front edge of the side view would intersect. Remember, the Ray command is located on the Line flyout of the Draw toolbar. On your own, make sure that Ortho is off.

Using Apparent Intersection

 You can use the Apparent Intersection object snap to find the intersection where two objects would meet if extended. You can also use it to find the intersection of two objects that appear to intersect on the screen, but in fact do not intersect in 3D space. It will also find actual intersections. You will use it to find the point where the bottom edge of the top view and the left edge of the side view would intersect. Refer to Figure 6.22. In order to select the short line segment in the side view, enlarge the views on the screen by using Zoom Window. Otherwise, if an actual intersection is inside the aperture box, it will be selected instead of the apparent intersection you are trying to locate.

Pick: **Ray icon**
From point: *(pick Snap to Apparent Intersection icon)*
appint of *(pick on line 1, making sure not to get the intersection of two lines inside the aperture box)*
and *(pick on line 2, making sure not to include any intersections inside the aperture box)*
Through point: *@3.5<45* ⏎
Through point: ⏎

The ray you will use for the miter line is added to the drawing, as shown in Figure 6.22.

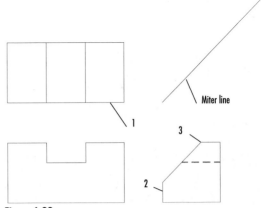

Figure 6.22

Next you will project the corner point (point 3 in Figure 6.22), created by the slanted surface in the side view, to the miter line by drawing a vertical ray from the intersection. You will use the Intersection running object snap to pick the intersection points.

On your own, make sure the Intersection running mode object snap and Ortho are turned on.

Pick: **Ray icon**
From point: *(pick point 3 in Figure 6.22)*
Through point: *(pick anywhere above point 3)*
Through point: ⏎

The ray is added to your drawing, extending vertically from point 3. Next, you will project another ray from where the vertical ray intersects the miter line.

Pick: **Ray icon**
From point: *(pick where the vertical ray meets the miter line)*
Through point: *(pick a point to the left of the miter line)*
Through point: ⏎

Your drawing will look like Figure 6.23.

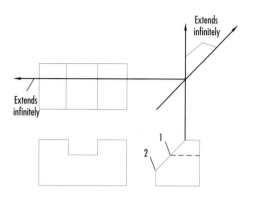

Figure 6.23

On your own, project point 1 to the top view and point 2 into the front view, using the same method as before. Now that you have projected the depth of the surfaces to the top view, you are ready to trim the lines in the top view. Remember, when rays are trimmed they become regular lines. On your own, use the Trim command and implied Crossing and trim the lines so that your drawing looks like Figure 6.24.

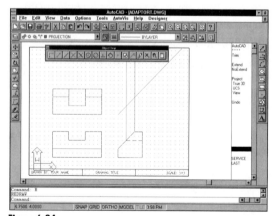

Figure 6.24

On your own, pick the Properties icon from the Object Properties toolbar and change the layer of the new lines in the top and front views to VISIBLE. When you have completed the visible lines, use the techniques you have learned

to set layer VISIBLE as the current layer and freeze layer PROJECTION. Leaving the projection lines frozen in the drawing is useful, because if you need to change something, you can just thaw the layer instead of having to create the projection lines over again.

> ■ *TIP* When you are drawing vertical and horizontal lines to intersect the miter line, the command Snap to Intersection will help to capture the intersection points between the vertical projection lines and the miter line. Turning Ortho on will ensure that only vertical and horizontal lines are drawn. ■

Sizing and Positioning the Drawing

Use Zoom XP, as you learned in Tutorial 5, to establish a relationship between the model space drawing and the size it is shown on the paper. When you use Zoom All, all of the drawing or the limits area is fit into the viewport; this does not give you any particular scale for the end drawing. You will use Zoom XP with a value of 1 to set one unit in paper space equal to one unit in model space. You will type the alias *Z* for the Zoom command.

Command: **Z** ⏎

All/Center/Dynamic/Extents/Left/Previous/Vmax/Window/
<Scale (X/XP)>: **1XP** ⏎

The drawing is zoomed so that one unit in paper space is equal to one unit in model space.

> ■ *TIP* You can check to see that Zoom XP worked by switching to paper space and using Distance from the Inquiry flyout on the Object Properties toolbar to measure the length of a model space line. (This is where you picked List in Tutorial 3.) Be sure to switch back to model space. ■

Next, use the Pan command on your own to position the views you drew inside the viewport. Position the views so that they appear centered in the viewport. When you have finished, your screen should look like Figure 6.25.

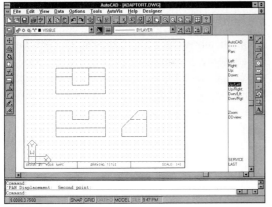

Figure 6.25

Turn the running mode Snap to Intersection off. Remember that leaving this mode turned on can cause difficulty with other commands. When you are finished projecting lines from intersections, you should always turn the running mode object snap back off in order to avoid problems.

■ **TIP** It is good practice to frequently save your drawings on your disk. If your system experiences a power failure and you lose the drawing in memory, you can always retrieve a recent version from your disks. ■

Drawing Holes

Figure 6.26 shows an object with two holes and the way in which they are represented in front and top views.

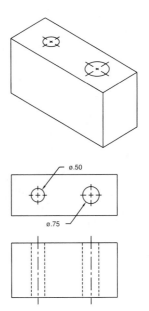

Figure 6.26

The *diameter symbol ø* indicates a diameter value. If no depth is specified for a hole, it is assumed that the hole goes completely through the object. No depth is specified for these holes, so the hidden lines in the front view go from the top surface to the bottom surface.

You will add two holes to the orthographic views you have drawn to represent the adapter, as shown in Figure 6.27.

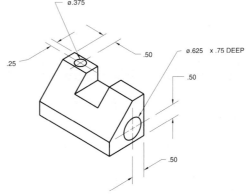

Figure 6.27

In Figure 6.27, the .625-diameter hole has a depth specification of .75. This means that the hole is .75 deep, drilling from the surface of the object. A 120° conical point is added to the bottom of any hole that does not go completely through an object. This is because twist drills, the type used most often to drill holes, have a conical point. The 120° drill point is not included as part of the depth of the hole.

Centerlines for holes must be included in all views. A centermark and four lines extending beyond the four quadrant points are used to define the center point of a hole in its *circular view* (the view where the hole appears as a circle). A single centerline, parallel to the two hidden lines, is used in the other views, called *rectangular views* because the drill hole appears as a rectangle. Centerlines should extend beyond the edge of the symmetrical feature by a distance of at least 3/8" on the plotted drawing.

You will start with the top view and add the circular view of the .375 diameter hole, as specified in Figure 6.27. As indicated in the figure, the hole's center point is located .5 from the left surface of the view and .25 from the back surface.

Use the Drawing Aids dialog box to change the Snap spacing to .25; make sure that it is on.

> *Pick:* **Circle Center Diameter icon**
>
> 3P/2P/TTR<Center point>: **2.5,5.25** ⏎
>
> Diameter: **.375** ⏎

Drawing Centerlines and Centermarks

Next, you will draw the centerlines for the circle. Engineering drawings use two different thicknesses of lines: thick lines are used for visible lines, cutting plane lines, and short break lines; thin lines are used for hidden lines, centerlines, dimension lines, section lines, long

break lines, and phantom lines. In AutoCAD, you use color to tell the plotter or printer which thickness to use when printing lines.

Since centerlines in the circular view should be thin, you need to draw them on a new layer. Make layer THIN current. You could also use layer CENTERLINE, but it has a CENTERLINE linetype. Because the Center Mark command automatically creates the dashes at the center of the circle by drawing short lines, centermarks will usually look better if they are not drawn with a linetype that already contains a dash (like the centerline linetype). This is why you will use layer THIN, which has a continuous linetype.

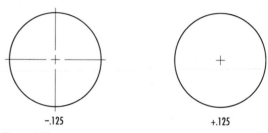

Figure 6.28

The Dimcen variable controls the size and appearance of the centermark that is added to the circle. The *absolute value* of the Dimcen variable determines the size of the centermark. The *sign* (positive or negative) determines the style of the centermark. Figure 6.28 shows the different styles of centermarks that you can create by setting Dimcen to a positive or negative value. Setting Dimcen to zero will cause no centermark to be drawn. (This is useful when using dimensioning commands like Radial, which automatically add a centermark. You will learn more about dimensioning and adding centermarks in Tutorial 7.) Usually the cross at the center should appear 1/8" wide, but since this hole is relatively small, you will need a smaller centermark, so you will use the value –.05.

Command: **DIMCEN** ⏎

New value for DIMCEN <0.0900>: **−.05** ⏎

Next, show the Dimensioning toolbar so that you can select the Center Mark command.

Pick: **Tools, Toolbars, Dimensioning**

The Dimensioning toolbar appears on your screen. It is shown in Figure 6.29. Position it so that it is located conveniently for you to select commands.

Center Mark

Figure 6.29

 Next, you will use Center Mark to draw the circle's centerlines on layer THIN. Use the Center Mark icon from the Dimensioning toolbar whenever you are drawing a centerline in the circular view (where the hole appears round).

Pick: **Center Mark icon**

Select arc or circle: *(**pick on the outer edge of the circle**)*

■ *TIP* Use Zoom Window and turn Snap off if you need to make it easier to target the objects and intersections. ■

You should see the circular view centerlines appear in the drawing. Next, you will use hidden lines and the Intersection object snap to project the width of the hole into the front view. On your own, set the current layer to layer HIDDEN_LINES. Now turn the running mode Intersection object snap on. Because you are going to project straight lines, make sure you have Ortho turned on.

Pick: **Line icon**

From point: *(**pick the intersection of the circle's horizontal centerline with the left edge of the circle**)*

To point: *(**pick Snap to Perpendicular icon**)*

per to *(**pick any point on the lowest horizontal line in the front view**)*

To point: ⏎

Repeat this procedure to draw the projection line for the right side of the hole and for the centerline. (Use Zoom Previous or Zoom All if needed to return your screen to full size.)

Now that you have finished this step, your screen should look similar to Figure 6.30.

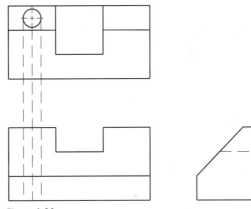

Figure 6.30

On your own, use the Properties icon from the Object Properties toolbar to change the layer for the middle of the three lines you just drew to layer CENTERLINE and save your drawing. Next, you will break the line so that it is the right length for a centerline in the front view.

Breaking the Centerline

You will break the middle vertical line so that you can use it as a centerline by selecting the 2 Points Select option from the Break flyout on the Modify toolbar. Centerlines should extend at least .375″ past the edge of the cylindrical feature when the drawing is plotted. You will extend the centerline .5" past the bottom of the front view to meet this criterion.

Pick: **Break 2 Points Select icon**

Select object: *(select the middle vertical line)*

The first point of the break is the point that you want to use as the top of the centerline. The second point is past the end where the line extends into the top view, so that all of the centerline extending into the top view is broken off. Refer to Figure 6.31. On your own, make sure that Snap is turned on to make selecting easier.

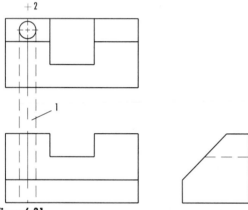

Figure 6.31

Enter first point: *(select break point 1 on the line, .5 units away from the top of the object in the front view)*

Enter second point: *(select point 2, past the top end of the line)*

> ■ *TIP* If you pick on the end of the line, you may not get a location past the end of the line for the second point of the break; instead, short segments of the line may remain and look like random points in your drawing. ■

Now use the hot grips with the Stretch command to stretch the centerline down .5" past the bottom of the front view on your own. Refer to Tutorial 4 if you need to review using hot grips and the Stretch command.

Your drawing should look like Figure 6.32.

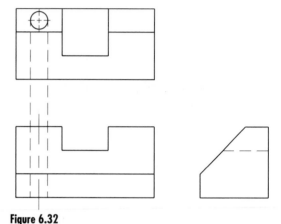

Figure 6.32

Before continuing, trim the two hidden lines where they extend past the front view of the object on your own.

To complete the hole in the side view, you will project the location of the hole into the side view. Then you will use hot grips with the Move, Copy option to copy the lines from the front view to the side view.

On your own, thaw layer PROJECTION, set it as the current layer, and draw a projection line from the top view, where the vertical centerline intersects the circle, and project it out past the miter line. Use the Snap to Intersection running mode and Ortho to help you. Make sure that Snap is off. Refer to Figure 6.33. Then project the intersection of the horizontal projection line with the miter line into the side view, creating a vertical projection line.

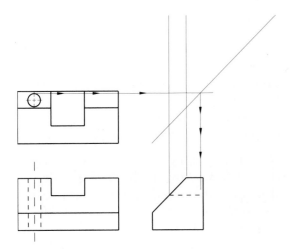

Figure 6.33

Because holes are symmetrical, the side view of the hole appears the same as the front view, except for the location. You will use hot grips with the Move, Copy option to copy the two hidden lines and the centerline from the front view to the side view.

Activate the hot grips for the two hidden lines and the centerline by picking on them on your own.

> *Pick: (the grip on the top endpoint of the right-hand hidden line as the base grip)*
>
> ****STRETCH****
>
> <Stretch to point>/Base point/Copy/Undo/eXit: ⏎
>
> ****MOVE****
>
> <Move to point>/Base point/Copy/Undo/eXit: **C** ⏎

A faint copy of the lines you selected appears, attached to the crosshairs. (You should still have the running mode Intersection object snap active.) Target the intersection between the top line in the side view and the projection line locating the side of the hole, as you see in Figure 6.34. You may want to use Zoom Window to help you choose the correct intersection. The three lines are copied to this location. Press ⏎ to exit the Copy command.

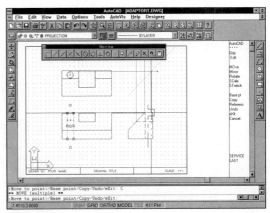

Figure 6.34

Leave the projection lines and the miter line; you will need them to project the other hole from the side view. Your drawing will look like Figure 6.35. Redraw your screen before you continue.

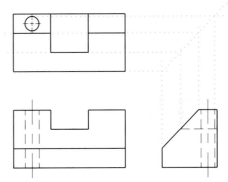

Figure 6.35

Next you will add the circular view of the .625 hole to the side view and then project the width and centerline of the hole into the front view, as presented above. Before you continue, set layer VISIBLE as the current layer.

The center point of the hole is 6.75,2 and the diameter is .625. Draw the circle on your own.

Now make layer THIN current and use the Center Mark command to add centerlines. Return to the layer HIDDEN_LINES and use the Intersection object snap to project the hole's edge lines into the front view on your own.

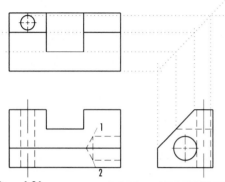

Figure 6.36

Add the .75 depth of the hole by using Offset to create a vertical line parallel to the right surface of the front view on your own. If you need help using Offset, refer to Tutorial 3. Then use the Trim command to remove the excess lines. Refer to Figure 6.36. Next, change the line you offset to layer HIDDEN_LINES on your own by using the Properties icon from the Object Properties toolbar.

Next you will add the 120° conical point to both views. First you will enlarge the area you are working on so that it fills your screen.

Pick: **Zoom Window icon**

Zoom the front view so that you can add the conical point. Make sure that Ortho is off.

Pick: **Line icon**

From point: *(target intersection 1 in Figure 6.36)*

To point: **@.75<240** ⏎

To point: ⏎

Command: ⏎

Line from point: *(target intersection 2)*

To point: **@.75<120** ⏎

To point: ⏎

Use the Trim command to remove any excess lines on your own, so your drawing looks like Figure 6.36.

Pick: **Zoom Previous icon**

to return to your original display screen area.

On your own, change the current layer to layer CENTERLINE. Project a line from the side view, where the centerline crosses the edge of the circle, to the front view, at least .5 inches beyond the drill point.

Notice that the horizontal centerline in the front view coincides with an edge line. Refer to the section at the beginning of this tutorial and Figure 6.16 on line precedence. When you are printing, the object's edge line takes precedence over the centerline, but on the screen, AutoCAD displays the last line drawn on top of the others.

On your own, use Break to remove the excess centerline, as you learned earlier, so that only a .5-inch tail remains at each end of the hole. The practice of showing this short tail of centerline is optional. If you feel that it makes the drawing difficult to interpret, it may be left off entirely. If it is shown, there must be a gap of about 1/16" on the plotted drawing between the visible (or hidden) line and the short centerline tail.

To get rid of the excess blip marks,

Pick: **Redraw View icon**

Now project the center of the hole from the side view to locate it in the top view so that you can copy the hole from the front view. Make sure layer CENTERLINE is current and that Snap to Intersection running mode and Ortho are on.

Pick: **Line icon**

From point: *(select the point where the centerline touches the top of the hole in the side view)*

To point: *(select a point slightly past the miter line)*

To point: ↵

Command: ↵

Line from point: *(select the point where the projected line intersects with the miter line)*

To point: *(select a point left of the top view)*

To point: ↵

Setting the Global Linetype Scaling Factor

You can change the linetype scale factor for all linetypes at once by setting the global linetype scaling factor. All standard AutoCAD linetypes are defined in the file *acad.lin*. As you recall from Tutorial 4, each linetype is defined by the distance to draw each dash, gap, and dot. Because these are defined using specific distances, you may need to adjust the lengths of the dashes and gaps for use in your drawing. The Ltscale command lets you adjust all of the linetype lengths by the scaling factor you specify. A setting of 2 for Ltscale makes the dashes and gaps twice as long as the original pattern; a setting of .5 makes them half as long. When you are plotting your drawing to a particular scale, you usually set the Ltscale factor to the reciprocal of the plot scale. For instance, if you are going to plot the drawing at 1"=10", then you set Ltscale to 10. Ltscale affects all linetypes in your drawing at the same time (although you can adjust differently for paper space and model space).

■ **Warning:** When you are working in metric units, the linetypes sometimes may not appear correctly. A line may appear to be the correct color, but not the correct pattern. This is because the lengths defined in the linetype file are defined in terms of inches. Incorrectly shown metric lines may have changed pattern, but the spacing is so small it can't be seen. Use Ltscale to adjust the spacing. For metric units, try a value of 25.4. ■

You may need to adjust your lines with Ltscale to make the CENTERLINE linetype visible and the HIDDEN linetype have shorter dashes.

Command: **LTSCALE** ↵

New scale factor <1.0000>: **.65** ↵

Your drawing should look like Figure 6.37.

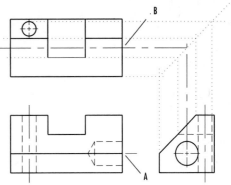

Figure 6.37

Now you are ready to use hot grips and Move, Copy to copy the hidden lines from the front view into the top view. You will use a procedure similar to last time, except that you will change the base point. When prompted for the base point, you will select the intersection of the centerline with the right edge of the object in the front view. Refer to Figure 6.37.

Pick: *(on the hidden lines forming the hole and drill point in the front view to activate the hot grips, but not the centerline)*

Pick: *(any grip as the base grip)*

Press: ↵ *(or the return button or the spacebar, to cycle past Stretch to the Move command)*

** MOVE **

<Move to point>/Base point/Copy/Undo/eXit: **B** ↵

Base point: *(target point A)*

<Move to point>/Base point/Copy/Undo/eXit: **C** ↵

You will see a faint copy of the lines you have selected attached to the crosshairs. Move the crosshairs to the point in the top view where the projection line for the center of the hole intersects with the right side of the top view. You will target this point and press the pick button to select it as the location for the copy.

<Move to point>/Base point/Copy/Undo/eXit:
(select point B)

You will see the copy appear in the top view, as shown in Figure 6.38. Press ⏎ to end the command.

On your own, use the Break command again to shorten the projected centerline in the top view so that it extends about .5" past the edge of the hole. Erase the centerline that is projected from the side view. Then freeze layer PROJECTION. Redraw your screen.

Your screen should look like Figure 6.38.

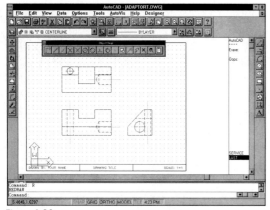

Figure 6.38

Saving Your Drawing

You have completed this orthographic drawing. Save the drawing on your disk as *adaptort.dwg*.

Projecting Slanted Surfaces on Cylinders

Next you will work with projecting slanted surfaces on cylinders. You will begin by opening the existing drawing *cyl1.dwg* from the data files for this manual.

■ **TIP** Your data files should be installed on the hard drive. Do not open files directly from your floppy disk. Doing so can cause you to encounter several problems, such as running out of disk space or AutoCAD temporary files being left on your disk if you do not exit properly or the program crashes. If you must open a file from a floppy, it is a good practice to pick New and then use the file you want to open as a prototype. You should specify the new drawing file name so that it is created on the hard drive. This will prevent many problems and the likelihood of corrupted drawing files. ■

Begin a new drawing called *cyl1orth.dwg* from the prototype drawing *cyl1.dwg*. Your screen should look similar to Figure 6.39.

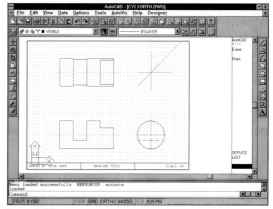

Figure 6.39

On your own, add a slanted surface to the front view as shown in Figure 6.40. The top of the slanted surface is located 1.5 inches from the right end of the cylinder, and the bottom of the surface is at the horizontal centerline. The top point of the slanted surface, the intersection of the vertical centerline and the edge of the cylinder in the side view, is labeled 1 in Figure 6.40. The bottom edge of the slanted surface is labeled 2,3 and is located directly on the horizontal centerline in the side view.

Next, remove all excess lines left from the shallow surface in the front view. Remove the same surface from the top view and side view; you will be creating a different surface in its place. Your drawing should look similar to Figure 6.40.

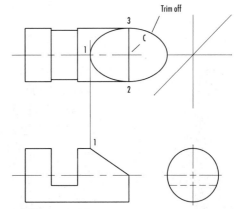

Figure 6.41

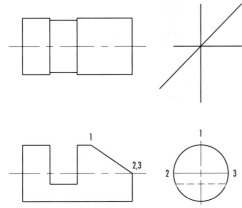

Figure 6.40

What is the shape of the slanted surface in the top view? The front view of the surface appears as a straight slanted line. The side view is a semicircle, but in the top view the shape is an ellipse. Refer to Figure 6.41. A circular shape seen from an angle other than straight on is an ellipse. Since the locations of points 1, 2, and 3 on the ellipse are known, you can use the Ellipse command to draw the shape in the top view.

Using Ellipse

The Ellipse command draws ellipses. AutoCAD provides four different ways to specify an ellipse. Notice the various methods identified in Figure 6.42, where the Ellipse flyout is shown. The Ellipse flyout is located on the Draw toolbar.

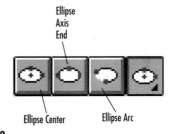

Figure 6.42

 Next you will practice the Ellipse command by drawing some ellipses off to the side of the drawing. You will erase them when you are done practicing. To draw an ellipse by specifying three points,

Pick: **Ellipse Axis End icon**

Arc/Center/<Axis endpoint 1>: *(select a point)*

Axis endpoint 2: *(select a point)*

<Other axis distance>/Rotation: *(select a point)*

An ellipse is created on your screen, using the three points that you selected. The ellipse has a *major axis*, the longest distance between two points on the ellipse, and a *minor axis*, the shorter distance across the ellipse. AutoCAD determined which axis was major and which was minor by examining the distance between the first pair of endpoints and comparing it to the distance specified by the third point.

One way to describe an ellipse is to create a circle and then tip the circle away from your viewing direction by a rotation angle. This method requires you to specify the angle of rotation, instead of the endpoint of the second axis.

You will draw an ellipse, using two endpoints and an angle of rotation. The rotation angle must be between 0 and 89.4°.

See Figure 6.43 to determine the location of the points.

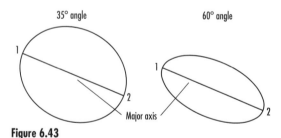

Figure 6.43

Pick: **Ellipse Axis End icon**

Arc/Center/<Axis endpoint 1>: *(select point 1)*

Axis endpoint 2: *(select point 2)*

This construction method defines the distance between points 1 and 2 as the major axis (diameter) of the ellipse. This circle will be rotated into the third dimension by the specified rotation angle.

<Other axis distance>/Rotation: **R** ⏎

Rotation around major axis: **35** ⏎

On your own, erase the ellipses you created from your screen.

The other two construction methods for ellipses are analogous to the two methods already described, but they use radius values rather than diameter values. That is, the center point of the ellipse is known and is used as a starting point instead of an endpoint. Next, you will draw an ellipse by specifying a center point and two axis points; refer to Figure 6.44 for the locations of the points.

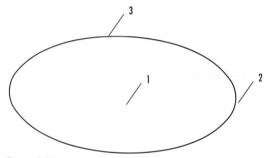

Figure 6.44

Command: ⏎ *(to restart the command)*

Arc/Center/<Axis endpoint 1>: **C** ⏎

The center of the ellipse is the intersection of the major and minor axes. You can enter a coordinate or use the cursor to select a point on the screen.

Center of ellipse: *(select point 1)*

Next, you must provide the endpoint of the axis. The angle of the ellipse is determined by the angle from the center point to this endpoint.

Axis endpoint: *(select point 2)*

Now you must specify either the distance measured from the center of the ellipse to the endpoint of the second axis (measured perpendicular to the first axis), or a rotation angle.

<Other axis distance>/Rotation: *(select point 3)*

AutoCAD used the point that you selected to decide whether the first axis was a major or minor axis. The ellipse is drawn on your screen.

 Next you will draw an ellipse by specifying a center point, one axis point, and an angle of rotation. (See Figure 6.45.) When you select the Ellipse Center icon, you pick one of the two basic approaches, eliminating the need to specify Center in another step.

Pick: **Ellipse Center icon**

Center of ellipse: *(pick the point marked 1 in Figure 6.45)*

Axis endpoint: *(pick point 2)*

<Other axis distance>/Rotation: **R** ↵

Rotation around major axis: **30** ↵

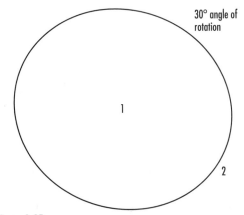

Figure 6.45

Now that you have practiced with the Ellipse command, you are ready to add the ellipse to your drawing. Erase the ellipses you drew when practicing before you go on.

On your own, project point 1 into the top view. To do this, change to layer PROJECTION. The Intersection running mode object snap should still be on. Use it and the Line command to draw a vertical line from point 1 in the front view extending up into the top view past the center. When you are finished, change the current layer back to layer VISIBLE.

Pick: **Ellipse Center icon**

Center of ellipse: *(pick the point marked C in Figure 6.46)*

Axis endpoint: *(pick point 1 in the top view, where your projection line crosses the top line)*

<Other axis distance>/Rotation: *(pick point 2 in the top view)*

Your ellipse should look like the one in Figure 6.46.

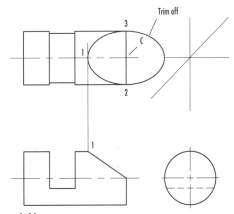

Figure 6.46

Use the Trim command to remove the unnecessary portion of the ellipse. The remaining curve represents the top view of the slanted surface. Remove the projection line from the screen by freezing layer PROJECTION. Your drawing should appear similar to Figure 6.47.

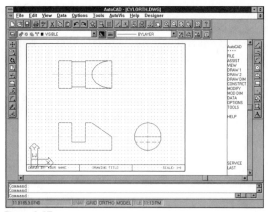

Figure 6.47

Circular shapes appear as ellipses when tipped away from the direction of sight. Not all surfaces are uniform shapes like circles and ellipses; some are irregular. You can create irregularly curved surfaces using the Polyline command and the Spline option that you learned in Tutorial 3. To project an irregularly curved surface to the adjacent view, identify a number of points along the curve and project each point. Use the Polyline command to connect the points. Then use the Pedit, Spline option to create a smooth curve through the points.

Save the drawing as *cylorth.dwg* and print the drawing from paper space. You have completed Tutorial 6. Turn off any extra toolbars you may have open, leaving only the Draw and Modify toolbars on the screen. Exit AutoCAD.

KEY TERMS

- absolute value
- circular view
- diameter symbol
- edge view
- hidden line
- inclined surface
- intersection
- limiting element
- major axis
- minor axis
- miter line
- normal surface
- orthographic view
- precedence
- project
- projection lines
- rectangular view
- sign

KEY COMMANDS

- Center Mark
- Construction Line
- Dimcen
- Ellipse
- Ortho
- Snap to Apparent

Draw front, top, and right-side orthographic views of the following objects. The letter M after an exercise number means that the problem's dimensional values are in millimeters (metric units). If no letter follows the exercise number, the dimensions are in inches.

6.1 Base Block

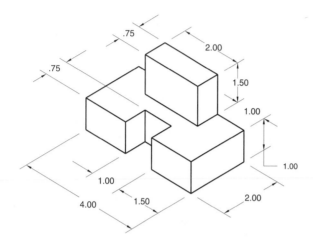

6.2M Shaft Guide

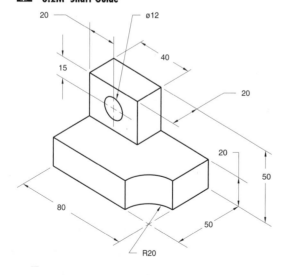

6.3 Piston Guide

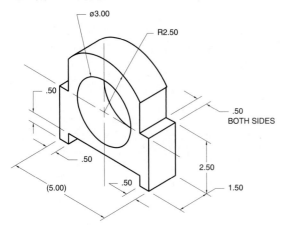

ø3.00
R2.50
.50
.50 BOTH SIDES
.50
.50
2.50
(5.00)
1.50

6.4 M Lock Catch

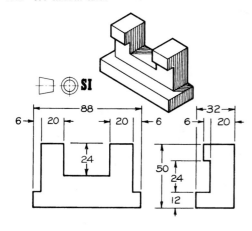

SI

88
6
20
20
6
6
32
20
24
50
24
12

6.5 Bearing Box

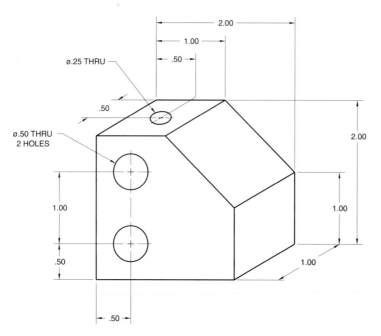

ø.25 THRU
2.00
1.00
.50
.50
ø.50 THRU
2 HOLES
2.00
1.00
1.00
.50
1.00
.50

6.6M Bushing Holder

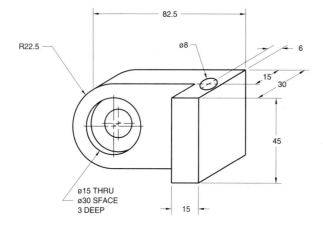

82.5
R22.5
ø8
6
15
30
45
ø15 THRU
ø30 SFACE
3 DEEP
15

6.7 Shaft Support

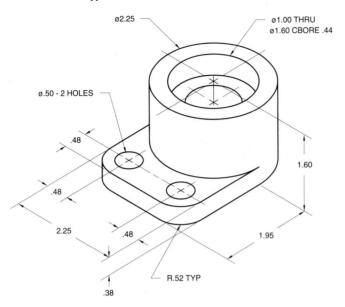

ø2.25

ø1.00 THRU
ø1.60 CBORE .44

ø.50 - 2 HOLES

.48

.48

1.60

2.25

.48

1.95

R.52 TYP

.38

6.8M Stop Plate

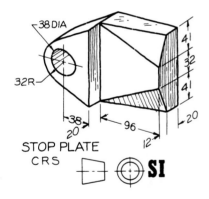

38 DIA

32R

41

32

41

38
20

96

12

20

STOP PLATE
C R S

SI

6.9 Lift Guide

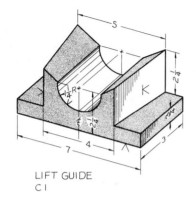

LIFT GUIDE
C1

6.10M Clamp

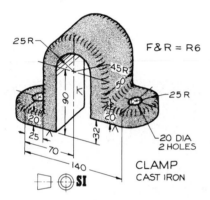

F&R = R6

CLAMP
CAST IRON

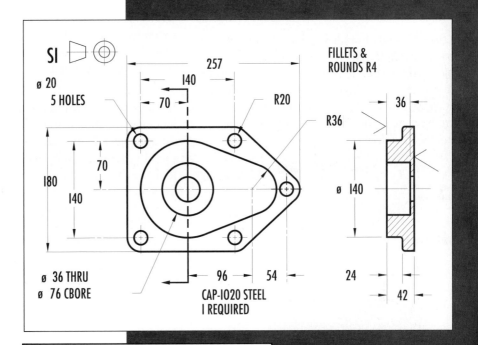

SI

ø 20
5 HOLES

FILLETS &
ROUNDS R4

257

140

70

R20

R36

36

70

180

140

ø 140

ø 36 THRU
ø 76 CBORE

96

54

24

42

CAP-1020 STEEL
1 REQUIRED

TUTORIAL 7

Basic Dimensioning

Objectives

When you have completed this tutorial, you will be able to

1. Understand dimensioning nomenclature and conventions.

2. Control the appearance of dimensions.

3. Set the dimension scaling factor.

4. Locate dimensions on drawings.

5. Set the precision for dimension values.

6. Dimension a shape.

7. Use baseline and continued dimensions.

8. Save a dimension style and add it to the prototype drawing.

9. Use associative dimensioning to create dimensions that can update.

Introduction

In the previous tutorials you have been learning how to use AutoCAD to define the *shape* of an object. Dimensioning is used to show the *size* of the object in your drawing. The *dimensions* you specify will be used in the *manufacture* and *inspection* of the object. Figure 7.1 shows a dimensioned drawing.

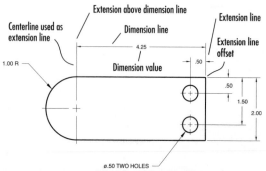

Figure 7.1

For the purpose of inspecting the object, a *tolerance* must be stated to define to what extent the actual part may vary from the given dimensions and still be acceptable. In this tutorial you will use a general tolerance note to give the allowable variation for all dimensions. Later, in Tutorial 16, you will learn how to specify tolerances for specific dimensions and geometric tolerances.

Nomenclature and Conventions

Dimensions are used to accurately describe the details of a part or object so that it can be manufactured. In engineering drawing, dimensions are always placed outside of the object outline, unless placing the dimension on the object would result in a drawing that is easier to interpret. *Extension lines* relate the dimension to the feature on the part. There should always be a gap of 1/16" on the plotted drawing between the feature and the beginning of the

extension line, called the *extension line offset*, as shown in Figure 7.1. Centerlines can also be extended across the object outline and used as extension lines, without leaving a gap where they cross object lines.

Dimension lines are drawn between extension lines and have arrowheads at each end to indicate how the dimension relates to the feature on the object. Dimensions should be grouped together around a view and be evenly spaced to give the drawing a neat appearance. The numbers on the dimension line should never touch the outline of the object. *Dimension values* are usually placed near the midpoint of the dimension line, except when it is necessary to stagger the numbers from one dimension line to the next so that all of the values do not line up in a row. Staggering the numbers makes the drawing easier to read.

Since dimension lines should not cross extension lines or other dimension lines, begin by placing the shortest dimensions closest to the object outline. Place the longest dimensions farthest out. This way you avoid dimension lines that cross extension and other dimension lines. It is perfectly acceptable for extension lines to cross other extension lines.

When you are selecting and placing dimensions, think about the operations used to manufacture the part. When possible, provide *overall dimensions* that give the greatest measurements that exist for each dimension of the object, because this tells the manufacturer the starting size of the material used to make the part. The manufacturer should never have to add shorter dimensions or make calculations to arrive at the sizes needed for anything in the drawing. All necessary dimensions should be specified in your drawing, but no dimensions should be duplicated, as this may lead to confusion, especially when determining whether a part meets the specified tolerance.

AutoCAD's Semiautomatic Dimensioning

AutoCAD's semiautomatic dimensioning feature does much of the work in creating dimensions for you. You can create the extension lines, arrowheads, dimension lines, and dimension values automatically. In order to get the most power from the dimensioning feature, use associative dimensions. Associative dimensions are linked to their locations in the drawing by information stored on the layer DEFPOINTS that AutoCAD creates. Because of this, associative dimensions automatically update when the drawing is modified. You can also use nonassociative dimensions, but this is not a good practice because these dimensions are created from separate line, polyline, and text objects, and do not contain information allowing them to update. Associative dimensioning is the default in AutoCAD.

You should create drawing dimensions in model space. It is possible to add dimensions to paper space, but dimensions placed in paper space are associated to paper space locations, while the drawing should be in model space. If the drawing in model space is updated or moved, dimensions placed in paper space will not move with it, making it very hard to update your drawing.

Starting

Start by launching AutoCAD Release 13 for Windows. The AutoCAD drawing editor should appear on your display screen. You will begin this tutorial by drawing an object and dimensioning it. To begin a new drawing,

Pick: **New icon**

Name your drawing *obj-dim.dwg*. Start the drawing from the prototype drawing provided with the data files, *proto.dwg*. To select the prototype drawing, pick on the button with the word Prototype. Use the dialog box that

appears to change the directory to the one in which your data files are stored, and pick *proto.dwg*. When you have finished selecting the prototype and have typed in the name *obj-dim.dwg* for your new drawing, pick OK to exit the dialog box.

You return to the drawing editor with the border and settings you created in the prototype drawing on your screen.

Dimensioning a Shape

Review the object in Figure 7.1. You will draw this shape and then dimension it.

Layer VISIBLE should be the current layer in the drawing. If it is not, set layer VISIBLE as the current layer. Set Grid for .5 spacing and Snap for .25 spacing. Make sure that both are turned on.

Use the commands you learned in the previous tutorials to draw the object shown in Figure 7.1, according to the specified dimensions, on your own. It is not necessary to draw the centerlines now. Create the drawing geometry *exactly* in order to derive the most benefit from AutoCAD's semiautomatic dimensioning capabilities. Draw the object now. When you have finished, your screen should look like Figure 7.2.

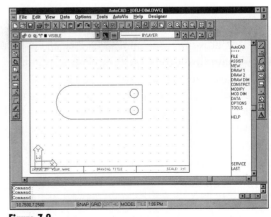

Figure 7.2

Using the DIM Layer

In order to produce a clear drawing of a subject, you draw the outline and visible lines of the object with a thick width pen when printing or plotting. The dimension, hidden, center, and hatch lines are plotted or printed with a thin width pen. This way your eye is drawn first to the bold shape of the object and then to the details of its size and other features.

As you recall from previous tutorials, in AutoCAD the plotted or printed pen width is set according to the color of the object on your drawing screen. In order for you to be able to select a thin pen for the dimension lines, they must be a different color than you used for object lines (to which you will assign a thick pen when plotting). Also, you will frequently want to turn off all of the dimensions in the drawing. Having dimensions on a separate layer makes color assignment and turning off dimensions easy tasks.

On your own, use the Object Properties toolbar to set DIM as the current layer.

Dimension Standards

There are standards and rules of good practice that specify how the dimension lines, extension lines, arrowheads, text size, and various aspects of dimensioning should appear in the finished drawing. Mechanical, electrical, civil, architectural, and weldment drawings, among others, each have their own standards. Professional societies and standards organizations publish drawing standards for various disciplines. The American National Standards Institute (ANSI) publishes a widely used standard for mechanical drawings. It contains various rules of good practice to help you create clear drawings that others can easily interpret.

For example, there should be a 1/16" space between the extension lines and the feature from which they are extended. The extension line should extend 1/8" past the last dimension line. Arrowheads and text should be approximately 1/8" tall on 8.5" × 11" drawings. The dimension line closest to the object outline should be at least 3/8" away from the object outline on your plotted drawing. Each succeeding dimension line should be at least 1/4" away from the previous dimension line.

Drawing Scale and Dimensions

When you are adding dimensions to your drawing, it is important to have already determined the scale at which you will plot your drawing on the sheet of paper so that you will know how to set up the dimension features to produce the correct sizes on the final plot. If you are going to plot from paper space, you should set the model space Zoom XP scaling factor *before* you dimension your drawing. This is already set to 1XP in *proto.dwg*, from which you started the current drawing. It is a good idea to check this by repeating the Zoom XP scale command option or by using Views, Floating Viewports, MV Setup from the pull-down menu and selecting Scale viewports.

On your own, switch to model space if you are not already there and make sure your drawing is zoomed to 1XP before continuing.

■ *TIP* Dimensions take a lot of drawing space, especially between the views of a multiview drawing. Often you may need to zoom your drawing to half-size (.5XP) or some other smaller scale to have room to add dimensions. Keep this in mind for future drawings. Do not change your XP scale factor for this tutorial. ■

Turning on the Dimensioning Toolbar

You can quickly select the dimensioning commands from the Dimensioning toolbar. You will show the Dimensioning toolbar by picking the Tools menu bar option.

Pick: **Tools, Toolbars, Dimensioning**

The Dimensioning toolbar floats on your screen, as shown in Figure 7.3.

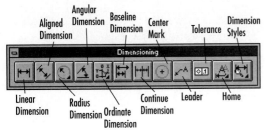

Figure 7.3

Using Dimension Styles

 Many features controlling the appearance of the dimensions are set by *dimension variables (dim vars)*, which you can set using the Dimension Styles dialog box from the Data option on the menu bar, or by typing Ddim at the command prompt. You can also call this dialog box using the Dimension Styles icon from the Dimensioning toolbar. Dimension variables all have names and you can set each one by typing its name at the command prompt. However, using the dialog box allows you to set many dimensioning variables at once, so you will select it from the Dimensioning toolbar.

Pick: **Dimension Styles icon**

The Dimension Styles dialog box appears on your screen, as shown in Figure 7.4.

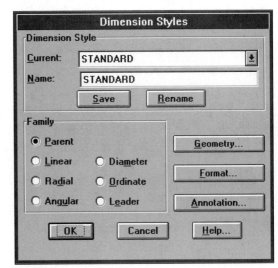

Figure 7.4

Creating a Named Dimension Style

The Dimension Styles dialog box allows you to easily change the dimension variables that control the appearance of the dimensions. Notice that the current style name is STANDARD. This basic set of features is provided as the default.

You can create your own *dimension style* with a name that you specify. This way you can save different sets of dimension features that will be useful for different types of dimensioning standards, for example, mechanical or architectural. You will create a dimension style named MECHANICAL by typing it in the box to the right of Name in place of the text STANDARD. Highlight the text STANDARD and to replace it,

Type: **MECHANICAL**

Pick: **Save (from the buttons below where you typed in MECHANICAL)**

Notice that MECHANICAL now appears as the current dimension style at the top of the dialog box, replacing STANDARD. Near the bottom of the dialog box you will see the message *Created MECHANICAL from STANDARD.*

MECHANICAL is now the current style and will be affected by changes that you make to the dimension variable settings.

Dimension styles can work three ways: as parent styles, children styles, or overrides. A parent style is the default selection when creating a new style. Setting the features of a parent style sets the characteristics for all of its children, until a different selection is made for the child.

After you create a parent style and set its characteristics, you can pick on a child type, such as linear, radial, angular, diameter, ordinate, or leader (listed in the Family section of the Dimension Styles dialog box), and then change the features for that child type. This allows you to vary the style for that particular type of dimension. For example, if you set up a child style for radial dimensions, it can be different from the parent style. With that parent style current, whenever you create a radial dimension, the child style for radial is automatically applied to vary the appearance of that radial dimension from the parent style.

When you change a style and pick the Save button, the changes are applied to all of the existing dimensions in the drawing that were created using that style. If you make changes to a style and do not pick the Save button, the changes are used as overrides that take effect on newly created dimensions, but not on existing dimensions using that style.

You will continue creating your parent style for MECHANICAL. (Make sure that the Parent radio button is still selected.) To set the dimension variables, you will use the selections for Geometry, Format, and Annotation on the right-hand side of the dialog box. The Geometry button allows you to control how the dimension lines, extension lines, arrowheads, and centermarks are drawn.

Pick: **Geometry**

The Geometry dialog box appears on your screen on top of the previous Dimension Styles dialog box. It should look like Figure 7.5.

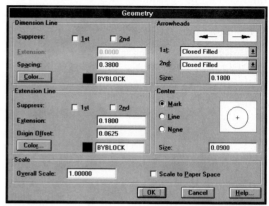

Figure 7.5

Setting the Features

In this dialog box you can change many features of the dimensions. An important dimensioning variable is the scaling factor that you see near the bottom of the dialog box. You should set this variable first before going on to set other size features, as it controls the sizes of many features at once. If you later determine that a different setting would be better, you can change the scaling factor at any time.

The Dimension Scaling Factor

There are essentially two ways to scale the dimension features: relative to model space or relative to paper space. Think about a drawing that is to be plotted at a scale of 1 plotted inch to 2 model units. If you want your arrowheads and text to be 1/8" on the plotted sheet, then at this scale the arrows need to be 1/4" tall in the drawing, so that when reduced by half they will still appear 1/8". To do this you could double all of the settings that control sizes of the dimension elements, such as arrowhead size, but as there are many of them, this would be time-consuming. Instead, you can automatically scale all of the size features at once by setting the overall scale.

The selection for overall scale near the bottom of the dialog box provides a scaling factor to use for all of the dimension variables that control the sizes of dimension elements. Its value appears in the box to the right of Overall Scale. Setting the overall scale to 2 doubles the size of all the dimension elements in the drawing, so that when plotted half-size, your arrowheads, text, and other elements will maintain the correct proportions on the final plot. (You can also type Dimscale at the command prompt and enter the value by which the dimension elements are multiplied.)

Picking Scale to Paper Space sets the scaling for dimension elements so that the values you enter in the dialog box are the sizes on the drawing when paper space is plotted at a scale of 1=1. The effect of doing this is that all of the dimension size features are multiplied by the reciprocal of the zoom XP factor.

■ *TIP* Using paper space scaling for dimensions can be tricky, but effective. If you zoom in on the view and add a dimension, the dimension elements will be smaller than the dimensions you add when you are not zoomed in. This happens because when you are using Scale to Paper Space, AutoCAD uses the zoom scaling factor to determine how to scale the dimension elements. When you zoom back out, the dimensions appear smaller. You can force the dimension to update by picking a dimension editing command. But it is hard to control the placement of dimensions when they update in this way. ■

For this tutorial, you will select a default overall scale factor of 1 for the dimensions, and you will not scale to paper space. Because you will plot your drawing full-size, zoomed to 1XP, the dimension features will be shown the same size you set them in the dialog box. When overall scale factor is used to size the dimension features, zooming does not affect the size of dimension elements.

Next, you will continue to use the Geometry dialog box to set the standard sizes for dimension features used in engineering drawings. The settings that you will make are based on an 8.5" × 11" sheet size. Larger drawing sheets use larger sizes.

Dimension Line Visibility and Size

In the Dimension Line area at the upper left of the dialog box, you can control the appearance of the dimension lines. You can choose to suppress the 1st or 2nd dimension line when the dimension value divides the dimension line into two parts. You can use the box to the right of Extension to set the value for a distance that the dimension line should extend beyond the extension line. This area is grayed out and cannot be selected, unless you are using *oblique stroke arrows* as the type of arrowhead. The value entered in the box to the right of Spacing controls the distance between successive dimensions added using baseline dimensioning. The default value, .38, will work for now. (The ANSI standard states that successive dimensions should be at least 1/4" or .25" decimal, so .38 will meet this criteria.) Next you will use the Color selection to set the color by layer.

Pick: **Color**

The standard Select Color dialog box appears on your screen.

Pick: **BYLAYER**

Pick: **OK**

You are returned to the Geometry dialog box.

Arrow Size and Style

To set the arrow size, use the upper right area of the dialog box. On your own, change the value in the box to the right of the word Size. Use the text box to replace the default value, 0.1800, with 0.125, the decimal equivalent of 1/8". Refer to Figure 7.6.

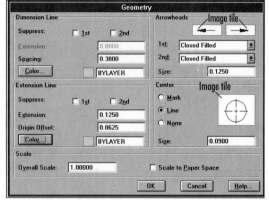

Figure 7.6

You can set the style of the arrow two different ways. One way is to pick on the selection Closed Filled, or on the downward-pointing arrow to the right of the Closed Filled selection. A list of the available arrow styles from which you can choose appears on your screen. You can also pick on the *image tile*, or active picture of the arrow, near the top of the Arrowheads area and scroll through the available arrow styles.

Pick: *(on the image tile of either arrow)*

The tile changes to show an oblique tick for the arrow on that side. Notice that Oblique now appears in the box to the right of whichever arrow you picked. (If you picked the first arrow, both changed). On your own, continue picking on the picture of the arrow until it has returned to the style Closed Filled.

Continue to refer to Figure 7.6 for the remaining selections.

Extension Line Visibility and Size

In the Extension Line area of the dialog box, you can suppress the first, second, or both of the extension lines while you are dimensioning. To do this you would pick in the box to the left of 1st to suppress the first extension line, for example. Leave this setting unchanged. The value in the box to the right of the word Extension sets the distance that extension lines extend past the dimension line. The usual setting for this on 8.5" × 11" paper is .125". Use the same method you used when setting the arrow size to change the value to .125. The box to the right of Origin Offset controls the distance from the edge of the selected feature that the extension line starts when dimensioning. As this is commonly 1/16", leave this value, .0625 (the decimal equivalent for 1/16"), unchanged. Next you will set the color for extension lines so that they are drawn the color of the layer on which they appear.

Pick: **Color**

Use the Select Color dialog box to pick the BYLAYER button on your own. Pick OK to return to the Geometry dialog box. When you have returned to the Geometry dialog box, you will notice that the color shown in the box to the right of Color is now cyan and the box to its right says BYLAYER.

Centermark Size and Style

When you create radius or diameter dimensions, you can use the Center area of the Geometry dialog box to control the size and style of the centermark, or to suppress it so that no centermark is created. If the value for centermark size is too large, AutoCAD will not draw full centermarks; instead it will draw just the tick mark in the center of the circle. Picking the radio button to the left of Line will cause full centermarks to be drawn, and picking the radio button to the left of None will cause no center-marks to be drawn. Picking Mark will cause a tick to be drawn at the center.

Pick: **Line**

The button will become filled in and the image tile will show a full set of centermarks, as shown in Figure 7.6. You can also set the cen-termark style by using the image tile on the right of this area.

Pick: (on the image tile showing the full centermarks)

The centermarks disappear from the circle and the radio button for None becomes high-lighted. You will pick on the picture again to cycle back so that just a centermark or tick will be drawn.

Pick: (on the image tile where there are now no centermarks)

A tick or mark appears at the center of the circle.

Pick: (on the image tile again so that full centermarks are shown)

Leave the selection set to show the full center-marks. You set the size of the centermark using the box to the right of the word Size. The value for Size sets the distance from the center to the end of one side of the center cross of the cen-termark. To make a smaller center cross, set it to a smaller value; for a larger center cross, set the size larger. Leave the value for your center-marks set to the default value, .09, for now. Center marks can also be controlled by typing Dimcen at the command prompt as you did in Tutorial 6. When you have finished making the selections as shown in Figure 7.6, you are ready to return to the previous dialog box.

Pick: **OK**

You return to the Dimension Styles dialog box. Next you will set up the format for the appear-ance of text and dimensions added to the drawing.

Pick: **Format**

The Format dialog box shown in Figure 7.7 appears on your screen. It contains selections that allow you to control the placement of the text, arrows, and leader lines relative to the dimension and extension lines.

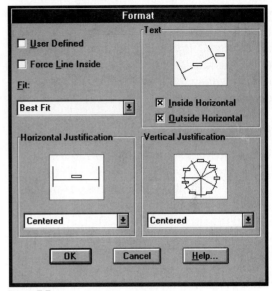

Figure 7.7

Fit

The upper left area of the dialog box contains selections for user-defined text placement, and for forcing a line to be drawn between the exten-sion lines even when text and arrows are placed outside the extension lines. If you select User Defined, when you pick or type the location for the dimension line, the dimension value will be located at that position. The default option for Fit, Best Fit, places *both* the text and arrows inside the extension lines, if possible. If that is not possible, *either* the text or the arrows will be placed inside the extension lines. If neither will fit, the arrows and text are both placed outside the extension lines. Generally these are all accepted practices as long as the dimension can be correctly interpreted. Leave this selection set to Best Fit.

Text Orientation

The Text area of the Format dialog box controls whether text always reads horizontally (i.e., is unidirectional text) or whether it will be aligned with the dimension line. You can control this individually for dimension values placed inside the extension lines or outside the extension lines. Use the image tile for this area to scroll through these options to see the effect each would have. When you are finished experimenting, both settings should be oriented horizontally.

Horizontal Justification

Using the Horizontal Justification area of the dialog box, you can choose to have text centered, placed near the first extension line or second extension line, or over the first or second extension line. The default choice, Centered, is usually selected for mechanical drawings. Pick on the image tile in the Horizontal Justification area so that you can see the effects of these selections. When you are finished, return it to the default choice, Centered.

Vertical Justification

The final area of the Format dialog box controls the vertical justification of text relative to the dimension line. The options for vertical justification are Centered, Above, Outside, and JIS (Japanese Industrial Standard). For the ANSI standard mechanical drawings, select Centered to create dimension values that are centered vertically, breaking the dimension line. The ISO (International Standards Organization) standard usually uses the options for text above or outside the dimension line. The JIS option orients the text parallel to the dimension line and to its left. (To see a picture of this, pick JIS from the menu and pick the Help button at the bottom right of the dialog box. Help for Format appears on your screen. Pick on the

highlighted option Vertical Justification to see its help screen. Remember to exit Help when you are done.) When you are finished examining this portion of the dialog box, return its setting to Centered on your own.

Pick: **OK**

You return to the Dimension Styles dialog box. Next you will set up the appearance of annotation, or text for the dimensions.

Pick: **Annotation**

The Annotation dialog box shown in Figure 7.8 appears on your screen.

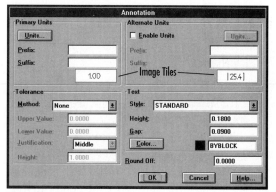

Figure 7.8

Primary Units

You use the Primary Units area of the dialog box to set up the number of decimal places, prefix, and suffix for dimensions. When you are dimensioning a drawing, it is important to consider the precision of the values used in the dimensions. Specifying a dimension to four decimal places, which is AutoCAD's default, implies that accuracies of 1/10,000th of an inch are appropriate tolerances for this part. Standard practice is to specify decimal inch dimensions to two decimal places (an accuracy of 1/100th of an inch) unless the function of the part makes a tighter tolerance desirable.

Next you will set up the display of units in dimension values.

Pick: **Units**

The Primary Units dialog box shown in Figure 7.9 appears on your screen. You can use it to select the type of units and set the number of decimal places for dimension values. It also contains selections to suppress leading or trailing zeros in the dimension value. You can also suppress dimensions of 0 feet or 0 inches. In decimal dimensioning, it is common practice to show leading zeros in metric dimensions and suppress them in inch dimensions when the dimension is less than 1 unit. For this tutorial you will leave the type of units set to Decimal.

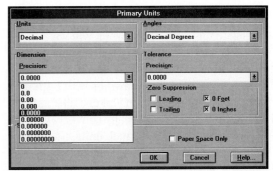

Figure 7.10

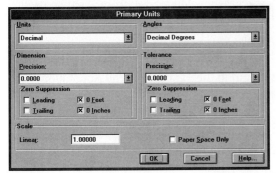

Figure 7.9

Setting Dimension Precision

Pick: (on 0.0000 shown in the Dimension area below Precision)

A list of decimal places pulls down from the selection, as shown in Figure 7.10. The current selection shows four places after the decimal, which is AutoCAD's default, but not good engineering practice in dimensioning. Pick on the selection 0.00 to set the number of decimal places to two, which is common practice when greater dimension precision is not required. You will learn about using tolerances in Tutorial 16. For now, leave the tolerance precision set as it is.

Angles

The type of units used to dimension angles can be set using the Angles area of the Primary Units dialog box. To select, pull down the list and select from the types of units shown. (You can further control the appearance of angular dimensions by selecting the Angular child style in the Dimension Styles dialog box.)

Zero Suppression

Zero Suppression is available for both dimensions and tolerances, in four categories: leading 0s, trailing 0s, 0 feet, and 0 inches. Selecting the box to the left of any of these words suppresses those zeros; an X appears in the box when it is selected. Pick the boxes to the left of Leading to suppress leading 0s for both dimensions and tolerances.

Scale

The Scale area of the Primary Units dialog box is used to specify a scaling factor by which all of the actual dimension values are multiplied. This is different than the overall scaling factor because it is the number shown as the dimension that is affected, not the sizes of the arrowheads and other elements. Leave the value to the right of Linear in the Scale area set to 1.0000.

Pick: **OK** *(to exit the Primary Units dialog box)*

Alternate Units

Alternate units are useful when you want to dimension a drawing using more than one system of measurement. For example, it is fairly common to dimension drawings with both metric and inch values. When you use alternate units, AutoCAD automatically converts your drawing units to the alternate units, using the scaling factor you provide. You can enable alternate units by picking the box to the left of Enable Units in this portion of the Annotation dialog box. Until you select Enable Units, the Units button is grayed out. To set the scale for the alternate units, pick Enable Units, pick Units, then use the linear scale selection in the Alternate Units dialog box that appears. You can also type a prefix and suffix for the alternate units in the appropriate boxes. The prefix you enter will appear in front of the dimension value; the suffix, after the dimension value. You can control the appearance of the alternate units by picking on the image tile, where you see the value [25.4]. The default is that the alternate units are placed inside square brackets following the normal units. You will not enable alternate units now.

Tolerance Method

You use the Tolerance portion of the Annotation dialog box to specify limit or variance tolerances with the dimension value. You can set the tolerance method to Symmetrical, Deviation, Limits, Basic, or None. The values shown in the boxes identified in Figure 7.8 are image tiles that display the units and tolerance method selected. Pick on the image tile just above the Tolerance box (in the Primary Units area) to scroll through the selections for tolerance so that you can see the effect of these settings. For now leave the setting for the tolerance method set to None.

Text

You can control the text style, height, color, and the gap between the end of the dimension line and start of the text with the Text area of the Annotation dialog box. You will select style MYTEXT.

Pick: (on STANDARD)

On your own, use the list of text styles to select MYTEXT as the style for dimension text. Remember that in order for styles to be listed here, they must have been previously created in the drawing with the Style command. Style MYTEXT was previously created in the prototype drawing you started from. Next you will set the text height to .125, the standard height for 8.5" × 11" drawings.

On your own, highlight the value to the right of Height in the Text area of the Annotation dialog box.

Type: .125

Now you will set the color for the dimension text to BYLAYER.

Pick: Color

On your own, use the dialog box that appears to select BYLAYER to set the color for dimension text and return to the Annotation dialog box.

Pick: OK (to exit the Annotation dialog box)

Now you have set up your basic sizes for the dimensions.

Saving the Parent Style

In order to save your changes to dimension style MECHANICAL, you must be sure to pick the Save button near the top of the Dimension Styles dialog box, just under the name of the style. If you do not pick Save to save the changes to the dimension style and then you pick OK to close the dialog box, the changes will be applied as overrides. When a change is used as an override, any new dimensions that are created will use those changes, but dimensions that have already been created with that style name will not be affected.

Pick: **Save**

Note the message *Saved to MECHANICAL* at the bottom of the dialog box. The selections you have made apply to dimension style MECHANICAL only. As you dimension your drawing, you can return to this dialog box and create a new style if you need to create dimensions with a different appearance.

Pick: **OK (to exit the Dimension Styles dialog box)**

Using Dimstyle

As you have seen, there are many settings that you can change to affect the appearance of the dimensions you add to your drawings. You can set these variables either through the Dimension Styles dialog box or by setting each individual dimension variable at the command line. You can use the Dimstyle command to list and set dimension styles and variables at the command line. If any dimension variable overrides are set, they will be listed after the name of the dimension style. There should not be any overrides set for style MECHANICAL because you have saved your settings to the parent style.

Command: **DIMSTYLE** ⏎

Dimension Style Edit (Save/Restore/STatus/Variables/ Apply/?)<Restore>: **ST** ⏎

The dimension variables and their current settings are listed in the text window on your screen. On your own use the scroll bars to scroll up to see entries and follow the *Return* prompts to look through the entries.

Associative Dimensioning

The Dimaso variable controls whether *associative dimensioning* is turned on or off. Associative dimensioning means that each dimension is inserted as a block or group of drawing objects relative to the points selected in the drawing. If the drawing is scaled or stretched, the dimension values automatically update. This is a very useful feature and Dimaso generally should be on. Also, dimensions created with Dimaso on are automatically updated if you make a change to their dimension styles. When Dimaso is turned on (which is the default), the entire dimension acts as one object in the drawing. Dimensions created with Dimaso turned off cannot be updated, but their individual parts, such as arrowheads or extension lines, can be erased or moved. Make sure that Dimaso is on when you view your dimension style variables.

When you are finished, use the Windows Control box to close the text window.

Now you are ready to start adding dimensions to your drawing. Look at the status bar. You will notice that the coordinates still display the cursor position with four decimal places of accuracy, even though you have selected to display only two decimal places in your dimensions. AutoCAD can keep track of your drawing and settings you have made in the drawing database to a precision of at least 14 decimal places, but the dimensions, which you will add to your drawing in the next steps, will only be shown to 2 decimal places as set in the current style. Display of the decimal places on the status bar is set independently of the dimension precision (using the Units command you learned in Tutorial 1). This is useful because you can still create and display an accurate drawing database while working on the design, but dimension according to the often lower precision required to manufacture acceptable parts.

Adding Linear Dimensions

 The Linear Dimension command measures and annotates a feature with a horizontal or vertical dimension line. The value inserted into the dimension line is the perpendicular distance between the extension lines. If you use the Linear Dimension command and dimension a line drawn at an angle on the screen, the value AutoCAD returns is just the X or Y axis component of the length. The dimension is adjusted to horizontal or vertical, depending on the points selected.

Make sure Snap is on to help you locate the dimensions .5 units away from the object outline (thus meeting the criterion that a dimension must be at least 3/8", or .375, away from the object outline). On your own, turn the Intersection running mode on. You will use this to select the exact intersections in the drawing for the extension lines so that the dimensions are drawn accurately.

You will dimension the horizontal distance from the end of the block to the center of the upper hole. Refer to Figure 7.11 for your selections.

Pick: **Linear Dimension icon**

First extension line origin or RETURN to select: *(pick point 1 at the top right-hand corner)*

Second extension line origin: *(pick Snap to Quadrant icon)*

qua of: *(pick on the upper small hole, near point 2)*

Dimension line location (Text/Angle/Horizontal/ Vertical/Rotated): *(pick a point two snap units above the top line of the object)*

The dimension should appear in your drawing, as shown in Figure 7.11.

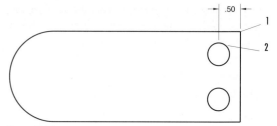

Figure 7.11

Next you will create the diameter and radius dimensions in your drawing. Usually you will not want the text and arrows inside the dimension when creating diameter and radius dimensions. Using the child settings makes it easy to make certain types of dimensions vary somewhat from the parent style. You will activate the Dimension Styles dialog box using the Dimensioning toolbar.

Pick: **Dimension Styles icon**

The Dimension Styles dialog box appears on your screen. You will use it to set the child styles for radial and diameter dimensions to be different from the parent style MECHANICAL. The style MECHANICAL should be listed as the current style at the top of the Dimension Styles dialog box.

Setting a Child Style

To create a child style for the diameter that is different than the parent,

Pick: **Diameter**

as shown in Figure 7.12.

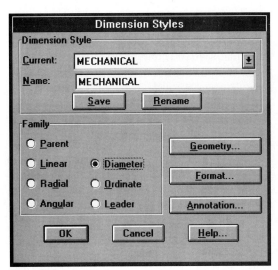

Figure 7.12

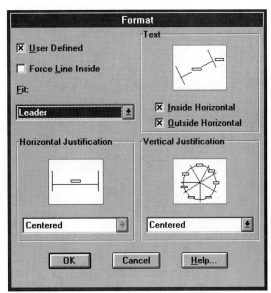

Figure 7.13

You want most of the settings to remain the same as the parent style; you will only change the items that should be different between radial dimensions and other dimensions. You can think of parent styles and child styles as similar to inherited genetic characteristics. Children inherit their appearance from their parents, but each child may appear different from the parents. That is why you set the general appearance first, using the Parent selection. Then you set up the specific differences for each child's appearance after the parent style has been created. *Once a child style has been set, changing the parent style will no longer change the child style.*

Pick: **Format**

Use the methods that you have learned to complete the dialog box that appears so that it looks like Figure 7.13. Pick User Defined for the placement of text, and Leader for the fit. These selections will allow you to place a leader to locate the dimension value off the object when you add a diameter dimension. When you have finished this step on your own, pick OK to return to the Dimension Styles dialog box.

Adding Special Text-Formatting Characters

When you are using AutoCAD's shape fonts (like Roman Simplex) to create your text, you can type in special characters during any text command and when entering the dimension text.

You can add special text characters to the dimensioning text when you are using AutoCAD's shape fonts by preceding the text with the code %% (double percent signs). The most common special characters have been given letters so that they are easier to remember. Otherwise, you can use any special character in a text font by typing its ASCII number. (Most word processor documentation includes a list of ASCII values for symbols in an appendix.)

The most common special characters and their codes are listed in the table below:

Code	Character	Symbol
%%C	diameter symbol	Ø
%%D	degree symbol	°
%%P	plus/minus sign	±
%%O	toggles on and off the overscore mode	Example
%%U	toggles on and off the underscore mode	Example
%%%	draws a single percent sign	%
%%N	draws special character number *n*	*n*

You will use the special character %%C to draw the diameter symbol ahead of the diameter dimensions.

Next you will type in the special character %%C as a prefix in the Annotation dialog box. This way, whenever you create a diameter dimension, the Ø symbol will automatically precede the value. Adding the Ø symbol ahead of the dimension value for diameter dimensions (and R for radial dimensions) is common in metric dimensioning. It is a practice that is becoming common in decimal inch dimensioning. Inch dimensions often have the abbreviations DIA. (for diameter) and R (for radius) following the value.

Pick: **Annotation**

In the Primary Units area of the dialog box, use the text entry box for Prefix and type in the special character code.

Type: **%%C (in the text box to the right of Prefix)**
Pick: **OK (to return to the previous dialog box)**

Saving the Changes

You must save the changes to the style in order to use it. You cannot use child styles as overrides. To save the changes,

Pick: **Save**

Next you will set up the radial child type in a similar way.

Pick: **(the radio button to the left of Radial)**
Pick: **Format**

On your own, set the dialog box so that User Defined is used for the placement of text, and Leader for fit. When you have finished this step, pick OK to return to the Dimension Styles dialog box.

Pick: **Annotation**

In the Primary Units area of the dialog box, for the prefix,

Type: **R**
Pick: **OK (to exit the Annotation dialog box)**
Pick: **Save**
Pick: **OK (to exit the Dimension Styles dialog box)**

Now you are ready to add the diameter dimensions for the two small holes.

Creating a Diameter Dimension

The Diameter Dimension selection adds the centermarks and draws the *leader line* from the point you select on the circle to the location you select for the dimension value. The leader line produced is a radial line. This means that if extended, it would pass through the center of the circle. Keep in mind that this only happens when you set the dimension variables so that you control the placement of the text to use a leader and be user defined, as you did when setting up the child styles for the diameter and radial dimensions.

You will pick the Diameter Dimension icon from the Dimensioning toolbar. It is on the Radius Dimension flyout shown in Figure 7.14. After you have entered the dimension, you will use the Text option of the Diameter Dimension command to add an additional line of text below the dimension value. Refer to Figure 7.16 for the placement of the dimension.

Radius Dimension flyout

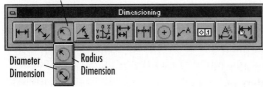

Diameter Dimension   Radius Dimension

Figure 7.14

Pick: **Diameter Dimension icon**

Select arc or circle: **(pick on the lower of the two small circles)**

■ *TIP* Make sure Snap and Ortho are off to make it easier to select the circles and draw angled leader lines. ■

Dimension line location (Text/Angle): **T** ⏎

The Edit MText dialog box shown in Figure 7.15 appears on your screen. You may see a message that the font you have selected is not available in Windows; if so, substitute a different font for use in the Edit MText dialog box on your own.

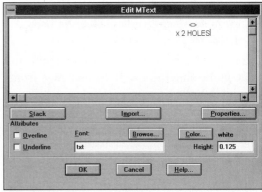

Figure 7.15

You will use the dialog box to add the text *x 2 HOLES* below the <> brackets indicating the default text. When the dimension is added to your drawing, the <> brackets will be replaced with the diameter value for the hole contained in your drawing database.

On your own, position the cursor after the <> brackets and press ⏎ to create a new line for text.

Type: **x 2 HOLES**
Pick: **OK**

■ *TIP* You can replace the angle brackets with text, or with a value for the dimension. If you do this, however, your dimension values will not automatically update when you change your drawing. It is a better practice to create your drawing geometry accurately and accept the default value that AutoCAD provides represented by the angle brackets. ■

Continue creating the diameter dimension. Next you will be prompted for the dimension line location.

Dimension line location (Text/Angle): *(pick a point below and to the left of the circle at about 7 o'clock, .5 outside the object outline)*

Use the grid as a visual reference to help you position the dimension text .5 units away from the object outline.

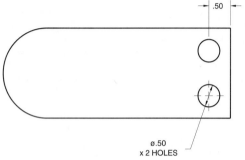

Figure 7.16

> ■ *TIP* When dimensioning a full circle of 360° (as opposed to an arc), always use the diameter command, rather than the radius. This is important because equipment used to manufacture and inspect holes and cylinders is designed to measure diameter, not radius. The machinist who is making the part should never have to calculate any of the dimensions. This includes not having to double the radius to arrive at the diameter. It is very difficult to measure the radius of a hole, because its center has already been drilled out. ■

Next, you will add a centermark for the upper of the two small holes.

Pick: **Center Mark icon**

Select arc or circle: *(pick on the edge of the top circle)*

The centermark is added to the drawing, as shown in Figure 7.17.

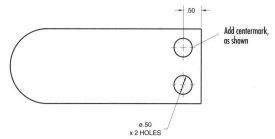

Figure 7.17

 Now you will use the Radius Dimension command to create the radius dimension and centermarks for the rounded end. You will use the Radius Dimension flyout on the Dimensioning toolbar to select the command.

Pick: **Radius Dimension icon**

Select circle or arc: *(pick on the rounded end)*

Dimension line location (Text/Angle): *(pick a point above and to the left of the circle at about 10 o'clock, .5 outside the object outline)*

The radial dimension and centermark for the arc are added. Your drawing should look like Figure 7.18.

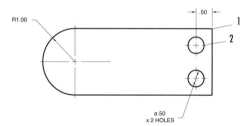

Figure 7.18

Next you will add a linear dimension for the vertical distance from the upper edge of the part to the center of the top hole. Refer to Figure 7.18 as you make selections.

Pick: **Linear Dimension icon**

First extension line origin or RETURN to select: *(pick point 1 at the top right-hand corner)*

Second extension line origin: *(target point 2)*

Dimension line location
(Text/Angle/Horizontal/Vertical/Rotated): **(pick a point two snap units to the right of the object)**

The dimension added to your drawing should look like Figure 7.19.

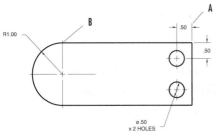

Figure 7.19

Next you will use baseline dimensioning to add the horizontal dimension between the edge of the part and the rounded end.

Baseline Dimensioning

Baseline and chained dimensioning are two different methods of relating one dimension to the next. In *baseline dimensioning*, as the name suggests, each succeeding dimension is measured from one extension line, or baseline. In *chained* or *continued dimensioning*, each succeeding dimension is measured from the last extension line of the previous dimension. Baseline dimensioning can be more accurate, because the tolerance allowance is not added to the tolerance allowance of the previous dimension, as it is in chained dimensions. However, chained dimensioning may often be preferred because the greater the tolerances allowed, the cheaper the part should be to manufacture. The more difficult the tolerance is to achieve, the more parts will not pass

inspection. Figure 7.20 depicts the two different dimensioning methods. Notice that if a tolerance of +/− .01 is allowed, the major size of the baseline dimensioned part can be as large as 4.26 or as small as 4.24. However, using chained dimensions, an acceptable part could be as large as 4.27, or as small as 4.23.

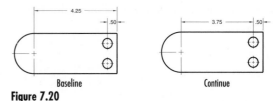

Figure 7.20

AutoCAD provides commands to make baseline and chained dimensioning easy.

Use the Baseline Dimension command to create the next dimension. It is preferable to add dimensions with Baseline or Continue, because adding a second dimension using only the Linear Dimension command will draw the extension line a second time, which will give a poor appearance to your drawing when it is plotted. Refer to Figure 7.19 as you make selections. You will use the *RETURN to select* option to tell AutoCAD which dimension you want to use as the base dimension. Since you have added several dimensions since the horizontal location for the upper small hole, you will need to specify that it is the base dimension before you can create the baseline dimension. If you were continuing from the previous dimension, you would not need to do this.

Pick: **Baseline Dimension icon**

Second extension line origin or RETURN to select: ⏎

Select base dimension: **(pick A for the base dimension)**

Second extension line origin or RETURN to select: **(pick**
   intersection B)

Second extension line origin or RETURN to select: ⏎

Select base dimension: ⏎

The new dimension should appear, as shown in
Figure 7.21. Notice that AutoCAD automatically selected the location for the dimension
based on the settings of your dimension style
variables.

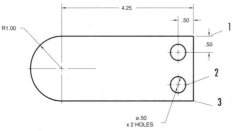

Figure 7.21

Next you will add baseline dimensions for the
vertical location of the lower hole and for the
overall height. Refer to Figure 7.21 for the
points to select.

Pick: **Baseline Dimension icon**

Second extension line origin or RETURN to select: ⏎

Select base dimension: **(pick 1 for the base dimension)**

Second extension line origin or RETURN to select: **(pick**
   intersection 2)

Second extension line origin or RETURN to select: **(pick**
   intersection 3)

Second extension line origin or RETURN to select: ⏎

Select base dimension: ⏎

Your drawing looks like Figure 7.22.

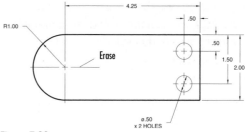

Figure 7.22

Using Xplode

The Xplode command lets you change dimensions, blocks, polylines, and other grouped
objects back into their individual components
and at the same time to control the color, layer,
and linetype of the components. You will use
Xplode so that you are able to erase just the
single right line of the centermark for the
rounded end. You will select the Layer option
of the Xplode command and the default layer,
DIM, for the exploded objects to be on.

AutoCAD also has a command named Explode,
which doesn't have any options. It simply
changes the grouped objects back into individual objects, regardless of the X, Y, and Z scale
factors. (In previous versions of AutoCAD the
X, Y, and Z scaling factors of a block had to be
the same or it could not be exploded.) Refer to
on-line help to further investigate the differences between these two commands on your
own.

Because the rounded end is an arc and not a
full circle, its centermark should only extend
for the half-circle shown. The right portion of
the centermark needs to be erased. Because
associative dimensions are created as blocks,
they are all one object; no part can be erased

singly. To eliminate the extra portion of the centermark on the rounded end, next you will type the Xplode command.

Command: **XPLODE** ↵

Select objects to XPlode.

Select objects: **(pick on the dimension for the rounded end)**

Select objects: ↵

All/Color/LAyer/LType/Inherit from parent block/<Explode>: **LA** ↵

XPlode onto what layer? <DIM>: ↵

You will see the message *Object exploded onto layer DIM* at the command prompt. There is no change in appearance of the drawing, but now you are able to erase the individual line at the right of the centermark for the rounded end.

On your own, erase the extra line from the centermark for the rounded end and save your drawing at this time.

Switch to paper space on your own and add the note *ALL TOLERANCES ± .01 UNLESS OTHERWISE NOTED*. Use the %%P special character to make the ± sign. Below the tolerance note add notes indicating the material from which the part is to be made, such as *MATERIAL: SAE 1020*, and a note stating *ALL MEASUREMENT IN INCHES*. Use the Edit Text command to change the text in your title block, noting the name of the part, the drafter's name, and the scale, to complete the drawing. Save your drawing on your own at this time.

Saving As a Prototype

Any drawing you have created can be used as a prototype from which you can start new drawings. You will use Save As and save your *obj-dim.dwg* with a new name so that you can use it as a prototype to start other drawings.

On your own, erase the object and all dimensions and set layer VISIBLE as the current layer. Use Save As to save this drawing as

dimproto.dwg. Drawings started from *dimproto.dwg* will already have the dimension styles you created earlier in this tutorial available to use.

Dimensioning the Adapter

You will continue to work with the orthographic views of the adapter from Tutorial 6 by adding dimensions to the drawing. Pick the New icon from the Standard toolbar and start a new drawing. In the dialog box, pick on Prototype and select the file *adapter1.dwg* from the data files as the prototype drawing. In the drawing name box, type the name *adapt-dm.dwg*.

Your screen should look like Figure 7.23. This is the same drawing you did in Tutorial 6, except that the holes have not been added and the views have been moved apart to make room for dimensions.

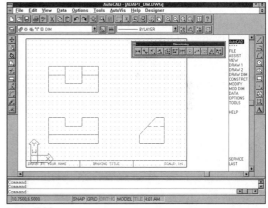

Figure 7.23

Dimensions are usually placed between views when possible, except for the overall dimensions, which are often placed around the outside.

Pick: **Dimension Styles icon**

Use the dialog box to select dimension style MECHANICAL. To select it, pick on the name STANDARD, as shown in Figure 7.24, and then select MECHANICAL from the list of dimension styles that pulls down. Its name should then appear as the current style. If it is not the current style, pick it before selecting OK.

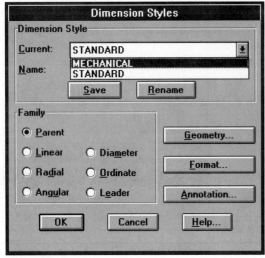

Figure 7.24

Pick: **OK** *(to exit the dialog box)*

You will check to see that associative dimensioning is turned on. Later in the tutorial you will use the feature to automatically update dimensions, and dimensions cannot be updated unless they were created with Dimaso turned on. You will type Dimaso at the command prompt.

Command: **DIMASO** ⏎

New value for DIMASO <On>: *(if the default value is Off, then type* **On** *⏎; otherwise, accept the default)* ⏎

Turn on the Intersection running mode. Check the Object Properties toolbar to see that you are on layer DIM. If not, make it the current layer on your own.

Add the horizontal dimension that shows the width of the left-hand portion of the block. The shape of this feature shows clearly in the front view, so add the dimension to the front view, placed between the views. You will use the

RETURN to select option, which allows you to pick an object from the screen instead of specifying the two extension line locations. AutoCAD will automatically locate the extension lines at the extreme ends of the object you select.

Pick: **Linear Dimension icon**

First extension line origin or RETURN to select: ⏎

Select object to dimension: *(pick the top left-hand line in the front view)*

Dimension line location (Text/Angle/Horizontal/Vertical/Rotated): *(pick .5 units above the object outline)*

Your drawing should appear similar to Figure 7.25.

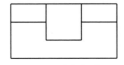
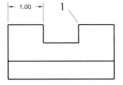

Figure 7.25

Using Continue Dimension

Now use AutoCAD's Continue Dimension command to add a chained dimension for the size of the slot. You will use the *RETURN to select* option again to choose which dimension is to be continued. The Baseline Dimension command has a similar option so that if you have added other dimensions, you can go back and select a different dimension to act as the base dimension.

When using this feature, pick near the extension line you want to have continued (or be the baseline). Refer to Figure 7.25.

Pick: **Continue Dimension icon**

Second extension line origin or RETURN to select: ⏎

Select continued dimension: *(pick the right extension line of the existing dimension)*

Second extension line origin or RETURN to select: *(pick point 1)*

Second extension line origin or RETURN to select: ⏎

Select continued dimension: ⏎

The chained dimension should appear in your drawing, as shown in Figure 7.26.

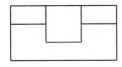

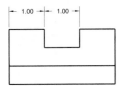

Figure 7.26

Adding the Angular Dimension

Before you add the angular dimension, you will use the Dimension Style dialog box to change the style for the Angular child dimension.

Pick: **Dimension Style icon**

Pick: (the radio button next to Angular)

Pick: **Annotation**

Pick: **Units** *(in the Primary Units area)*

Change the precision for the dimension by picking the box that shows 0.00, then selecting 0. This will cause AutoCAD to display the angle's dimension to zero decimal places.

Pick OK to exit the Units dialog box, then pick OK again to exit the Annotation dialog box. In the Dimension Style dialog box, be sure to pick

Save before you exit the dialog box so changes are saved to the MECHANICAL dimension style.

 Add the angular dimension for the angled surface in the side view, referring to Figure 7.26.

Pick: **Angular Dimension icon**

Select arc, circle, line or RETURN: *(pick line 1)*

Second line: *(pick line 2)*

Dimension arc line location (Text/Angle): *(pick near point 3)*

The angular dimension is added to the side view, as shown in Figure 7.27. You also have the option during the command of pressing ⏎ instead of selecting the first line. If you do so, you will be prompted for three points to define the angle. You do not have to use the %%D special character to make the degree sign; AutoCAD inserts it automatically unless you override the default text.

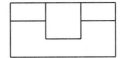

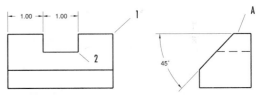

Figure 7.27

Add the dimension for the depth of the top surface in the side view.

Pick: **Linear Dimension icon**

First extension line origin or RETURN to select: ⏎

Select object to dimension: *(pick the short horizontal line labeled A in Figure 7.27)*

Dimension line location (Text/Angle/Horizontal/Vertical/Rotated): *(pick .5 units above the top line of the side view)*

Now add the dimension for the height of the slot in the front view. Refer to Figure 7.27.

Pick: **Linear Dimension icon**

First extension line origin or RETURN to select: *(pick the intersection labeled 1)*

Second extension line origin: *(pick the right-hand bottom corner of the slot, labeled 2)*

Dimension line location (Text/Angle/Horizontal/Vertical/Rotated): *(pick .5 units away from the right-hand side of the front view)*

The dimension for the slot should appear in your drawing, as shown in Figure 7.28.

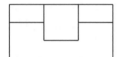

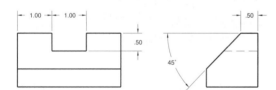

Figure 7.28

On your own, add the overall dimensions to the outsides of the views, so that your drawing looks like Figure 7.29.

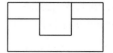

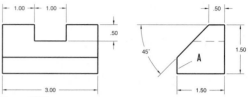

Figure 7.29

■ *TIP* Notice that you do not need dimensions for the short vertical line in the side view, labeled A. This is because its length is already defined by the overall dimension for the height of the part, the .50 distance across the top surface, and the 45° angle. To include this dimension would be an example of over-dimensioning. You will learn more in the advanced dimensioning tutorial about why over-dimensioning must be avoided. If you want to give this dimension, include the value in brackets, or followed by the word REF, to indicate that it is a reference dimension only. ■

Adding the Tolerance Note

Use the Pan command on your own to move the views of the object up in the viewport to make room near the bottom for drawing notes. Next, switch to paper space. Use the Object Properties toolbar to make layer TEXT current.

Now use the text command to add .125" text stating *ALL TOLERANCES ARE ± .01 UNLESS OTHERWISE NOTED*. Use the %%P special text character to create the symbol. Add a second note below the first, stating *MATERIAL: SAE 1020* and a third that says, ALL MEASUREMENTS ARE IN INCHES. Edit the drawing title.

If you want, plot your drawing while in paper space. Plot the drawing limits at scale 1=1.

On your own, switch back to model space.

Save your drawing *adapt-dm.dwg* before you continue.

Your finished drawing should appear similar to Figure 7.30.

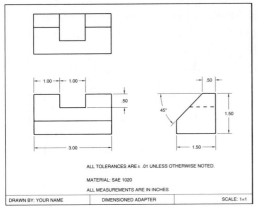

Figure 7.30

■ *TIP* You can trim and extend linear and ordinate dimensions using the Trim and Extend commands. When trimmed or extended, the dimension acts as though there were a line between the endpoints of the extension line origins. Trimming and extending takes effect on the dimension at that location. ■

Automatically Updating Dimension Values Created with Associative Dimensioning

Dimensions created with associative dimensioning (Dimaso) turned on are an AutoCAD block object. Blocks are a group of objects that behave as one. If you try to erase a dimension that was created with associative dimensioning on, the entire dimension is erased, including the extension lines and arrowheads. Because these dimensions have the special properties of a block, they also can be updated automatically.

Using Stretch

 The Stretch command is used to stretch and relocate objects. It is similar to stretching with hot grips which you did in Tutorial 4. You *must* use implied Crossing, Crossing, or Crossing Polygon methods to select the objects for use with the Stretch command. When selecting, objects entirely enclosed in the crossing window specified will be moved to the new location rather than stretched. Keep this in mind and only draw the crossing box around the portion of the object you want to stretch, leaving the other portion unselected to act as an anchor.

You will use the Stretch command to make the adapter wider. When you do this, the appropriate dimensions automatically update to the new size. Notice that the overall width of the front view is currently 3.00. Refer to Figure 7.31.

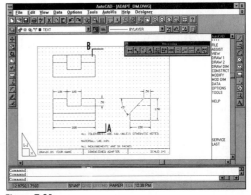

Figure 7.31

Pick: **Stretch icon**

Select objects to stretch by crossing-window or -polygon...

Select objects: *(use Crossing and select point A)*

Other corner: *(select point B)*

Select objects: ⏎

Base point or displacement: *(target the bottom right-hand corner of the front view)*

Second point of displacement: *(move the cursor over .5 units to the right and pick)*

> ■ *TIP* If you have trouble stretching, double-check to make sure that you are not still in paper space. Also, the Stretch command only works when you use the Crossing option. If you type the command, you must make sure to select Crossing or type *C* and press ⏎ afterwards, or use implied Crossing (by drawing the window box from right to left), as you did in the previous step. ■

Notice that the overall dimension now reads 3.50. It has updated automatically. The result is shown in Figure 7.32.

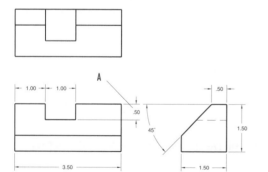

Figure 7.32

Modifying Dimensions

You can modify individual dimensions and override their dimension styles using the Properties icon on the Object Properties toolbar. Refer to Figure 7.32.

Pick: **Properties icon**

Select objects: *(pick on the vertical dimension labeled A)* ⏎

The Modify Dimension dialog box shown in Figure 7.33 appears on your screen. As with other objects, you can modify the layer, color, linetype, thickness, and linetype scale of a dimension. Notice that near the bottom center of the dialog box is a selection for style that contains the name MECHANICAL. You can pick on the name to pull down the list of styles that are already created in the drawing. If you want to change the dimension so that it uses a different style, you can select the new style using this area.

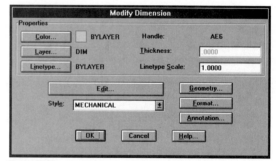

Figure 7.33

Near the center of the dialog box is a button labeled Edit. You pick the Edit button to select the Edit MText dialog box to change the value for the dimension.

You will select Format and change the text orientation so that it is parallel to the dimension line.

Pick: **Format**

Use the dialog box that appears to orient the text parallel to the dimension line. To do this, pick on the image tile at the upper right of the dialog box until it is oriented as shown in Figure 7.34.

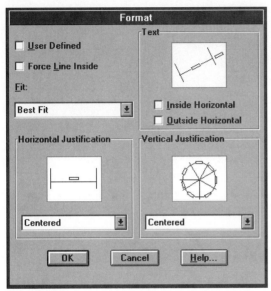

Figure 7.34

Pick: **OK**

Pick: **OK**

The value for the .50 vertical dimension changes so that it is parallel to the dimension line, as shown in Figure 7.35. The dimension style for MECHANICAL has not changed; only that single dimension was overridden to produce a new appearance.

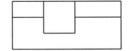

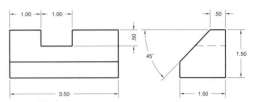

Figure 7.35

Updating Dimensions

When you change a dimension style, all of the dimensions that were created using that style automatically change to take on the new appearance. (This only works for dimensions created with associative dimensioning on.) You will change the arrow style to a new appearance for all of the dimensions created with style MECHANICAL. Use the Dimension Styles dialog box to change the arrow style.

Pick: **Dimension Styles icon**

The Dimension Styles dialog box appears on your screen. You will use the Geometry option. Make sure that the radio button for Parent is highlighted. Then,

Pick: **Geometry**

Pick: **(Right Angle in the Arrowheads section of the dialog box)**

Pick: **OK (to exit the Geometry dialog box)**

Pick: **Save**

Pick: **OK (to exit the Dimension Styles dialog box)**

The dimensions in the drawing update automatically to reflect this change. The arrows are now right-angled arrows instead of filled arrows. Your drawing should appear similar to Figure 7.36.

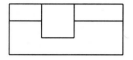

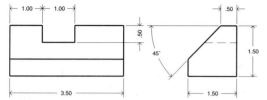

Figure 7.36

Now switch to paper space and use the Edit Text (Ddedit) command to edit the text in the title box to make corrections to the date, name, and scale if necessary. Title your drawing *ADAPTER*. Return to model space when you are done and save the drawing on your own.

Close any open toolbars except for the Draw and Modify toolbars, which should be docked to the edge of the graphics window. Exit AutoCAD. You have completed this tutorial.

> ■ *TIP* If you are not sure what dimension style and overrides may have been used when creating a particular dimension, use the List command and pick on the dimension. You will see a list of the dimension style and any overrides that were used for the dimension. ■

KEY TERMS

associative dimensioning
baseline dimensioning
chained (continued) dimensioning
dimension line
dimension style
dimension value
dimension variables (dim vars)
dimensions
extension line
extension line offset
image tile
inspection
leader line
manufacture
overall dimensions
tolerance

KEY COMMANDS

Angular Dimension
Baseline Dimension
Continue Dimension
Diameter Dimension
Dimension Status
Dimension Styles
Linear Dimension
Radius Dimension
Stretch
Xplode

Draw and dimension the following shapes. The letter M after an exercise number means that the units are metric. Add a note to the drawing saying *METRIC: All dimensions are in millimeters* for metric drawings. Specify a general tolerance for the drawing.

7.1M Stop Plate

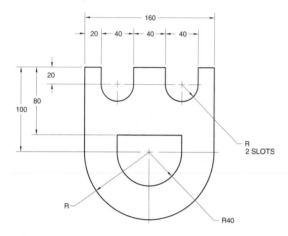

7.2M Hub

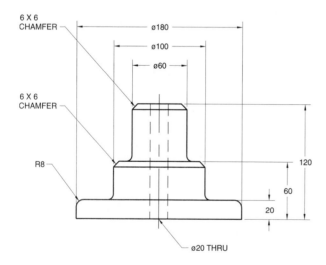

7.3 Guide Block

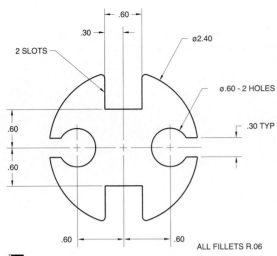

2 SLOTS

.60

.30

.60

ø2.40

ø.60 - 2 HOLES

.30 TYP

.60

.60

.60

.60

ALL FILLETS R.06

7.4 Angle Bracket

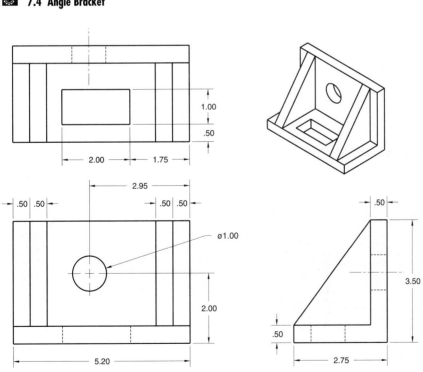

1.00

.50

2.00

1.75

2.95

.50 | .50

.50 | .50

ø1.00

2.00

5.20

.50

3.50

.50

2.75

7.5 Interchange

Draw and dimension this intersection, then mirror the circular interchange for all lanes.

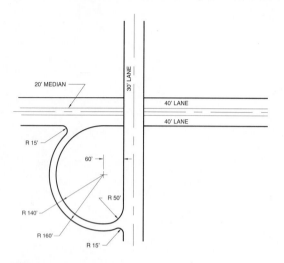

7.6 Bearing Box

Draw and dimension the following shape. Specify a general tolerance for the drawing.

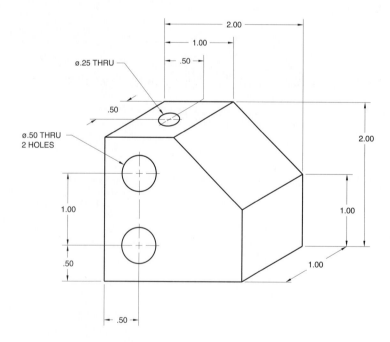

7.7 Plot Plan

Draw and dimension using the civil engineering dimensioning style shown, and find the missing dimensions.

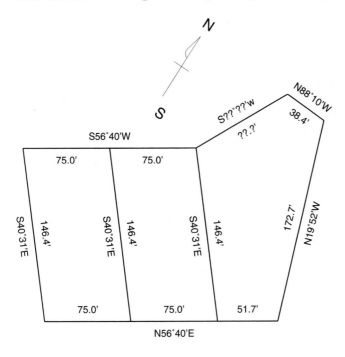

7.8 Hub

Draw and dimension the hub shown. Use one grid unit equal to .125" for a decimal inch drawing or 25 mm for a metric drawing.

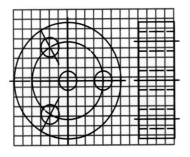

7.9 Floor Plan

Use architectural units to draw and dimension the floor plan, making good use of layers.

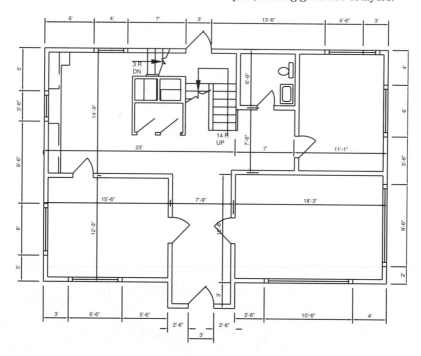

7.10 Support

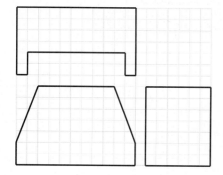

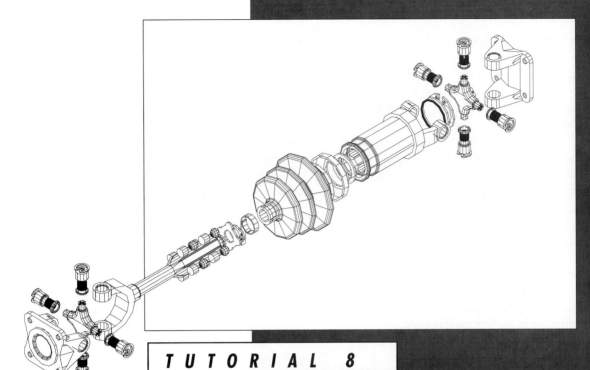

Using Blocks & Customizing Toolbars

Objectives

When you have finished this tutorial, you will be able to

1. Use the Block, Write Block, Insert Block, and Donut commands.

2. Create your own toolbar.

3. Add your block symbols to buttons.

4. Create a block library drawing.

5. Create a logic circuit diagram using blocks.

6. Create named groups.

Introduction

You can unify a collection of drawing objects into a single symbol using AutoCAD's Block command. Using blocks can help you organize your drawing and make it easy to update symbols. Blocks can have associated text objects called attributes that you can even make invisible. You can extract this attribute information to an external database or spreadsheet for analysis or record keeping. You will learn more about using attributes in Tutorial 10. There are many advantages to using blocks. You can construct a drawing by assembling blocks that consist of small details. You can draw objects that appear often once, then insert them as needed, rather than drawing them repeatedly. Inserting blocks rather than copying basic objects results in smaller drawing files, which saves time in loading and regenerating drawings. Blocks can also be nested, so that one block is a part of another block.

In this tutorial you will create symbols used in drawing the electronic logic gates *Buffer, Inverter, And, Nand, Or,* and *Nor.* You will learn to use the Block and Insert Block commands to easily add the symbols into your drawing. Using Write Block, you will make the symbols into separate drawings available to be added to any drawing. You will learn how to create a block library drawing, which you can use to insert a group of blocks into your current drawing. Finally, you will learn to create a custom toolbar to which you will add your block symbols.

You can also create named groups in AutoCAD, using the Object Group command. A *group* is a named selection set. You can make *selectable groups*. When a group is selectable, picking on one of its members selects all the members of the group that are in the current space, and not on locked or frozen layers. An object can belong to more than one group.

Groups are another powerful way of organizing the information in your drawing. Some CAD programs use only groups to organize drawing information, instead of layers. In many ways groups are like blocks, but groups differ because of their special use for selecting objects by named sets.

One case where you might use blocks and groups together is in an architectural drawing. You could create each item of furniture—a table, chair, etc.—as a separate block and then insert it multiple times into the drawing. Each block could be drawn using several different layers. You could then name all of the tables as a group called TABLES, and all of the chairs as a group called CHAIRS. The TABLES and CHAIRS groups could belong to a group named FURNITURE. Unlike blocks, which act as a single object, you can still move or modify items in groups individually, but you can also quickly select the group (using the Select Group selection method) and modify it as a unit.

Starting

Start AutoCAD so that AutoCAD's drawing editor is on your display screen. On your own, start a new drawing and name it *eblocks.dwg,* using the file *electr.dwg,* which was provided with the data files, as a prototype. When you have started the new file from the prototype, it should appear on your screen, as shown in Figure 8.1.

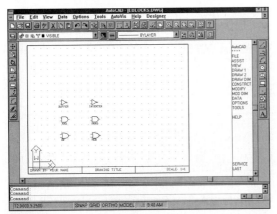

Figure 8.1

Figure 8.1 shows six symbols used in drawing electronic logic circuits: the Buffer, Inverter, And, Nand, Or, and Nor gates. You will make each symbol into a block.

Using the Block Command

A *block* is a set of objects formed into a compound object or symbol. You define a block from a set of objects in your current drawing. You specify the block name and then select the objects you want to be part of the block. The *block name* is used whenever you insert the compound object (block) into the drawing. Each insertion of the block into the drawing, the *block reference*, can have its own scale factors and rotation. A block is treated as a single object by AutoCAD; you select the block for use with commands like Move or Erase simply by picking any part of the block. You can explode blocks into their individual objects using the Explode or Xplode commands, but doing so removes the blocks' special properties.

A block can be composed of objects that were drawn on several layers, with several colors and linetypes. The layer, color, and linetype information of these objects is preserved in the block. Upon insertion, each object is drawn on its original layer, with its original color and

linetype, no matter what the current drawing layer, object color, and object linetype happen to be. There are three exceptions to this rule:

1. Objects that were drawn on layer 0 are generated using the current layer when the block is inserted, and they take on the characteristics of that layer.

2. Objects that were drawn with color BYBLOCK inherit the color of the block (either the current drawing color when they are inserted or the color of the layer when color is set by layer).

3. Objects that were drawn with linetype BYBLOCK inherit the linetype of the block (either the current linetype or the layer linetype, as with color).

> ■ *TIP* Block names within a drawing can be up to 31 characters long and can contain letters, numbers, and the following special characters: $ (dollar sign), - (hyphen), and _ (underscore). AutoCAD converts all letters to uppercase. If a block with the name you choose already exists, AutoCAD tells you that the name already exists. You can either exit without saving, so that you do not lose the original block, or you can choose to redefine the block. ■

Blocks defined using the Block command are stored only in the current drawing and you can use them only in the drawing in which they were created. To save a block so that it can be transferred to another drawing, you need to use the Write Block command or the Windows Clipboard Write Block. You will learn more about Write Block later in this tutorial. You can also use blocks in other drawings by copying vectors (graphical objects) to the Windows Clipboard and pasting them into a drawing, which results in a block. In addition, you can insert the drawing containing a block into another drawing. When you insert a drawing, as you did in Tutorial 4 with the subdivision drawing, *subdivis.dwg*, any blocks it contains are then defined in the drawing in which it is inserted.

On your own, zoom up the area of the drawing showing the And gate so that your screen appears similar to Figure 8.2.

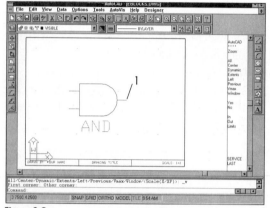

Figure 8.2

 You will use the Block icon from the Block flyout on the Draw toolbar to create a block from the objects making up the And gate on the screen. Make sure to select Block and not Insert Block on the same flyout. Refer to Figure 8.2 to help you select point 1.

Pick: **Block icon**

Block name (or ?): **AND** ⏎

Insertion base point: *(pick Snap to Endpoint icon)*

endp of *(pick endpoint 1)*

Select objects: *(use implied Crossing to select all of the objects making up the And gate, including the name AND)* ⏎

The objects that were selected as part of the block are erased from the screen. In order to restore these objects to the drawing, you will type *OOPS* at the command prompt. (You can

also pick the Oops command from the Miscellaneous toolbar when it is turned on.)

Command: **OOPS** ⏎

The objects that were selected for the block are restored in your drawing.

■ *TIP* Remember, Oops restores only the last objects that were erased in your drawing. ■

On your own, use Zoom Previous to restore the original view of the drawing, where you can see each symbol, and create a block of each of the other logic gates in the drawing. Use Snap to Endpoint to select the right-hand endpoint of the lead from each electronic logic gate (as you did for And). Use Oops to restore each set of objects to the screen as you create the block. Name the blocks NAND, OR, NOR, INVERT, and BUFFER.

To see how you can use blocks within the current drawing, you will insert the AND block you created.

Inserting a Block

 The Insert Block command inserts blocks or other drawings into your drawing using a dialog box, as you saw in Tutorial 4. When inserting a block, you specify the block name, the insertion point (the location for the base point you picked when you created the block), the scale, and the rotation for the block. A selection of 1 for the scale causes the block to remain the original size. The scale factors for the X, Y, and Z directions do not have to be the same. However, using the same size ensures that the inserted block has the same proportions, or aspect ratio, as the original. A new feature in AutoCAD Release 13 is that even blocks that are inserted with different X, Y, and Z scaling factors can be exploded back into their original objects, using the Explode command. A rotation of 0° ensures the same orientation as the original. You can also use Insert Block by typing *INSERT* at the command prompt.

You will select Insert Block from the Block fly-out on the Draw toolbar and use the dialog box to insert the And gate into your drawing.

Pick: **Insert Block icon**

The Insert dialog box appears on your screen, as shown in Figure 8.3.

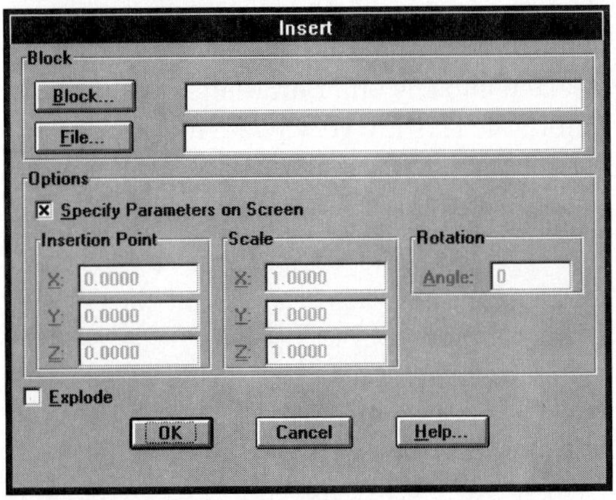

Figure 8.3

Select the Block button from the dialog box to bring up the Defined Blocks dialog box, which lets you select from the blocks that are defined in your drawing. The Defined Blocks dialog box looks like Figure 8.4.

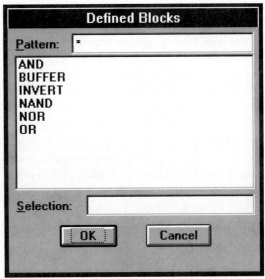

Figure 8.4

Pick: **AND**

Pick: **OK**

You return to the Insert dialog box. Make sure that the box to the left of Specify Parameters on Screen has an X in it. This allows you to pick the location for the block from your screen after you exit the dialog box. Otherwise, you must type in the exact coordinates for its location in the dialog box, as well as the rotation angle and scale. To exit the dialog box and begin specifying the insertion parameters on screen,

Pick: **OK**

The AND block appears, attached to the crosshairs by its endpoint, the insertion base point you specified. Continuing with the Insert command,

Insertion point: *(pick a point anywhere to the right of the logic symbols)*

X scale factor <1>/Corner/XYZ: ⏎

Y scale factor (default = X): ⏎

Rotation angle <0>: ⏎

The AND block is added to your drawing. On your own, try erasing one of the lines of the newly added block. Notice that the entire block behaves as one object and is erased all together. Cancel the Erase command without erasing anything. In Tutorial 7, when you added associative dimensions to your drawing, they were also blocks. This is why they behaved as a single object, even though they were a collection of lines, polylines, and text. Before you can change individual pieces of a block, you must use the Explode command, as you did with Xplode in Tutorial 7. However, exploding takes away all of a block's special properties.

Using Explode

 AutoCAD treats blocks as a single object. The Explode command replaces a block reference with copies of the simple objects composing the block. Explode also turns 2D and 3D polylines, associative dimensions, and 3D mesh back into individual objects. When you explode a block, the resulting image on the screen is identical, except that the color and linetype of the objects may change. This occurs because properties such as the color and linetype of the block return to the settings determined by their original method of creation, either BYLAYER or the set color and linetype with which they were created. There will be no difference unless the objects were created in layer 0 or with the color and linetype set BYBLOCK, as specified in the list at the beginning of this tutorial.

You will use the Explode command to break the newly inserted AND block into individual objects for editing.

Pick: **Explode icon**

Select objects: **(pick any portion of the newly inserted AND block)** ⏎

The block is now broken into its component objects. On your own, try erasing a line from the block. You will see that it is now possible to erase one line at a time, rather than the whole block.

Using Write Block

The block you just created is only defined in the current drawing. You can use the Write Block command to make the block into a separate drawing so that it is available for insertion into other drawings. A drawing file made with the Write Block command is the same as any other drawing file. You can call it up and edit it like any other drawing you have created. Any AutoCAD drawing can be used as a block and inserted into any other drawing, as you did in Tutorial 4. This is a very powerful feature when creating drawings in AutoCAD.

■ **TIP** You can use Write Block to export a part of any drawing to a separate file. This can speed up your computer's response time by keeping file sizes smaller. When you are working in an extremely large file, but just need to work on a certain area of it, use Write Block to export that area to a new file. Work on the new, smaller-sized file that contains only the part of the drawing you need. When you are done, use Insert Block to insert the finished work back into the large drawing again. You will have to explode it once it is inserted before you can edit it. ■

The Write Block command has two steps. The first step is to specify the new name for the file that will be created on your default drive. After you have specified a name for the new AutoCAD drawing file, the second step allows you to select the block in four ways:

1. Type the name of a block you have previously created with the Block command in the current drawing at the command prompt and press ⏎. The objects composing the specified block are written to the drawing file you created in the first step.

2. Type an equal sign (=) and press ⏎. This is a shortcut you can use when the block you previously created with the Block command has the same name as the AutoCAD drawing file you just specified.

3. Type an asterisk (*) and press ⏎. This saves the *entire current drawing* to the name specified as the Write Block file name (as the Save command does), except that unused named objects, layers, and other definitions are eliminated.

4. Press ⏎. This null response is followed by prompts that allow you to specify the objects and the insertion base point, as in the Block command prompts. AutoCAD writes the selected objects to the drawing file name you specified and deletes them from the current drawing, just as when you use the Block command. You can use the Oops command to retrieve them if necessary.

You will use the Write Block command to write the AND block you created to a new AutoCAD drawing file called *and-gate.dwg*.

Command: WBLOCK ⏎

If you have AutoCAD configured to use dialog boxes for file operations (the default), the Create Drawing File dialog box shown in Figure 8.5 appears on the screen. (If not set, Filedia equal to 1 on your own and try again.)

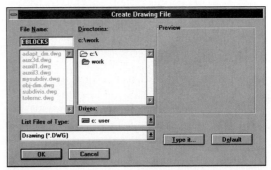

Figure 8.5

You will replace the current drawing name, *c:\work\eblocks.dwg*, shown in the box at the upper left of the dialog box with the name for the new file that will be created. You do not need to include the file extension; the *.dwg* extension is automatically added. The name must have only eight characters to be a legal DOS file name. (Make sure to use the *c:\work* directory for the new file; this will keep your hard drive organized.) Highlight *eblocks.dwg* in the input box below File Name.

Type: **AND-GATE**

When you have finished,

Pick: **OK**

■ *Warning:* Do not give the write block the same name as an existing drawing. If you do this, you will see the AutoCAD message *The specified file already exists. Do you want to replace it?* Select No unless you are certain that you no longer want the old drawing file. ■

Next, AutoCAD will prompt you for the block name. Type in the name of the AND block you created previously.

Block name: **AND** ⏎

The block you created, named AND, has been saved to the new AutoCAD drawing file *and-gate.dwg*.

> ■ **TIP** Write Block does not save views, User Coordinate Systems, viewport configurations, unreferenced symbols (including block definitions), unused layers or linetypes, or text styles. This compresses your drawing file size by getting rid of unwanted overhead. ■

On your own, use the Write Block command to create separate drawing files for each of the other blocks you created. Name them *or-gate, nor-gate, nnd-gate, inverter,* and *buffer.*

Creating Block Libraries

You can build block libraries of standard symbols for electronic symbols, standard mechanical fasteners, furniture, landscaping symbols, and many other items you might frequently use. You can create a library of blocks and use it for all the standard parts needed for any application. You can also order block libraries from third-party sources. There are many third-party vendors that offer disks of standard shapes saved in block form.

To create a block library, start a new drawing with the name for the library. Insert each of the blocks into the library drawing. You can also add text below each block indicating where the insertion point for each block is and the exact block name. Save the file. To use the block library, insert the library file into the current drawing, using the Insert Block command and selecting File as the type of object to insert. Pick OK. At the prompt for the insertion point, pick Cancel. Now each of the blocks that was inserted into the library file is available in the current drawing. This is because when you insert a drawing, all of its block definitions (and groups too) are inserted with it. Even though you cancel the Insert command, the blocks are still defined. Now you can insert each block you want to use from within the drawing. Using

block libraries helps keep your hard drive organized. Group like symbols together into a library file. Print out the drawing of it showing the block insertion points and names of the blocks and hang it up near your computer for quick access to block information. This also helps inform coworkers about the blocks that you have created that they could be using.

Creating Custom Toolbars

One of the ways you can *customize* your AutoCAD software is by creating new toolbars and programming the existing toolbars so they contain the commands you use most frequently. You can also reprogram the individual buttons on the AutoCAD toolbars. Next, you will create your own toolbar for the logic symbols you created. You will program the buttons on the new toolbar to insert the logic symbols. To create a new toolbar,

Pick: **Tools, Customize Toolbars**

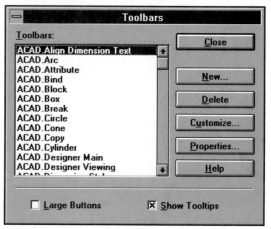

Figure 8.6

The Toolbars dialog box appears on your screen, as shown in Figure 8.6. You will pick New to create a new toolbar.

Pick: **New**

Figure 8.7 shows the New Toolbar dialog box.

Figure 8.7

You will use the New Toolbar dialog box to type in the name of the toolbar you want to create. You can also use it to select the menu group you want to customize. You will add your toolbar to the ACAD menu group. Pick in the text box below Toolbar Name.

Type: **LOGIC**

Pick: **OK**

■ *TIP* The Toolbar dialog box and its associated dialog boxes are case-sensitive. What you type here will appear on your toolbar *exactly* as you have typed it. If you want your toolbar to look like those that are used by AutoCAD, use an initial capital letter, Logic, to name your toolbar. ■

You will see a small toolbar with the name Logic at the top appear on your screen. It should look similar to Figure 8.8.

Figure 8.8

The Toolbars dialog box should still be on your screen. You will use the Customize selection to add buttons to your Logic toolbar.

Pick: **Customize**

Now the Customize Toolbars dialog box appears on your screen. You can use it to select standard button icons to add to the toolbar. The button icons are grouped by their functions under the category heading. In the Categories area, use the list that appears and,

Pick: **Custom**

When you have Custom selected, the dialog box appears as shown in Figure 8.9.

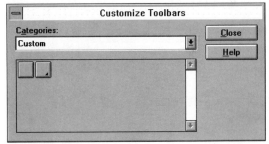

Figure 8.9

Notice that there are two types of custom buttons: regular buttons and flyout buttons. To add a button to your toolbar, pick on the button icon you want to add, and keeping the pick button held down, drag a copy of that button to the toolbar. On your own, drag a regular button over to your Logic toolbar. Now drag a regular button to the toolbar three more times so that you have a total of four regular buttons on the toolbar. (Note that the Logic toolbar expands to accommodate the buttons as you add them.) Next, drag a flyout button to the toolbar. When you have finished adding the five buttons to the Logic toolbar,

Pick: **Close**

Now you are ready to add commands to your custom buttons. To edit the button properties, you will pick on the custom button you want to modify, using the return button on your pointing device or mouse; this opens the Button Properties dialog box. If the Toolbars dialog box is closed, picking once with the return button on a toolbar button opens the Toolbars dialog box. Picking with the return button a second time on the toolbar button opens the Button Properties dialog box. On your own, pick on the left-most regular button on the Logic toolbar, using your return button to open the Button Properties dialog box. Figure 8.10 shows the Button Properties dialog box.

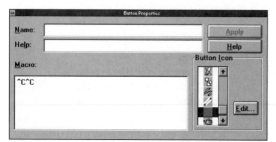

Figure 8.10

The input box to the right of Name allows you to specify the name of the button that you are creating. This name will appear, exactly as you type it, as the tool tip for the button. You use the input box to the right of Help to enter the help line for the button, which will appear in the status bar at the bottom of the screen. The area below the heading Macro lets you type in the command selection you want AutoCAD to use when the button is picked. Do not add any spaces between ^C^C and the first command you type in. AutoCAD uses a space to represent pressing ⏎ or ⎵SPACEBAR⎵.

You will program these buttons to insert the blocks you have created into your current drawing when picked. When you pick the button, the command will use the Insert Block command automatically to insert the specified block.

■ **TIP** The block or file you are inserting must be in the current path or you must tell AutoCAD where the block is located by adding the path name ahead of the block name. (You can change the current path in the Preferences dialog box.) The back-slash has a special meaning in this dialog box; it means pause for user input. Therefore, when you add a path to a macro, you use the forward slash instead of the backslash to separate the directories (e.g., *c:/work/and-gate.dwg*). Finally, you can create a block library drawing and insert it into the drawing so that all of the blocks are available in the current drawing prior to using the Logic toolbar you just created. ■

Using Special Characters in Programmed Commands

There are some special characters that you can use in programming the toolbars. One of these is ^C. If you put ^C in front of the command name you are programming, it will cancel any unfinished command when you select its button from the toolbar. In programming the menus and buttons, it is often a good idea to put cancel twice before a command, in case you have left a subprompt active (so that it would take two cancels to return to the command prompt). Putting cancel before the Insert command is useful because you cannot select the command during another command. Do not add cancel before the toggle modes or before transparent commands (they must be preceded by '), because if you execute cancel before you use them, they cannot be used during another command.

A space or ⏎ is automatically added to the end of every command you program for the buttons. This way the command is entered when the button is selected. You can enter a string of commands, or a command and its options by separating them with a space to act as ⏎. Below is a list of special characters that you can use to program the toolbar buttons.

Character	Meaning
space	⏎
;	⏎
;	at the end of a line, suppresses the addition of a space to the end of a command string
+	at the end of a command string, allows it to continue to the next line
\	pause for user input
/	separates directories in path names, since \ [backslash] pauses for user input
\n	new line
^B	Ctrl-B Snap mode toggle
^C	Ctrl-C Cancel
*^C^C	at the beginning of a line, automatically restarts command sequence (repeating)
^D	Ctrl-D Coordinates toggle
^E	Ctrl-E Isoplane toggle
^G	Ctrl-G Grid mode toggle
^O	Ctrl-O Ortho mode toggle
_	an underscore preceding a command or option automatically translates AutoCAD keywords and command options for use with foreign-language versions

Pick: (in the input box to the right of Name)

Type: **AND**

Pick: (in the input box to the right of Help)

Type: **INSERTS AND GATE SYMBOL**

Pick: (to the right of ^C^C, below Macro)

Type: **INSERT AND-GATE**

Pick: **Edit** *(to the right of the button icon pictures)*

The Button Editor dialog box appears on your screen. (If you double-click with the right button on an icon *when the Toolbars dialog box is not yet open*, the Toolbars dialog box opens and then the Button Editor dialog box opens.) The button editor works similarly to many paint-type drawing programs. Each little box in the button editor presents a pixel, or single dot, on your screen. You can select the drawing tool from the buttons at the top of the editor. The color of the drawing tool is selected from the colored buttons at the right of the editor. Use the button editor to draw a picture of the And gate, as shown in Figure 8.11. If you have trouble using the button editor, pick the Help button at the bottom right of the editor.

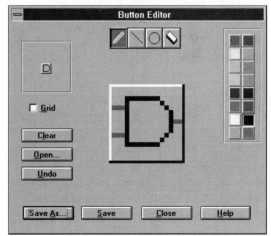

Figure 8.11

When you have finished drawing your button icon,

Pick: **Save As**

On your own, use the Save As dialog box that appears to name the icon *and.bmp* in the *c:\work* directory.

> ■ *TIP* If you wish to save your *.bmp* files (or any other files) to a subdirectory within *c:\work* or in any other directory, you must add that subdirectory to your environment support path. If you do not do this, AutoCAD will not be able to find your *.bmp* files upon loading. ■

In order to test an icon, all your dialog boxes must be closed; when you close the dialog boxes the necessary files are updated so that you can use your new icons. When you are finished using the button editor,

Pick: **OK**

Pick: **Close**

Pick: **Apply**

When you pick Apply, your selections are applied to the icon. You should now see the And gate icon you made appear on the Logic toolbar that you created. On your own, use the Windows Control box to close the Button Properties dialog box. To close the Toolbars dialog box,

Pick: **Close**

All of the dialog boxes should now be closed. On your own, test the new And button you created on your toolbar.

Pick: **And icon**

The And gate should appear attached to the crosshairs, ready to insert into your drawing. Position it as you like in your drawing and press the pick button. Press ⏎ for the remaining prompts for scale and rotation. The And gate should appear in your drawing.

Next you will program the Nand gate button for the toolbar so that it inserts the NAND block when you pick it. Use the Button Properties dialog box by double-clicking the return button while the arrow cursor is positioned over the button second from the left on the Logic toolbar. Do this on your own now.

The Button Properties dialog box appears on your screen. Type the entries listed below. (Notice that there is no space between ^C^C and Insert.)

Name: **NAND**

Help: **INSERTS NAND GATE SYMBOL**

Macro: **^C^CINSERT NND-GATE**

On your own, pick on the And icon you created from the icons displayed in the Button Icon area of the dialog box and pick Edit. The button editor appears on your screen, with the icon for the And gate already showing. Pick the Save As button and save it to the new name *nand.bmp*.

> ■ *TIP* If you do not want to create the bit maps for each button, the files have been provided with the data files. To use one, when the button editor is on your screen, pick Open and switch to the *c:\datafile* directory. The files are named *and.bmp*, *nand.bmp*, *or.bmp*, *nor.bmp*, *inverter.bmp*, and *buffer.bmp*. Once the appropriate file is selected, okayed, and appears in the button editor, pick Save As and save it to the *c:\work* directory. ■

Now use the button editor to change it so it looks like a Nand gate, not an And gate, by adding pixels at the right end to look like the small circle on the Nand gate. (It is difficult to make fine details; just a block at the end will look OK on the button.) When you have finished drawing it,

Pick: **Save**

Pick: **Close**

Pick: **Apply**

You will add the Or and Nor gates to the next two buttons. The entries for them are listed below. Use the Button Editor dialog box on your own to create button icons for them.

Name: **OR**

Help: **INSERTS OR GATE SYMBOL**

Macro: **^C^CINSERT OR-GATE**

Name: **NOR**

Help: **INSERTS NOR GATE SYMBOL**

Macro: **^C^CINSERT NOR-GATE**

When you are finished, close all of the dialog boxes. On your own, test your new icons.

Creating the Inverter Flyout

Next, you will create a flyout item for the Buffer and Inverter symbols. When you create a flyout, you essentially create a small toolbar that appears when the flyout button is held down. First you will create a new toolbar named Inverter. Then you will add the buttons for the Inverter and the Buffer gates to it. Once the Inverter toolbar is finished, you will use it as a flyout on the Logic toolbar.

Pick: **Tools, Customize Toolbars**

Pick: **New**

Type: **INVERTER** *(in the Toolbar Name text box)*

Pick: **OK**

The new toolbar appears on your screen without any buttons.

Pick: **Customize**

Use the Customize Toolbars dialog box that appears to select the category Custom and add two regular blank buttons to the Inverter toolbar. If you need help adding buttons to a toolbar, refer to adding the buttons to the Logic toolbar, earlier in this tutorial. (Sometimes you may need to drag your toolbar to a location where

you can reach it easily.) When you have added two buttons to the Inverter toolbar, close the Customize Toolbars dialog box on your own.

Now open the Button Properties dialog box for the left button on the Inverter toolbar, using the return button. Use the dialog box to make the following entries.

Name: **INVERTER**

Help: **INSERTS INVERTER SYMBOL**

Macro: **^C^CINSERT INVERTER**

Name: **BUFFER**

Help: **INSERTS BUFFER SYMBOL**

Macro: **^C^CINSERT BUFFER**

Create your own button icons for the inverter and the buffer and save them to the appropriate *.bmp* names. When you are finished modifying the button properties, don't forget to pick Apply to apply these settings to the buttons. Close the dialog boxes on your own.

Now you are ready to add the Inverter toolbar as a flyout shown on the Logic toolbar.

Adding a Flyout

Double-click on the empty flyout button on the Logic toolbar, using the return button. The Flyout Properties dialog box appears on your screen. Use it to make the selections you see in Figure 8.12. Make sure to turn on Show This Button's Icon at the bottom of the dialog box.

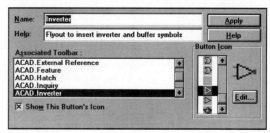

Figure 8.12

Pick: **Apply**

Now the Inverter flyout is added to the Logic toolbar.

On your own, close the Flyout Properties dialog box and then use the Customize selection from the Toolbars dialog box to add the Explode flyout to the Logic toolbar you are creating. To add the Explode flyout, all you need do is pick the category Modify and drag a copy of the Explode flyout that appears over to your Logic toolbar. When you have finished, close all the dialog boxes and the Inverter toolbar.

> ■ **TIP** Remember, when there is no Close button in a dialog box, use the Windows Control box to close it. ■

Editing Your Menu Files

When you finish creating a custom toolbar and close the dialog box, AutoCAD writes to the *.mnr* and *.mns* menu files. It may not, however, write the correct *.bmp* files to your new buttons. To save the correct *.bmp* files to your new buttons you may need to edit your *.mns* file and copy the changes to your *.mnu* file. If your new toolbar does not show correctly, save your drawing now and exit AutoCAD on your own. (If you do not have difficulty with the toolbar buttons showing correctly, skip to Drawing the Half-Adder Circuit.)

From the Program Manager, start your word processor and open your menu file. If you followed the customization procedure in the Getting Started section of this manual, your menu file is *c:\r13\win\support\acad.mns*.

Find the ***TOOLBARS section, and within that section find **LOGIC. Below the Logic Toolbar entry should be **INVERTER. Edit these two sections so that they look like the section below:

**LOGIC

ID_Logic	[_Toolbar("Logic", _Floating, _Show, 400, 50, 1)]
ID__0	[_Button("And-gate", and.bmp, and.bmp)] ^C^CInsert and-gate
ID__1	[_Button("Nand-gate", nand.bmp, nand.bmp)]^C^CInsert nndgate
ID__2	[_Button("Or-gate", or.bmp, or.bmp)]^C^CInsert or-gate
ID__3	[_Button("Nor-gate", nor.bmp, nor.bmp)]^C^CInsert nor-gate
ID__4	[_Flyout("Inverter", inverter.bmp, inverter.bmp, _OwnIcon, ACAD.INVERTER)]
ID_Explode_0	[_Flyout("Explode", ICON_16_EXPLOD, ICON_32_EXPLOD, _OtherIcon, ACAD.TB_EXPLODE)]

**INVERTER

ID_Inverter	[_Toolbar("Inverter", _Floating, _Hide, 400, 50, 1)]
ID__5	[_Button("Inverter", Inverter.bmp, Inverter.bmp)]^C^CInsert inverter
ID__6	[_Button("Buffer", Buffer.bmp, buffer.bmp)]^C^CInsert buffer

ID_Logic gives the information for the Logic Toolbar, that its name is Logic; it is floating, not docked; and AutoCAD will open with it showing. ID_0 gives the information for the first icon, And gate. Its icon is *and.bmp* and the command it invokes is Insert and-gate. As you can see in the line defining the Explode icon, you can provide two different file names for the icons; the first is a 16-bit file and produces a small icon. In the Toolbars dialog box there is an option with which you can show large icons, the second file name in the menu file is the file used for the large icon. The large icons use *.bmp* files which have an image that is 32 pixels square. In the case of the icons you just created, if you chose to show large icons, your And gate, for instance, would appear in the upper left of the button, and AutoCAD would fill the rest of the button with some default pattern. To create *.bmp* files for use as large buttons, use a paint program, such as Windows Paintbrush to create and save your own files. Make the image 32 pixels by 32 pixels.

Save your file *acad.mns* on your own. Copy the **LOGIC and **INVERTER sections of this file. Close *acad.mns*. Open *acad.mnu*. At the bottom of the ***TOOLBARS section, paste the **LOGIC and **INVERTER sections that you copied from *acad.mns*. Save *acad.mnu* on your own. Exit your word processor and start AutoCAD again.

Your Logic Toolbar should appear exactly as you left it, and you can use it to create your half-adder circuit. You can choose Hide from the Toolbar Properties dialog box, or use the Windows control box to remove this toolbar from your screen at any time.

Drawing the Half-Adder Circuit

Now you have learned to customize the AutoCAD toolbars. You can create your own blocks and add them to other toolbars. You can also program your own commands onto tool-

bars and icons, or modify existing toolbars and icons, using the techniques you have learned. Next you will use the new toolbar you created to draw a half-adder circuit using electronic logic symbols.

On your own, start a new drawing named *hfadder.dwg* from the prototype drawing *proto.dwg*, which was provided with the data files. Drag the Logic toolbar to the upper right of the screen (or some other convenient location). Now switch to paper space and use the Edit Text command to change the text in the title bar to *HALF-ADDER CIRCUIT* and type in your own name. Specify NONE for the scale, as electronic diagrams are not usually drawn to scale. Switch back to model space when you are finished.

■ *TIP* The Minsert command lets you insert a rectangular array of blocks. It can be useful when laying out drawings that have a pattern, like rows of desks, or sometimes rows of electronic components. The Divide and Measure commands also allow you to insert blocks. The Divide command divides an object you select into the number of divisions you specify. You can either insert a point object where the divisions will be (as you did in Tutorial 5), or you can select the Block option and specify a block name. The Measure command works in a similar way, except that instead of specifying the number of divisions you want to use, you specify the lengths of the divisions. Use the on-line help facility to get more information about these commands. ■

On your own, set the current layer to THIN and Snap to .1042 so that it will align with the spacing between the leads on the logic symbols you created. Use the toolbar on your own to insert the logic symbols, as shown in Figure 8.13. Use the Line and Arc commands with

Snap to Endpoint to connect the output from one logic circuit to the inputs of the next as shown in Figure 8.13.

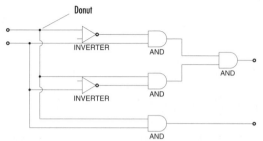

Donut

INVERTER

AND

AND

AND

INVERTER

AND

Figure 8.13

You will use the Donut command to make the filled and open circles for connection points.

Using Donut

 The Donut command draws filled circles or concentric filled circles. See Figure 8.14.

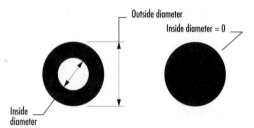

Outside diameter

Inside diameter = 0

Inside diameter

Figure 8.14

The Donut command is on the Draw toolbar on the Circle flyout. The Donut command requires numerical values for the inside and outside diameters of the concentric circles, as well as a point location for the center point. If you enter a zero value for the inside diameter, a solid dot appears on the screen. The diameter of the dot will equal the stated outside diameter value. Next you will draw some donuts off to the side of your logic circuit drawing.

Pick: **Donut icon**

Inside diameter<0.5000>: (↵)

Outside diameter<1.0000>: (↵)

Notice that a doughnut-like shape is now moving with your cursor. The *Center of doughnut* prompt is repeated so that you can draw more than one donut. Select points on your screen:

Center of doughnut: *(select a point off to the side of your drawing)*

Center of doughnut: *(select another point)*

To exit the command when you are done practicing drawing donuts of the same size, you will press (↵).

Center of doughnut: (↵)

On your own, erase the donuts you drew off to the side of the drawing. Restart the Donut command and set the inside diameter for the donut to 0. Set the outside diameter to .075. Draw the solid connection points for the logic circuit that you see in Figure 8.13 on your own.

Now add the open donuts at the ends of the lines. First, start the Donut command and set the inner diameter to .0625 and the outer diameter to .075. Create four donuts on the side of your drawing, then place them using the Move command so that each donut is attached at its outside edge, not its center. Use Snap to Quadrant to select the basepoint on the donut, then Snap to Endpoint to select the endpoint of the line. When you have drawn and placed all the connection points, your drawing should look like Figure 8.13.

Freezing the TEXT Layer

The drawing *electr.dwg*, which you used to create the blocks for the logic gates, used different layers for the text, lines, and shapes of the logic gates. You can freeze or turn off layers to control the visibility within parts of a block. You will freeze the text in the blocks, because it is there to help identify the blocks while you are inserting them, but would not ordinarily be shown on an electronic circuit diagram. The text in the blocks is on layer TEXT, as is the text in your title block. Since you do not want to freeze the text in the title block, you will first move it onto a new layer.

On your own, switch to paper space. Make a new layer, TEXT1, and then use the Properties icon from the Object Properties toolbar to move all three portions of the title block text onto layer TEXT1.

Next you will use Layer Control to freeze the layers that have the text showing the name of the type of gate. Switch back to model space on your own.

Pick: **Layer Control icon**

Use the Layer Control list that appears on your screen to freeze layer TEXT. The names of the gates should disappear from the screen.

Save your drawing on your own before continuing.

Your drawing should look similar to Figure 8.15.

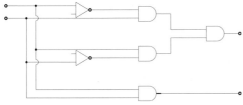

Figure 8.15

Creating Object Groups

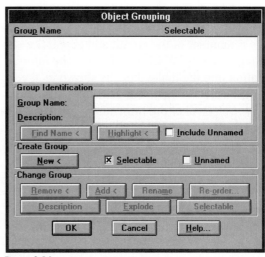

Figure 8.16

The Object Grouping dialog box appears on your screen, as shown in Figure 8.16. You will create a new group named EXISTING-CIRCUIT. For the existing circuit, you will select the upper Inverter and And gate and the two leads into each of them. Groups can be selectable, not selectable, named, or unnamed. When a group is selectable, the entire group becomes selected if you pick a member of the group when selecting objects. You can return to the dialog box and turn Selectable off at any time. Type the information listed below into the appropriate areas of the dialog box.

Group Name: **EXISTING-CIRCUIT**

Description: **Portions of the Existing Circuit**

Pick: **New**

Select objects: *(pick the upper inverter and upper And gate and the two leads to the left of each of them)*

Select objects: ⏎

Another powerful method of organizing your drawing is to use named groups. You can select these groups for use with other commands. Object Group is on the Standard toolbar.

Pick: **Object Group icon**

The dialog box returns to your screen so you can continue creating groups. You will group the remainder of the circuit and give it the group name NEW-CIRCUIT. Type the information into the dialog box, as shown in Figure 8.17. When you are finished, pick New. Select all of the remaining objects to form the NEW-CIRCUIT group. Press ⏎ when you are done selecting. Pick OK to exit the dialog box.

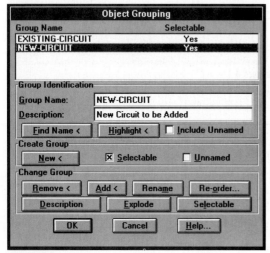

Figure 8.17

Now you are ready to try your groups with drawing commands. Remember that whenever you are prompted to select objects, you can pick the Select Group icon from the Selection flyout on the Standard toolbar and then pick the group name, or you can pick on any member of a selectable group to select the entire group. You will change the linetype property for the NEW-CIRCUIT group to a HIDDEN linetype.

Pick: **Properties icon**

Select objects: *(pick on one of the symbols from the NEW-CIRCUIT group)*

The entire group becomes selected, as shown in Figure 8.18.

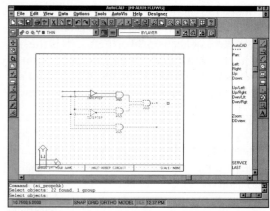

Figure 8.18

Select objects: ⏎

The Object Properties dialog box appears on your screen.

Pick: **Linetype**

Pick: **Hidden**

Pick: **OK (twice)**

The wiring for the new portion of the circuit becomes dashed.

Use Undo from the Standard toolbar on your own to reverse the linetype change you just made before you go on.

To complete your drawing, you will erase the unused leads from the inverters and reposition the remaining lead. In order to be able to select each object individually, return to the Object Grouping dialog box on your own and turn Selectable off.

Use the Explode command you added to your Logic toolbar to explode the inverters and erase the unused lead from them. Use the Stretch command with hot grips to move the remaining lead so that it begins in the center of the inverter as shown in Figure 8.19. Save and plot your drawing.

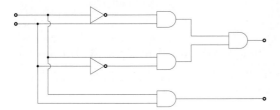

Figure 8.19

■ *TIP* You can quickly change blocks by redefining them. If you redefine a block, all of the blocks with that name will update to the new appearance. To redefine a block, make the changes you want to the block. Then use the Block command to save it as a block again, keeping the file name the same as in the past. Insert the block again, or type *INSERT* ⏎ at the command prompt. When prompted for the block name, type the old block name followed by = and then the new block name. Do not include spaces because the spacebar has the same effect as ⏎. All of the existing blocks which have the same name as the redefined block will update to the new appearance. ■

You have completed this tutorial. Save your file and close any toolbars you may have open, except for the Draw and Modify toolbars. Exit AutoCAD.

8.1 Stress Test Circuit

Create this circuit board layout using Grid and Snap set to 0.2. Use donuts and polylines of different widths to draw the circuit.

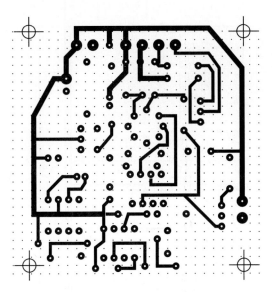

8.2M Decoder Logic Unit

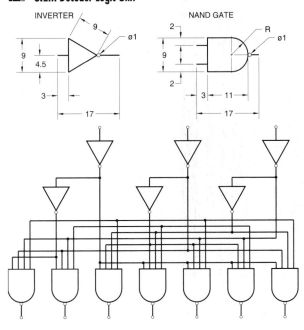

▥ 8.3 Floor Plan

Draw the floor plan for the first floor house plan shown. Create your own blocks for furniture and design a furniture layout for the house.

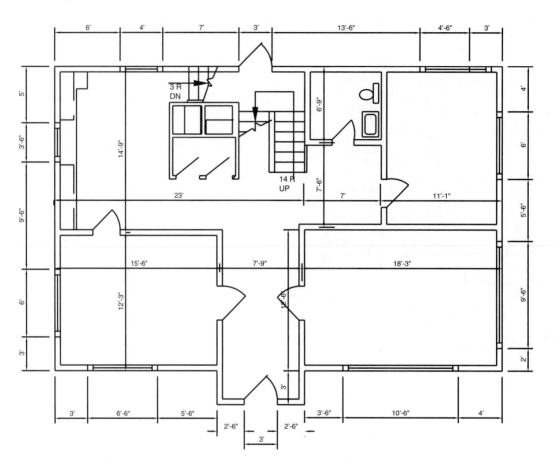

8.4M Arithmetic Logic Unit

Create blocks for the components according to the metric sizes shown. Use the blocks you create to draw the circuit shown.

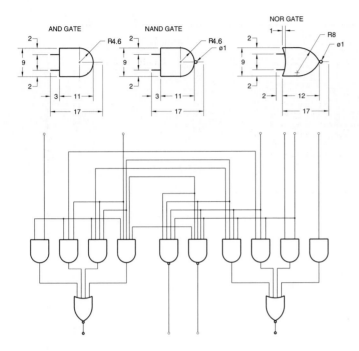

8.5 Amplifier Circuit

Use the grid at the top to draw and create blocks for the components shown. Each square equals –0.0625. Use the blocks you create to draw the amplifier circuit.

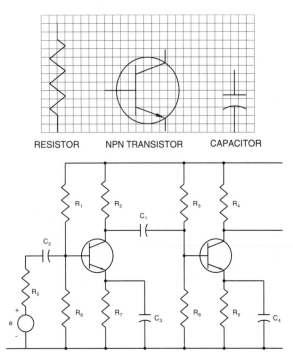

RESISTOR   NPN TRANSISTOR   CAPACITOR

8.6 Office Plan

Draw the office furniture shown and create blocks for DeskA, DeskB, ChairA, ChairB, and the window. Use the floor plan shown as a basis to create your own office plan.

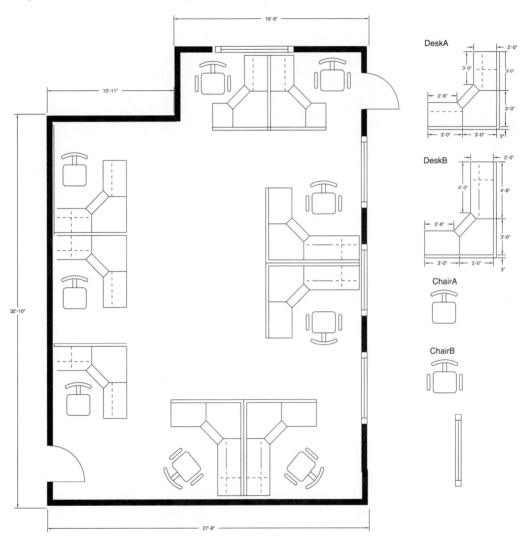

8.7 Piping Symbols

Draw the following piping symbols to scale; the grid shown is 0.2 inches (you do not need to show it on your drawing). Use polylines 0.03 wide. Make each symbol a separate block and label it. Insert all of the blocks into a drawing named *pipelib.dwg*.

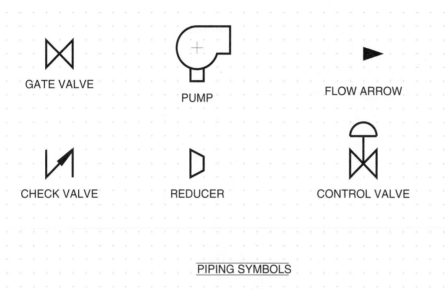

GATE VALVE

PUMP

FLOW ARROW

CHECK VALVE

REDUCER

CONTROL VALVE

PIPING SYMBOLS

8.8 Utilities Layout

Construct this representation of a group of city blocks' utilities layout (grid spacing is at 0.25). Using Polylines 0.05 wide, construct and label the three main symbols shown. Then scale the symbols down to 1/4 size, make each one a block, and insert them in the proper location.

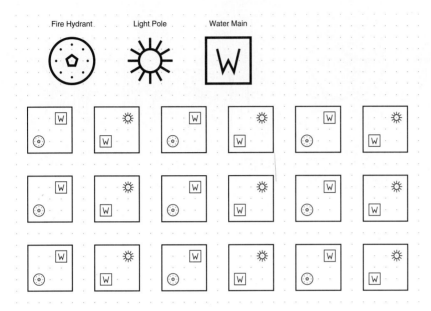

8.9 Mechanical Fasteners

Draw the hex head bolt, square head bolt, regular hex nut, and heavy hex nut according to the proportions shown. Make each into a block. Base the dimensions on a major diameter (D) of 1.00. This way when you insert the block you can easily determine the scaling factor that will be necessary to produce the correct major diameter. Insert each block into a drawing named *fastenrs.dwg*.

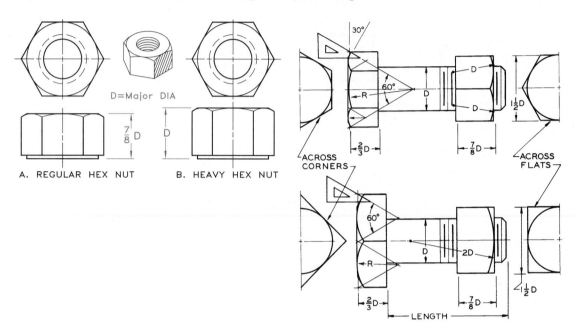

8.10 Wood Screws

Create blocks for each of the wood screws shown. The grid gives the proportions for drawing them. Make the diameter of each 1.00 so that you can easily determine the scale to insert them for other sizes.

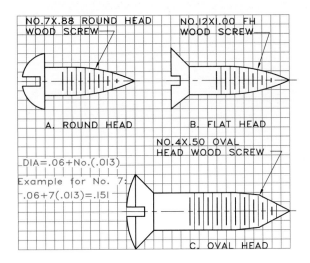

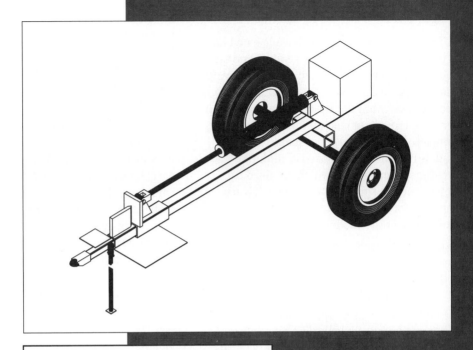

TUTORIAL 9

Introduction to Solid Modeling

Objectives

When you have completed this tutorial, you will be able to

1. Change the 3D viewpoint.

2. Work with multiple viewports.

3. Set individual limits, grid, and snap for each viewport.

4. Create and save User Coordinate Systems.

5. Set Isolines to control model appearance.

6. Create model geometry, using primitives, extrusion, and revolution.

7. Use Boolean operators to add, subtract, and intersect parts of your model.

8. Use region modeling.

9. Shade and plot your models.

10. Convert AME models.

Introduction

In Tutorials 1 through 8 you have learned how to use AutoCAD to model three-dimensional objects. When you create a multiview drawing, it is a model of the object that contains enough information within the various views to give the person interpreting it an understanding of the complete three-dimensional shape. Your models have been composed of multiple two-dimensional views, which are used to convey the information. Now you will learn to use AutoCAD to create actual three-dimensional (3D) models.

AutoCAD allows three types of 3D modeling: *wireframe, surface,* and *solid modeling.* Wireframe modeling uses 3D lines, arcs, circles, and other graphical objects to represent the edges and features of an object. It is called wireframe because it looks like a sculpture made from wires. Surface modeling takes 3D modeling one step further to add surfaces to the wireframe so that the model can be shaded and hidden lines removed. A surface model is like an empty shell: there is nothing to tell you how the inside behaves. A solid model is most like a real object because it represents not only the lines and surfaces, but the volume contained within. In this tutorial you will first learn to create solid models and then surface models.

Using the computer, it is possible to represent three dimensions in the drawing database; that is, drawings are created using X, Y, Z coordinates. Solid modeling is the term for creating an accurate 3D model of the drawing that goes beyond representing just the lines that form the edges of the surfaces, to describing a volume that is contained by the surfaces and edges making up the object. When you create a solid modeled drawing database, in some ways it is as if the actual object is stored inside the computer. Some of the benefits of solid models are: they are easier to interpret; they

can be rendered (shaded) so that someone unfamiliar with engineering drawing can visualize the object easily; two-dimensional views can be generated directly from them; and their *mass properties* can be analyzed. The need to create a physical prototype of the object may even be eliminated.

Creating a solid model is somewhat like sculpting the part out of clay. You can add and subtract material with *Boolean operators* and create parts by *revolution* and *extrusion.* Of course, when modeling with AutoCAD you can be much more precise than when modeling with clay. At first, AutoCAD's solid modeling will look much like wireframe or surface modeling. This is because wireframe representation is usually used even in solid modeling and surface modeling to make many operations quicker to draw on your screen. When you want to see a more realistic representation, you can shade or hide the back lines of solid models and surface models.

Starting

Launch AutoCAD. The AutoCAD drawing editor should be on your screen. Begin a new drawing on your own, based on the default prototype drawing *acad.dwg* that is provided in the AutoCAD software. Call your new drawing *block.dwg.*

When you are working in 3D, the grid is very useful. It can help you to relate visually to the current viewing direction and coordinate system. On your own, make sure you are in model space and that the grid is turned on.

You should see the regularly spaced grid dots appear at a spacing of 1 unit.

3D Coordinate Systems

AutoCAD defines the model geometry using precise X, Y, Z coordinates in space called the *World Coordinate System* (*WCS*). The WCS is fixed, and drawing geometry in both 2D and 3D is stored in the database using the WCS. Its default orientation on the screen is a horizontal X axis, with positive values to the right, and a vertical Y axis, with positive values above the X axis. The Z axis is perpendicular to the computer screen, with positive values in front of the screen. The default orientation of the axes is shown in Figure 9.1.

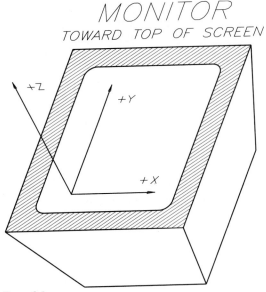

Figure 9.1

While you have been creating 2D geometry, you have been using this default WCS. You have been looking straight down the Z axis, so that a line in the Z direction appears as a point. When you do not specify the Z coordinate, AutoCAD uses the default elevation, which is zero. This has made it easy for you to create and save 2D drawings.

Setting the Viewpoint

When you are creating 3D geometry, it is useful to establish a different direction, or several directions, from which to view the XY plane, so that you can see the object's height along the Z axis. You can do this with the Vpoint command. Using the Vpoint command, you can view the coordinate system from any direction you want. You will select Tripod from the Views menu bar to set the *viewpoint*.

Pick: **View, 3D Viewpoint, Tripod**

Rotate/<View point> <0.0000,0.0000,1.0000>:

An X, Y, Z *coordinate system locator* and *globe* appear on your screen, as shown in Figure 9.2.

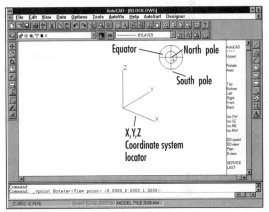

Figure 9.2

You will use the globe to select the viewing direction. You can think of the center point of the globe as the view looking straight down on the top of the XY plane. The center circle is like the equator. Any point inside the center circle shows the view looking down on the object from the top. Points selected outside the center circle show the view looking up at the object from below. A horizontal line divides the globe into front and back. If you pick a point below the line, you are viewing the object from the front; above the line and you are viewing the object from the back. Table 9.1 summarizes the use of the globe.

Table 9.1

	Inside of center circle	Outside of center circle
Above horizontal line	Top rear view	Bottom rear view
Below horizontal line	Top front view	Bottom front view

Move your pointing device so that the crosshairs are inside the center circle in the lower right-hand quadrant. When you have them positioned as shown in Figure 9.2, press the pick button to select the viewing direction. The location you have selected produces a viewpoint that shows the coordinate system as though you are looking from the above right.

Your screen should appear similar to Figure 9.3, depending on the exact point you selected. Notice that your view has changed. You are no longer looking straight down on the XY plane, which is represented by the grid area. Your view shows it from an angled direction.

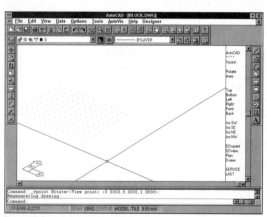

Figure 9.3

User Coordinate Systems

A *User Coordinate System (UCS)* is a set of X, Y, Z coordinates that you can define to help make it easy to create your 3D models. You can

define your own User Coordinate System, which may have a different origin and rotation from, and can be tilted at any angle with respect to, the World Coordinate System. UCSs are helpful because your mouse moves in only two dimensions, and UCSs let you orient the basic drawing coordinate system at any angle in the 3D model space of the drawing, so that you can still draw using your mouse or other pointing device. You can define any number of UCSs, give them names, and save them in a drawing; however, only one can be active at a time.

The UCS Icon

The UCS icon appears in the lower left of the screen. It will help you orient yourself when looking at views of the object. Since the monitor screen is essentially flat, even if the object is a 3D solid model in the database, only 2D views of it can be represented on the monitor. Because wireframe models look the same from front and back, or from any two opposing viewpoints, it is especially important to keep track of what view you are seeing on the screen.

The UCS icon is always drawn in the XY plane of the current UCS. The arrows at the X and Y ends always point in the positive direction of the X and Y axes of the current UCS. A W appearing in the UCS icon tells you that the UCS is currently lined up with the WCS. The box in the lower corner of the icon indicates that you are viewing the UCS from above. A plus sign in the lower left of the icon indicates that the icon is positioned at the origin of the current UCS. Notice that there is no plus sign in the lower left of the UCS now. When you start a drawing from the *acad.dwg* prototype, the origin (or 0,0,0) of the drawing is in the lower left of the screen. If the UCS were at the origin, it would be partially out of the view, so the default is that the UCS icon is not at the origin. You can use the Ucsicon command to reposition the icon so that it is at the origin of the X, Y, Z coordinate system. When you do this, a plus sign appears in the icon.

A special symbol may appear instead of the UCS icon to indicate that the current viewing direction is viewing the UCS edgewise. (Think of the X, Y coordinate system of the UCS as a flat plane like a piece of paper; in this case, the viewing direction is set so that you are looking directly onto the edge of the paper.) When this happens you cannot use most of the drawing tools, so the icon appears as a box containing a broken pencil. Take special notice of this icon.

The *perspective icon* replaces the UCS icon when perspective viewing is in effect. It appears as a cube drawn in perspective. Many commands are also limited when perspective viewing is in effect. The Dview command turns on perspective viewing. You will use the Dview command in Tutorial 15, when you create auxiliary views.

Figure 9.4 shows the different icons.

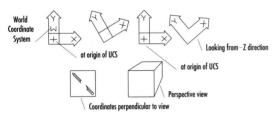

Figure 9.4

On your own, create three new layers. One layer will contain the solid model, another will contain viewports, and the third will contain the drawings border. On your own, use the Layer Control dialog box to create the following layers:

MODEL	Magenta	Continuous
VPORT	White	Continuous
BORDER	White	Continuous

Set layer MODEL as the current layer.

Showing the Solids Toolbar

On your own, use the Tools selection from the menu bar and select the Toolbars option. From the list of toolbars that appears, pick to show the Solids toolbar. It will appear on your screen, as shown in Figure 9.5. Use the methods you have learned to position it in an area of your screen where it will be accessible, but not in the way of your drawing.

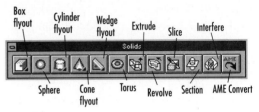

Figure 9.5

Creating an Object

AutoCAD has three methods you can use to create model geometry: primitives, extrusion, and revolution. *Primitives* are basic shapes, like boxes, cylinders, and cones, that can be joined together to form more complex shapes. You will create a solid three-dimensional box. Later in this tutorial you will learn how to join the shapes together with Boolean operators, and how to create other shapes with extrusion and revolution.

The Box command draws a rectangular prism solid model. You can draw a box by specifying the corners of its base and its height, its center and height, or its location and length, width, and height. Figure 9.6 shows the information you specify to draw a box using these methods. The default method of defining a box is specifying two corners across the diagonal in the XY plane and the height in the Z direction. By selecting the Box Center icon on the Box fly-out, you can specify the center of the box and the length, width, and height, instead of the corners of the diagonal and the height. Or you can pick one corner and then select the Length option at the command prompt, and you will see further prompts for the length, width, and height. You can also type the Cube option at the command prompt once the Box command is started, so you can specify that all dimensions are equal. You will then only be prompted for one length. On your own, use the Help command to get help for the Solids toolbar and the Box command.

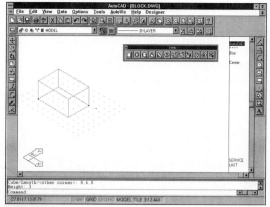

Figure 9.7

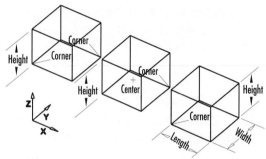

Figure 9.6

 You will use the Box flyout on the Solids toolbar to select the Corner method and draw a solid box by specifying the corners of the base and the height of the box.

Pick: **Box Corner icon**

The prompt for the box primitive appears. You will type in the coordinates.

Center/<Corner of box> <0,0,0>: **2,2,0** ⏎

Cube/Length/<other corner>: **8,6,0** ⏎

Height: **3** ⏎

Your screen should appear similar to Figure 9.7.

Creating Multiple Viewports

Now you have a 3D solid object on your screen. You will next create viewports to show several views of the object on your screen at the same time. This way you can create just one solid model and produce a drawing with the necessary two-dimensional orthographic views directly from that 3D model. Having several views also makes it easier to create the model. You have already used a single paper space viewport for printing and plotting in Tutorial 5. Now you will create four viewports and then change them so that they contain four different views of the model. Going back to the Rose Bowl analogy from Tutorial 5, you can think of the four separate viewports as four separate TV screens, each one of them showing the Rose Bowl game. Each TV set can show a different view of the game (like when the same event is on several different stations at once). Each cameraperson is looking at the game from a different angle, producing a different view on each TV set. Yet there is only one Rose Bowl being played. Multiple viewports in AutoCAD work in the same way. You can have many viewports that each contain a different view of the model, but there is only one model space and each object in it need only be created once.

On your own, set layer VPORT as the current layer before you continue.

Next you will switch to paper space by selecting it from the View menu. When you do this, AutoCAD automatically sets Tilemode to 0 if needed. Once paper space is enabled, do not change Tilemode. Doing so disables paper space and can become confusing when you are using paper space viewports.

Pick: **View, Paper Space**

You switch from model space into paper space. Notice the triangular paper space icon in the lower left of your screen. As you remember from Tutorial 5, paper space allows you to arrange views of your model, text, and other objects as you would on a sheet of paper. You cannot see model space until you create a viewport in paper space (like turning on your TV). You use the Mview command to create viewports. You will pick it from View on the menu bar.

Pick: **View, Floating Viewports, 4 Viewports**

> ■ *TIP* Floating Viewports will be grayed out unless you are in paper space. ■

The Mview command is echoed in the command prompt area, and its prompts appear. You will place the viewports inside the area you want to plot full size, centered on your page, as you did in Tutorial 5. You may find the values used below work for your printer. If you determined a different setting in Tutorial 5, use the one that works for your printer.

Fit/<First Point>: **.25,.25** ⏎
Second point: **10.25,7.75** ⏎

Four viewports appear on your screen, each one containing an identical view of the object. Now you will set the paper space limits to the size of the sheet of paper you will use when plotting. Each viewport, and paper space, can have its own Limits, Grid, and Snap settings.

The default limits in AutoCAD are set to 12,9. Because the paper size that you will print on is only 11×8.5, the default limits are too large. It is a good practice to set the limits in paper space close to the paper size, because that way you can see how much area you actually have to draw on the sheet. Starting the limits at 0,0 usually makes it easier to position your plots on the page where you want them when you are printing. Also, this way, if you use Zoom All, the graphics window fills your screen and not a larger area (which would make your viewports smaller on the screen and harder to work in). While still in paper space,

Pick: **Data, Drawing Limits**

ON/OFF/<Lower left corner> <0.0000,0.0000>: ⏎
Upper right corner <12.0000,9.0000>: **10.5,8** ⏎

To fill the screen area with the new drawing limits, you will pick the Zoom All icon from the Standard toolbar.

Pick: **Zoom All icon**

On your own, set layer MODEL as the current layer. Your screen should appear similar to Figure 9.8.

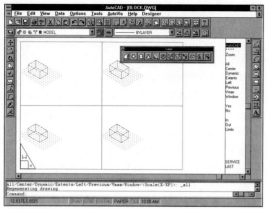

Figure 9.8

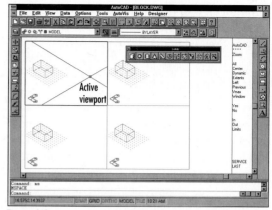

Figure 9.9

Next you will change the viewpoint for each viewport so that together they show a top, front, right-side, and isometric view of the model.

Remember, when you are in paper space, you cannot change things that are in model space. The original box was created in model space, where you create and make changes to your model. To switch back to model space,

Command: **MS** ⏎

You can tell that you are in model space when you see that the UCS icon has reappeared. It is now displayed in *each* of the viewports.

You will make the top left-hand viewport active and change it to show the view of the model looking straight down from the top.

Selecting the Active Viewport

To make the upper left viewport active, you will move the arrow cursor until it is positioned in the top left-hand viewport and then press the pick button.

Pick: *(anywhere in the upper left viewport)*

You will see the crosshairs appear in the viewport, as shown in Figure 9.9, indicating that this is now the active viewport.

You can only draw in and pick points from a viewport when it is active. After you create something, whatever you create is visible in all other viewports showing that area of the WCS, unless the layer that the object is on is frozen in another viewport. (Using the Vplayer command, you can freeze layers in specific viewports.) Keep in mind that while you are using a certain viewport to *access* model space, there *is* only one model space. Because of this, you can start drawing something in one viewport and finish drawing it in another.

■ *TIP* You can press Ctrl-R to toggle the active viewport. This is helpful when you are unable to pick in the viewport for some reason. ■

Changing the Viewpoint

Now you will change the viewpoint for the upper left-hand viewport so that it shows the top view of the object. By controlling the viewpoint using the Vpoint command, you can create a view from any direction. The numbers you enter in the Vpoint command are the X, Y, Z coordinates of a point that defines a *vector*, or directional line. The other point defining the vector is the origin point, 0,0,0. Your line of sight, or viewpoint, is defined by this vector, or imaginary line, toward the origin from the coordinates of the point you enter. The actual *size* of the number you enter does not matter, only the *direction* it establishes. Thus entering *2,2,0* is the same as entering *1,1,0*, as only the direction of sight is determined by this number. See Figure 9.10.

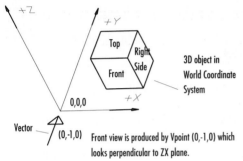

Figure 9.10

You can also select the Rotate option and establish the viewing direction by specifying the rotation angle in the XY plane from the X axis, and then the rotation above the XY plane for the viewpoint. See Figure 9.11. Experiment with these options on your own.

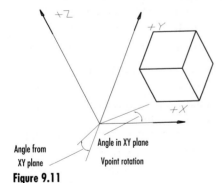

Figure 9.11

Using the View Flyout

 The View flyout located on the Standard toolbar is shown in Figure 9.12. You can use it to quickly pick different viewing directions for the active viewport. The selections on the View flyout use the Vpoint command to select common viewing directions. You will use it to pick the Top View icon.

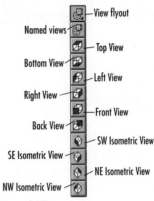

Figure 9.12

Pick: **Top View icon**

The view in the upper left viewport changes. Now you are looking straight down on top of the box. The top view of the box is too large to fit entirely within view. Your screen should look like Fig 9.13, which shows the 3D box viewed from the top in the upper left viewport. Notice that the UCS icon is viewed straight on now that you have changed the viewing direction.

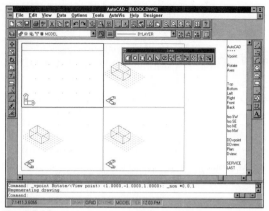

Figure 9.13

> ■ *TIP* You can turn on the View toolbar using the selection Tools, Toolbars, View from the menu bar; it contains the same selections as the View flyout. You may find it convenient to turn on this toolbar so that the selections for changing the view direction are always available on the screen and easy to pick. ■

You will use the Zoom command with the Scale XP (meaning relative to paper space) option to move the viewing distance farther away so that the object appears smaller and will fit into the viewport.

 The Left option of the Zoom command allows you to specify the coordinates of the point that will be placed in the lower left-hand corner of the viewport when the zoom is performed. Using Zoom Left will help make the views align between viewports. You will type the alias for the Zoom command. instead of using the icon.

Command: **Z** ⏎

All/Center/Dynamic/Extents/Left/Previous/Vmax/Window/
    <Scale(X/XP)>: **L** ⏎

Lower left corner point: **0,0,0** ⏎

Magnification or Height <4.5000>: **.5XP** ⏎

The top view of the box should fit into the viewport at half paper space scale. Your drawing should appear similar to Figure 9.14.

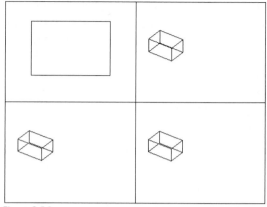

Figure 9.14

Each viewport can contain its own setting for Grid, Snap, and Zoom. Now set the grid and snap to .5 in the upper left viewport on your own. Then, make the lower left viewport active by moving the arrow cursor into that viewport and pressing the pick button. The crosshairs appear in the lower left viewport.

 You will use the View flyout on the Standard toolbar to pick the Front View icon, and change the view of the lower left viewport.

Pick: **Front View icon**

Now the view in this viewport is looking onto the front of the object, as though you took the top view and tipped it 90° away from you. (Imagine that the original Y axis is projecting straight into your monitor.) The object is again too large to fit in the viewport. Think of the distance of the object from the viewport or drawing screen as determined by the Zoom command. Use Zoom Left and Zoom Scale XP to zoom out so that the entire view will fit in the viewport. This time you will specify the coordinates 0,0,–1 to be located in the lower left corner of the viewport. You will do this because, since the model starts at the Z coordinate 0, if 0,0,0 is in the lower left corner for the front view, the object will be touching the bottom of the viewport. Using –1 for the Z coordinate will locate the view in a better position.

Pick: **Zoom Left icon**

Lower left corner point: **0,0,–1** ⏎

Magnification or Height <4.5000>: **.5XP** ⏎

Now you see the entire front view of the object in the viewport. Your screen should appear similar to Figure 9.15.

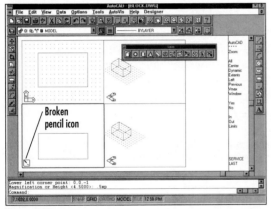

Broken
pencil icon

Figure 9.15

Notice the broken pencil icon in the lower left corner. This icon tells you that your current XY drawing plane or UCS is parallel to the viewing direction. If you were to try to draw, you couldn't keep track of what was being created. Every shape you drew would appear as a straight line because you are viewing your XY drawing surface edgewise. Think about the results you would get if you looked edgewise at a piece of paper at eye level and then tried to

draw on it. You may also have seen the message *Grid too dense to display* when you originally changed your viewpoint. If the coordinate system is viewed edge on, the grid is, of course, a solid line of dots, making it too dense to appear on the screen in a meaningful fashion.

Creating User Coordinate Systems

You can have more than one coordinate system in AutoCAD. The object is stored in the World Coordinate System, which stays fixed, but you can use the UCS command to create a User Coordinate System oriented any way you want. The UCS command allows you to position a User Coordinate System anywhere with respect to the WCS. You can also use it to change the origin point for the coordinate system, and save and restore named coordinate systems.

You will create a new UCS that is rotated 90° around the WCS X axis, so that the XY plane of the new UCS is parallel to the front viewing plane. This will provide a coordinate system that will make it easy to model objects in the front view. AutoCAD uses the right hand rule to determine the positive direction for rotation around an axis. To determine the direction of rotation, point your thumb in the positive direction of the axis in question. Your fingers curl in the positive rotation direction. You can use the Standard toolbar UCS flyout shown in Figure 9.16 to quickly pick the options of the UCS command.

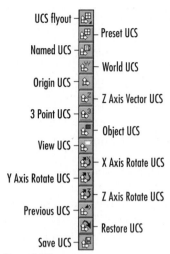

UCS flyout

Preset UCS

Named UCS

World UCS

Origin UCS

Z Axis Vector UCS

3 Point UCS

Object UCS

View UCS

X Axis Rotate UCS

Y Axis Rotate UCS

Z Axis Rotate UCS

Previous UCS

Restore UCS

Save UCS

Figure 9.16

Pick: X Axis Rotate UCS icon

Rotation angle about X axis <0>: **90** ⏎

The grid in your drawing now appears parallel to the front view (lower left viewport). The broken pencil now appears in the top view (upper left viewport). Refer to Figure 9.17.

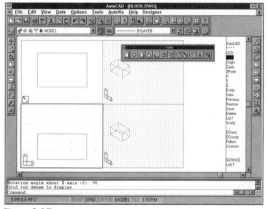

Figure 9.17

Now the XY plane of the current coordinate system is lined up with the front view, but not the top view. You will save this UCS so you can return to it later when you want to work in the front view. You will press ⏎ or the return button on your pointing device to repeat the UCS command, pick the Save option, and name the coordinate system FRONT because you see it in the front view. To restart the UCS command,

Command: ⏎

UCS Origin/ZAxis/3point/OBject/View/X/Y/Z/Prev/
    Restore/Save/Del/?/<World>: **S** ⏎

?/Desired UCS name: **FRONT** ⏎

Set the grid and snap spacing for this viewport to .5 units on your own now.

> ■ *TIP* The UCS command and the View command work well in conjunction with one another. After you have established a User Coordinate System, you can set up a view that aligns with it by using the Plan command and selecting the UCS option. This allows you to easily define a view that corresponds to any defined UCS. The View command lets you save a named view in the drawing. Use the View command to save the view with the same name as the UCS with which it is aligned. Save named views of any important viewing directions in your model. ■

Now that you have completed setting the viewing direction and UCS for the front view, make the lower right viewport active on your own. Now the crosshairs appear in the lower right viewport. Next, you will create a side view of the model in this viewport.

 You will use the Right View icon from the View flyout to select the viewing direction for this view.

Pick: Right View icon

The view changes to show a right-side view of the object. Again, the entire object will not fit in the viewport at the current zoom factor. You will use the Zoom Left command and the Zoom XP option to zoom the view relative to paper space, as you did with the first two views.

Command: **Z** ⏎

All/Center/Dynamic/Extents/Left/Previous/Vmax/Window/ <Scale(X/XP)>: **L** ⏎

Lower left corner point: **0,–1,0** ⏎

Magnification or Height <4.5000>: **.5XP** ⏎

Now the right-side view will fit in the lower right viewport. But this viewport has the broken pencil icon now. You will create another UCS parallel to this view to make it easy to work in this viewport.

 The View UCS icon creates a UCS that is parallel to the viewing plane in the active viewport and at the origin of the Z axis.

Pick: **View UCS icon**

The grid appears in the lower right-hand viewport and the broken pencil icon appears in the top and front views. Save this User Coordinate System so you can return to it when you want to draw in the right-side view. Press ⏎ to repeat the previous UCS command.

Command: ⏎

Origin/ZAxis/3point/OBject/View/X/Y/Z/Prev/ Restore/Save/Del/?/<World>: **S** ⏎

?/Desired UCS name: **SIDE** ⏎

■ *TIP* If you have trouble lining up the views, use MVSETUP and use the Align option. Select Horizontal or Vertical alignment. Then pick a point in one viewport (using Snap to Endpoint or some other object snap) and pick a point that should align in the other viewport (again using an object snap). The view in the viewport that you pick second will be shifted to line up with the point you selected in the first viewport. You can use Pan in the first viewport to locate the view where you want it before you begin the MVSETUP command. ■

On your own, set the grid and snap for the lower right viewport to .5 units. When you are finished, your screen should look like Figure 9.18.

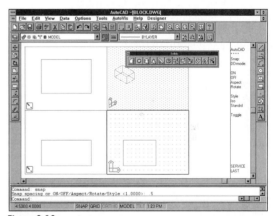

Figure 9.18

Make the upper left viewport active on your own before you continue. To restore the World Coordinate System to the top left viewport,

Command: **UCS** ⏎

Origin/ZAxis/3point/OBject/View/X/Y/Z/Prev/ Restore/Save/Del/?/ <World>: ⏎ *(to accept the default of World)*

This returns the drawing to the original World Coordinate System. Now the UCS icon shows the letter W and appears parallel in the top view. In AutoCAD the default XY plane is thought of as the *plan view* of the WCS. A plan view is basically a top view. It may sound familiar if you have worked with architectural drawings. Any time you want to restore the original coordinate system, set the UCS equal to World Coordinates with the command you used above.

Make the upper right-hand viewport, in which the box appears as its 3D shape, active by moving the arrow cursor to this viewport and pressing the pick button. You will see the crosshairs appear in the viewport when it is selected. On your own, pick the SE Isometric view from the View flyout to change the view for this viewport. Use Zoom .8X to make the view 80% of the previous size so that it fills the upper right viewport.

In the next portion of the tutorial, you will familiarize yourself with the basic solid modeling primitives that you can use to create drawing geometry. You will add and subtract them with Boolean commands to create more complicated shapes.

Creating Cylinders

 Next you will add a cylinder to the drawing. Later, you will turn the cylinder into a hole by using the Subtract command. You create cylinders by specifying the center of the circular shape in the XY plane and the radius or diameter, then giving the height in the Z direction of the current UCS. You can use the Baseplane option to change the height above the XY plane that the circular shape is drawn in. The Elliptical option allows you to specify an elliptical shape instead of circular, and then go on to give the height.

The Center of other end option of the Cylinder command allows you to specify the center of the other end, by picking or typing coordinates, instead of giving the height. The Cylinder flyout is shown in Figure 9.19.

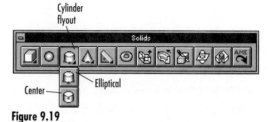

Figure 9.19

■ *TIP* If you do not specify a Z coordinate, it is assumed to be your current elevation, which is presently 0. ■

Pick: **Cylinder Center icon**

Elliptical/<center point> <0,0,0>: **4,4** ⏎

Diameter/<Radius>: **.375** ⏎

Center of other end/<Height>: **3** ⏎

Next you will change the color of the cylinder. For many drawings it is the best practice to make objects on separate layers and use the layer to determine the color, but this is not always the case with solid modeling. When you use a Boolean operator to join two objects, the result of that operation will always be on the current layer, unless you change it later. Sometimes it is easier to see features if they are different colors, so you may want to set the color for an object before you join it with another using Boolean operations. To change the color of the cylinder, which you will use to create a hole,

Pick: **Properties icon**

Select objects: *(pick the cylinder)*

Select objects: ⏎

The Modify 3DSolid dialog box appears on the screen, as shown in Figure 9.20. You will use the Color option to set the color for the cylinder to blue while leaving it on layer MODEL (which will still remain magenta). Currently the color is set to BYLAYER.

Figure 9.20

Pick: **Color**

From the color chart that appears on your screen, select the color blue from the band of standard colors across the very top of the box. It is color number 5, the fifth color from the left.

Pick: **OK** *(to accept the color selection)*

Pick: **OK** *(to exit the Modify 3DSolid dialog box)*

The cylinder should be changed to the new color. Your screen should look like Figure 9.21.

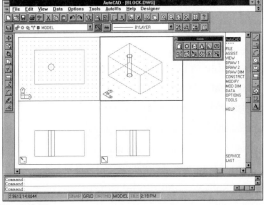

Figure 9.21

Now you have a box and a cylinder, each occupying the same space. Although this could not be done in the real world, in the drawing database, these two objects are both occupying the volume inside the cylinder.

Setting Isolines

Before you continue, you will set the variable called Isolines. This variable controls the wireframe appearance of the cylinders, spheres, and tori (like a 3D donut) on the screen. You can set the value for Isolines between 4 and 2047, to control the number of tessellation lines used to represent rounded surfaces. *Tessellation lines* are lines displayed on a curved surface that help you better visualize the surface. The number of tessellation lines you set will be shown on the screen representing the contoured surface of the shape. The higher the value for Isolines, the better the appearance of rounded wireframe shapes will be. The default setting of 4 looks very poor, but saves time in drawing. The highest setting may look the best, but it takes more time for the calculations, especially when processing a complex drawing.

You will set the value to 12. You can change the setting for Isolines and regenerate the drawing at any time. You will use the Regenall command to regenerate all of the viewports to show the new setting for Isolines.

Command: **ISOLINES** ⏎

New value for ISOLINES <4>: **12** ⏎

Command: **REGENALL** ⏎

The cylinder in the upper right viewport should look similar to Figure 9.22.

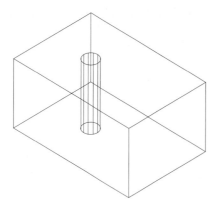

Figure 9.22

Next you will use the Boolean operator Subtract to remove the cylinder from the box so that it forms a hole.

Building Complex Solid Models

You can create complex solid models using Boolean operators, which find the *union* (addition), *difference* (subtraction), and *intersection* (common area) of two or more sets. These operations are named for Irish logician and mathematician George Boole, who formulated the basic principles of set theory. In AutoCAD the sets can be 2D areas (called *regions*), or they can be 3D solid models. Often *Venn diagrams* are used to represent sets and Boolean operations. Figure 9.23 will help you understand how Union, Subtract, and Intersection work. The order in which you select the objects is only important when subtracting (i.e., A subtract B is different from B subtract A).

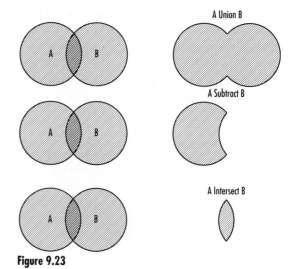

Figure 9.23

The Boolean operators are on the Explode flyout shown in Figure 9.24.

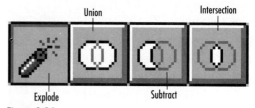

Figure 9.24

 You will use Subtract to remove the volume of the cylinder from the box, thereby forming a new solid model with a hole in it.

Pick: **Subtract icon**

Select solids and regions to subtract from...

Select objects: *(pick the box and press* ⏎*)*

Select solids and regions to subtract...

Select objects: *(pick the cylinder and press* ⏎*)*

The resulting solid model is a rectangular prism with a hole through it. If you selected the items in the wrong order (i.e., picked the cylinder and subtracted the box from it) the result would be a null solid model, one where there is no volume. If you picked both the box and the cylinder and then pressed ⏎, then the command would not work unless you picked something else to subtract. If you make a mistake, you can back up by typing U ⏎ at the command prompt to use the Undo command, and then try it again.

Next you will shade your model in the top right-hand viewport. You should have the AutoVis selection available on the menu bar. If you have not loaded AutoVision or have not modified your menus, please refer to the Getting Started section of this manual.

Pick: **AutoVis, Render**

The Render dialog box appears on your screen, as shown in Figure 9.25.

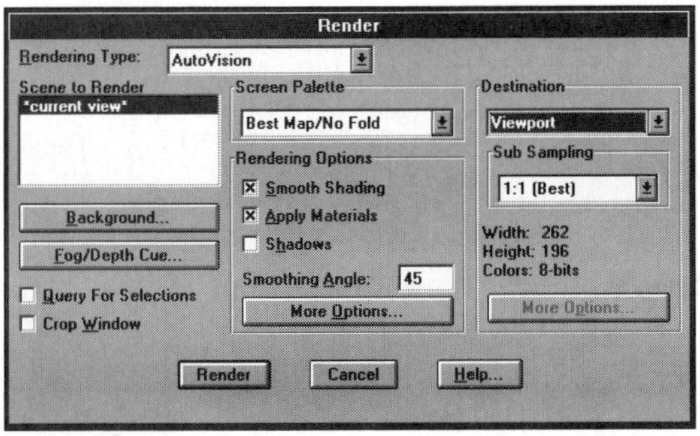

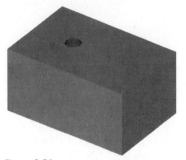

Figure 9.25

You will accept the defaults, using AutoVision as the type of rendering to perform. You will learn more about using AutoVision and its features in Tutorial 17. If you do not have AutoVision loaded, you can use AutoCAD's Shade or Render command. Use on-line help to get more information about these commands if necessary.

Pick: **Render**

The model in the active viewport becomes shaded. Notice that you can now tell that the cylinder has formed a hole in the block. The color inside the hole is blue and the rest of the block is magenta.

Your drawing in the upper right viewport should appear similar to Figure 9.26.

Figure 9.26

■ *TIP* The Facetres variable adjusts the smoothness of shaded objects and objects from which hidden lines have been removed. Its value can range from 0.5 (the default) to 10. Viewres controls the number of straight segments used to draw circles and arcs on your screen. Viewres can be set between 1 and 20000. You can type these commands at the command prompt and set their values higher to improve the appearance of rendered objects. ■

You will regenerate your drawing to eliminate the shading so you can continue to work on it by typing *REGEN* at the command prompt on your own before you continue. (You can't select objects when they are shaded.)

Command: **REGEN** ⏎

Save your drawing, *block.dwg*, on your own before continuing. Saving periodically will prevent you from losing your drawing in the event of a power failure or other hardware problem. Also, it is useful to save after you complete a major step in your work, before you go on to the next thing. That way, if you want to return to the previous step, you can open the previous version of the drawing, discarding the changes you have made.

Next you will use the wedge primitive to create a wedge and subtract it from the block.

Creating Wedges

The Wedge flyout has two selections for drawing solid wedges: Wedge Center and Wedge Corner, shown in Figure 9.27. The Wedge Center selection asks you to specify the center point of the wedge you want to draw. As with other solid primitive commands, you can use the Baseplane option to select a different height, or Z level, at which to draw.

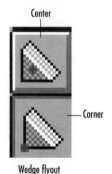

Center

Corner

Wedge flyout
Figure 9.27

 The Corner selection for the wedge primitive starts by drawing the rectangular shape from the first point you select to the second point you select as the diagonal of the rectangular base. The height given starts at the first point you select and gets smaller in the X direction toward the second point. You can also type the option

letters during the command to enter length, width, and height. You will use Wedge Corner to continue modifying your shape.

Pick: **Wedge Corner icon**

Center/<Corner of Wedge> <0,0,0>: **8,6** ⏎

Cube/Length/<other corner>: **6,2** ⏎

Height: **3** ⏎

Now you will use the Boolean operators to subtract the wedge from the object.

Pick: **Subtract icon**

Select solids and regions to subtract from...

Select objects: *(pick the box with the hole and press ⏎)*

Select solids and regions to subtract...

Select objects: *(pick the wedge and press ⏎)*

With the wedge subtracted, your drawing should look similar to Figure 9.28.

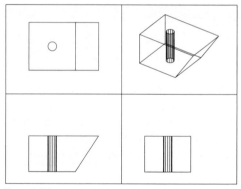

Figure 9.28

Next you will use Hide to remove hidden lines from the upper right-hand view. Make sure the upper right viewport is active. Then,

Command: **HIDE** ⏎

You will see the 3D object on your screen with the hidden lines removed. The shape in the upper right viewport should look like Figure 9.29.

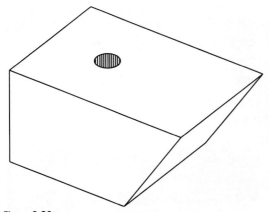

Figure 9.29

Regenerate your drawing display from the drawing database, so that you can continue to work using wireframe.

Command: **REGEN** ⏎

Creating Cones

The cone primitive creates a solid cone defined by a circular or elliptical base tapering to a point perpendicular to the base. It is similar to the cylinder primitive that you have already used. The circular or elliptical base of the cone is always created in the XY plane of the current UCS. The height is along the Z axis of the current UCS. The Cone flyout is shown in Figure 9.30.

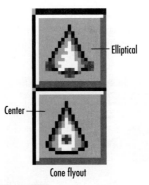

Elliptical

Center

Cone flyout

Figure 9.30

Next you will create a cone, using the cone primitive, and then subtract it from the block to make a countersink for the hole. It will have a circular base and a negative height. Specifying a negative height causes the cone to be drawn in the opposite direction from a positive height (i.e., in the negative Z direction).

Pick: **Cone Center icon**

Elliptical/<center point> <0,0,0>: **4,4,3** ⏎

Diameter/<Radius>: **.625** ⏎

Apex/<Height>: **–.75** ⏎

Use the Boolean operator to subtract the cone from the block to make a countersink for the hole.

Pick: **Subtract icon**

Select solids and regions to subtract from...

Select objects: *(pick the block and press* ⏎*)*

Select solids and regions to subtract...

Select objects: *(pick the cone and press* ⏎*)*

Next, shade the object in the upper right viewport, using the AutoVision Render command as you did before.

Pick: **AutoVis, Render**

Pick: **Render**

The upper right viewport now displays a view of the shaded object. Your screen should look like Figure 9.31.

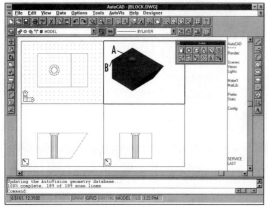

Figure 9.31

Next you will use the cylinder primitive to add a rounded surface to the top of the block. On your own, use Regenall to regenerate the views.

Selecting a Named Coordinate System

 First you will restore the User Coordinate System SIDE that you saved earlier for drawing in the side view. This will make it easier for you to create objects that are parallel to the side view. You will use the Standard toolbar to pick the UCS flyout selection for Named UCS.

Pick: **Named UCS icon**

The UCS Control dialog box appears on your screen, as shown in Figure 9.32. You will use it to select the User Coordinate System for the side view, which you saved earlier in this tutorial.

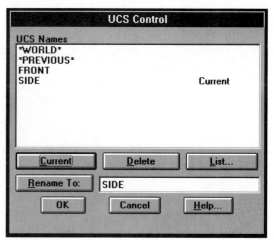

Figure 9.32

Pick: **SIDE**

Pick: **Current**

Pick: **OK**

The grid appears in the side view. The broken pencil icon appears in the top and front views. In the isometric view in the upper right viewport, the grid changes to a different angle, since it now lines up with the side and not the top of the object.

On your own, make sure the upper right viewport is active.

Next you will use the Cylinder flyout and choose to draw a cylinder by specifying its center. Refer to Figure 9.31 for the points to select. You will use the Midpoint object snap to pick your points.

Pick: **Cylinder Center icon**

Elliptical/<center point> <0,0,0>: *(pick the Snap to Midpoint icon)*

mid of *(pick the back top line, labeled A)*

Diameter/<Radius>: *(pick the Snap to Endpoint icon)*

endp of *(pick the endpoint of the edge, labeled B)*

Center of other end/<Height>: **1** ⏎

Your drawing should look like Figure 9.33.

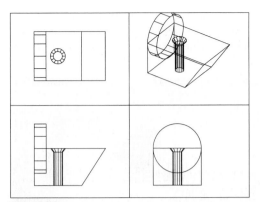

Figure 9.33

 Next you will use the Boolean operator Union to join the cylinder to the block.

Pick: **Union icon**

Select objects: **(pick the new cylinder and the block)**

Select objects: (↵)

When the objects are unioned, your drawing should appear similar to Figure 9.34.

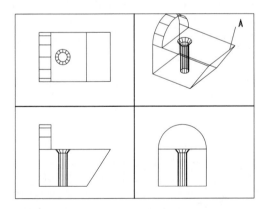

Figure 9.34

On your own, use the Pan command to pan the view of the object in the upper right viewport so that the entire object fits in the available space. Save your completed *block.dwg* drawing.

Using Fillet

AutoCAD's Fillet command lets you add concave or convex rounded surfaces between plane or cylindrical surfaces on an existing solid model. You have already used it to created rounded corners between 2D objects. Now use the command to create a rounded edge for the front, angled surface of the object. Refer to Figure 9.34 for your selection of point A.

Pick: **Fillet icon**

Polyline/Radius/Trim<Select first object>: **(pick on line A)**

Enter radius <0.0000>: **.5** (↵)

Chain/Radius<Select Edge>: (↵)

Your drawing with the rounded corner added should look like Figure 9.35.

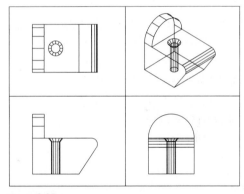

Figure 9.35

Next you will undo the fillet, and then use the Chamfer command to add an angled surface.

Command: **U** (↵)

The fillet that you added to your drawing is eliminated.

Using Chamfer

The Chamfer command works on solid models in a way similar to the Fillet command, except that it adds an angled surface instead of a rounded one. Selecting the Loop option allows you to add a chamfer all around a surface. Selection of the correct surfaces will be easier if you make the upper left-hand viewport active.

On your own, make sure the upper left viewport is active.

Pick: **Chamfer icon**

Polyline/Distance/Angle/Trim/Method/<Select first line>: **D** ⏎

Enter first chamfer distance<0.0000>: **.75** ⏎

Enter second chamfer distance<0.7500>: ⏎

Command: ⏎

Polyline/Distance/Angle/Trim/Method/<Select first line>: *(pick line A in Figure 9.34)*

Select base surface/Next/ <OK>: *(if the top surface of the object is highlighted, press ⏎; if not, type N so that the next surface becomes highlighted)*

Enter base surface distance<0.7500>: ⏎

Enter other surface distance<0.7500>: ⏎

Loop/Select edge: *(pick line A)* ⏎

■ *TIP* When you are picking on the solid model, it is often difficult to tell at first which surface will be selected because each wire of the wireframe represents not one surface, but the intersection between two surfaces. Many commands have the Next option to allow you to move surface by surface until the one you want to select is highlighted. If the surface you want to select is not highlighted, use the Next option until it is. When the surface you want to select is highlighted, press ⏎. ■

On your own, render the drawing in the upper right viewport. Your screen with the chamfer should look like Figure 9.36.

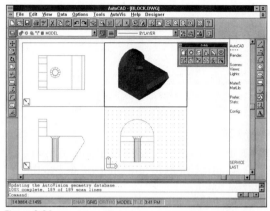

Figure 9.36

Plotting Solid Models from Paper Space

Next, you will plot your multiview drawing of the block. Change to paper space on your own by typing *PS* ⏎ at the command prompt. When you have done so, the paper space icon replaces the UCS icon.

You will use the Mview command with the Hideplot option so that when you plot your drawing from paper space, the back surface lines are automatically removed. When selecting the objects for the Hideplot option of the Mview command, remember that you are selecting which *viewports* will be shown with hidden lines removed. To select a viewport, pick on its border, not inside it.

Command: **MVIEW** ⏎

ON/OFF/Hideplot/Fit/2/3/4/Restore/<First Point>: **H** ⏎

ON/OFF: **ON** ⏎

Select objects: *(pick on the border of the upper right viewport)*

Select objects: ⏎

Nothing noticeable happens when you finish the command, but when you plot your drawing, the back surface lines will automatically be removed from this viewport.

On your own, make BORDER the current layer and freeze layer VPORT. This way the viewport borders will not print on your drawing. Use the Line command to draw a border around all of the viewports while you are in paper space. Add a title bar and notes if you wish as you did in Tutorial 5.

Make sure that you are in paper space, and plot the drawing limits at a scale of 1=1. The views you have drawn should be exactly half-size on the finished plot, because the Zoom XP scale factor was set to .5. Your plotted drawing should be similar to Figure 9.37.

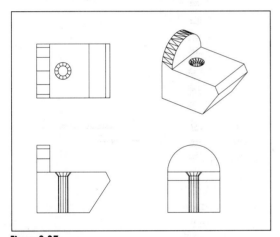

Figure 9.37

Saving Your Multiview Setup as a Prototype

Switch your drawing back to model space.

The UCS icons are restored to the viewports.

On your own, thaw layer VPORTS and set layer MODEL as the current layer in the drawing.

Now you will erase the object and save the basic settings to use as a prototype drawing when creating new 3D drawings. Type the alias for the Erase command at the prompt.

Command: **E** ⏎

Select objects: **(pick the solid block object you have drawn)**

Select objects: ⏎

It is erased from all viewports.

 Next you will restore the World Coordinate System, using the World UCS icon from the UCS flyout.

Pick: **World UCS icon** ⏎

Use Saveas to save this drawing to file name *solproto.dwg*.

Creating Solid Models with Extrude and Revolve

Next you will learn how to create new solid objects, using the extrusion and revolution methods. Start a new drawing, using the New icon from the Standard toolbar. Use the prototype drawing *solpro_d.dwg*, provided with your data files or that you just created, *solproto.dwg*. Call the new drawing *extrusn.dwg*.

Your screen should appear similar to Figure 9.38.

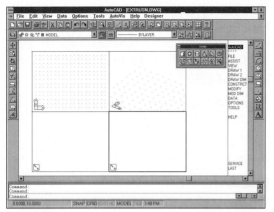

Figure 9.38

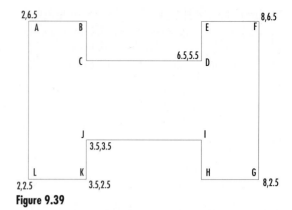

Figure 9.39

Since you started your new drawing from the solid modeling prototype, you will notice that the viewports are already created. The UCSs called FRONT and SIDE that you saved previously are still available with the UCS Restore option. Layer MODEL should be the current layer.

On your own, pick in the upper left viewport to make it active. Check to see that Grid and Snap are turned on. If they are not, turn them on.

You will draw the I shape in Figure 9.39, using the 2D Polyline command. The coordinates have been provided in Figure 9.39 to make it easier for you to select the points.

Pick: **Polyline icon** *(not 3D Polyline)*

From point: *(pick point A)*

Current line-width is 0.0000

Arc/Close/Halfwidth/Length/Undo/Width/<Endpoint of line>:
(pick points B–L in order)

Arc/Close/Halfwidth/Length/Undo/Width/<Endpoint of line>:
C ⏎

You will set the fillet radius to .25 and then use the Fillet command's Polyline option to round all of the corners of the polyline.

Command: **FILLET** ⏎

Polyline/Radius/Trim/<Select first object>: **R** ⏎

Enter fillet radius <0.0000>: **.25** ⏎

Command: ⏎

FILLET Polyline/Radius/Trim/<Select first object>: **P** ⏎

Select 2D polyline: *(pick the polyline you just created)*

12 lines were filleted

■ *TIP* If you did not close your object by typing *C*, you will get the message *11 lines were filleted.* In this case, undo the last two commands and redraw your I shape, closing the polyline with the Close option. ■

On your own, use the Snap grid to draw the two circles shown in Figure 9.40.

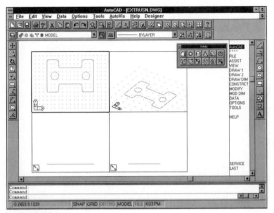

Figure 9.40

Region Modeling

You can also use the Boolean operators with the closed 2D shapes made by circles, ellipses, and closed polylines. You can convert closed 2D shapes to regions, which are essentially 2D solid models, or areas. Use the Region command from the Polygon flyout of the Draw toolbar. Region modeling and extrusion can be combined very effectively to create complex shapes. Turn Snap off on your own to make it easier to select.

Pick: **Region icon**

Select objects: *(use implied Crossing to pick the polyline and both circles; then press ⏎)*

3 regions created

Now you are ready to subtract the two circles from the polyline region to form holes in the area.

Pick: **Subtract icon**

Select solids and regions to subtract from...

Select objects: *(pick the polyline region and press ⏎)*

Select solids and regions to subtract...

Select objects: *(pick the circles and press ⏎)*

Nothing noticeable happens; however, now there is only one region, which has two holes in it. You can confirm this by selecting the object: notice that the circles become highlighted as part of the outline of the object.

Extruding a Shape

Now you can extrude this shape to create a three-dimensional object. Extrusion is the process of forcing material through a shaped opening to create a long strip that has the shape of the opening. AutoCAD's extrusion command works in a similar way to form the shape. Closed 2D shapes, such as splines, ellipses, circles, donuts, polygons, regions, and polylines can be given a height, or extruded.

Polylines must have at least 3 vertices and not more than 500 vertices in order to be extruded. You can specify a taper angle if you want the top of the extrusion to be a different size than the bottom. You can use the Path option to extrude a shape along a path curve that you have previously drawn. Path curves can be lines, splines, arcs, elliptical arcs, polylines, circles, or ellipses. You may encounter problems when trying to extrude along path curves with a high amount of curvature and where the resulting solid would overlap itself. The Extrude icon appears on the Solids toolbar.

Pick: **Extrude icon**

Select objects: *(pick the region and press ⏎)*

Path/<Height of Extrusion>: **2** ⏎

Extrusion taper angle <0>: ⏎

Now render the object in the upper right viewport. The solid object on your screen should look like Figure 9.41.

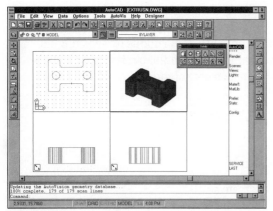

Figure 9.41

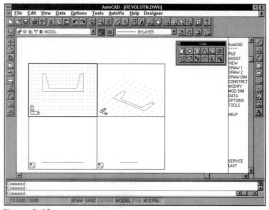

Figure 9.42

■ *TIP* Using the Pedit command to join objects into a closed polyline is an effective way to create 2D objects to extrude. ■

On your own, save your drawing *extrusn.dwg* and start a new drawing from the prototype provided with your data files, *solpro_d.dwg*. Call your new drawing *revolutn.dwg*.

Creating Solid Models by Revolution

Creating a solid model by revolution is similar in some ways to creating an extrusion. You can use it to sweep a closed 2D shape about a circular path to create a symmetrical solid model that is basically circular in cross section.

Use the 2D Polyline command on your own to draw the closed shape you see in Figure 9.42.

 Now you will revolve the polyline about an axis to create a solid model. You don't need to draw the axis line; you can specify it by two points. The Revolve icon is on the Solids toolbar.

Pick: **Revolve icon**

Select objects: *(pick the polyline and press* ⏎*)*

Axis of revolution - Object/X/Y/<Start point of axis>:
   (pick one endpoint of the bottom line)

<End point of axis>: *(pick the other endpoint)*

Angle of revolution <full circle>: ⏎

Next, you will use the Pan command in the lower left and lower right viewports so that all of the object shows in the views. On your own, pick to make the lower left viewport active. Then,

Pick: **Pan Point icon**

Displacement: **0,0,0** ⏎

Second point: **0,0,2** ⏎

The view should move up two units in the Z direction. Now repeat this on your own for the lower right viewport. When you have finished moving the view in the lower right viewport up, use the Pan command again in the upper right viewport so that the view looks similar to Figure 9.43.

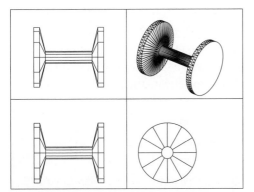

Figure 9.43

Save your drawing and then pick the Save As command to save the drawing with a new name. Call your new drawing *intsct.dwg*. On your own, erase the solid model you just made using the Revolve command.

Using the Boolean Operator Intersection

Like the Union and Subtract Boolean operators you learned to use earlier in the tutorial, Intersection lets you create complex shapes from simpler shapes. Intersection finds only the area that is common to the two or more solid models or regions you have selected. Next you will create the shape you see in Figure 9.44 by creating two solid models and finding their intersection.

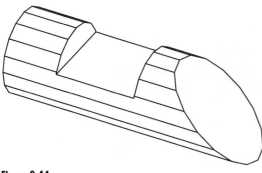

Figure 9.44

 First create the shape in the front view of a surface that you will extrude to create the angled face and notch. To restore the saved coordinate system aligned with the front view,

Pick: (in the lower left viewport to make it active)

Pick: **Restore UCS icon** (⏎)

?\Name of UCS to restore: **FRONT** (⏎)

The grid appears in the lower left viewport, parallel to the front view. Use the Polyline command to create a polyline that defines the shape of the object in the front view, as shown in Figure 9.45.

Pick: **Polyline icon**

From point: **2.5,0** (⏎)

Current line-width is 0.0000

Arc/Close/Halfwidth/Length/Undo/Width/<Endpoint of line>: **2.5,1.5** (⏎)

Arc/Close/Halfwidth/Length/Undo/Width/<Endpoint of line>: **3.5,1.5** (⏎)

Arc/Close/Halfwidth/Length/Undo/Width/<Endpoint of line>: **3.5,1** (⏎)

Arc/Close/Halfwidth/Length/Undo/Width/<Endpoint of line>: **5,1** (⏎)

Arc/Close/Halfwidth/Length/Undo/Width/<Endpoint of line>: **5,1.5** (⏎)

Arc/Close/Halfwidth/Length/Undo/Width/<Endpoint of line>: **5.5,1.5** (⏎)

Arc/Close/Halfwidth/Length/Undo/Width/<Endpoint of line>: **7,0** (⏎)

Arc/Close/Halfwidth/Length/Undo/Width/<Endpoint of line>: **C** (⏎)

Once you have drawn the shape of the object in the front view, you may need to use the Pan command so that you can see it in the top and side views.

Pick: (in the upper left viewport)

Pick: **Pan Point icon**

Displacement: **0,0,0** (⏎)

Second point: **0,0,−2** (⏎)

On your own, repeat these steps for the lower right viewport. When you have finished, your drawing will look like Figure 9.45.

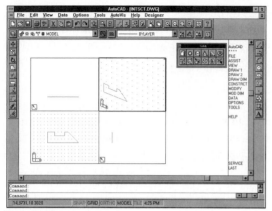

Figure 9.45

Now extrude this shape to form a solid model.

Pick: **Extrude icon**

Select objects: *(pick the polyline)* ⏎

Path/Height of extrusion: **−3** ⏎

Extrusion taper angle <0>: ⏎

The solid model that you see in Figure 9.46 is created.

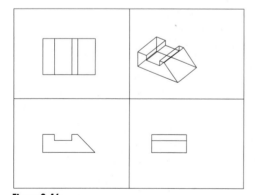

Figure 9.46

Next you will create the circular shape of the object in the side view, as shown in Figure 9.47.

On your own, pick in the lower right viewport to make it current. Restore the named UCS called SIDE to activate the grid and coordinates parallel to the side view.

Pick: **Circle 2 Point icon**

First point on diameter: **1.5,0** ⏎

Second point on diameter: **1.5,1.5** ⏎

In the top view, front view, and view of your entire object, the circle is drawn a distance away from the previously drawn solid model. You will use the Extrude command to elongate the circle into a cylinder.

Pick: **Extrude icon**

Select objects: *(pick the circle)* ⏎

Path/Height of extrusion: **10** ⏎

Extrusion taper angle <0>: ⏎

The isometric view of your drawing in the upper right viewport should look like Figure 9.47.

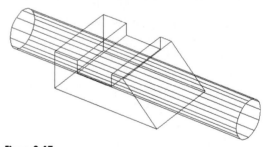

Figure 9.47

 Now you are ready to use the Intersection command to create a new solid model from the overlapping portions of the two solid models you have drawn.

Pick: **Intersection icon**

Select objects: *(pick the cylinder and the extruded polyline)* ⏎

The solid model is updated. Your resulting drawing will look like Figure 9.48.

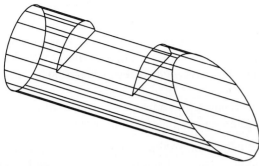

Figure 9.48

Save your *intsct.dwg* drawing.

> ■ *TIP* The Delobj variable controls whether the 2D object is automatically deleted after being extruded or revolved. A value of 1 means delete the object that was used to create the other object (the circle that created the cylinder, for example); a value of 0 means do not delete the object. The default setting in AutoCAD is 1 (delete). ■

On your own, experiment by typing *DELOBJ* at the command prompt and changing the setting to 0. Create a polyline and extrude it in one direction, using a taper angle. The original polyline will still remain. Select it again and extrude it in the other direction, using the same taper angle. This way you can create a surface that tapers in both directions.

On your own, hide the Solids toolbar, by picking on its close box.

Converting AME Solid Models

The Ameconvert command allows you to convert solid models which were created using AutoCAD's AME 2 or 2.1 solid modeler to ACIS solid models that are used in AutoCAD Release 13. To convert an AME model, type

AMECONVERT ⏎ at the command prompt and then select the objects to convert. AutoCAD ignores objects that are not AME 2 or 2.1 solid models. You may notice a difference in the models once they are converted as the ACIS modeler used by Release 13 is more accurate and may interpret some surfaces on the AME model as not meeting exactly, or holes as not going all of the way through the model.

The new ACIS modeler has several advantages over the models created using AME. The ACIS models and mass property calculations performed on them are more accurate than AME models. ACIS models can be created with more complex fillets and chamfers, and with extrusion along a path curve to easily create a wider variety of shapes. AME models had the advantage that because they were stored using Constructive Solid Geometry (CSG), the individual primitives could be edited even after being joined together in Boolean operations. ACIS models are stored in the drawing database using Boundary Representation (B-Rep) and therefore the primitives are not available for editing after being joined with Boolean operations.

Creating Surface Models

You can also create surface models using AutoCAD. AutoCAD's surface models are composed of a faceted polygonal mesh that approximates curved surfaces. Surface modeling is more difficult to use than solid modeling and provides less information about the object, as only the surfaces and not the interior volumes of the object are described in the drawing database. However, surface modeling is well suited to applications like modeling 3D terrain for civil engineering applications.

In general you should not mix solid modeling, surface modeling, and wireframe modeling, as these modeling methods cannot be edited in the same ways to create a cohesive single structure. Select the single method that is best for your application.

You can convert solid models to surface models using the Explode command. Surface models can be likewise converted to wireframe models. However, because they do not contain the same information, you cannot go from wireframe to surface models, or from surface models to solid models.

Next you will show the Surface toolbar. On your own use the Tools menu bar selection and,

Pick: **Toolbars, Surfaces**

The Surfaces toolbar appears on the screen as shown in Figure 9.49.

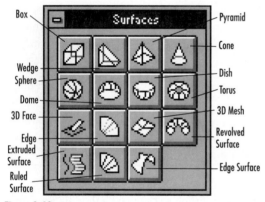

Figure 9.49

The Surface toolbar contains selections for Box, Wedge, Pyramid, Cone, Sphere, Dome, Dish, Torus, 3D Face, Edge, 3D Mesh, Revolved Surface, Extruded Surface, Ruled Surface, and Edge Surface. Notice that many of the selections are similar to the selections on

the solids toolbar. However, the objects created using surface modeling are like an empty shell. They do not contain information about the volume and mass properties of the object, as solid models do. An important consideration is that you cannot use Boolean operators to join surface models. You must edit the mesh which creates surface models differently.

As using Box, Wedge, Pyramid, Cone, Sphere, Dome, Dish, Torus, Revolved Surface, and Extruded Surface are basically similar to the methods you used earlier in the tutorial to create solid models, they will not be covered here. Instead, on your own, refer to AutoCAD's Help command for further information about creating these shapes.

The 3D Face selection lets you add a 3 or 4 sided surface anywhere in your drawing. It can be used to create surfaces on top of wireframe drawings.

The Edge command lets you change the visibility of an edge of a 3D face, which you created with the previous command. You can use it to hide edges that join with other 3D faces, or to hide back edges which may not hide correctly using the Hide command.

In this tutorial, you will learn to create a 3D Mesh, an Edge Surface, and a Ruled Surface. Understanding how to create these will allow you to create a wider variety of shapes that are especially useful for modeling 3D terrain, like mountainous topography.

On your own, start a new drawing from the prototype drawing *contrdat.dwg* which was provided with the data files. Name your new drawing *surf1.dwg*. When you have finished, your screen should look like Figure 9.50.

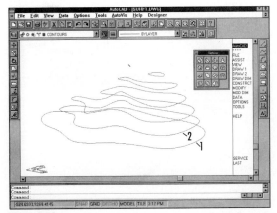

Figure 9.50

On the screen are 2D splines representing contours (lines of equal elevation on a contour map). The splines have been moved along the Z axis so that they are at different heights. The bottom line is at Z 200, the next at Z 220; the next at Z 240 and so forth. You can create these contour lines on your own by drawing splines (or polylines) through your data points, joining points of equal elevation. Then use Move to relocate the resulting spline. Specify a base point of 0,0,0 and a displacement of 0,0,200 (or whatever your elevation may be). If needed, change the viewpoint and use Zoom so that you see the lines clearly.

On your own, make layer MESH current.

Creating a Ruled Surface

Using the Ruled Surface selection from the Surfaces toolbar, you can create a surface mesh between two graphical objects. The objects can be lines, points, arcs, circles, ellipses, elliptical arcs, 2D polylines, 3D polylines, or splines. The objects can either be both open (like lines) or both closed (like circles). Points can be connected to either open or closed objects.

Controlling Mesh Density

The mesh comprising a surface is defined in terms of a *matrix* of M and N vertices. Two system variables control the density of the mesh for creating surfaces. M and N specify the column and row locations of vertices. Surftab1 controls the density of the mesh in the M direction. Surftab2 controls the density of the mesh in the N direction. The larger the value for Surtab1 and Surftab2, the more tightly the generated mesh will fit the initial objects selected. As with other commands, increasing the density of the mesh increases the time for calculation and display of the object.

The Surftab1 variable controls the density of the mesh for Ruled Surfaces, because they are always a 2 × Surftab1+1 mesh. To set the value for Surftab1,

Command: **SURFTAB1** ⏎
New value for SURFTAB1 <6>: **40** ⏎

 Next you will create a ruled surface between the two bottom splines. Use the Surfaces toolbar to select Ruled Surface. Refer to Figure 9.50 for the lines to select.

Pick: **Ruled Surface icon**
Select first defining curve: *(pick 1)*
Select second defining curve: *(pick 2)*

The ruled surface will be added between the two splines as shown in Figure 9.51.

Figure 9.51

> ■ *TIP* When you are picking the defining curves, pick in a similar location on each curve. If you pick on the opposite end of the second curve, the generated mesh will be intersecting. ■

On your own, repeat the command for Ruled Surfaces between curves 2 and 3, 3 and 4, 4 and 5, and 5 and 6 shown in Figure 9.51. You may need to zoom in so that you can pick the red splines. (Remember Zoom is transparent and can be used during the command.) If you pick on the blue mesh, you will not be able to use it as a defining curve. When you have finished, your drawing will look like Figure 9.52. Notice that for the upper curves, you can tell that half of the mesh for the ruled surface is connected to each curve. To create an accurate model, sometimes you must change Surftab1 to vary the mesh refinement so that there is not as much interpolation.

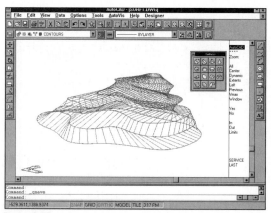

Figure 9.52

On your own, freeze layer CONTOURS to remove the original splines from the screen and save your drawing.

Start a new drawing on your own from the prototype *edgsurf.dwg*. Name your new drawing *surf2.dwg*. On your own, make MESH the current layer. Your screen should look like Figure 9.53.

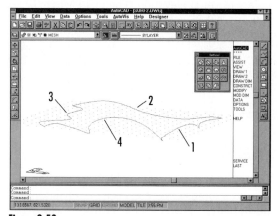

Figure 9.53

You will use the four 3D splines shown on the screen to define an edge surface.

Creating an Edge Surface

 The Edge Surface selection from the Surfaces toolbar creates a mesh defined by four edges you select. The mesh density for an edge surface is controlled by the Surftab1 and the Surftab2 variables. The Edge Surface selection creates a *Coons patch* mesh. A Coons patch is a *bicubic surface* interpolated between the four edges. First set the values for Surftab1 and Surftab2 to control the mesh density. Then you will select Edge Surface from the Surfaces toolbar. Refer to Figure 9.53 for the splines to select.

Command: **SURFTAB1** ⏎
New value for SURFTAB1 <6>: **20** ⏎
Command: **SURFTAB2** ⏎
New value for SURFTAB2 <6>: **20** ⏎
Pick: **Edge Surface icon**
Select edge 1: *(pick 1)*
Select edge 2: *(pick 2)*
Select edge 3: *(pick 3)*
Select edge 4: *(pick 4)*

The edge surface appears in your drawing as shown in Figure 9.54.

Figure 9.54

Save your drawing on your own.

Next you will define a rectangular mesh. You will erase the objects from the screen and use Save As to give the drawing a new name.

Pick: **Erase icon**
Select objects: **ALL** ⏎
Select objects: ⏎

All of the objects are erased from the screen. On your own, use the Save As command to save the drawing with the name *mesh.dwg*. The name *mesh.dwg* appears in the titlebar at the top of the current drawing.

Creating 3D Mesh

 The 3dmesh command lets you construct rectangular polygon meshes by entering the X, Y, and Z coordinates of the points in the mesh. First you define the number of columns (M) and then the number of rows (N) for the mesh matrix. The values for M and N must fall between 2 and 256. After M and N are defined, you are prompted to type in the values for the vertices defining the mesh. You can often use a lisp program or script in conjunction with the 3dmesh command to automate creation of the mesh.

Pick: **3D Mesh icon**
Mesh M size: **3** ⏎
Mesh N size: **3** ⏎
Vertex (0, 0): **0,0,0** ⏎
Vertex (0, 1): **15,0,1** ⏎
Vertex (0, 2): **30,5,2** ⏎
Vertex (1, 0): **0,15,0** ⏎
Vertex (1, 1): **18,18,3** ⏎
Vertex (1, 2): **30,15,0** ⏎
Vertex (2, 0): **0,30,2** ⏎
Vertex (2, 1): **15,30,0** ⏎
Vertex (2, 2): **30,34,3** ⏎

The mesh appears in your drawing. You will change the size of your drawing so that you can see it in its entirety. You will use Zoom Extents to do this.

 You will change your view so that you can see the mesh clearly in your drawing.

Pick: **Zoom Extents icon**

Pick: **SW Isometric View icon**

Your drawing should appear similar to Figure 9.55, which has the coordinates of the vertices noted.

Now you know how to establish viewports and viewing directions and create and join the basic shapes used in solid modeling. With these tools you can create a wide variety of complicated shapes. In the next tutorials, you will learn how to apply more of the power of solid modeling to change the solid models and analyze the mass properties. Practice creating shapes and working with the User Coordinate Systems on your own.

Save your drawing and close toolbars before exiting AutoCAD. You are now ready to begin the next tutorial.

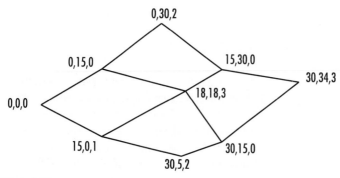

Figure 9.55

bicubic surface
Boolean operators
Coons patch
coordinate system locator
difference
extrusion
globe
intersection
mass properties

matrix
perspective icon
plan view
regions
revolution
solid modeling
surface modeling
tessellation lines
union

User Coordinate System
(UCS)
vector
Venn diagrams
viewpoint
wireframe modeling
World Coordinate System
(WCS)

3D Mesh
Box Center
Box Corner
Cone Center
Cylinder Center
Delobj
Edge Surface
Extrude
Facetres
Front View
Hide
Intersection
Isolines
Model Space
Named User Coordinate
System
Paper Space

Plan
Regenerate All
Region
Render
Restore User Coordinate
System
Revolve
Right View
Ruled Surface
Shade
Snap to Midpoint
Southwest Isometric View
Subtract
Surftab1
Surftab2
Top View
Union

User Coordinate System
User Coordinate System
icon
View
View User Coordinate
System
Viewpoint
Wedge Center
Wedge Corner
World User Coordinate
System
X Axis Rotate User
Coordinate System
Zoom Extents
Zoom Left

Use solid modeling to create the parts shown according to the specified dimensions.

9.1 Connector

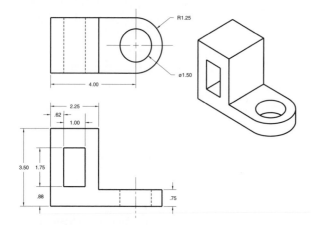

9.2 Angle Link

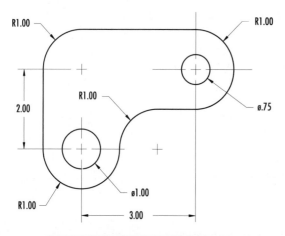

9.3 Support Base

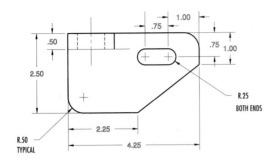

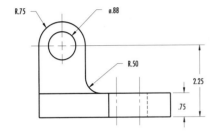

9.4 Chess Piece

Create the rook chess piece body by revolving a polyline. Use Subtract to remove box primitives to form the cutouts in the tower. Add an octagon for the base. Extrude it to a height of .15 and use a taper angle of 15°. (Use your *solproto.dwg* as the prototype from which you start the rook.)

Polyline used for revolution

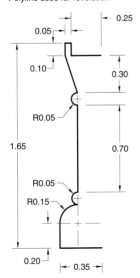

Top view of rook

Cut outs
for tower
are 36
degrees

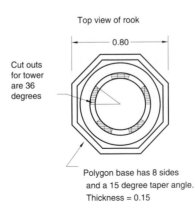

Polygon base has 8 sides
and a 15 degree taper angle.
Thickness = 0.15

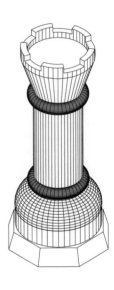

Draw the following shapes using solid modeling techniques. The letter M after an exercise number means that the units are in millimeters.

9.5M Bushing Holder

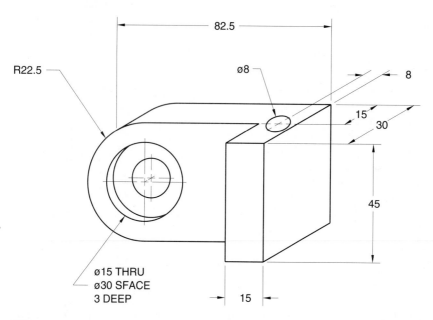

9.6 Shaft Support

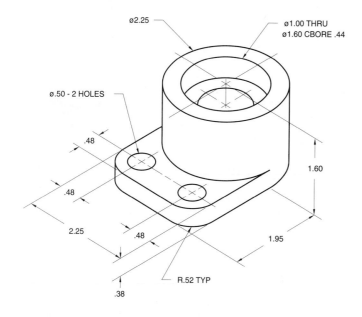

9.7 Balcony

Design a balcony like the one shown here, or a more complex one, using the solid modeling techniques you have learned. Use "two-by-fours" (which actually measure 1.5″ × 3.5″) for vertical pieces. Use Divide to insure equal spacing.

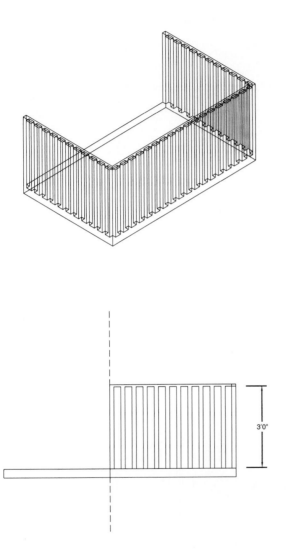

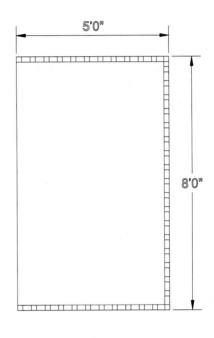

5'0"

8'0"

3'0"

9.8 Bridge

Create a bridge as shown below. First determine the size of the bridge that you want and the area that you would like left open with the arc. Use extrusion to create the arc. Add the structure of the bridge, making sure that the supports are evenly spaced and of sufficient height. (You can enter one half of the supports and then use Mirror.)

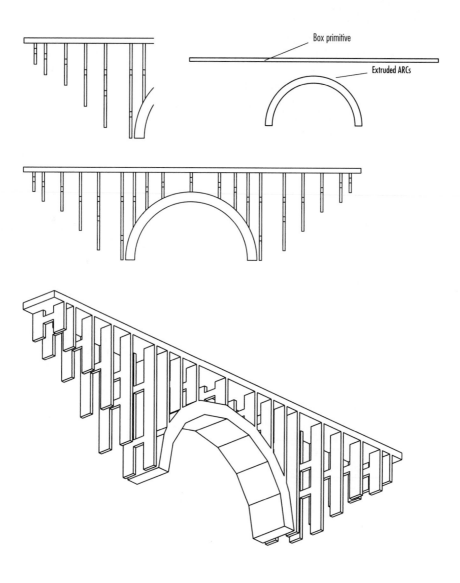

Box primitive

Extruded ARCs

9.9 Hold Down

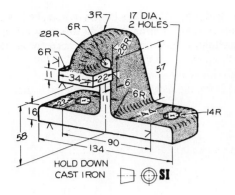

HOLD DOWN
CAST IRON

9.10 Rock Plan

Use the information in the file *rockplan.pts* to create the plan and profile drawings for the railroad bed as shown below. The file has the information in the following order: point number, northing (the y coordinate to the north from the reference point), easting (the x coordinate to the east from the last reference point), elevation, and description. Start by locating each point at the northing and easting given. Extrapolate between elevations and draw polylines for the contour lines at 1' intervals. Make the contour lines for 95' and 100' wide, so that they stand out. Create a 3D mesh using the information in the file *rockplan.pts* modeling the terrain shown.

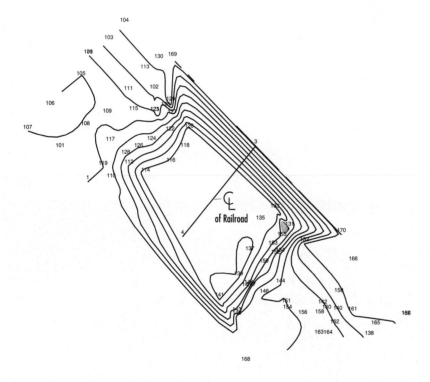

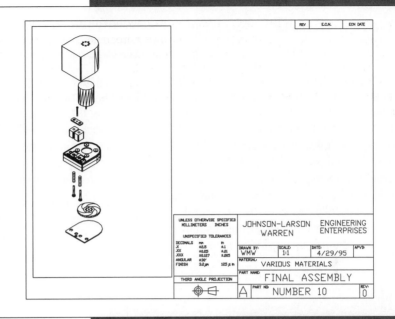

	REV	E.C.N.	ECN DATE

UNLESS OTHERWISE SPECIFIED
MILLIMETERS   INCHES

UNSPECIFIED TOLERANCES

DECIMALS	mm	in
.X	±0.5	±.1
.XX	±0.25	±.01
.XXX	±0.127	±.005
ANGULAR	±30'	
FINISH	3.2μm	125 μin

THIRD ANGLE PROJECTION

JOHNSON–LARSON   ENGINEERING
WARREN   ENTERPRISES

DRAWN BY: WMW	SCALE: 1:1	DATE: 4/29/95	APVD:

MATERIAL: VARIOUS MATERIALS

PART NAME: FINAL ASSEMBLY

A   PART NO: NUMBER 10   REV: 0

TUTORIAL 10

Creating Assembly Drawings from Solid Models

Objectives

When you have completed this tutorial, you will be able to

1. Create an assembly drawing from solid models.

2. Use external references.

3. Check for interference.

4. Analyze the mass properties of solid models.

5. Create an exploded isometric view.

6. Create and insert ball tags.

7. Create and extract attributes.

8. Create a parts list.

9. Use Windows Object Linking and Embedding (OLE).

Introduction

In previous tutorials you have created solid models of objects and projected 2D views to create dimensioned part drawings. To make a set of working drawings using solid modeling, you will create an assembly drawing of a clamp, similar to the one in Figure 10.1. The parts for the assembly drawing have been created as solid models and are included with the data files. You will proceed directly to creating the assembly drawing. The purpose of an assembly drawing is to show how the parts go together, not to fully describe the shape and size of each part. Assembly drawings usually do not show dimensions or hidden lines.

you use external references to *attach* a part drawing to the assembly drawing, AutoCAD automatically updates the assembly drawing whenever you make a change to the original part drawing.

You will also learn more about how to use the Insert command, as you did in Tutorial 8 when inserting the electronic logic symbols. Objects added to a drawing with the Insert command are *included* in the new drawing, but if you update the original drawing, the new drawing is not automatically updated.

The parts in Figures 10.2, 10.3, 10.4, and 10.5 are the parts provided with the data files. The insertion points are identified by 0,0 in the drawings.

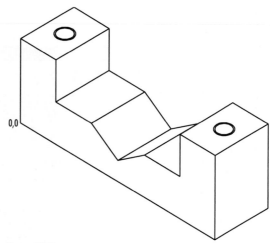

Figure 10.2

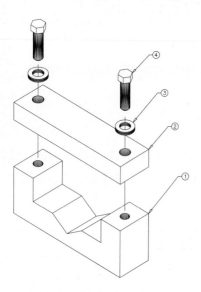

Figure 10.1

Attaching Solid Models

In this tutorial you will learn how to use the External Reference commands to attach part drawings to create an assembly drawing. When

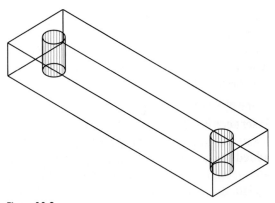

Figure 10.3

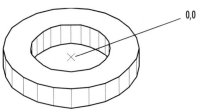
0,0

Figure 10.4

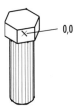
0,0

Figure 10.5

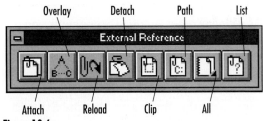

Figure 10.6

Position the toolbar in an area of the screen where it will be out of your way while you are drawing. Remember, you can move it to a different location at any time. Your screen should appear similar to Figure 10.7.

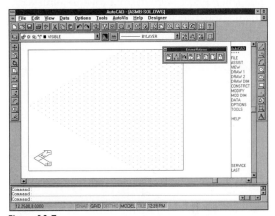

Figure 10.7

Starting

Start AutoCAD and from the prototype drawing called *prot-iso.dwg* provided with the data files, begin a new drawing, *asmb-sol.dwg*, on your own.

The AutoCAD drawing editor appears on your screen, with the settings from a prototype drawing for 3D isometric views. On your own, turn on the External Reference toolbar, shown in Figure 10.6, by selecting it from the Tools menu.

Using External References

 The External Reference command lets you use another drawing without really adding it to your current drawing. The advantage of this is that if you make a change to the referenced (original) drawing, then any other drawing to which it is attached updates automatically the next time that drawing is opened. Like blocks, external references can be attached anywhere in the drawing; they can be scaled and rotated. You can quickly select the External Reference command options from the External Reference toolbar. As with any other AutoCAD command, you can also type the command name, *XREF*, and select the option you want. You will begin your drawing by attaching the base of the clamp to your assembly drawing as an external reference, using the Attach icon from the External Reference toolbar.

Pick: **Attach icon**

The Select file to attach dialog box appears on your screen, as shown in Figure 10.8. On your own, use this dialog box to select the file *base-3d.dwg*. (Since the working directory is set to *c:\work*, the files you will see are the files you have created in the previous tutorials. Change the directory to *c:\datafile* and select the file named *base-3d.dwg* from the list at the left.)

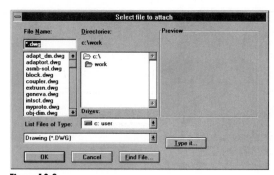

Figure 10.8

When you have the *base-3d.dwg* selected in the dialog box,

Pick: **OK**

You will return to your drawing for the following selections.

Insertion point: **0,0,0** ⏎
X scale factor <1>/Corner/XYZ: ⏎
Y scale factor (default = X): ⏎
Rotation angle <0>: ⏎

You will set the Isolines variable to a larger value to produce more tessellation lines and improve the appearance of the rounded surfaces.

Command: **ISOLINES** ⏎
New value for Isolines <4>: **12** ⏎
Command: **REGEN** ⏎

The base should now appear in your drawing. On your own, use the Zoom command with the XP option to set the paper space zoom factor to .5. When you have finished these steps, your screen should appear similar to Figure 10.9.

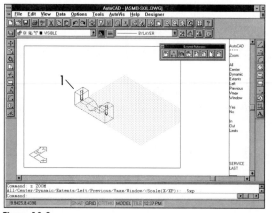

Figure 10.9

Drawings that are attached using the External Reference command, as *base-3d.dwg* was, cannot be modified in the new drawing, nor can you analyze the mass properties of a referenced object.

You can use the object snaps to locate new objects in relation to the referenced object's geometry.

Layers and External References

When you attach a drawing to another drawing using External Reference Attach, all of its subordinate features, such as layers, linetypes, colors, blocks, and dimensioning styles, are attached with it. You can control the visibility of the layers that are attached with the referenced drawing, but you cannot modify anything or create any new objects on these layers. The Visretain system variable controls the extent to which changes in the visibility, color, and linetype are made. If Visretain is set to 0, any changes you make in the new drawing apply only to the current drawing session, and will be discarded when you end the drawing session, or reload or attach the original drawing.

Use Layer Control from the Object Properties toolbar to pull down the list of available layers. Scroll up to the top of the list if necessary to see the result of attaching the base as an external reference. The list should appear similar to Figure 10.10.

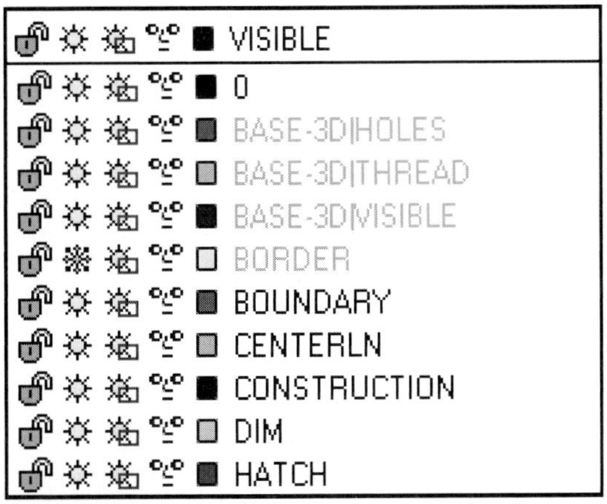

Figure 10.10

Notice that the layer names that were attached with the external reference have the name of the external reference drawing in front of them.

Pick: (to freeze layer BASE-3D|VISIBLE)

Notice that the base externally referenced to the drawing disappears from the screen.

Pick: (to thaw layer BASE-3D|VISIBLE)

You can also control which portions of an external reference are visible by using the External Reference Clip command. This command creates a paper space viewport that you use as a window to select the portion of the external reference that you want to have visible in the drawing. Use the Help command to find more information about the Xrefclip command on your own.

Inserting the Solid Part Drawings

Next you will insert the solid model drawing *cover-3d.dwg* into the assembly drawing as a block. You will use the Insert command, as you did in Tutorial 8 when inserting the electronic logic symbols. Use the Draw toolbar to select the icon for Insert Block.

Pick: **Insert Block icon**

Pick: **File**

Use the Select Drawing File dialog box that appears to select the drawing *cover-3d.dwg* from the directory containing the data files provided. When you are done selecting the file,

Pick: **OK** *(to exit the Select Drawing File dialog box)*

Pick: **OK** *(to exit the Insert Dialog box)*

You return to the command prompt for the remaining prompts.

Insertion point: *(pick Snap to Endpoint icon)*

endp of *(pick the upper left corner of the base, labeled 1 in Figure 10.9)*

X scale factor <1> /Corner/XYZ: ⏎

Y scale factor (default=X): ⏎

Rotation angle <0>: ⏎

The solid model of the clamp cover appears in your drawing, as shown in Figure 10.11. Save your drawing on your own.

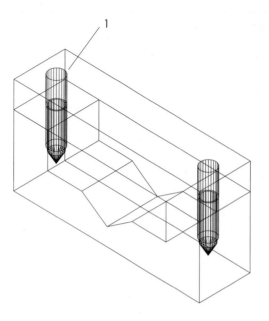

Figure 10.11

Now you are ready to insert the washers into the assembly. You will select the Attach icon from the External Reference toolbar. On your own, make sure Snap is off and use Zoom Window to enlarge the view to make selection easier.

Pick: **Attach icon**

On your own, use the dialog box that appears to select the drawing *washr-3d. dwg*, making sure to select the correct drive and directory. Pick OK to exit the Select File to Attach dialog box. You return to your drawing for the following selections.

Insertion point: **(pick Snap to Center icon)**

cen of **(target the top edge of hole 1, indicated in Figure 10.11)**

X scale factor <1>/Corner/XYZ: ⏎

Y scale factor (default = X): ⏎

Rotation angle <0>: ⏎

Insert the washer for the right-hand side of the part on your own.

You will notice the message *Xref WASHR-3D has already been loaded. Use XREF Reload to update its definition.* If you knew that drawing *washr-3d.dwg* had changed and you wanted to update it without having to exit the drawing, you could use the Reload option of the Xref command. Reload is especially useful if you are working on a networked system and sharing files between members of a design team. At any point you can use the Reload option to update the definition of the referenced objects in your drawing. In this case, you want to insert another copy of the same drawing, so you can just ignore this message. (You could also use the Copy command to create the other washer.)

Once you have the washers inserted, your drawing should appear similar to Figure 10.12.

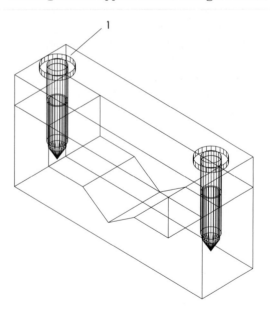

Figure 10.12

Next, you will add the two screws to the assembly to finish inserting the parts you will need. You will use Insert to add the two screws to your drawing. You will use Insert and not

External Reference so that you can use other AutoCAD features, for example checking interference, later in this tutorial.

Pick: **Insert Block icon**

On your own, use the Insert dialog box that appears on your screen to select *screw-3d.dwg* from the data files that were provided. Close the Select Drawing File dialog box by picking OK when you have the correct file selected. Make sure that Specify Parameters on Screen is checked in the Insert dialog box, and pick OK to return to the command prompt for the remaining selections.

Insertion point: *(pick Snap to Center icon)*

of *(pick the top edge of the washer labeled 1 in Figure 10.12)*

X scale factor <1> /Corner/XYZ: ⏎

Y scale factor (default=X): ⏎

Rotation angle <0>: ⏎

The screw should appear in your drawing. Repeat this process on your own for the right-hand screw. Now your drawing should show the parts completely assembled, as in Figure 10.13.

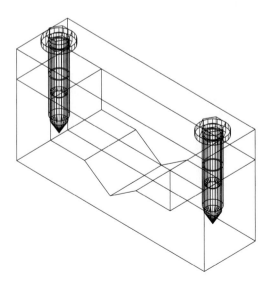

Figure 10.13

Checking for Interference

When you are designing a device, you often want to know whether the parts will fit together correctly. You can use AutoCAD's Interfere command to check and see whether two parts overlap. If they do overlap, Interfere creates a new solid, showing where the two objects were overlapping. You can use this command to determine whether the objects are fitting together as intended.

You will check to see whether the screws in the assembly will fit into the holes in the cover. The Interfere command will only analyze solids. Objects which have been inserted into the drawing are blocks. To return the cover and screws that you inserted to solids, you will explode them. When you explode something, it removes one level of grouping at a time, so exploding the block will return it to a solid. If you explode the solid, it will become a surface model. On your own, list the cover and two screws. Notice that they are block references. To explode the cover and screws,

Pick: **Explode icon**

Select objects: *(pick the cover and the two screws)*

Select objects: ⏎

The cover and two screws become exploded. Now list them again on your own. You should see that they are now listed as 3dsolid objects. Next you will check to see whether the screws interfere with the cover. From the Solids toolbar,

Pick: **Interfere icon**

Select the first set of solids: *(pick on the cover)* ⏎

Select the second set of solids: *(pick on the left-hand screw)* ⏎

■ *TIP* You may need to use Zoom Window to select the screw. ■

Comparing 1 solid against 1 solid.

Solids do not interfere.

The solids do not interfere, so you know that the left hand screw fits through the cover. If the solids do interfere you will see a message similar to *Interfering solids: 2 Interfering pairs: 1 Create interference solids ? <N>:*. You can type *Y*⏎ if you want AutoCAD to create a new solid on the current layer of the overlap between the interfering solids.

Determining Mass Properties

 AutoCAD enables you to inquire about the mass properties of an object. One of the advantages of the solid modeler used in Release 13 is that volumes and mass properties are calculated very accurately, unlike the AME modeler used in Release 12. Pick the Mass Properties icon from the Inquiry/List flyout on the Standard toolbar. Like Interfere, this command works only on solids, not on external references.

Pick: **Mass Properties icon**

Select objects: *(pick the clamp cover)*

Select objects: ⏎

The AutoCAD text window opens displaying the mass property information, as shown in Figure 10.14.

```
                    AutoCAD Text Window
 Edit
Command: _massprop
Select objects: 1 found

Select objects:

---------------   SOLIDS   ---------------

Mass:                  0.2659
Volume:                0.2659
Bounding box:      X:  0.3706  --  0.4901
                   Y:  4.8989  --  6.0166
                   Z:  0.0000  --  1.9893
Centroid:          X:  0.4303
                   Y:  5.4577
                   Z:  0.9947
Moments of inertia: X: 8.2980
                   Y:  0.4003
                   Z:  7.9969
Products of inertia: XY: 0.6245
                   YZ: 1.4433
                   ZX: 0.1138
Radii of gyration: X:  5.5866
                   Y:  1.2270
                   Z:  5.4843

Press RETURN to continue:
```

Figure 10.14

Close the AutoCAD text window on your own. AutoCAD provides the opportunity to write the mass properties information to a file. Once you close the text window, you will see the prompt,

Press RETURN to continue: ⏎

Write to a file ? <N>: ⏎

Creating an Exploded Isometric View

Exploded views are often used to show how parts are assembled. Exploded views show the parts in the assembly moved away from their assembled positions yet still aligning, as though the assembly had exploded. In this tutorial, you will create an exploded view from your assembly drawing by moving the parts away from each other. In an exploded view, the parts still should align with their assembled positions, so you will move them only along one axis. You will use the Move icon on the Modify toolbar to move the screws out of their assembled positions.

Pick: **Move icon**

Select objects: *(pick the two screws)* ⏎

<Base point or displacement>: **0,0,0** ⏎

Second point of displacement: **0,0,5** ⏎

The screws move up in the drawing and off the view. You will use the Zoom Extents command so that the entire drawing fits on the screen.

Pick: **Zoom Extents icon**

Your drawing should appear similar to Figure 10.15.

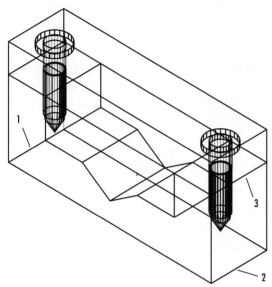

Figure 10.15

You will continue with the Move command and relocate the washers in line with their present position, but 3 units above, along the Z axis.

Pick: **Move icon**

Select objects: *(pick the two washers)* ⏎

<Base point or displacement>: **0,0,0** ⏎

Second point of displacement: **0,0,3** ⏎

The washers line up below the screws, still in line with the holes on the object. Next, move the clamp cover up 2 units along the Z axis.

Command: ⏎

Select objects: *(pick the cover)* ⏎

<Base point or displacement>: **0,0,0** ⏎

Second point of displacement: **0,0,2** ⏎

When you create an exploded view, it is customary to add thin lines that show how the parts assemble.

On your own, create a new layer called ALIGN that is color cyan and linetype CONTINUOUS. Set it as the current layer.

 To create lines in the same plane as the objects' centers, you need to define a UCS that aligns through the middle of the objects. Zoom in, if necessary, to pick the correct lines on the part.

Pick: **3 Point UCS icon**

Origin point <0,0,0>: *(pick Snap to Midpoint icon)*

mid of *(target the middle of line 1 in Figure 10.15)*

Point on positive portion of the X axis
   <1.0000,0.7500,0.0000>: *(pick Snap to Midpoint icon)* ⏎

mid of *(target line 2)*

Point on positive-Y portion of the UCS XY plane
   <1.0000,0.7500,0.0000>: *(pick Snap to Midpoint icon)*

mid of *(target line 3)*

The UCS icon changes to line up with the plane through the middle of the parts. If it does not, repeat the command and try again. On your own, make sure that Snap is on. Draw the lines on your own, indicating how the parts align, by picking points on the Snap. Save your drawing before you continue.

Creating a Block for the Ball Tags

Ball tags identify the parts in the assembly. They are made up of the item number in the assembly enclosed in a circle, hence the name ball tags. To make it easy to add the ball tags to the drawing, you will make a simple block with one visible attribute and several invisible attributes. An *attribute* is basically text information that you can associate with a block. It can be used very effectively for adding information to the drawing. You can also extract attribute information from the drawing and import it into a database or word processing program for other uses.

Figure 10.16 shows ball tags like the ones you will create.

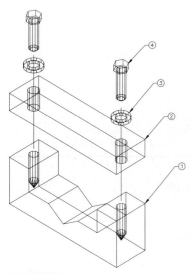

Figure 10.16

On your own, set layer TEXT as the current drawing layer. Select the View UCS to align the UCS with the view so that you can create text that appears straight on.

You will draw the circle first. Then you will add the attributes and make the circle and attribute objects into a block.

Pick: **Circle Center Diameter icon**

On your own, draw a circle of diameter .25 above and to the right of the assembly drawing.

Defining Attributes

 Use the Toolbars selection from the Tools menu to turn on the Attribute toolbar. You can use the Define Attribute icon from this toolbar to create the special attribute text objects.

Pick: **Define Attribute icon**

The Attribute Definition dialog box appears on your screen. You will use it to make the selections shown in Figure 10.17.

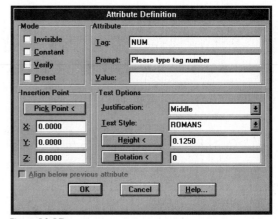

Figure 10.17

Attribute Modes

Attributes can have the special modes, or properties, listed below.

Mode	Definition
Invisible	The attribute does not appear in the drawing, but can still be used for purposes such as extracting to a database.
Constant	The value of the attribute is set at the beginning of its definition, instead of AutoCAD prompting you for a value on insertion of the block.
Verified	You can make sure the value is correct after typing the value.
Preset	You can change the attribute later, but you will not be prompted for the value when you insert it.

You will now use the Attribute Definition dialog box to add the number in the center of the ball tag as a visible attribute. You will also create attributes for the other information commonly shown in the parts list, such as the part name, part number, material, and quantity, as invisible attributes.

Leave each of the Mode boxes unchecked to indicate that you want this attribute to be visible, variable, not verified, and typed in, not preset.

Attribute Tag

The *attribute tag* is a variable name that is replaced with the value that you type when prompted as you insert the attribute block. The tag appears when the block is exploded, or before the attribute is made into a block. You will use NUM for the attribute tag. On your own, type *NUM* in the Tag text box.

Attribute Prompt

The *attribute prompt* is the prompt that appears in the command area when you insert the block into a drawing. Make your prompt descriptive, so that it is clear what information needs to be typed in, and how it should be typed. For instance, *Enter the date (dd/mm/yy)* is a prompt that specifies not only what to enter—the date—but also the format—numeric two-digit format, with day first, then month, then year.

On your own, type *Please type tag number* in the Prompt text box.

> ■ *TIP* The prompt appears with a colon (:) after it to make it clear that some entry is expected. AutoCAD includes the colon automatically; you do not need to type a colon in the dialog box now. ■

Default Attribute Value

The Value text box defines a default value for the attribute. Leaving the box to the right of Value empty results in no default value for the tag. Since every tag number will be different, there is no advantage to having a default value for the ball tag attribute. If the prompt is often answered with a particular response, that response would be a good choice for the default value.

Attribute Text Options

The attribute uses the same types of options that regular text uses. You can center, fit, align it, and so on, just like any other text. Select the Middle option from the Justification pull-down menu. This means that the middle of the text will appear in the center of the ball tag when you type it.

The text style romans is the default, and this is acceptable. You do not need to change the Text Style setting.

Set the text height to .125.

Leave the rotation set at 0° for this attribute. You do not want the attribute rotated at an angle in the drawing.

Attribute Insertion Point

You can use the Pick Point button in the Insertion Point section to select the insertion point on your screen, or you can type the X, Y, Z coordinates into the appropriate boxes.

> *Pick:* **Pick Point**

Your drawing returns to the screen.

> Start point: **(pick Snap to Center icon)**
>
> cen of **(pick on the edge of the circle)**

AutoCAD returns you to the Attribute Definition dialog box.

> *Pick:* **OK**

The tag NUM is centered in the circle. The circle and the attribute tag should look like Figure 10.18. If the attribute is not in the center of the circle, use the Move command and move it so that it is centered in the circle on your own.

Figure 10.18

Next you will make the attributes for part name (description), part number, material, and quantity. These attributes will be invisible; they will not show up in the drawing and make it look crowded, but they can still be extracted for use in a spreadsheet or database.

To restart the Define Attribute command,

> *Press:* ⏎

The Attribute Definition dialog box appears on your screen. On your own, select the invisible mode. When it is selected, an X appears in the box to its left. Continue to use the dialog box to create an invisible attribute with the following settings:

Tag: **PART**

Prompt: **Please type part description**

Value: **(leave Value blank so that there is no default value)**

Justification: **Left**

Text Style: **ROMANS**

Height: **.125**

Rotation: **0**

When you have entered the this information into the dialog box, you are ready to select the location in the drawing for your invisible attribute.

Pick: **Pick Point**

Your drawing returns to the screen.

Start point: **(pick a point to the right of the ball tag circle)**

Pick: **OK (to exit the Attribute Definition dialog box)**

The tag PART should appear at the location you selected.

On your own, use this method to create the three invisible attributes listed below. Choose the box to the left of Align below previous attribute to locate each new attribute below the previous attributes.

Tag: **MATL**

Prompt: **Please type material**

Value: **Cast Iron**

Justification: **Left**

Text Style: **ROMANS**

Height: **.125**

Rotation: **0**

Tag: **QTY**

Prompt: **Please type quantity required**

Value: **(leave blank)**

Justification: **Left**

Text Style: **ROMANS**

Height: **.125**

Rotation: **0**

Tag: **PARTNO**

Prompt: **Please type part number, if any**

Value: **(leave blank)**

Justification: **Left**

Text Style: **ROMANS**

Height: **.125**

Rotation: **0**

When you are finished, you should see each of the tags in the drawing.

Defining the Block

Now you will make the circle and all the attributes into a block. (It is important to remember that first you create the shapes for the block and the attributes, and then you define them into a block.)

Pick: **Block icon**

Block name (or ?): **BALLTAG** ⏎

Insertion base point: **(pick Snap to Quadrant icon)**

qua of **(pick on the left side of the circle to select its left quadrant point)**

Select objects: **(use implied Windowing to pick the circle and the tag names NUM, PART, MATL, QTY, PARTNO)** ⏎

The circle and the tag names disappear from your screen. Now you can insert the necessary ball tags into the drawing, using the block you have made.

> **■ TIP** You may want to use the Write Block command (Wblock) so that the block is available for other drawings. Remember, without Write Block, the block is defined only in the present drawing. **■**

Inserting the Ball Tags

 You will use the Leader icon on the Dimensioning toolbar to draw the arrow and the lines to the ball tag. Then you will insert the block BALLTAG to add the circle and tags.

 You will select Snap to Nearest and use it as an override to pick exactly on the edge of the parts you select for the beginning of the leader line. When you place the aperture box on an object, Snap to Nearest finds the point on the object nearest to the crosshairs.

You will use Figure 10.16 as your guide as you insert the ball tags.

On your own, make sure that Ortho is off. Turn off the Attribute and External Reference toolbars. Then turn on the Dimensioning toolbar.

> **■ TIP** Because block BALLTAG has its insertion point on the left, you want to draw the leaders towards the right. **■**

Pick: **Leader icon**

From point: *(pick Snap to Nearest icon)*

nea to *(pick a point on the right edge of part 1, the base)*

To point: *(pick a point that is above and to the right of part 1)*

To point: *(Turn on Ortho and pick a point that is about .125 to the right)*

To point (Format/Annotation/Undo)<Annotation>: ⏎

Annotation (or RETURN for options): ⏎

Tolerance/Copy/Block/None/<Mtext>: **B** ⏎

Block name (or ?): **BALLTAG** ⏎

Insertion point: *(pick Snap to Endpoint icon)*

Pick: (endpoint of leader)

X scale factor <1>/Corner/XYZ: ⏎

Y scale factor (default=X): ⏎

Rotation angle <0>: ⏎

Please type tag number: **1** ⏎

Please type part number, if any: **ADD1** ⏎

Please type quantity required: **1** ⏎

Please type material<Cast Iron>: **STEEL** ⏎

Please type part description: **BASE** ⏎

> **■ TIP** Remember the space bar does not act as ⏎ when you are entering text. Because text may include spaces, when entering text commands press ⏎ to end the command. **■**

The circle and number 1 appear on your screen. Your prompts for attribute information may have appeared in a different order. If you made a mistake, you can correct it in the next section.

Continue to create the leader lines for parts 2 (the cover), 3 (the hex head), and 4 (the washers) on your own. The leader can start anywhere on the edge of the part. Try to pick points and leader line angles that place the ball tags in a location in the drawing where they are accessible and easy to read. When adding the ball tags, refer to the table on the next page showing the information for each part.

Item	Name	Material	Quantity	Number
1	Base	Steel	1	ADD1
2	Clamp Top	1020ST	1	ADD2
3	.438 x.750 x.125 Flat Washer	Stock	2	ADD3
4	.375-16 UNC x.125 Hex Head	Stock	2	ADD4

When you are done, your drawing should look similar to Figure 10.16. Save your drawing before you continue.

Changing an Attribute Value

 The Edit Attribute command is very useful for changing an attribute value. If you have mistyped a value, or you want to change an existing attribute value, use the Edit Attribute icon from the Attribute toolbar. If necessary show the Attribute toolbar on your own.

Pick: **Edit Attribute icon**

Select block: **(pick on one of the ball tags)**

The Edit Attributes dialog box appears, as shown in Figure 10.19. (Your attributes may appear in a different order.)

```
                    Edit Attributes
Block Name:   BALLTAG

Please type tag number       4
Please type material         STOCK
Please type part descript    .375-16UNC x .125 HEX HEAD
Please type part number,     ADD4
Please type quantity requ    2



    OK      Cancel     Previous    Next    Help...
```

Figure 10.19

It contains the attribute prompts and a text box to the right of each prompt with the value you entered. To change a value, pick its text box and type the new value. Pick OK to exit the dialog box.

Changing an Externally Referenced Drawing

The major difference between blocks and externally referenced drawings is that external references are not *really* added to the current drawing. A *pointer* is established to the original drawing (external reference) that you attached. If you change the original drawing, the change is also made in any drawing to which it is attached as an external reference. You will try this feature by opening drawing *base-3d.dwg* and filleting its exterior corners. On your own, open the file *base-3d.dwg*.

The drawing of the base appears on your screen. On your own, use the commands that you have learned to add a fillet of radius .25 to all four exterior corners of the base. Save the changes to the drawing *base-3d.dwg* so that they will occur in your assembly drawing.

Now reopen the assembly drawing *asmb-sol.dwg* and observe that the changes you made to *base-3d.dwg* appear in that drawing also.

The change to the base has been made in the assembly drawing, as shown in Figure 10.20. The changes took place when this drawing was opened and the newly changed external referenced drawing *base-3d.dwg* was loaded into *asmb-sol.dwg*, automatically updating the assembly.

On your own, make sure that *asmb-sol.dwg* is your current drawing before continuing. Its name should appear in the title bar near the top of the screen.

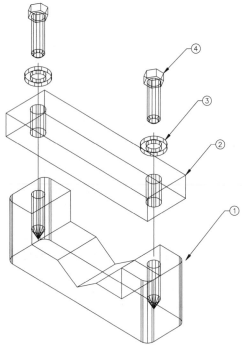

Figure 10.20

■ **TIP** You must keep careful track of your drawings when using external references. If you did not intend to change the assembly, but made a change to its externally referenced drawing, the drawing will change. If you want to change an externally referenced drawing but not the assembly, use Save As and save the changed drawing to a new file name. If you move an externally referenced drawing to a different directory, you will have to Reload the external reference and supply the new path in order for AutoCAD to find it. ■

Creating the Parts List

The next task is to create a *parts list* for your drawing. The required headings for the parts list are Item, Description, Material, Quantity, and Part No. The item number is the number that appears in the ball tag for the part on your assembly drawing. Sometimes a parts list is created on a separate sheet, but often in a small assembly drawing, like the clamp assembly, it is included in the drawing. This is preferable because having both in one drawing keeps the parts list from getting separated from the drawing it should accompany. The parts list is usually positioned either near the title block or in the upper right-hand corner of the drawing.

There is no standard format for a parts list; each company can have its own standard. You can loosely base dimensions for a parts list on one of the formats recommended in the MIL-15 Technical Drawing Standards, or in the ANSI Y14-2M Standards for text sizes and note locations in technical drawings.

Extracting Attribute Information

You will extract the attribute information that you created and then import it into a spreadsheet or word processing program to format. Afterwards you can insert it back into the drawing using Windows Object Linking and Embedding (OLE). You will type the dynamic dialog command for attribute extraction, Ddattext.

Command: **DDATTEXT** ⏎

The Attribute Extraction dialog box appears on your screen, as shown in Figure 10.21. You can use it to set the format of the attribute information you will extract.

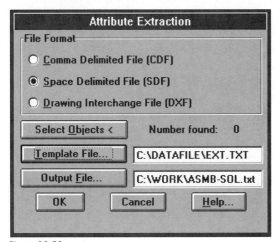

Figure 10.21

You can extract attributes to three different file formats:

Comma Delimited File (CDF) creates a file containing one record for each block reference, with the individual attributes separated by commas. The character fields in the record are enclosed in single quotation marks.

Space Delimited File (SDF) creates a file containing one record for each block reference. The fields of each record are a fixed width; no commas or other separators are used.

Drawing Interchange File (DXF) creates a file like the one used to exchange drawings between systems (DXF), except that it has only block reference, attribute, and end-of-sequence objects. This format does not require the use of a template file.

On your own, make sure the space delimited format is selected. The radio button to the left of Space Delimited File (SDF) should be filled in.

The *template file* tells AutoCAD how to structure the extracted attribute information. You will use the template file provided with the data files, *ext.txt*. If you want to see the structure of the template file, use the DOS editor to open *ext.txt*. It is a text file that matches the block name, X and Y coordinates, and attribute names with the types and lengths of the fields to which the information will be extracted. The field information begins with either C (for character) or N (for numeric), indicating the type of field, followed by a three-digit number indicating the field length and three more places indicating the number of decimal places; for example, N007001 or C020000.

On your own, use the Attribute Extraction dialog box to select the template file *ext.txt*. Pick the Template File button. Use the dialog box that appears to select *ext.txt* as the template file. You will need to change to the *c:\datafile* directory in order to see *ext.txt* listed.

The *output file* has the same prefix, *asmb-sol*, as the AutoCAD file in which you are working, with the file extension *.txt*. Leave it set to the default name, *asmb-sol.txt*.

Next you will select the attributes to extract from your drawing. You will pick the Select Objects button to return to your drawing for the selections.

Pick: **Select Objects**

Select objects: *(pick all four ball tags)* ⏎

Pick: **OK** *(to exit the Attribute Extraction dialog box)*

You will see the message *4 records in extract file* appear in the command window. The file *asmb-sol.txt* contains four lines.

Switching to the Spreadsheet or Word Processing Program

Press: ⎇ALT ⎋ESC *(to return to the Program Manager)*

> ■ **TIP** You can also use ⎇ALT - ⭾TAB to switch between tasks using Windows. ■

Pick: (the appropriate icon to start your spreadsheet or word processing program)

> ■ **TIP** You may need to close AutoCAD in order to open the spreadsheet. Having too many applications open may cause you to run out of memory. ■

Figure 10.22

On your own, use your knowledge of your spreadsheet or word processing program to open or insert the extracted file, *asmb-sol.txt*, and clean up the formatting. Figure 10.22 shows an example of the file open and formatted using Microsoft Word. The titles for the parts list have been added as the top row. On your own, delete the XY coordinate columns, because you do not need them for the parts list. (The coordinates for the attributes can be useful in a number of ways for inserting information or drawing objects back into your drawing.) Save the changes you made to *asmb-sol.txt*.

In your Windows spreadsheet or word processor, highlight the parts list text on your own. Use the spreadsheet or word processor's Edit, Copy command sequence to copy the selection to the Windows Clipboard. Press ⎇ALT-⎋ESC again until you see your AutoCAD drawing return to the screen. (If you had to close AutoCAD in order to start your spreadsheet program, restart it and open *asmb-sol.dwg*.)

Pasting from the Clipboard

AutoCAD's drawing editor should be on your screen, showing the assembly drawing created earlier in the tutorial. Be sure that you have completed the previous step of highlighting and copying the selection from your spreadsheet to the Clipboard. Then,

Pick: **Edit, Paste**

The selection that you copied to the Windows Clipboard from the spreadsheet appears in your graphics window. On your own, resize and position the pasted text until it is similar to Figure 10.23. Resize the pasted item by picking on the dark boxes that appear at the corners of the selection. The cursor changes to a double arrow when positioned over the boxes, and you can drag those boxes to move the object's outline. When you have the cursor positioned over the object, but not at the edges where you can resize the selection, you will see the four-way arrow cursor. Position the pasted selection by picking on it when you see the four-way arrow cursor and hold the pick button down while dragging it to a new location.

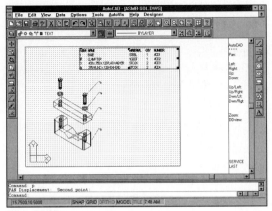

Figure 10.23

> ■ *TIP* You may want to paste text into paper space, depending on how your drawing is organized. ■

If you cannot extract the attributes, use any Windows-compatible word processor or spreadsheet to create the parts list. Once you have typed in the information for the parts list, highlight all of the text information that you have entered. Select the Edit menu, Copy option to copy the selection to the Windows Clipboard. Switch back to the AutoCAD program and use the Edit menu, Paste option to add the parts list to your AutoCAD drawing.

On your own, use the hot grips with Move, if necessary, to locate the parts list in the upper right of the drawing. Save your drawing.

Using Windows Object Linking and Embedding

Object Linking and Embedding (OLE) is a Windows feature that you can use to transfer information between Windows based applications. You can use several different forms of information inside a single file using OLE. Objects that can be embedded in Windows applications are things like spreadsheets, drawings, charts, or text. For example, you can embed the text for the parts list, which you exported and formatted in the steps above, in your AutoCAD drawing. Embedding a linked object is very similar to attaching an external reference drawing in AutoCAD. When you edit the original source document, the changes automatically appear in the document or drawing where it is embedded. This is because the data that actually makes up the linked object resides only in the source document. (Pasting a copied selection into the drawing is similar to inserting a block.) To embed a Windows object into your AutoCAD drawing, you would use the AutoCAD menu bar.

Pick: **Edit, Insert Object**

The Insert New Object dialog box appears on your screen as shown in Figure 10.24.

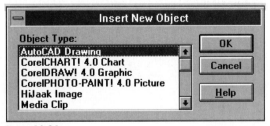

Figure 10.24

Use the scroll bars to scroll down the list of object types available on your system. Selecting an object type from the list automatically launches the application with which that object type is associated. For example if you pick MS Word 6.0 Document, MS Word 6.0 will automatically start. You can then open the file you have previously saved, or create a new file if you want. When you are finished using the application (like MS Word 6.0) you would pick File, Close and Return to Object from the menu bar.

Pick: **Cancel** *(to cancel the Insert New Object dialog box)*

Experiment on your own embedding objects in your AutoCAD drawing and embedding AutoCAD drawings in your other documents. When objects are embedded in a *destination application*, double-clicking on them automatically starts the *source application* where the object was created.

On your own, switch to paper space and plot your drawing. Close any spreadsheet or word processing programs that you may have left open.

■ *TIP* If you are going to send a drawing to someone on a disk, you must make sure to send any externally referenced drawings too. Otherwise the recipient will not be able to open the referenced portions of the drawing. A good way to avoid this problem is to use the External Reference command with the Bind option at the command prompt to link the two drawings. This makes the referenced drawing a part of the drawing to which it is attached. Also, it prevents you from unintentionally deleting the original drawing. However, once you have done this, the external reference will no longer update. To use the External Reference Bind option, after you have attached the drawing, type *XREF* at the command prompt, select the Bind option, and type the names of all the externally referenced files you want to bind to your assembly (separated by commas, but no spaces). One drawback to binding the files is that the file size for the assembly drawing will be larger. ■

Hide any extra toolbars you may have on your screen. Leave the Draw and Modify toolbars shown. Exit AutoCAD. You have completed Tutorial 10.

10.1 Clamping Block

Prepare assembly and detail drawings, and a parts list based on the information below using solid modeling techniques. Create solid models of Parts 1 and 2. Join Part 2 to Part 1 using two .500–3UNC x 2.50 LONG HEX-HEAD BOLTS. Include .500–13UNC NUTS on each bolt. Locate a .625 x 1.250 x.125 WASHER under the head of each bolt, between Parts 1 and 2, and between Part 1 and the NUT. (Each bolt will have 3 washers and a nut.)

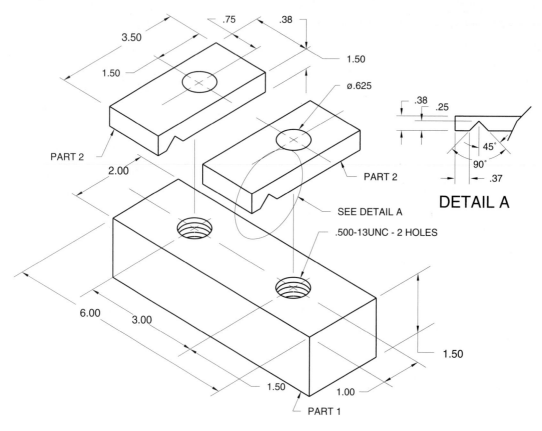

10.2 Pressure Assembly

Prepare assembly and detail drawings, and a parts list of the pressure assembly using solid modeling techniques. Create solid models of the BASE, GASKET, and COVER shown below. Use four .375–16UNC x 1.50 HEX-HEAD SCREWS to join the COVER, GASKET, and BASE. The GASKET assembles between the BASE and COVER.

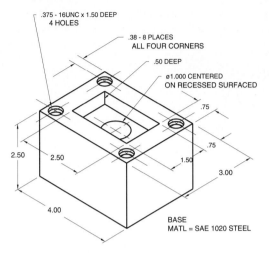

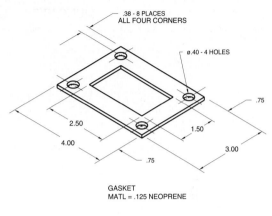

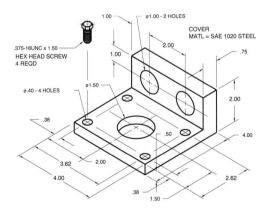

10.3 Hub Assembly

Create an assembly drawing for the parts shown. Use 8 bolts with nuts to assemble the parts.

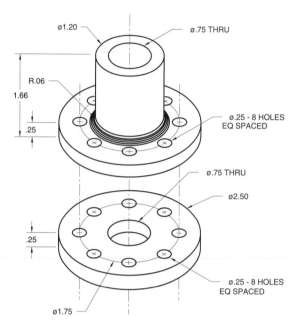

10.4 Turbine Housing Assembly

Create detail drawings for the two parts shown below and save each as a Wblock. Insert each block to form an assembly.

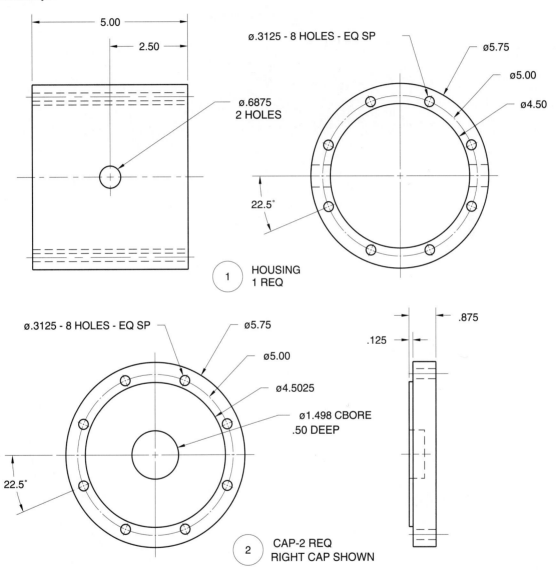

10.5 Geneva Wheel Assembly

Create an assembly drawing for the parts shown, using solid modeling techniques. Experiment with different materials and analyze mass properties.

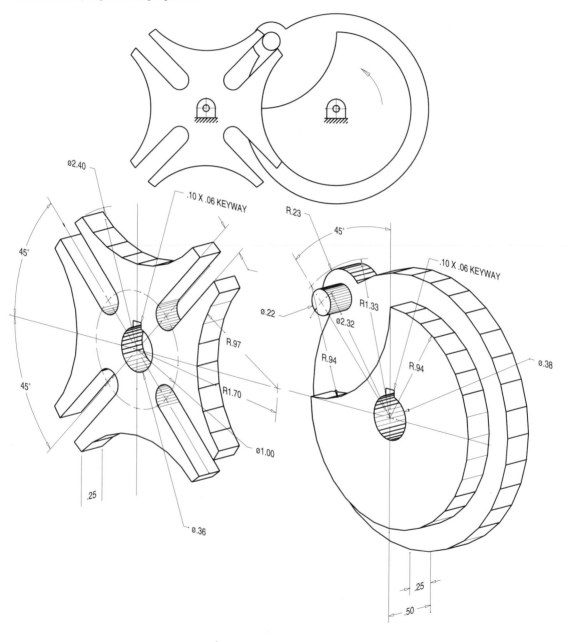

10.6 Staircase

Use solid modeling to generate this staircase assembly detail.

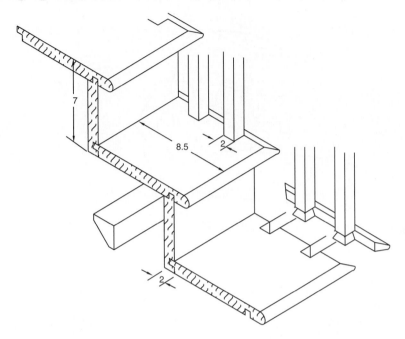

Create an assembly drawing for the parts shown, using solid modeling techniques. Analyze the mass properties of the parts.

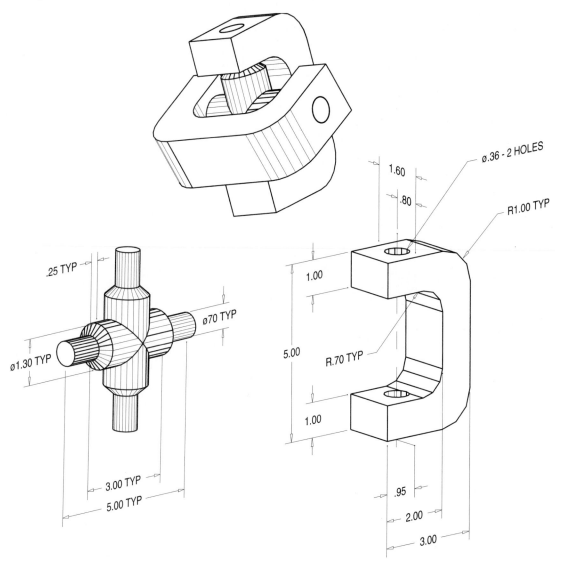

10.8M Step Bearing

Use solid modeling to create an assembly drawing of this step bearing.

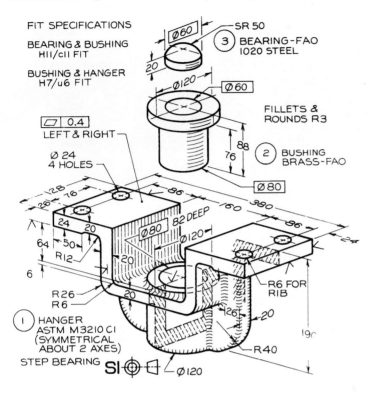

FIT SPECIFICATIONS

BEARING & BUSHING
H11/c11 FIT

BUSHING & HANGER
H7/u6 FIT

$\boxed{\varnothing 60}$ ─ SR 50

③ BEARING-FAO
1020 STEEL

$\varnothing 120$ ── $\boxed{\varnothing 60}$

FILLETS &
ROUNDS R3

$\boxed{\square\ 0.4}$
LEFT & RIGHT

$\varnothing 24$
4 HOLES

② BUSHING
BRASS-FAO

88
76

$\boxed{\varnothing 80}$

128
76
26
24 20
64 50
R12
6

86
160   390
82 DEEP
$\varnothing 80$
$\varnothing 120$
86
24

R6 FOR
RIB

R26
R6
20
20

126   20

① HANGER
ASTM M3210 CI
(SYMMETRICAL
ABOUT 2 AXES)

190

STEP BEARING   SI ⊕ ⊟ ─ $\varnothing 120$

R40

10.9M Tensioner

Create an assembly drawing of this tensioner using solid modeling.

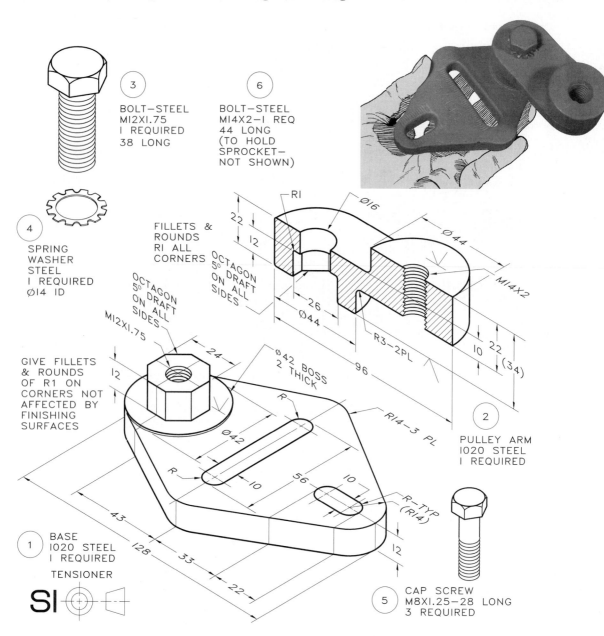

3
BOLT–STEEL
M12X1.75
1 REQUIRED
38 LONG

6
BOLT–STEEL
M14X2–1 REQ
44 LONG
(TO HOLD
SPROCKET–
NOT SHOWN)

4
SPRING
WASHER
STEEL
1 REQUIRED
Ø14 ID

FILLETS &
ROUNDS
R1 ALL
CORNERS

OCTAGON
5° DRAFT
ON ALL
SIDES

R1

Ø16

Ø44

M14X2

22

12

26

Ø44

R3–2PL

96

10

22

(34)

2
PULLEY ARM
1020 STEEL
1 REQUIRED

OCTAGON
5° DRAFT
ON ALL
SIDES

M12X1.75

24

Ø42 BOSS
2 THICK

R

GIVE FILLETS
& ROUNDS
OF R1 ON
CORNERS NOT
AFFECTED BY
FINISHING
SURFACES

12

Ø42

R14–3 PL

R

10

56

10

R–TYP
(R14)

12

1
BASE
1020 STEEL
1 REQUIRED

43

128

33

22

5
CAP SCREW
M8X1.25–28 LONG
3 REQUIRED

TENSIONER

SI

Make an assembly drawing of this trailer hitch using solid modeling.

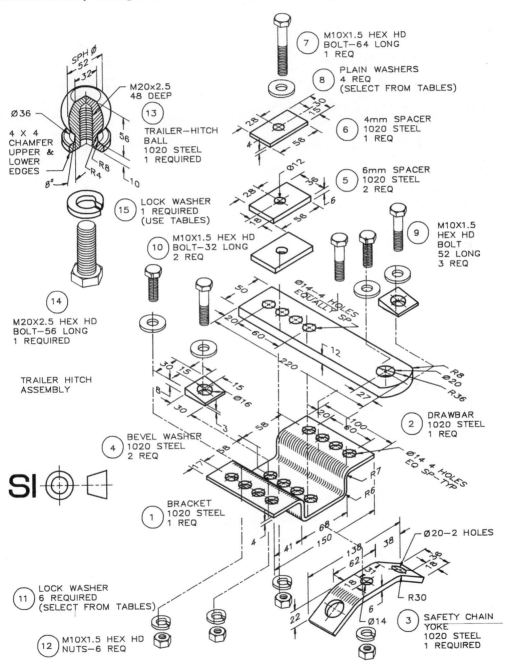

SPH Ø
52
32

Ø36
4 X 4
CHAMFER
UPPER &
LOWER
EDGES
R8
R4
8°
10

M20x2.5
48 DEEP

⑬

TRAILER—HITCH
BALL
1020 STEEL
1 REQUIRED

56

⑦ M10X1.5 HEX HD
BOLT—64 LONG
1 REQ

⑧ PLAIN WASHERS
4 REQ
(SELECT FROM TABLES)

30
13
28
58
4

⑥ 4mm SPACER
1020 STEEL
1 REQ

Ø12
28
36
6
58
8

⑤ 6mm SPACER
1020 STEEL
2 REQ

⑨ M10X1.5
HEX HD
BOLT
52 LONG
3 REQ

⑮ LOCK WASHER
1 REQUIRED
(USE TABLES)

⑩ M10X1.5 HEX HD
BOLT—32 LONG
2 REQ

⑭

M20X2.5 HEX HD
BOLT—56 LONG
1 REQUIRED

TRAILER HITCH
ASSEMBLY

Ø14—4 HOLES
EQUALLY SP
50
20
60
12
220
27
R8
Ø20
R36

② DRAWBAR
1020 STEEL
1 REQ

30
15
15
8
30
Ø16
3
58
20
100
60
Ø14 4 HOLES
EQ SP—TYP
R7
R6

④ BEVEL WASHER
1020 STEEL
2 REQ

17
8

① BRACKET
1020 STEEL
1 REQ

SI ⊙⊏

⑪ LOCK WASHER
6 REQUIRED
(SELECT FROM TABLES)

⑫ M10X1.5 HEX HD
NUTS—6 REQ

4
41
150
68
138
62
4
22
6
Ø14
18
31
38
38
18
R30
Ø20—2 HOLES

③ SAFETY CHAIN
YOKE
1020 STEEL
1 REQUIRED

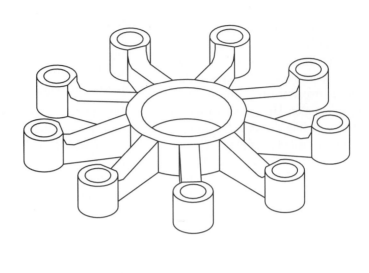

Introduction to AutoCAD Designer

Objectives

When you have completed this tutorial, you will be able to

1. Create models of parts parametrically, using AutoCAD Designer.

2. Sketch initial part shapes.

3. Add constraints to define part geometry.

4. Add parametric dimensions to control part size.

5. Create 3D parts from sketched geometry.

6. Change dimensions and update the size of 2D profiles and 3D features.

7. Use equations linking part geometry.

8. Generate correctly drawn orthographic drawings automatically.

9. Analyze mass properties.

Introduction

In this tutorial you will learn to create 3D solid models and engineering drawings using AutoCAD Designer. AutoCAD Designer is a *parametric* solid modeling program. Parametric modeling allows you to input not only sizes for a part's features, but also the relationships between the features, to create an "intelligent" model. The term "parametric" basically means "using variables." Through the use of parametric modeling, dimensions in AutoCAD Designer can act as variables or contain equations. This allows you to build intelligence into your drawing. Each feature "knows" how it relates to other features in the model. You can update drawing geometry created using parametric modeling automatically by changing specified sizes and relationships.

As you are working through this tutorial, keep the engineering design process in mind. In the beginning phases of design development, all of the relationships may not be completely determined; rough sketches are used to define the basic shapes. In subsequent stages, the design is refined further and the initial designs are updated. Additional analysis may be performed to determine the suitability of different design alternatives. When the final design is determined, drawings are produced to specify the information necessary to manufacture the part.

The process of parametric modeling parallels the engineering design process. In the initial stages, you roughly define shapes and sizes in the model by specifying *constraints*, which define relationships between the elements that create the model, such as perpendicularity, tangency, and parallelism. You add dimensions specifying the sizes of the features that make up the part. As in the design process, when new information becomes available, you can change the sizes and relationships of the features. You can evaluate the mass properties of the model and use the resulting information in making design decisions. AutoCAD Designer automatically updates the model by interpreting the size and constraint parameters. The process of defining the constraints that will be used to build the model is central to the parametric modeling process.

When you are using AutoCAD Designer, initially you sketch rough shapes, using AutoCAD's Line, Polyline, Circle, and Arc commands. AutoCAD Designer adds constraints to the rough sketch by applying rules; for example, that lines within 4° of horizontal or vertical are intended to be true horizontal or true vertical, respectively. (In the engineering design process, a colleague trying to interpret your rough sketch would also apply these rules.) The sketch is turned into a 3D *feature* using extrusion, revolution, or *sweeping*, similar to the way you created solid models, starting in Tutorial 9. (Sweeping is similar to extrusion, except that the path can be a curved or a straight line.) Features are the simpler 3D building blocks from which parts are created.

Next, you combine features and establish relationships between them to create a part. You learned in the preceding tutorials to use AutoCAD to create 2D and solid models. It was important to create geometry to exact sizes when beginning your drawing in order to end up with a useful database. AutoCAD Designer allows you to easily resize features, and add and change constraints defining the relationships between the features of the part. You can also specify equations relating various features of the model. If you change these equations, the parametric model updates. Just as creating an accurate drawing database was important in creating useful models using 2D and solid modeling, in AutoCAD Designer it is important to constrain and dimension your model in a way that accurately reflects your design intentions.

Because of the intelligence built in through the parametric modeling process, you can generate fully dimensioned orthographic views from the model, with hidden lines shown correctly when the part is designed to your satisfaction. If you make a change to the model, the orthographic views update automatically. Likewise, you can change the orthographic views and have the model automatically update. This ability to change either the model or the drawings and have the other update is referred to as *bidirectional associativity*.

AutoCAD and the Designer Database

The information for the parameters and constraint information in Designer is stored in your drawing file, using blocks on special frozen layers. AutoCAD Designer generates layers for storage of the AutoCAD Designer database, dimensions, work planes, axes, and points. Changing the contents of these layers can corrupt or destroy the AutoCAD Designer database.

■ **Warning:** Information for the AutoCAD Designer database is stored in the layers ADP_FRZ, ADP_WORK, ADD_VIEWS, ADD_DIMS, ADV_#_VIS, and ADV_#_HID. Making changes to these layers can corrupt or destroy the parametric modeling database. *Leave these layers frozen and do not edit their contents.* Do not use commands, like Explode, Mirror, or Scale, that can change or destroy block information. ■

It is always a good idea to save your work often; if you use Save As and save to a new name, you can restore the previous file if you have made an error that you cannot correct

easily or if your drawing database becomes corrupt. AutoCAD's Audit command, which reports the integrity of your drawing, will not work correctly with AutoCAD Designer files. It will always report errors because of the extended entity data required by AutoCAD Designer.

Starting

Before you begin, make sure that you have installed and configured AutoCAD Designer. If you have not installed AutoCAD Designer, refer to the section on configuring AutoCAD Designer in Getting Started. Make sure that you have set up your AutoCAD program so that the AutoCAD Designer menu is automatically loaded.

Launch AutoCAD. Make sure that AutoCAD Designer is loaded. You should see the AutoCAD drawing editor on your screen with the Designer pull-down menu available. When you pick on the Designer menu bar selection, the Designer pull-down menu should appear at the far right of the pull-down menus, as shown in Figure 11.1.

Figure 11.1

Designer has two toolbars available: the Designer Main toolbar and the Designer Viewing toolbar. Turn on the Designer toolbars, using the Designer pull-down menu.

Pick: **Designer, Utilities, Toolbars, Designer Main**

The Designer Main toolbar appears on the screen.

Repeat this process on your own for the Designer Viewing toolbar (called Designer View in the Designer menu). When you have finished these steps, the Designer toolbars appear on your screen, as shown in Figure 11.2. Use the techniques you have learned to move them near the top of the screen out of the way of your graphics window.

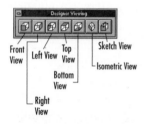

Front View
Left View
Top View
Sketch View
Isometric View
Bottom View
Right View

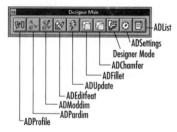

ADList
ADSettings
Designer Mode
ADChamfer
ADFillet
ADUpdate
ADEditfeat
ADModdim
ADPardim
ADProfile

Figure 11.2

Using AutoCAD Designer Help

Like the AutoCAD Help command you learned in Tutorial 1, AutoCAD Designer has its own help files. To get help for Designer commands,

Pick: **Designer, Designer Help**

Figure 11.3

The Designer help window appears on your screen, as shown in Figure 11.3. It has the same basic appearance and function as regular AutoCAD help, except that only Designer commands appear. The Designer commands are not a part of the standard AutoCAD package and are not found under the regular AutoCAD Help command. You can use the Designer help window to get help for either commands or system variables. You can also use the buttons for Search, Contents, Back, etc., as you do with AutoCAD help.

Pick: **Designer Commands**

The list of Designer commands appears in the Designer help window, as shown in Figure 11.4. All Designer commands start with the letters AD followed by the remainder of the command name. You will pick on the ADProfile command to get help.

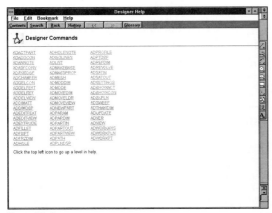

Figure 11.4

Pick: (on ADPROFILE)

Help for the ADProfile command appears in the help window, as shown in Figure 11.5. If you have difficulty with commands during the next tutorials, refer to Designer help. Designer help does not contain information about where to select commands from the toolbars. If you are unsure of the command name for a particular selection, refer to the Designer Command Summary provided in this guide.

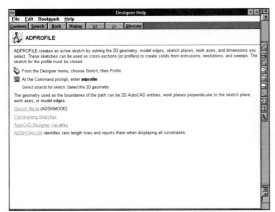

Figure 11.5

On your own, close the Designer help window by double-clicking its Windows Control box.

The part you will create in this tutorial, shown in Figure 11.6, is a clevis, a common mechanical device used for attachment.

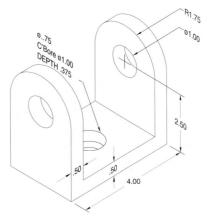

Figure 11.6

Provided with the AutoCAD Designer program are two prototype drawings that contain basic settings useful in AutoCAD Designer, which you can use in starting new drawings. The drawing *adesign.dwg* contains basic settings for ANSI standard drawings; drawing *adiso.dwg* mainly follows the ISO standard for drawings. Both prototype drawings are found in the *c:\ad\com\sup* directory. Use the AutoCAD Designer prototype drawing *adesign.dwg* to begin your new drawing, *clevis.dwg*, on your own now.

On your own, create a new layer named CLEVIS. Assign it color blue and linetype CONTINUOUS and set it as the current layer.

Creating Geometry as a Sketch

Using AutoCAD Designer, you can enter the basic shapes very roughly. In the process of constraining the sketch, endpoints will be aligned and lines will be straightened. Like doing rough sketches by hand early in the design process, you begin by showing the basic shapes and not worrying too much about lining things up perfectly until later.

Choose a major surface of the object to sketch first. Other, smaller features will be added to the first feature. The first feature you create acts as the *base feature*. All other features are related to the base feature by the constraints you add. The way the model updates, and the ease of adding features, is determined by the choice of the base feature.

On your own, use the AutoCAD Line and Arc commands to sketch the shapes that make up the rounded shape of the clevis, as shown in Figure 11.7. Notice that the lines do not have to be perfectly straight and endpoints do not need to meet exactly. The exact lengths are not important. You can specify a new value when you constrain and dimension the sketch in the next steps. This basic shape cannot contain any internal features, such as holes. Lines, arcs, circles, and non-spline fit polylines can all be used to create sketch geometry, as well as work planes that are perpendicular to the sketch plane, work axes, and part edges. (You will learn about these later.) Your sketch should look similar to Figure 11.7.

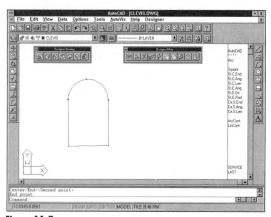

Figure 11.7

In the next steps, you will use AutoCAD Designer to apply constraints that will establish the final relationships between the lines and arcs that will make up the features of your clevis. Later you will refine this shape and then use an extrude command, as you did when creating solid models, to make a 3D solid feature. You will use AutoCAD Designer to create the hole feature after you create the initial feature.

Setting AutoCAD Designer Variables

 You can use the Designer Settings dialog box, available from the Designer Main toolbar, to control the constraints that are applied in solving your sketch.

Pick: ADSettings icon

The dialog box shown in Figure 11.8 appears on your screen. You can use it to set the angular tolerance and the pickbox size, and to determine whether constraints are applied to your geometry.

Figure 11.8

The default setting for rule mode is on; this tells AutoCAD Designer to apply constraints that realign sketch geometry. Turning rule mode off means that AutoCAD Designer will not realign geometry, except for aligning the endpoints of objects. Similarly, sketch mode tells AutoCAD Designer to assume that the sketch is inaccurate, and to apply constraints to realign sketch geometry. The default setting for sketch mode is also on. Turning sketch mode off tells AutoCAD Designer to assume that your sketch is accurate and not to change the geometry.

The Angular Tolerance area of the dialog box determines the variation that lines can have before AutoCAD Designer considers them to be horizontal, vertical, or parallel. Initially the Angular Tolerance variable is set to $4°$.

The size of the AutoCAD pickbox tells AutoCAD Designer whether endpoints that do not touch should be joined. You can use the Pickbox Size area of the dialog box to change the pickbox size, or you can zoom out on the drawing so that the gaps in the sketch will fit into the existing pickbox.

The Constraint Display Size selection in the dialog box allows you to change the size of the symbols that are used to display constraints on the screen. You will see how to display the constraints later in this tutorial, and if you want to change the size, you can return to this dialog box.

You can also specify the linetypes that you will use to draw the objects in the sketch, using the Sketch Linetypes area of the dialog box. Objects drawn with linetypes other than the one specified will be considered construction lines when it is time to apply constraints to solve the sketch.

You can also select the Drawing Settings dialog box to set drawing variables controlling various aspects of the way orthographic drawings are created from the model.

> Pick: **Cancel (to exit the Designer Settings dialog box without saving any changes)**

Sketch Menu

Selecting Sketch from the Designer pull-down menu shows the Sketch selections that you use to add information such as constraints and dimensions to your sketches.

The Sketch submenu contains the following selections:

Profile	applies constraints to the sketch to create a 2D profile
Path	creates a constrained path for use in sweeping operations
Sketch Plane	defines the active sketch plane and its orientation
Constraints	shows, adds, or deletes the constraints in a sketch
Fix Point	changes the location of the sketch's fixed point
Add Dimension	adds dimensions to the active sketch

When using AutoCAD Designer, you create a sketch, constrain it, and make it into a feature by extrusion, revolution, or sweeping. Only one sketch can be active at a time. When a feature is complete, you can define a new orientation for the sketch plane and start a new sketch. You can link the new sketch to the existing feature and then make it into a new feature.

Menu Names and Designer Commands

When you use the pull-down menu options and the toolbars in AutoCAD Designer, the command name that you can type is echoed in the command window, just as it is with AutoCAD.

■ TIP Once you become familiar with AutoCAD Designer, you may want to use shortened command names, which you can type quickly. AutoCAD provides a bonus LISP program, *adbonus1.lsp*, to enable you to use shortened command names for several Designer commands. The bonus commands have not been fully tested, so be sure to save your work before you load them. To load *adbonus1.lsp*, at the command prompt type (*load "ADBONUS1"*) ⏎. You have to type it exactly as shown, with parentheses and quotation marks. Once you load *adbonus1.lsp*, you can use the shortened commands listed below:

ad	prompts for operation to perform on parts
adf	prompts for operation to perform on features
adw	prompts for operation to perform on work planes
adp	for ADProfile command; creates a sketch
ada	for ADPardim command; adds dimensions
adc	for ADModdim command; changes dimensions
adm	for ADMode command; flips viewing mode
add	for ADDimdsp command; displays dimensions
adco	for ADShowcon command; prompts to Display, Add or Remove
adv	for ADPartview command; prompts for type or view ■

Creating a Profile

A *profile* is a closed 2D shape that you can extrude, revolve, or sweep. AutoCAD Designer solves the active sketch geometry by applying rules to form closed 2D profiles, or cross-sectional shapes. You can create a profile by picking ADProfile from the Designer Main toolbar. A similar command, selected from the Designer menu, Sketch, Path, creates paths of open or closed planar shapes that you can use in sweeping operations to create solid models. (A sweeping operation is similar to extrusion, except that the path can be curved, instead of always a straight line.)

Pick: **ADProfile icon**

Select objects: *(use implied Crossing to select all of the objects in your sketch)*

Select objects: ⏎

When Designer applies constraints to the drawing, the lines and arcs of your sketch are neatened up. You will see a box and Xs marking locations on the sketch. The box indicates that the enclosed point is fixed with respect to the World Coordinate System (WCS). As you add constraints and update your model, this point stays at the same WCS location and the other points are moved to meet the constraint criteria. The Xs mark the locations of temporary points in the sketch that are added as AutoCAD Designer is solving the sketch. Your drawing should appear similar to Figure 11.9.

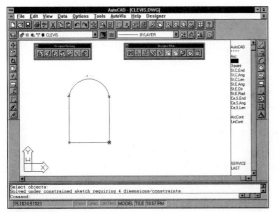

Figure 11.9

> ■ **TIP** To select a different point to act as the fixed point for the sketch, select Designer, Sketch, Fix Point or use the command ADFixpt. Follow the prompts to select a new point to be fixed in the WCS. ■

Showing the Constraints

Creating a profile applies the following constraints:

Symbol	Constraint
H	Lines sketched nearly horizontal are horizontal.
V	Lines sketched nearly vertical are vertical.
T	Two arcs or an arc, a circle, and a line sketched nearly tangent are tangent.
N	Two arcs or circles whose centers are sketched nearly coincident are concentric.
C	Two lines sketched nearly overlapping along the same line are colinear.
P	Lines sketched nearly parallel are parallel.
L	Lines sketched nearly perpendicular are perpendicular. The lines must be attached for perpendicularity to be inferred automatically.
R	Any arcs and circles sketched with nearly the same radius have the same radius. (One of the arcs or circles must include a radius dimension before AutoCAD Designer applies this rule.)
none	Objects are attached using the endpoint of one object and the near point of the other object.

You can see what constraints AutoCAD Designer has applied to the profile by selecting Designer, Sketch, Constraints, Show from the pull-down menu.

Pick: **Designer, Sketch, Constraints, Show**
All/Select/Next<eXit>: A ⏎

The constraints that were applied appear on your screen, as shown in Figure 11.10. Each sketched object is identified by a number. The constraints are shown by a code letter and number matching the object to which the constraint applies.

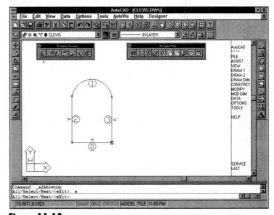

Figure 11.10

Refer to the list of constraints to help you interpret the constraints in your drawing. The constraints shown in your drawing may not be exactly the same as shown in the figure, depending on how you created your sketch. When you are finished examining the constraints, press ⏎ on your own to end the command and remove the constraints from your screen.

In the next step you will learn to add more constraints to the drawing. Adding constraints allows you to control the geometry in your drawing. These constraints tell AutoCAD Designer how to control the appearance of the drawing when a change is made to a feature. You can also delete constraints that AutoCAD Designer has created.

You will add two tangent constraints so that the rounded end is tangent to the two vertical lines. If your drawing already has these constraints, you can skip this step. If your lines are not horizontal and vertical, you will need to add those constraints before you add the tangent constraints. To add new constraints,

> Pick: **Designer, Sketch, Constraints, Add**
>
> Hor/Ver/PErp/PAr/Tan/CL/CN/PRoj/Join/XValue/Yvalue/Radius/<eXit>: **T** ⏎
>
> Select first item to make tangent: **(pick on the arc)**
>
> Select second item to be tangent to first: **(pick on the right vertical line)**
>
> Hor/Ver/PErp/PAr/Tan/CL/CN/PRoj/Join/XValue/Yvalue/Radius/<eXit>: **T** ⏎
>
> Select first item to make tangent: **(pick on the arc)**
>
> Select second item to be tangent to first: **(pick on the left vertical line)**
>
> Hor/Ver/PErp/PAr/Tan/CL/CN/PRoj/Join/XValue/Yvalue/Radius/<eXit>: ⏎

If you try to add a constraint that already exists, you will see a message stating *Constraint already exists.* Do not duplicate existing constraints. If necessary, restart the command and continue on your own to add any constraints your sketch may be missing. On your own, use the Designer, Sketch,

Constraints, Show selection from the menu bar to show the constraints you have added. When you are finished, your drawing should have the constraints shown in Figure 11.11.

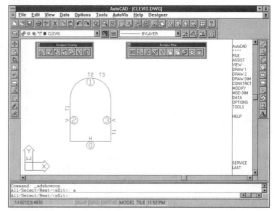

Figure 11.11

Deleting Constraints

If you have constraints in your drawing that are incorrect, you can delete them by picking Designer, Sketch, Constraints, Delete from the pull-down menu. Picking on the object from which you want to delete the constraint causes the constraints along that object to be shown. Next, pick on the symbol for the constraint you want to remove. Press ⏎ to end the command. Use this on your own if you have added any incorrect constraints.

Adding Parametric Dimensions

 Next you will add parametric dimensions to your drawing by picking ADPardim from the Designer Main toolbar. Constraints control the shape of the part you are creating; the parametric dimensions control the size. Because you already know that the rounded end is tangent to the vertical lines, only two dimensions are necessary to determine the size of the part: the radius of the rounded end and the distance from the horizontal line to the center of the arc (the length of the vertical side).

When you are adding parametric dimensions to a part, consider carefully the function of the part being designed, and dimension the features accordingly.

Designer uses the dimensions you place to change the size of the sketch, and the feature created from it. When you override the default value with a different one as you are dimensioning, your drawing will be updated to match the new value.

It makes sense to locate dimensions from the fixed point, just as you would when creating a fixture to hold the part during inspection. Because the fixed point will not move on the WCS, it is a good practice to dimension other locations from this point. Refer to Figure 11.12 when picking points.

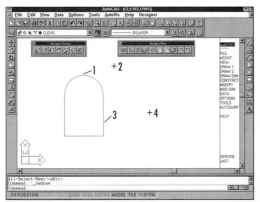

Figure 11.12

Pick: **ADPardim icon**
Select first item: *(pick on the rounded end near point 1)*
Select second item or place dimension: *(pick near point 2)*
Undo/Dimension value <xxxx>: **1.75** ⏎

Select first item: *(pick on the right-hand vertical line identified as point 3)*
Select second item or place dimension: *(pick near point 4)*
Undo/Hor/Ver/Align/Par/Dimension value <xxxx>: **2.5** ⏎
Select first item: ⏎

The dimensions appear in your drawing, as shown in Figure 11.13, and the drawing is updated automatically to reflect the sizes that you typed in.

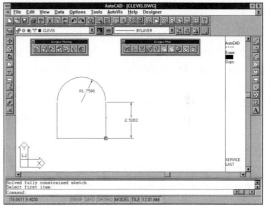

Figure 11.13

Parametric and Reference Dimensions

Do not worry if the dimensions shown during the modeling process are not in accordance with standard practices for engineering drawings. After you have modeled the part, you will generate correctly drawn views from the model. During this process, the dimensions will automatically be placed in accordance with rules of engineering practice. Once the dimensions appear in your drawing, you can edit them further to achieve the results you want. You can also place reference dimensions for the drawing, which are different from the parametric dimensions you just created. Parametric dimensions (or feature dimensions) control the model. Using reference dimensions allows you to correctly dimension the final drawing. Reference dimensions update automatically when the model is changed, but they do not control the model.

■ *TIP* Use the Pan and Zoom commands, if necessary, to move your view on the screen so that you can see the entire drawing. ■

Notice the message in the command prompt area that says *Solved fully constrained sketch*. This indicates that you do not need any additional dimensions or constraints; the profile is fully defined. Like a machinist trying to manufacture a part from the dimensions shown, AutoCAD Designer cannot fully define a parametric model if some dimensions are left out. Likewise, if you provide too many constraints or parametric dimensions, you will get a message saying that the drawing is over-constrained. Remember the tutorials on dimensioning: you learned not to over-dimension drawings because of the confusion that might result, either because of difficulty in interpreting the tolerance or because two different dimension values are given for the same feature.

■ *TIP* If you need to change the values of dimensions after they are added to the sketch, pick ADModdim from the Designer Main toolbar. After changing dimensions or features, use ADUpdate to update the model's appearance to reflect the change. ■

If you add an incorrect dimension to the sketch, you can use AutoCAD's Erase command to remove it. Only use Erase to remove dimensions while you are working on the sketch. In the next step you will create the object as a 3D solid feature; once you do that you must use ADEditfeat from the Designer Main toolbar to change it.

Creating Features

The Features selection from the Designer menu contains options for creating features. AutoCAD Designer allows you to create features by extruding, revolving, or sweeping a 2D profile. You can also easily add features such as holes, fillets, and chamfers to your model through the specialized commands provided. The final items on the Features submenu are used to create work planes, work axes, and work points.

Work Planes, Work Axes, and Work Points

A *work plane* is a parametric plane that you can use to locate the sketch plane or other features. When you create a work plane, you link it to the current feature geometry by an edge, axis, vertex, or surface of the part. This way, features located on the work plane move and update when the parent feature is moved.

A *work axis* is similar to a work plane, except that it is a line at the centerline of the cylindrical, conical, or toroidal surface that you select. You can use the work axis to locate other features.

A *work point* is a point that you can place to locate holes. You can constrain and dimension the location of the work point on the sketch plane, using the sketch commands you have learned. Work planes, work axes, and work points are all located on the generated layer ADP_WORK. Do not edit the contents of this layer; doing so may corrupt or destroy your modeling database.

Creating an Extrusion

Now that the profile is fully constrained, you are ready to extrude it to form a 3D solid feature. As you learned in Tutorial 9, extrusion is named for the manufacturing process in which material is forced through a shaped opening. In AutoCAD Designer, extrusion gives a profile thickness, thus creating a feature. To extrude the profile, you will pick from the Designer menu.

Pick: **Designer, Features, Extrude**

You will use the Designer Extrusion dialog box that appears on your screen, shown in Figure 11.14, to select the distance to extrude the part. Notice that the radio button for Base as the type of operation is already selected and cannot be changed. The first feature you create is always the base feature.

The Termination area of the dialog box controls the method for how the extrusion is ended. Selecting Blind extrudes the profile to the distance you specified in the Distance area. Mid Plane extrudes the profile equally in both directions ending at the overall depth specified in the Distance area. To Plane ends the extrusion at the planar face or work plane you pick. Through works only when cut or intersect is selected as the operation and goes all the way through the solid part.

In the Size area of the dialog box, you can set the distance for the extrusion and the draft angle, if you want. *Draft angle* is the term for the taper on a molded part that makes it possible to easily remove the part from the mold. You can use it to make parts that taper as they are extruded. You can specify negative draft angles to create parts that taper to a larger profile as they are extruded. You will set the distance to 3 and leave the other settings at their defaults.

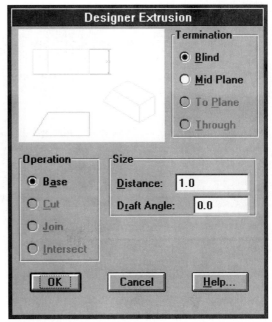

Figure 11.14

Type: **3** *(in the box to the right of Distance)*

Pick: **OK** *(to exit the dialog box)*

The part is extruded, creating a 3D solid. Because you are looking straight down the Z axis, you cannot see the thickness of the part.

Viewing the Part

 AutoCAD Designer allows you to change the viewing angle for the part and makes it easy to specify preset views by making selections from the Designer Viewing toolbar. You will select the Isometric View icon from the Designer Viewing toolbar so you can see the 3D shape of the part.

Pick: **Isometric View icon**

Your viewing direction changes so that your screen looks like Figure 11.15.

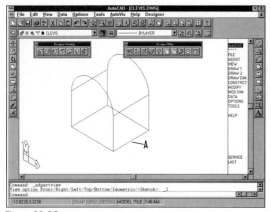

Figure 11.15

On your own, use the Zoom and Pan commands to position your view so that you can see the object well. (As you are working through the tutorial, use the Zoom and Pan commands as needed.)

Changing 3D Solid Features

 To change the dimensions for 3D solid features (the depth to which you extruded the profile), you will pick ADEditfeat from the Designer Main toolbar.

Pick: **ADEditfeat icon**

Sketch/<Select feature>: *(pick line A in Figure 11.15)*

The dimensions for the solid appear, as shown in Figure 11.16.

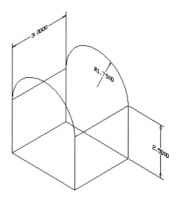

Figure 11.16

Select dimension to change: *(pick on the 3.0000 dimension)*

New value for dimension <3>: **4.5** ⏎

Select dimension to change: ⏎

 The dimensions disappear from your screen, but the size of the feature remains unchanged until you use ADUpdate to update it. Next you will select the ADUpdate icon from the Designer Main toolbar.

Pick: **ADUpdate icon**

The thickness of the object increases in the isometric view, as shown in Figure 11.17.

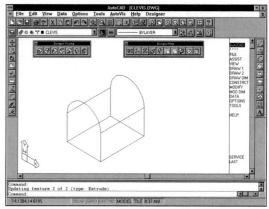

Figure 11.17

Adding the Hole

AutoCAD Designer provides commands that make it easy to add drilled, countersunk, or counterbored holes, as well as fillets and chamfers. These are found on the Features option of the Designer menu. You will add the hole to the rounded end of the clevis.

Pick: **Designer, Features, Hole**

Use the Designer Hole dialog box that appears on your screen (as shown in Figure 11.18) to select the type of hole to add. For this part, the operation will be a drilled hole; the placement will be concentric; the termination will be through (in other words, the hole goes completely through the object); and the diameter for the drill size of the hole will be 1. If you choose to create a through hole, the depth and point angle selections of the dialog box will be grayed out. This is because the hole will go completely through the object, so you would not see the drill point or need to specify the depth. If you change the thickness of the part, a through hole updates so that it still goes through the new thickness. For holes that are blind (do not go completely through), use the depth and point angle selections to specify the depth of the hole and the angle for the drill point.

If you select the C'Bore (counterbore) or C'Sink (countersink) radio buttons in the Operation area of the dialog box, the C'Bore/Sunk Size area of the dialog box is no longer grayed out and you can use it to input the appropriate sizes. Notice that the type of operation you select using the radio buttons is depicted in the image at the upper left of the dialog box. On your own, make the selections shown in Figure 11.18. Select a drilled through hole of diameter 1.00, using concentric placement. When you are finished, select OK to exit the dialog box.

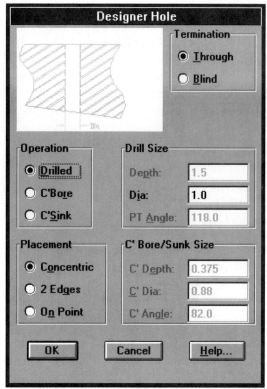

Figure 11.18

The options for placing the hole in your drawing are: locating it from a concentric edge, from two edges, or from a work point. When locating the hole from two edges, you are prompted to select objects to define the planar work face where the circular shape of the hole will be located, and then to specify the distance from each of the two edges that you select.

To locate the hole from a work point, you must first have placed a work point in the drawing by picking Designer, Features, Work Point. The selected work point will then be the center location for the hole.

If you locate the hole from a concentric edge, AutoCAD Designer prompts you to select the work face and then a rounded edge, and then locates the hole at its center. Selecting the right placement method is important because it will in part determine the way the model updates when you make changes to the dimensions or constraints.

You are returned to the drawing for the following selections.

Select work plane or planar face: **(pick on the rounded edge)**

Select concentric edge: **(pick the rounded edge again)**

The hole is added to your model, as shown in Figure 11.19.

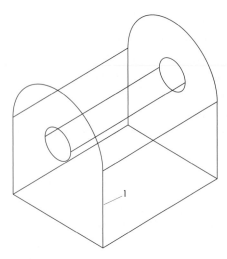

Figure 11.19

Save your work on your own before continuing.

Aligning the Sketch Plane

Next you will select a new sketch plane and create a sketch showing the shape of the clevis from that new direction. Aligning a sketch plane is much like orienting a User Coordinate System, which you have done in previous tuto-

rials. Remember, only one sketch plane can be active at a time. You will align the sketch plane by picking from the Designer menu.

Pick: **Designer, Sketch, Sketch plane**

Xy/Yz/Zx/Ucs/<Select work plane or planar face>: **(pick line 1 in Figure 11.19)**

Next/<Accept>: **N** ⏎

If the surface shown highlighted in Figure 11.20 is *not* highlighted on your screen, type *N* ⏎ in order to select the next surface on the object. Continue switching to the next surface until the correct surface is highlighted. When it is, press ⏎ to accept the selection. Refer to Figure 11.20 to make your selection.

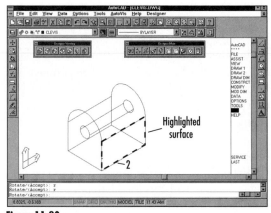

Figure 11.20

X/Y/Z/<Select work axis or straight edge>: **(pick line 2)**

Rotate/<Accept>: **R** ⏎

Rotate/<Accept>: **R** ⏎

You should now see the UCS icon aligned with the highlighted surface shown in Figure 11.20. If not, continue to type *R* and press ⏎ until the UCS is aligned with the previously highlighted surface. When the orientation matches, press ⏎ to accept it.

Rotate/<Accept>: ⏎

Sketching the Side View

The new sketch plane is now aligned with the highlighted surface, so that you can draw the clevis as it appears from the side. Once you draw the side view of the clevis, you can then extrude it and find the intersection of the two extrusions. Using AutoCAD Designer, you can perform both the extrude and intersection operations in a single command step.

Before sketching the side view of the clevis, set the color for object creation to red so that it is easier to see the new objects added to the drawing. On your own, use the Color control selection from the Object Properties toolbar and then use the Select Color dialog box to set the color to red.

Remember, usually it is a better practice to set the color using BYLAYER, because then changes to the layer are reflected in the color of the object. In this case, however, you will leave the layer set to CLEVIS and change the color for new objects created.

Notice the red color swatch that appears on the Object Properties toolbar. This is to alert you that the color red will be used for newly created objects. The layer color for layer CLEVIS remains blue. On your own, use the Line command to sketch the shape you see in Figure 11.21. Do not draw your lines exactly on top of the blue lines of the existing feature, because it will make them hard to select.

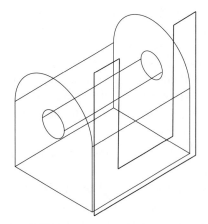

Figure 11.21

Creating the Profile

Next, you will create a profile from the sketched objects, as you learned earlier in the tutorial.

Pick: **Designer, Sketch, Profile**

Select objects: *(pick all of the new lines added to the new sketch)*

Select objects: ⏎

> ■ *TIP* You can also use implied Crossing or Window to select the objects. If you select the wrong item, you can type R ⏎ at the *Select objects:* prompt and select the items to remove from the selection set.

The lines that you drew are neatened up and you may see Xs drawn at the corners. Next, you will show the constraints that have been added to establish the relationships between the new lines. Select to show the constraints from the pull-down menu, as you did before:

Pick: **Designer, Sketch, Constraints, Show**

All/Select/Next/<eXit>: **A** ⏎

All/Select/Next/<eXit>: *(when you are finished examining the constraints shown on the screen, press ⏎)*

The constraints added to the drawing should appear similar to Figure 11.22. Your drawing may not be identical, depending on how you constructed each line. If you do not have horizontal constraints for the two bottom horizontal lines, add them on your own. If your vertical lines do not have vertical constraints, add them also by selecting Designer, Sketch, Constraints, Add.

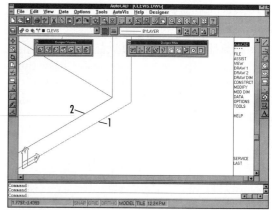

Figure 11.23

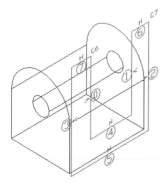

Figure 11.22

Next you will add a *projected* constraint for the endpoint of the bottom horizontal line. Adding a projected constraint allows you to link the new sketch profile to the existing part. You will also add *colinear* constraints between the vertical line of the sketch and the vertical line of the previous object, and between the two short top horizontal lines.

On your own, zoom the lower right-hand corner of the object, so that your drawing appears similar to Figure 11.23. Your drawing may be slightly different, depending on the actual sketch you created. Don't be concerned if your drawing is not exactly the same as the figure at this point. After you add the constraints in the next steps, it should be the same.

Pick: **Designer, Sketch, Constraints, Add**

Hor/Ver/PErp/PAr/Tan/CL/CN/PRoj/Join/XValue/Yvalue/ Radius/<eXit>: **PR** ⏎

Specify end point to constrain to item: **(pick the horizontal red line of the new sketch, labeled 1 in Figure 11.23)**

Specify line, arc, or circle to constrain to: **(pick the horizontal blue line 2 on the 3D solid)**

The new red line should move to align with the selected blue line from the solid model. You are returned to the command prompt, where you can continue to add constraints. Next, you will make the right vertical red line of the sketch colinear with the right vertical blue line of the solid.

Hor/Ver/PErp/PAr/Tan/CL/CN/PRoj/Join/XValue/Yvalue/ Radius/<eXit>: **CL** ⏎

Specify first line: **(pick red line 3 in Figure 11.24)**

Specify second line: **(pick blue line 4)**

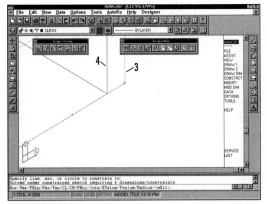

Figure 11.24

The right line of the sketch should now be co-linear with the solid. You still see the prompt options for adding constraints.

On your own, use the Zoom Previous command to return to the previous zoom factor and then zoom the lower left-hand corner of the object. You will need it enlarged so that you can add another colinear constraint for the left-hand vertical line in order to line it up with the solid.

Hor/Ver/PErp/PAr/Tan/CL/CN/PRoj/Join/XValue/Yvalue/ Radius/<eXit>: **CL** ⏎

Specify first line: *(pick the vertical red line of the new sketch)*

Specify second line: *(pick the vertical blue line on the 3D solid)*

Use Zoom Previous on your own to return to the original zoom factor so that you can add the co-linear constraint for the top two short horizontal lines if needed. If you already have this constraint in your drawing, you will see the message *Constraint already existed on selected items.*

Hor/Ver/PErp/PAr/Tan/CL/CN/PRoj/Join/XValue/Yvalue/ Radius/<eXit>: **CL** ⏎

Select first line: *(pick red line 5 in Figure 11.25)*

Select second line: *(pick red line 6)*

Hor/Ver/PErp/PAr/Tan/CL/CN/PRoj/Join/XValue/Yvalue/ Radius/<eXit>: ⏎

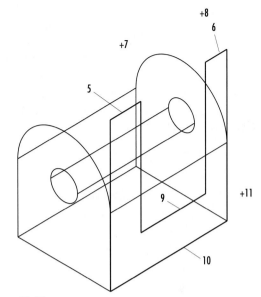

Figure 11.25

Now you are finished adding the constraints. The two vertical lines and the bottom line of the sketch should line up with the corresponding edges of the solid, as shown in Figure 11.25. The short horizontal lines at the top of the object should align with one another.

Dimensioning the New Profile

Next, you will add parametric dimensions to define the size of the object so that it is fully constrained, as you did with the previous profile. You will dimension the height of the profile so that it is equal to the height of the existing feature, using a simple equation. Use the Designer Main toolbar.

> ■ *TIP* If you have trouble selecting the lines to dimension, use Zoom Window to enlarge your work area. When you have finished making the selection, use Zoom Previous to restore the original zoom factor. ■

Pick: **ADPardim icon**

Select first item: *(pick on line 5 in Figure 11.25)*

Select second item or place dimension: *(pick near point 7)*

Undo/Hor/Ver/Align/Par/Dimension value <xxxx>: **.5** ⏎

Select first item: *(pick on line 6)*

Select second item or place dimension: *(pick near point 8)*

Undo/Hor/Ver/Align/Par/Dimension value <xxxx>: **.5** ⏎

Select first item: *(pick on line 9)*

Select second item or place dimension: *(pick on line 10)*

Specify dimension placement: *(pick near point 11)*

Undo/Hor/Ver/Align/Par/Dimension value <xxxx>: **.5** ⏎

Select first item: ⏎

When you have finished adding these dimensions, your drawing should look like Figure 11.26.

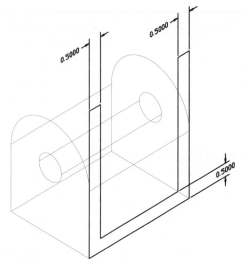

Figure 11.26

If you have added incorrect dimensions, use the Erase command to remove them on your own. Then try again to add the dimension.

AutoCAD Designer assigns a variable to each parametric dimension. You can change the way the dimensions are displayed by picking Designer, Display, Dim Display from the

pull-down menu. You will use it to display the dimensions as variables; in other words, in their parametric form.

Pick: **Designer, Display, Dim Display**

Parameters/Equations/<Numeric>: **P** ⏎

The dimensions on your screen change to display their variable names (yours may be different), as shown in Figure 11.27.

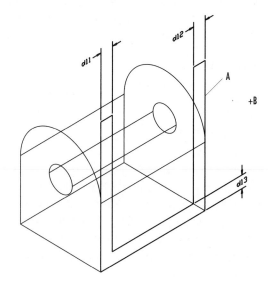

Figure 11.27

Next you will use the icon ADEditfeat from the Designer Main toolbar to display the dimensions for the previous sketch, so that you can determine the variable names of the rounded end and the height of the sketch. You will add an equation dimension for the height of the current sketch, so that it is the height of the first sketch plus the radius of the rounded end.

Pick: **ADEditfeat icon**

Sketch/<Select feature>: *(pick on the rounded end)*

Notice that the parametric dimensions for the previous sketch are now shown. On your own, review your drawing and determine the variable names for the radius of the rounded end (probably d1) and the height of the side (probably d2). Once you have this information, you will exit the command by pressing ⌇ESC⌇.

Select dimension to change: ⌇ESC⌇

Now you will add the equation dimension for the height of the side view sketch.

■ *Warning:* If your variable names are not the same, substitute the variable names from your drawing in the next step or your model will not work correctly. ■

Pick: **ADPardim icon**

Select first item: *(pick on line A in Figure 11.27)*

Select second item or place dimension: *(pick near point B)*

Undo/Hor/Ver/Align/Par/Dimension value <xxxx>: **d1+d2** ⌇←⌇

Select first item: ⌇←⌇

Your sketch updates so that the height is the same as the height of the original sketch.

On your own, change the dimension display back so that the numerical values are shown. When you are finished with this step, your drawing should look similar to Figure 11.28.

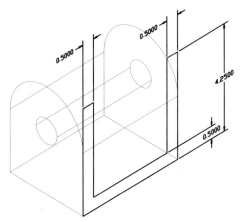

Figure 11.28

On your own, save your work so far.

Using AutoCAD Designer, you can extrude and find the intersection between two features in the same command step. Next, you will extrude the new sketch into a 3D solid object to find the intersection with the base solid you created earlier.

Pick: **Designer, Features, Extrude**

The Designer Extrusion dialog box appears on your screen, as shown in Figure 11.29. Notice that this time, since you have already created a base feature and only one is allowed per part, the radio button for Base is grayed out and cannot be selected. On your own, you will use the dialog box to create an intersection of the new extrusion with the previous feature. Use the following settings — Termination: Through, Operation: Intersect, and Draft Angle: 0.0. When you have finished, pick OK.

Figure 11.29

An arrow appears, indicating the direction in which the extrusion will proceed. If the arrow is pointing in the correct direction, so that the extrusion will go through the part, press ⌇←⌇ to accept. If not, type *DF* ↵ at the prompt to flip the direction, and then accept when the arrow is pointing the correct way.

Direction Flip/<Accept>: ↵

When the extrusion and intersection are complete, the resulting part should look like Figure 11.30. Use Redraw All, if necessary, to clean up your screen.

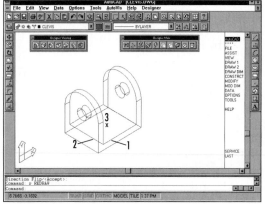

Figure 11.30

Adding the Counterbored Hole

Now you are ready to add the final hole and then create a multiview orthographic drawing from your model.

Pick: **Designer, Features, Hole**

On your own, use the Designer Hole dialog box to add a counterbored hole of diameter .75 through the object. Use the C'Bore/Sunk Size area of the dialog box to set the counterbore depth to .375 and the counterbore diameter to 1. You will place the hole using two edges. Refer to Figure 11.30 for your selections.

Select first edge: *(pick on line 1)*
Select second edge: *(pick line 2)*

Select hole location: *(place the corner of the box that rubberbands from the lines you selected by picking 3, near the center of the bottom surface of the object)*

Distance from first edge <xxxx>: **1.75** ↵
Distance from second edge <xxxx>: **1.75** ↵

The hole should be added to your model, as shown in Figure 11.31.

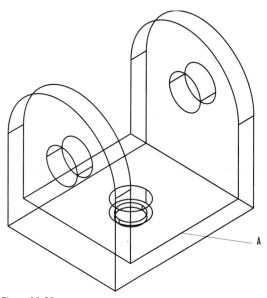

Figure 11.31

Now that the counterbored hole is added, it appears that there is not enough material in the base of the clevis for the depth of counterbore you chose, so you will make the base of the clevis thicker. To change the thickness of the base, use ADEditfeat from the Designer Main toolbar. Refer to Figure 11.31 for your selections.

Pick: **ADEditfeat icon**

Select feature: *(pick on line A)*

If the surfaces shown as dotted lines in Figure 11.32 are highlighted, press ↵ to accept the selection. Otherwise, type *N* ↵ on your own until the surfaces are highlighted, and then press ↵ to accept.

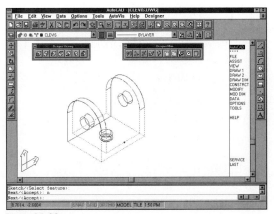

Figure 11.32

The dimensions for the side view of the part should appear on your screen.

> Select dimension to change: *(pick the .5 dimension between two bottom surfaces)*
>
> New value for dimension<.5>: **.75** ⏎
>
> Select dimension to change: ⏎

Next you will update the appearance of the model to show the change you have made.

> *Pick:* **ADUpdate icon**

Your model should now look like Figure 11.33.

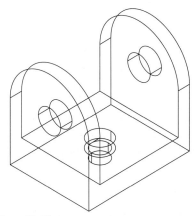

Figure 11.33

Save the model on your own.

Creating a Drawing from the Part

Now you have finished creating the model of the clevis. You are ready to have AutoCAD Designer create the dimensioned multiview orthographic drawing of the part. You create this drawing automatically by picking Create View from the Designer menu.

Before creating the drawing views, you will set the Color Control selection from the Object Properties toolbar so that colors are determined by layer before you create the drawings. Otherwise, all lines will continue to be drawn with red, regardless of what layer is used. Use the Object Properties toolbar to select the Color Control dialog box.

> *Pick:* **Color Control**

Use the Color Control dialog box to set the color to BYLAYER on your own. Be sure to pick OK when you have finished.

To create a drawing from the part,

> *Pick:* **Designer, Drawing, Create View**

The Designer Drawing View dialog box shown in Figure 11.34 appears on your screen. You will accept the defaults to create the base view. The base view is the main orthographic view from which the lines of sight for the other orthographic views are established; usually it is the front view of the object.

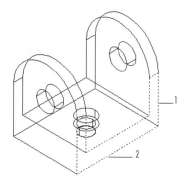

Figure 11.35

Figure 11.34

Pick: **OK**

You return to the drawing screen. You select the base view by picking a planar face, by picking a work plane, or viewing the object from the XY, YZ, ZX, or UCS planes. A planar face is any plane surface on the object, defined by picking on a straight edge. You can also select a work plane if you have previously created it. In addition you may type the command option letter to align the viewing direction for the base view with the coordinate axis pairs, or the UCS. You will pick a planar face.

Xy/Yz/Zx/Ucs/<Select work plane or planar face>: *(pick line 1 in Figure 11.35)*

Next/<Accept>: *(type N ⏎ until the surface is highlighted as shown)*

Next/<Accept>: ⏎

X/Y/Z/<Select work axis or straight edge>: *(pick line 2)*

Rotate/<Accept>: *(type R ⏎ until the UCS aligns with the highlighted surface)*

Rotate/<Accept>: ⏎

When you complete the selection of the planar face and the straight edge, you will be switched to paper space automatically. Once in paper space, you will specify a location for the view. Your screen should look like Figure 11.36. The title block you see is part of *adesign.dwg*, the prototype drawing you used to start drawing *clevis.dwg*.

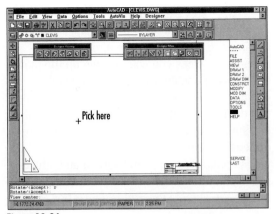

Figure 11.36

View center: *(pick near the point indicated in Figure 11.36)*

View center: ⏎

The front view and its dimensions are placed in your drawing, as shown in Figure 11.37. The dimensions may be too small to see clearly. This is because the prototype drawing, *adesign.dwg*, is set up to plot on a sheet of paper that is about 34″ by 22″. If you zoom the view, you should be able to read the dimensions. You will learn more about changing the appearance of the dimensions in the next section of this tutorial.

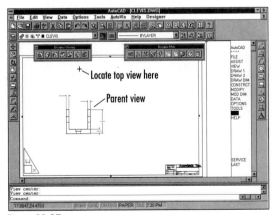

Figure 11.37

To restart the command,

Command: ⏎

The Ortho selection lets you create any orthographic view by picking a parent view and the placement for the new view. The placement determines the line of sight for the new view and AutoCAD Designer automatically unfolds the views to create the new view, according to standard practice. Orthographic views are always in alignment. On your own, select the box to the left of Ortho in the dialog box that appears on your screen to create an orthographic view.

When you are finished,

Pick: **OK**

Select parent view: *(pick in the center of the front view that you just placed)*

Location for orthographic view: *(pick above the front view, where you want the center of the top view located)*

Location for orthographic view: ⏎

> ■ **TIP** You can select either third angle projection (used mainly in the U.S., Canada and England) or first angle projection (used in most other countries) by setting the Designer variable ADProjtype. You can also select Designer, Settings and use the dialog box that appears to pick Drawing Settings. Leave the Drawing Settings dialog box set to select third angle projection for this tutorial.
> ■

On your own, repeat this process to create a right-side view. When you are finished, your screen should look like Figure 11.38.

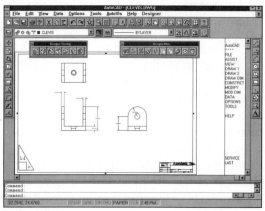

Figure 11.38

> ■ *TIP* It is easy to move the views around on the sheet at any time, so don't be too concerned about the placement at this point. If views overlap, you can move them by picking Designer, Drawing, Edit Views, Move. ■

Cleaning Up the Dimensions

When dimension views are created automatically, AutoCAD Designer uses the current dimension style for the dimensions it places in the drawing. If there is no dimension style active, a style called _AD0 is created. As additional dimension styles are needed, they are created and named _AD1, _AD2, and so on. If you create a dimension style and make it active before you create drawing views, that dimension style will be used for the dimensions, and additional _AD styles will not be created.

You can change the appearance of the dimensions in your drawing two ways. One way is by changing the dimension style characteristics, using the AutoCAD Dimension Style selection from the Data menu, and using the dialog box to change the dimension style. Refer to Tutorials 7 and 16 for information on using dimension styles. Because AutoCAD Designer

uses paper space and viewports to create the drawing views, choosing to use paper space scaling and then setting the dimension variables to the sizes you want shown on the plotted drawing usually works well. You will change the dimension style _AD0 so that only two decimal places are displayed and overall scaling is set to 2. Remember, it is unusual to dimension drawings to four decimal places unless accuracies of 1/10,000th of an inch are required. You will type the Ddim command to invoke the Dimension Styles dialog box.

Command: **DDIM** ⏎

The Dimension Styles dialog box appears on your screen. Dimension style _AD0 should be current. If it is not, make it the current style on your own.

Pick: **Geometry**

On your own, use the Geometry dialog box to set the overall scaling factor to 2. When you have finished, pick OK to return to the Dimension Styles dialog box. Next you will select Annotation and set the number of decimals for the display of dimensions.

Pick: **Annotation**

On your own, use the dialog box that appears to set the dimension precision for the primary units to two decimal places. When you have finished, pick OK to return to the Dimension Styles dialog box.

To save the changes to the style so that they take effect on the existing dimensions,

Pick: **Save**

Pick: **OK (to exit the Dimension Styles dialog box)**

The changes to the dimension style are reflected in your drawing. It should appear similar to Figure 11.39.

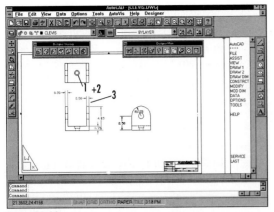

Figure 11.39

On your own, use hot grips to stretch the 4.25 dimension so that it does not overlap the shorter .75 dimension. Use hot grips again as necessary to stretch the other dimensions so that they are placed correctly.

The other way to change the appearance of the dimensions is to use the Designer menu to select Drawing, Dimension, Attributes.

Pick: **Designer, Drawing, Dimension, Attributes**

First you are prompted to select a dimension to edit, and the Designer Dimension Options dialog box shown in Figure 11.40 appears on your screen. You can use the dialog box to change the appearance of the selected dimension.

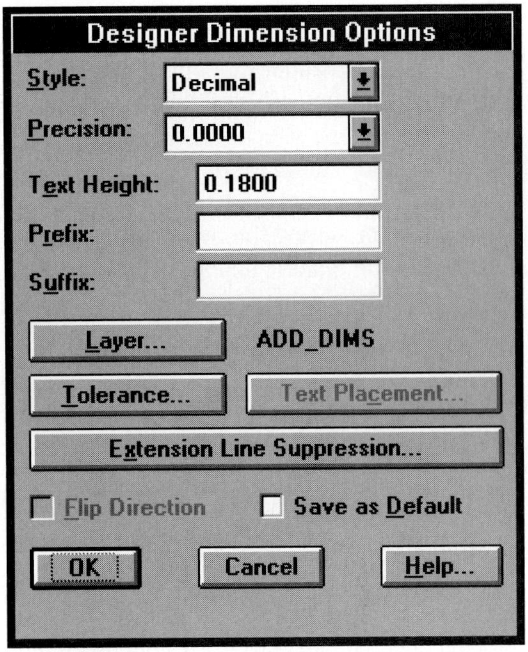

Figure 11.40

Pick: **Cancel** *(to exit without making changes)*

Next you will add the hole note in the top view for the counterbored hole.

Adding Annotation and Hole Notes

The Designer, Drawing, Annotation menu selection allows you to create, delete, and move an annotation, or text, that is associated with a view (so that if the view moves, so does the note). You can also use this menu to select the Hole Note command to create a standard hole note for the drawing.

Pick: **Designer, Drawing, Annotation, Hole Note**

Select arc or circle of hole feature: *(pick circle 1 in Figure 11.39)*

Location for hole note: *(pick point 2)*

Location for hole note: ⏎

The hole note is added to your drawing. The text is once again set up for a large sheet size.

AutoCAD Designer also provides commands for moving the dimensions in the drawing, including moving between drawing views; freezing and thawing individual dimensions; and adding and deleting reference dimensions (ones that do not affect the parametric model). Explore these commands on your own using the Designer, Dimension menu selection and Designer help.

On your own, clean up the appearance of the dimensions in your drawing.

Bidirectional Associativity

AutoCAD Designer has bidirectional associativity between the drawing and the model. If a change is made to the model, the drawing is automatically updated to reflect the change. Likewise, if a change is made to the parametric dimensions in the drawing, both the drawing and the model update automatically.

Next, you will change one of the dimensions in the drawing so that you can see this effect in your drawing. On your own, enlarge the front view of the clevis, using Zoom Window. Next, you will edit the dimension for the thickness of the right side of the clevis by picking ADModdim from the Designer Main toolbar. Then you will use the same toolbar to pick the Update icon.

Pick: **ADModdim icon**

Select dimension to change: *(pick on the .50 dimension for the right side of the clevis, identified as 3 in Figure 11.39)*

New value for dimension<.50>: **1** ⏎

Select dimension to change: ⏎

Pick: **ADUpdate icon**

Your screen briefly shows the model and then returns to the drawing, where it shows the changed thickness of the right side of the clevis. Your drawing and model are updated to reflect the change. You may have to stretch dimensions once again to relocate them. Your drawing should appear similar to Figure 11.41.

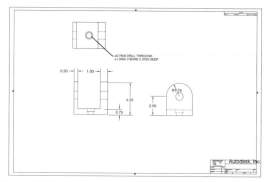

Figure 11.41

Next you will return to the view of the model.

Switching between Part and Drawing Mode

 The drawing views you have been working with are laid out in paper space. To return to the original part that was created in model space, you will select Designer Mode and toggle to part mode. You can also use this command to toggle to the drawing mode when the part is showing. Now, you will return to part mode, where you will analyze the mass properties of the part. Use the Designer Main toolbar to select the Designer Mode icon.

Pick: **Designer Mode icon**

You are returned to your model. Notice that the right side of the clevis on the model is now 1 unit thick.

Examining the Mass Properties

You can examine the mass properties of models created with AutoCAD Designer as you did in Tutorial 10 using solid modeling. To do so, you will pick the Mass Properties selection from the Designer menu.

Pick: **Designer, Utilities, Mass Properties**

ALl/Select/<ACtive>: ⏎

The Designer Mass Properties dialog box shown in Figure 11.42 appears on your screen, listing the mass properties information about your clevis. You can use the Density area of the dialog box to specify a different density for the material of the part. To use this dialog box effectively, you must keep track of your drawing units. The units you specify for the density should match the units you used when creating your model, even though no units are specified in the dialog box. Next, try changing the density. Pick in the text box to the right of Density and type in a new value, then press ⏎. Notice that changing the density changes the mass shown in the upper right corner of the dialog box. When you are finished examining the mass properties, change the Density setting back to 1, and pick OK to exit the dialog box.

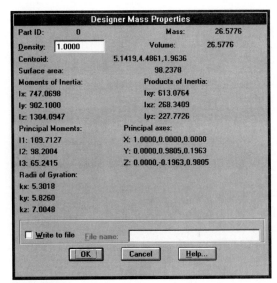

Figure 11.42

Creating a Section View

You will create a work plane and then add a section view of the clevis to your drawing. You will learn more about creating section views of 2D drawings and solid models in Tutorial 14.

A work plane is a parametric or non-parametric feature that orients the sketching plane, either to define other work planes or as a cutting plane for establishing cross-sectional views in the drawing. You will create a work plane to locate the cutting plane for the cross-sectional view you will create.

You will use the Features submenu from the Designer pull-down menu and select the Work Plane option to add the work plane. You will pick Planar Parallel to add a parametric work plane and then pick Offset to offset the parallel plane so that it is through the center of the clevis. Use the Designer Help command to find more information on creating work planes.

Pick: **Designer, Features, Work Plane**

The Designer Work Plane dialog box shown in Figure 11.43 appears on your screen.

Figure 11.43

Pick: **Planar Parallel** *(from the left side of the dialog box)*

Pick: **Offset** *(from the dialog box that appears)*

Pick: **OK**

Pick: **OK**

You will be returned to your drawing; make the following selections. Refer to Figure 11.44.

Xy/Yz/Zx/Ucs/<Select work plane or planar face>: *(pick line 1)*

Next/<Accept> edge: *(type N ⏎ if correct surface is not highlighted; otherwise press ⏎)*

Offset<1>: **1.75** ⏎

The work plane appears on your screen. If it appears toward the center of the clevis, as shown in Figure 11.44, press ⏎ to accept the position. If it is away from the clevis, type *F* and press ⏎ to flip the position of the work plane.

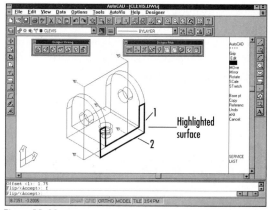

Figure 11.44

To add the section view to your drawing, you will select the Create View selection from the Designer menu.

Pick: **Designer, Drawing, Create View**

Use the Designer Drawing View dialog box that appears (shown in Figure 11.45) to select Base as the type of view and then Full Section. Pick in the box to the left of Hatch so that hatching will be created for your section view. Pick Pattern and then use the Hatch Options dialog box that appears to pick Pattern once again to get the Choose Hatch Patterns dialog box, and select the ANSI31 hatch pattern to use for the hatching. When you select ANSI31, you immediately return to the Hatch Options dialog box. Pick OK again to return to the Designer Drawing View dialog box. Type *A* in the text box for the Section Symbol label. Type *Section A-A* in the text box for the View Label. Pick OK when you have finished making these selections.

When you create a section view, you may specify whether to section through a point you specify on the view or use a work plane. Because you specified a full section, AutoCAD Designer assumes you will use a work plane as the cutting plane and prompts you for it.

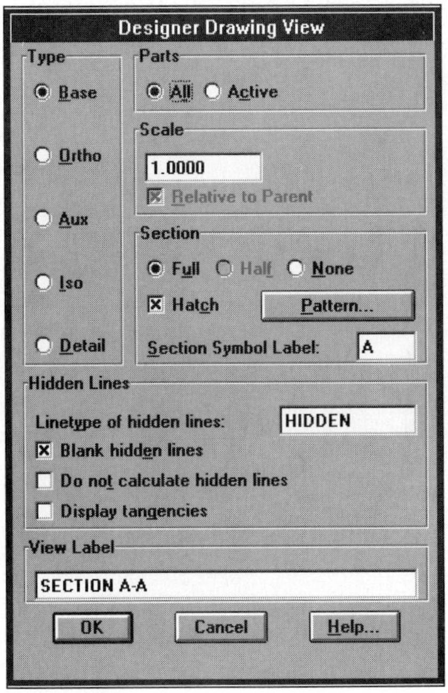

Designer Drawing View

Type
- ● Base
- ○ Ortho
- ○ Aux
- ○ Iso
- ○ Detail

Parts
- ● All ○ Active

Scale
`1.0000`
☒ Relative to Parent

Section
- ● Full ○ Half ○ None
- ☒ Hatch [Pattern...]
- Section Symbol Label: `A`

Hidden Lines
Linetype of hidden lines: `HIDDEN`
☒ Blank hidden lines
☐ Do not calculate hidden lines
☐ Display tangencies

View Label
`SECTION A-A`

[OK] [Cancel] [Help...]

Figure 11.45

Work plane will be the cutting plane.

Select work plane: *(pick the work plane)*

X/Y/Z/<Select work axis or straight edge>: *(pick line 2 in Figure 11.44)*

Rotate/<Accept>: **R** ⏎

Rotate/<Accept>: **R** ⏎

Rotate/<Accept>: ⏎

Type R ⏎ as needed until the UCS icon is oriented as shown in Figure 11.44, then press ⏎ to accept the orientation. You will return to paper space showing the drawing views for the next selections.

View center: *(pick the location shown in Figure 11.46)*

View center: ⏎

Save your drawing on your own.

The section view is generated automatically and the cutting plane added and labeled in the drawing.

You will move the views on the drawing so that they fit on the sheet better.

Pick: **Designer, Drawing, Edit View, Move**

Select view to move: *(pick the front view)*

View location: *(position the crosshairs above their current position to make more room for the section view on the sheet)*

Continue on your own to position the views until they fit on the sheet, as shown in Figure 11.46. To do this efficiently, position the parent view (in this case the front view) first, as its descendants move with it. Then position the top, side, and section views. Figure 11.46 shows the final drawing.

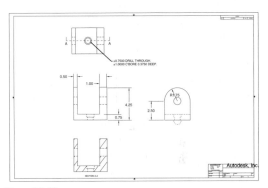

Figure 11.46

■ **TIP** The bonus program *adbonus1.lsp* contains the Adcleanview command, which provides tools to automatically clean up errors in AutoCAD Designer's hidden line drawing views. In order to produce a drawing that follows standard engineering drawing practices, many of AutoCAD Designer's drafting views have to be exploded and hand-edited. (Remember not to explode the actual model, just the drawing views.) You can use Adcleanview to automatically correct many of the errors that AutoCAD Designer makes when automatically generating views. To load the bonus programs, type *(load "ADBONUS1")* at the command prompt. It must be typed exactly with the parentheses and quotation marks. To run the Adcleanview command, type *ADCLEANVIEW* at the command prompt. You will see the prompt *Hide/Unhide/Blank/Convertpolys*. The Convertpolys option looks for 3D polylines and converts them to 2D polylines in the plane of the view. This option should be your first step in cleaning up a drawing view. Because 3D polylines always appear as continuous, no matter what their line style, converting them to 2D polylines often fixes lines that should appear hidden in the drawing, but are not shown correctly. The Hide option changes the layer of the selected segment to the hidden line layer and hidden line style. Unhide changes the layer of the selected segment to the visible line layer and ADP_FRZ layer, causing it to disappear from the display. You may encounter problems with the Adcleanview command. It turns on model space to let you make selections, and does not return you to paper space automatically. Each time you change the part, you must redo the changes you made with Adcleanview. You will see the response *nil* on the prompt line when the command is completed. You need not respond to this prompt. ■

Save *clevis.dwg* and close the AutoCAD Designer toolbars, and exit AutoCAD.

KEY TERMS

base feature
bidirectional associativity
colinear constraint
draft angle
feature
parametric
path
profile
projected constraint
sketch plane
sweeping
work axis
work plane
work point

KEY COMMANDS

ADCleanview
ADEditfeat
ADModdim
ADPardim
ADProfile
ADSettings
ADUpdate
Audit
Designer Help
Designer Mode
Dim Display
Display Work Plane
Fix Point
Hole
Hole Note
Isometric View
Show Constraints
Sketch Path

Create parametric models according to the dimensions for the parts shown below. Use AutoCAD Designer to generate a correctly shown dimensioned multiview drawing. On your own, change the dimensions and update the model. Does your model update to reflect the design considerations for the part? What are the extreme dimensions any feature can have and still allow the part to update correctly? An M following the problem number indicates that the dimensions are in metric units. For problems shown on a grid, use 1/4" (or use 10mm for metric) for each square.

 11.1 Clevis

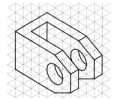

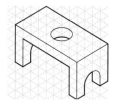 **11.4 Fixture Guide**

11.2 Corner Box

11.5 Journal

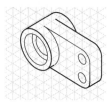

11.3 Angle Block

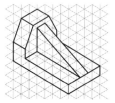

11.6 End Guide

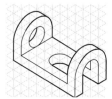

11.7 Saddle

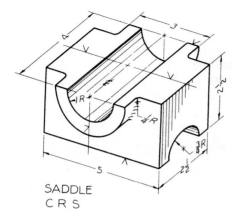

SADDLE
C R S

11.8M Stop Plate

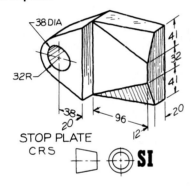

STOP PLATE
CRS

11.9M Clamp

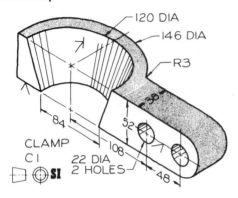

11.10 Floor Plan

Create a parametric model for the floor plan shown. Make it simpler to start with by leaving out the windows, doors, stairs, and furnishings. They can be added later. Consider how to update the model to reflect changes your client may want to make, such as a smaller square footage or a larger living room. Design flexibility into your model, so it can easily be changed to meet your client's requests.

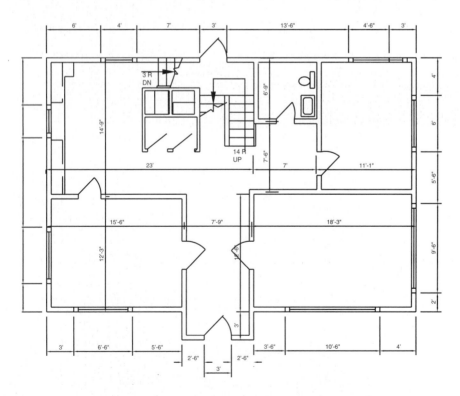

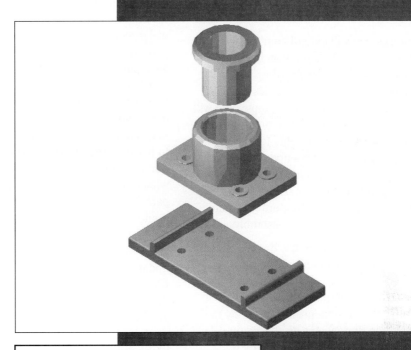

Creating Intelligent Parts with AutoCAD Designer

Objectives

When you have completed this tutorial, you will be able to

1. Use global parameters to create intelligent parts.

2. Update global parameters to change part drawings.

3. Add parametric fillets.

4. Edit parametric models.

Introduction

In this tutorial you will extend your knowledge of AutoCAD Designer to create an intelligent part for use in an assembly drawing. You will create *global parameters*, or variables that can be used from part to part. For example, you will define a global parameter for the hole size in a part. When you change the value for the global parameter, you will be able to update the part so that all of the holes dimensioned with that parameter update to the new size of the hole. You can also use global parameters to update across multiple drawings in an assembly.

To create intelligent parametric drawings keep the function of the part in mind. Think about the parts in an assembly. Which feature will act as the base? Which dimensions are critical dimensions? Which dimensions are likely to change? Which dimensions are based on relationships with other dimensions? Which dimensions are controlled by fit with mating parts? What are the tolerances involved in the type of fit? Answering as many of these questions as possible at the outset, before starting to draw, will save you time in the long run because you will create a parametric model that better represents your design intent. At the beginning, you may not be sure of all of the relationships, but you can use rough sketches to define the basic shapes. Later, as you understand the relationships better, you can change dimensions to global parameters. As the design changes, you may change the values for the global parameters and update the model to reflect the design changes. When the final design is determined, you can quickly produce dimensioned drawings to specify the information for manufacturing the part. Using parametric dimensions, you can create entire families of similar parts from one model.

Keep in mind when creating parametric models that the more constraints you apply to control drawing geometry, the fewer dimensions you will have to specify. This will help you create a parametric model that updates in the way you want it to. Many of the same considerations arise in creating a parametric model as in dimensioning a part for manufacture and inspection. Just as the dimensions you choose to place on the drawing control where inspection measurements will be made, and from what *datum surface*, so does the choice of a parametric dimension determine how the model will update. If you give a parametric dimension a new value, the fixed point remains at the same coordinate location and features update to reflect the new dimension value (depending on how they are related to the fixed point by the constraints and dimensions you have specified). The fixed point of the model is like the origin for the datum surfaces in the drawing.

■ *Warning:* Information for the AutoCAD Designer database is stored in layers ADP_FRZ, ADP_WORK, ADD_VIEWS, ADD_DIMS, ADV_#_VIS, and ADV_#_HID. Making changes to these layers can corrupt or destroy the parametric modeling database. *Leave these layers frozen and do not edit their contents.* Do not use commands, like Explode, Mirror, or Scale, that can change or destroy block information. ■

Maintaining the AutoCAD Designer Database

AutoCAD Designer stores information for the parametric model in blocks on frozen layers. Do not use the standard AutoCAD editing commands, which can change or corrupt the block information. AutoCAD Designer provides its own editing commands, which you can use to make changes to your model in a way that will keep the parametric database intact. If you are creating a parametric model, you must use AutoCAD Designer's commands to modify it in order for it to retain its parameter information.

The following AutoCAD editing commands can produce unexpected results or corrupt the parametric database when you use them on AutoCAD Designer data:

Array You can use the Array command successfully on AutoCAD Designer profiles and paths. Using Array on a profile or path arrays only the geometry and dimensions, not the profile constraint information. Using Array on an AutoCAD Designer part turns it into *a static base feature*. Static models do not have the parametric information that is necessary for them to automatically update, but you can add other features to a static model using AutoCAD Designer. You will learn more about static models in Tutorial 13.

Do not use the AutoCAD Scale or Mirror commands when you use Array on AutoCAD Designer parts; this results in invalid objects that are deleted as soon as you execute any AutoCAD Designer command.

Copy You can use the Copy command on AutoCAD Designer parts, profiles, and paths, but the copy becomes a static base feature. A copy of a profile does not include profile constraint information.

Erase Do not use Erase to delete AutoCAD Designer features or drawing objects. If you use Erase to remove work planes, work axes, or work points, they will reappear later when you display the work plane. You *can* use Erase to delete drawing objects while you are sketching.

Mirror Do not use the Mirror command on AutoCAD Designer parts; it results in invalid objects that are deleted as soon as you execute any AutoCAD Designer command. You can mirror profiles and paths, but keep in mind that the AutoCAD Designer data will not be mirrored. The mirrored profile or path will not include constraint information.

Scale Do not use the Scale command on AutoCAD Designer parts; it results in invalid objects that are deleted as soon as you execute any AutoCAD Designer command.

Explode Do not use Explode on AutoCAD Designer parts. Exploding an AutoCAD Designer part results in the complete loss of AutoCAD Designer's database.

Ellipses AutoCAD Release 13 ellipses are invalid objects to use for AutoCAD Designer profiles and paths.

Audit You will encounter errors due to the extended object data required by AutoCAD Designer if you use the Audit command in a drawing that contains AutoCAD Designer data.

In general, keep in mind that in AutoCAD Designer, you can use any standard AutoCAD commands when you are sketching and viewing your model, but once you have turned the sketch into a feature, you cannot use the standard AutoCAD commands to edit it.

Starting

Launch AutoCAD, making sure that AutoCAD Designer is loaded. If you have not installed AutoCAD Designer and set up your AutoCAD program so that the Designer menu is automatically loaded, refer to the Getting Started section of this manual.

Next you will turn on the Designer toolbars, using the Utilities selection from the Designer pull-down menu.

>   *Pick:* **Designer, Utilities, Toolbars, Designer Main**
>   *Pick:* **Designer, Utilities, Toolbars, Designer View**

The Designer Main and Designer Viewing toolbars appear on your screen. On your own, position them near the top of the screen.

On your own, start a new drawing from the AutoCAD Designer prototype drawing *c:\ad\com\sup\adesign.dwg*, which contains basic settings for ANSI standard drawings. Name your new drawing *c:\work\pivot.dwg*. When you are finished, pick OK to exit the dialog box and return to the AutoCAD drawing editor.

Sketching the Pivot Assembly Base

Figure 12.1 shows the pivot assembly that you will create in Tutorial 13. To create a parametric model of the pivot assembly, you must first examine the objects making up the assembly. Observe how the parts will fit together in the assembly.

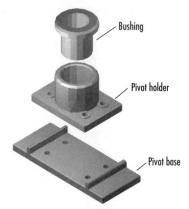

Figure 12.1

You will start by modeling the pivot base in this tutorial. Only one part at a time can be active in AutoCAD Designer. To begin modeling the pivot assembly, first you will need to determine which will be the base feature. Choose a major surface of the object to sketch as the base feature. Other, smaller features, such as the holes, can be added to the first feature. The large flat surface of the pivot base will serve as a good base feature because you can add the other features to it.

After you create the large flat surface, you can define the holes parametrically. Since the holes of the pivot base must mate with the holes in the pivot holder, you will use a global parameter for the hole size. You can control mating parts by specifying the same global parameters for the mating dimensions on each part. Then if you change the value for the global parameter, both parts update. You can use an equation to add a tolerance to the global parameter.

The holder part must always fit between the guides on the base part. Just as with the holes' global parameter, using a global parameter for this dimension will let you control the fit for both parts. You will define a global parameter for the size of the pivot holder and define the pivot base using the same global parameter, adding an allowance for a *clearance fit*. If you change the value of this global parameter, both parts will update so that they still mate.

Get ready to create the model for the base part. While you are designing it, keep in mind the fit between the parts in the pivot assembly.

Setting the Drawing Limits

Because the overall dimensions of the pivot are larger than 12,9, you will set the drawing limits in model space to a larger value to accommodate the size of the part.

Command: **LIMITS** ⏎

ON/OFF/<Lower left corner><0.0000,0.0000>: ⏎

Upper right corner <12.0000,9.0000>: **24,18** ⏎

Command: **Z** ⏎

All/Center/Dynamic/Extents/Left/Previous/Vmax/Window/ <Scale(X/XP)>: **A** ⏎

On your own, set the grid to 1 and the snap to .25. Create a layer named BASE with red as its color and linetype CONTINUOUS. Make BASE the current layer.

On your own, use the Line command to sketch the rectangle shown in Figure 12.2; it will make up the base surface. The lines do not have to be perfectly straight and the endpoints do not need to meet exactly, but often it is just as easy to sketch them with Ortho on and to make the endpoints meet. Remember, the basic shapes you sketch, which you will turn into features, cannot contain any internal features, such as holes.

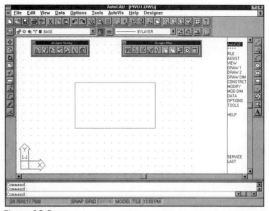

Figure 12.2

In the next steps, you will use AutoCAD Designer to make the shape you have drawn into a profile. From there, you will apply constraints and dimensions that will establish the final relationships between the lines. Then you will extrude the profile to make a 3D solid feature. After you create the initial feature, you will use AutoCAD Designer to create the holes and other features. Finally, you will add the fillets. You will do some of these steps on your own.

As you work through the next steps, refer to Tutorial 11 if you need help, or use the Designer Help selection from the Designer menu bar.

Now that you have sketched the shape on your own, you will make it into a profile.

Pick: **Designer, Sketch, Profile**

Select objects: **(pick all of the lines of the sketch)** ⏎

You will see the message *Solved under constrained sketch requiring 2 dimensions/ constraints*, or something similar. If the lines you sketched were not within 4° of horizontal or vertical, no horizontal or vertical constraints may have been applied. On your own, use the Designer, Sketch, Constraints, Show selection to show all of the constraints and check to see that there are horizontal constraints for the two horizontal lines and vertical constraints for the two vertical lines. Add these constraints if necessary, so that your sketch looks similar to Figure 12.3.

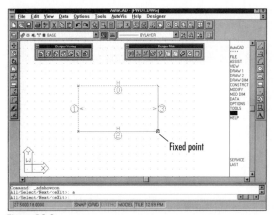

Figure 12.3

> ■ *TIP* Remember, you can use the Designer Settings dialog box (quickly selected from the ADSettings icon on the Designer Main toolbar) to control how constraints are initially applied in solving your sketch. ■

Notice the fixed point shown by the box on the drawing. This is the point that stays fixed on the coordinate system; all of the other drawing geometry updates around it. If the fixed point in your drawing is not the lower right corner, change it on your own. To show your current fixed point, or to select a different point to act as the fixed point for the sketch, select Designer, Sketch, Fix Point. Follow the prompts to select a new point to be fixed in the WCS, or press ⏎ to accept the fixed point as is.

Next you will create your global parameter scheme. You must create global parameters before you can use them. Once you have created them, you will use them to dimension the profile.

Creating Global Parameters

Global parameters are like variables that you can use in place of dimension values. You can also use them in equations. You can extract global parameters from a drawing and import them into another drawing. You can use this ability to link parameters from one drawing or part to another, creating an intelligent model that exists between parts even when they are stored in separate drawing files. When you change the value for a global parameter and export the parameters, all of the parts in the drawing update to show the new size.

To start, you should first think of a scheme for the names of the global parameters you will use in your drawing. Like global variables in programming, which are used across subroutines, global parameters in AutoCAD Designer are used across parts or drawings. Therefore, they must be named exactly the same everywhere they are used for them to work. When you are naming global parameters, don't use names that are too general (like "height" and "width"), because you would tend to repeat them for many different dimensions. Try to make names that are specific enough that you will not accidentally reuse the same name within the assembly. You can change the names of global parameters later to make them match, but you can prevent the need for a lot of changing by planning ahead.

Global parameters, because you can give them descriptive names, also help to keep the equations used in dimensions organized. Global parameter names can use any combination of letters and numbers and the underscore character (_). In AutoCAD Designer the global parameter names are not case sensitive, so that *HOLEDIST1* is the same as *holedist1*.

Look at Figure 12.4, which shows the pivot assembly with the names of global parameters listed. You will create these and set their values in the next steps. As you create the parameters you will see AutoCAD messages similar to *Parameter "basewidth" created: current value == 15.75* on the command line. You will stay in the Create command until you press ⏎ to exit.

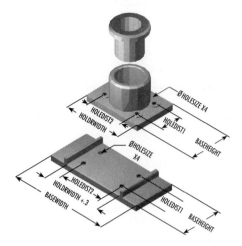

Figure 12.4

Pick: **Designer, Parameters, Create**

Enter equation: **BASEWIDTH=15.75**

Enter equation: **BASEHEIGHT=6.625**

Enter equation: **HOLEDIST1=3.15**

Enter equation: **HOLEDIST2=7.085**

Enter equation: **HOLESIZE=.75**

Enter equation: **HOLDRWIDTH=9.45**

Enter equation: ⏎

Create/Delete/List/Import/Export/<eXit>: **L** ⏎

The List option of the Parameters command opens a text window that lists the existing global parameters and what they are set to. Use Delete to delete any incorrect parameters. Use the Windows Control box to close the text window on your own, and then press ⏎ to end the Parameters command. (The ADList icon on the Designer Main toolbar performs a similar function to the Parameters List command. ADList will give you the parameters of a part, feature, or view, but it will not list the global parameters.)

■ *TIP* Pressing ⏎ to exit the Parameters command will not close the AutoCAD Text window. The window will hide, but not close. If this happens, hold down ⟨ALT⟩ and press ⟨ESC⟩ to switch between your Windows applications until you see the AutoCAD Text window. When it is active on your screen, use the Windows Control box to close it. ■

Next you will set Dimscale to a larger value so that the dimensions you create will be sized larger for the scale of the drawing. Refer to Tutorial 7 if you need help with dimension variables.

Command: **DDIM** ⏎

On your own, use the Dimension Styles dialog box that appears on your screen to select Geometry. Use the Geometry dialog box to set the Overall Scale factor to 2. Pick OK to exit the Geometry dialog box and return to the Dimension Styles dialog box. Next, pick Format and use the Format dialog box to pick User Defined and Leader for the Fit in the upper left corner of the dialog box. This will allow you to place radial and diameter dimensions where you want them on the screen. Pick OK to return to the Dimension Styles dialog box. (Make sure to pick Save to save the changes to dimension style STANDARD, or your changes will only be used as overrides to the dimension style.) Pick OK once again to exit.

Remember, parametric dimensions will not always appear on your screen according to standard engineering drawing practices. When you switch to drawing mode and place the views of the model, the dimensions are cleaned up and placed more in accordance with standard practice. You can relocate and stretch the dimensions placed around the views as needed. While creating the parametric model, you only need to strive for a clear placement of the dimensions so that you do not become confused by overlapping dimensions or extension lines that get in the way of your making other selections.

Next, you will assign BASEHEIGHT and BASEWIDTH as dimensions to the current sketch profile.

Pick: **Designer, Sketch, Add Dimension**

Select first item: *(pick on the right-hand vertical line)*

Select second item or place dimension: *(pick to the right of the same vertical line)*

Undo/Hor/Ver/Align/Par/Dimension value <8.0000>: **BASEHEIGHT** ⏎

Solved under constrained sketch requiring 1 dimensions/ constraints.

The dimension for the height of the pivot base appears on the screen. Now you will add the horizontal dimension:

Select first item: *(pick the bottom horizontal line)*

Select second item or place dimension: *(pick below the same line)*

Undo/Hor/Ver/Align/Par/Dimension value <12.5000>: **BASEWIDTH** ⏎

Solved fully constrained sketch.

Select first item: ⏎

The dimensions appear as shown in Figure 12.5. Notice the message that the sketch is fully constrained. Horizontal and vertical constraints define the lines that were not dimensioned.

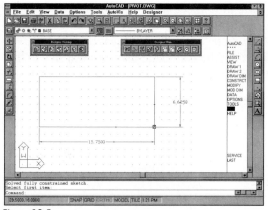

Figure 12.5

Now you are ready to extrude the profile to create a feature.

> ■ *TIP* It is always a good idea to save your work often; if you use the Save As command and save to a new name, you can restore the previous file if you make an error that you cannot correct easily. ■

Pick: **Designer, Features, Extrude**

Use the Designer Extrusion dialog box that appears to select blind termination and a distance of .865, as shown in Figure 12.6. Leave the draft angle set to 0.0. Because this is the first feature created, Base is automatically selected as the type of operation.

Figure 12.6

You cannot see the effect of the Extrude Feature command until you change the viewing angle. At present you are looking straight onto the XY plane and cannot see the depth in the Z direction. You will use the Designer Viewing toolbar to select the isometric view. Use the tool tips if you are unsure which icon to select.

Pick: **Isometric View icon**

The viewing direction changes so that your screen appears as shown in Figure 12.7.

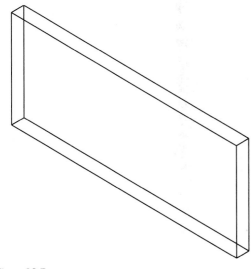

Figure 12.7

Next you will define the four holes. These holes must match the ones in the pivot holder, which you will draw in the next tutorial, in order for the pivot to assemble correctly. It does not matter exactly how far the holes are from the edge of the part, as long as the distance *between* the holes and their size are consistent between the base and the holder. The holder must also fit between the two guides on the base. Instead of dimensioning the holes from the edge of the part, you will create construction lines locating the center of the base and use global parameters to define these construction lines. You can then locate the holes from the construction centerlines, creating a model that will represent the design intent of the part.

On your own, create a new layer named CONSTRUCTION. Assign it the color blue and linetype HIDDEN. You will use this layer to create construction lines in the drawing. Set layer CONSTRUCTION as the current layer.

Construction Geometry

The default in AutoCAD Designer is that linetype CONTINUOUS is used for sketch geometry, which is then turned into features. You can use lines drawn with any other linetype as construction geometry to help create, constrain, and dimension the profile. When the profile is turned into a feature, the construction geometry is no longer displayed (although it still remains in case you need to edit the feature).

Pick: **ADSettings icon**

The Designer Settings dialog box appears. You will leave the settings as they are.

Notice the entry CONT* below the heading Sketch Linetypes. The wildcard * means that any linetype that starts with the letters CONT will be used for sketch geometry; this would be basically any continuous line. Any other linetypes in the profile will be considered construction geometry, so layer

CONSTRUCTION, to which you assigned a HIDDEN linetype, will work for creating construction geometry.

Pick: **Cancel *(to exit the dialog box)***

Next, use the Line command to draw the two construction lines shown in Figure 12.8. You do not have to locate them exactly, but if you use Ortho to draw them, you will be sure that horizontal and vertical constraints are applied. Notice that the UCS icon on your screen is aligned with the base surface that you extruded. Even though you are drawing in an isometric view, using Ortho still produces lines that are parallel to the X and Y axes. It helps a lot in creating AutoCAD Designer geometry, and 3D geometry in general, to work in a view where you can be sure which line you are selecting. Views where several lines are overlapping make it difficult to select the correct line.

On your own, draw the two construction lines.

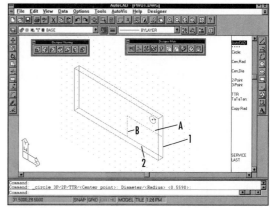

Figure 12.9

In the next step you will create a profile from both the construction lines and the circle. You will then constrain these objects to the existing feature and use global parameters to specify their relationships.

Pick: **Designer, Sketch, Profile**

Select objects: *(pick the circle and the two construction lines)* ⏎

You will see a message similar to *Solved under constrained sketch requiring 5 dimensions/constraints.* If your number of constraints listed is different, on your own, show the constraints. There should be constraints for the horizontal and vertical construction lines, and no others. If you get a message saying *Unable to construct continuous profile,* try sketching the lines again and redoing the profile. You can profile the sketch as many times as needed.

Next you will add projected constraints for the construction lines so that they will always line up with the edge of the part. Refer to Figure 12.9 for your selections.

Pick: **Designer, Sketch, Constraints, Add**

Hor/Ver/PErp/PAr/Tan/CL/CN/PRoj/Join/XValue/Yvalue/Radius/<eXit>: **PR** ⏎

Specify end point to constrain to item: *(pick near point A)*

Specify line, arc, or circle to constrain to: *(pick line 1)*

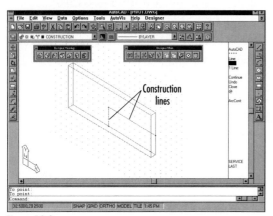

Figure 12.8

On your own, make layer BASE current and draw the circle shown in Figure 12.9. Do not worry about the exact placement and size. You will constrain and dimension the construction lines and the circle once they are sketched.

Solved under constrained sketch requiring 5 dimensions/constraints.

Hor/Ver/PErp/PAr/Tan/CL/CN/PRoj/Join/XValue/Yvalue/Radius/<eXit>: **PR** ⏎

Specify end point to constrain to item: *(pick near point B)*

Specify line, arc, or circle to constrain to: *(pick line 2)*

Solved under constrained sketch requiring 5 dimensions/constraints.

Hor/Ver/PErp/PAr/Tan/CL/CN/PRoj/Join/XValue/Yvalue/Radius/<eXit>: ⏎

Next you will add global parameters, so that the construction lines are always at the midpoints of the base, no matter what values you specify for BASEHEIGHT and BASEWIDTH. If you locate the construction lines by giving the BASEHEIGHT/2 and BASEWIDTH/2 dimensions for the distances from the edge of the part, the construction lines will always be at the center of the pivot base, no matter what the BASEHEIGHT global parameter is set to.

You can then dimension the holes from the construction lines. If you locate the first hole so that it is half the hole distance parameter away from the construction line, and locate the other holes the hole distance parameter away from the first hole, all of the holes will remain centered around the construction lines, even if the global parameter values are changed. Notice the difference between doing this and specifying an exact value, such as 7.8750 (one half of 15.75), for the location of the construction line. This value is currently half of the base width, but if you changed the global parameter for the base width to a new value, you would have to go back and change the dimension by which you located the center construction line, since *its* value will remain 7.8750.

Refer to Figure 12.10 for the points to select as you add the dimensions.

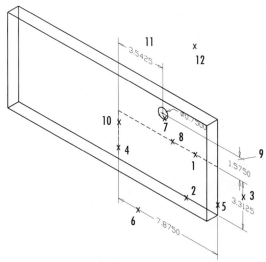

Figure 12.10

Pick: **Designer, Sketch, Add Dimension**

Select first item: *(pick point 1)*

Select second item or place dimension: *(pick point 2)*

Specify dimension placement: *(pick point 3)*

Undo/Hor/Ver/Align/Par/Dimension value <3.0000>: **=BASEHEIGHT/2**

Solved under constrained sketch requiring 4 dimensions/constraints.

Select first item: *(pick point 4)*

Select second item or place dimension: *(pick point 5)*

Specify dimension placement: *(pick near point 6)*

Undo/Hor/Ver/Align/Par/Dimension value <8.5000>: **=BASEWIDTH/2**

Solved under constrained sketch requiring 3 dimensions/constraints.

Select first item: *(pick point 7)*

Select second item or place dimension: *(pick point 8)*

Specify dimension placement: *(pick near point 9)*

Undo/Hor/Ver/Align/Par/Dimension value <1.4375>: **=HOLEDIST1/2**

Solved under constrained sketch requiring 2 dimensions/constraints.

Select first item: *(pick point 7)*

Select second item or place dimension: *(pick point 10)*

Specify dimension placement: *(pick near point 11)*

Undo/Hor/Ver/Align/Par/Dimension value <5.6250>:
=HOLEDIST2/2

Solved under constrained sketch requiring 1 dimensions/
constraints.

Select first item: *(pick point 7)*

Select second item or place dimension: *(pick near point 12)*

Undo/Dimension value <1.5000>: **=HOLESIZE**

Solved fully constrained sketch.

Select first item: (⏎)

When the dimensions are added, your drawing should look like Figure 12.10. On your own, save your drawing. You should see a message saying *Designer data saved.*

You will set the method for dimension display to parametric, so that the equations you type will appear instead of the dimension values.

Pick: **Designer, Display, Dim Display**

Parameters/Equations/<Numeric>: **E** (⏎)

Figure 12.11 shows the drawing with dimensions displayed in parametric form.

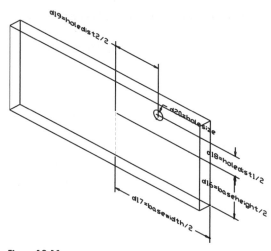

Figure 12.11

On your own, return the dimension display to numeric form by repeating the previous command and typing *N* (⏎) to select the Numeric option.

Pick: **Designer, Features, Extrude**

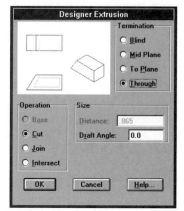

Figure 12.12

On your own, complete the Designer Extrusion dialog box so that it looks like Figure 12.12. Select Cut so that the hole feature is subtracted from the base feature when it is extruded. Pick Through so the hole goes all the way through the part. When the dialog box is correctly filled in, pick OK. You are returned to your drawing for the final prompt. If the direction arrow points in the same direction on your screen as in Figure 12.13, press (⏎) at the prompt *Direction Flip/<Accept>:*.

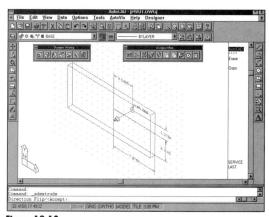

Figure 12.13

If not, type F ⏎ to flip the direction and then press ⏎ to accept. The hole is created through the part and the construction lines disappear.

On your own, draw the circle for the bottom right-hand hole shown in Figure 12.14 and turn it into a profile.

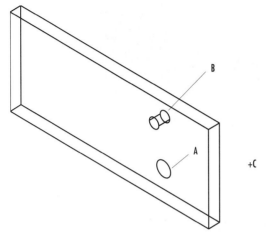

Figure 12.14

Next you will add an X value constraint to the new circle, so that its X value will always be equal to the X value of the circle you select. You will constrain the new hole to the existing hole. This way the two holes will always align. Refer to Figure 12.14 for your selections.

Pick: **Designer, Sketch, Constraints, Add**

Hor/Ver/PErp/PAr/Tan/CL/CN/PRoj/Join/XValue/Yvalue/ Radius/<eXit>: **XV** ⏎

Select first circle: *(pick on the new circle, A)*

Select second circle: *(pick on the original circle, B)*

Solved under constrained sketch requiring 2 dimensions/ constraints.

Hor/Ver/PErp/PAr/Tan/CL/CN/PRoj/Join/XValue/Yvalue/ Radius/<eXit>: ⏎

Notice that as you add constraints, AutoCAD Designer updates its message and lets you know that you now need one less dimension to

fully constrain the profile. Next you will add dimensions for the diameter of the circle and the vertical distance from the first hole to the second hole to fully constrain its position. The dimensions you add will be global parameters, letting you control the distance between the holes on both the base and the holder part. Because the holes are parametrically dimensioned from the center construction lines, the holes will always remain centered if the global parameter for the hole distance is changed. This would not be the case if the holes were located from the edge of the part. You will use the Designer Main toolbar to select ADPardim to add a parametric dimension.

Pick: **ADPardim icon**

Select first item: *(pick on circle A in Figure 12.14)*

Select second item or place dimension: *(pick on the original circle, B)*

Specify dimension placement: *(pick to the right of the objects, near C)*

Undo/Hor/Ver/Align/Par/Dimension value <3.8875>: **HOLEDIST1** ⏎

Solved under constrained sketch requiring 1 dimensions/ constraints.

Select first item: *(pick on circle A)*

Select second item or place dimension: *(pick to the right of circle A)*

Undo/Dimension value <1.5000>: **HOLESIZE** ⏎

Solved fully constrained sketch.

Select first item: ⏎

The dimensions appear in your drawing, and the drawing is updated automatically to reflect the global parameters that you typed in. Now the sketched profile should be fully constrained and you are ready to extrude it to create a feature.

On your own, extrude the new circular profile, using Cut and Through to form a hole through the object, as you did for the first circle. When you have finished, your drawing should appear similar to Figure 12.15.

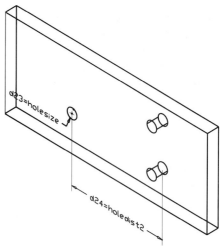

Figure 12.16

Now extrude the circular profile on your own to form the hole.

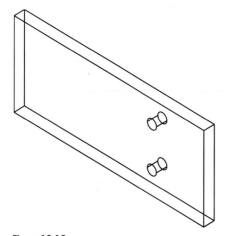

Figure 12.15

On your own, add a circle to the lower left corner of the part. Make it into a profile and add a constraint for the Y value so that it is the same as the hole in the lower right corner of the part. Add dimensions, using global parameters for the hole size and the hole distance, as shown in Figure 12.16.

Create the final hole on your own and make it a profile. You will only need to add two constraints and one dimension. Constrain the X value to be equal to that of the lower left hole, and the Y value to be that of the upper right hole. Add a dimension for the diameter of the circle that is the global parameter hole size. Finally, extrude the profile to cut through the part to form the feature. When you are finished, you should have four parametrically defined holes through the base plate, as shown in Figure 12.17. Save your drawing at this point on your own.

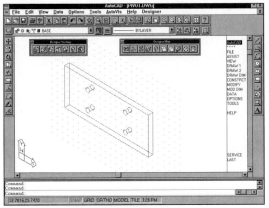

Figure 12.17

Next you will create a new sketch plane, aligned with the lower edge surface shown in Figure 12.18. You will then sketch the shape of the guides and extrude them to finish the base of the pivot assembly.

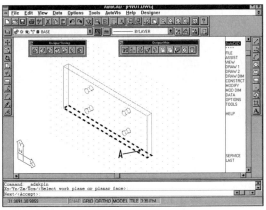

Figure 12.18

Pick: **Designer, Sketch, Sketch Plane**

Xy/Yz/Zx/Ucs/<Select work plane or planar face>: *(pick the face labeled A in Figure 12.18)*

Next/<Accept>: *(if the correct surface is highlighted, press ⏎; if not, type N ⏎ until it is and then press ⏎)*

X/Y/Z/<Select work axis or straight edge>: *(pick the straight edge B in Figure 12.19)*

Rotate/<Accept>: *(if the UCS icon appears as shown in Figure 12.19, press ⏎; if not, type R ⏎ until it is and then press ⏎)*

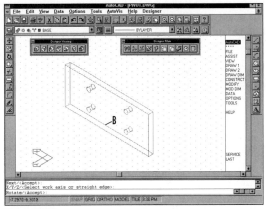

Figure 12.19

 You will change the viewing direction, using the Designer Viewing toolbar, so that you are looking directly onto the sketch plane.

Pick: **Sketch View icon**

The view should line up with the sketch plane, as shown in Figure 12.20.

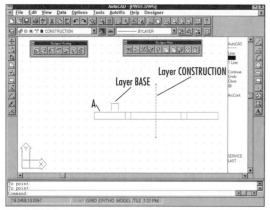

Figure 12.20

On your own, sketch the shapes you see in Figure 12.20. Make sure Snap is off before you begin. On layer BASE, use three lines for the profile of the guide. You will close the profile to the edge of the existing feature when you create the profile. Sketch the construction line near the middle of the part on layer CONSTRUCTION. When you are finished, your screen should look similar to Figure 12.20.

After you have sketched the construction line, change back to layer BASE. Next you will make the lines into a profile for the left-hand guide. You will close the profile to the edge of the existing feature. This way, even though you have not added a projected constraint, the profile will update along with the feature if it is updated. Use the Designer Main toolbar to select ADProfile so that you can create the profile.

> ■ *TIP* When selecting an edge or work feature in the ADProfile command, you will not see a message such as "1 found" to indicate that your selection was successful. After you press return, AutoCAD Designer will report your results. ■

Pick: **ADProfile icon**

Select objects for sketch.

Select objects: **(pick the three lines and the construction line you just created)** ⏎

Select edges or work features to close profile: **(pick edge A in Figure 12.20)**

Select edges or work features to close profile: ⏎

Solved under constrained sketch requiring 4 dimensions/constraints.

> ■ *TIP* If you see a message, *Unable to construct continuous profile*, you may not have drawn your lines so that they come close to touching the lines of the previous feature. For AutoCAD Designer to constrain lines as joining, they cannot be farther apart than the size of the AutoCAD pickbox. If your lines do not form a continuous profile, use Extend or Stretch so that the lines touch the edge of the base. Then profile the part again. You can use the ADProfile command as many times as needed. ■

Next you need to add dimensions to define this profile. Make sure BASE is the current layer, and then:

Pick: **ADPardim icon**

On your own, add the dimensions shown in Figure 12.21. Dimensioning the center construction line and the profile using global parameters will allow the model to be easily updated when you change any of the parameters. If you change the holder width, the guides will update so that they are still centered on the part and always have an allowance of .3(.15 × 2) more than the size of the holder.

When you are adding dimensions, make sure to pick on the part, and not on extension lines of dimensions that you added to the drawing previously. You must dimension to the part, not to other dimension lines. To help in selecting the lines of the part, use Zoom Window to enlarge the object and pick near the endpoint of the line where you will place the dimension. This way the extension lines will not extend across the lines of the object, obscuring them, when you want to place future dimensions. If you create an incorrect dimension, you can use the AutoCAD Erase command to remove it (if you have not yet made the sketch profile into a feature).

You can use parentheses to group operations within an equation being used as a dimension, but it is not necessary. If no parentheses are used, operations are performed in the following order: multiplication, division, addition, and finally subtraction. Often it is helpful to use parentheses to keep track of the order for yourself, even though they are not required by AutoCAD Designer.

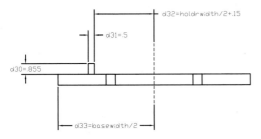

Figure 12.21

Notice that the message in the command prompt area says *Solved fully constrained sketch.* This indicates that you do not need any additional dimensions or constraints; the profile is fully defined.

■ *TIP* Pick ADModdim from the Designer Main toolbar to change the values of dimensions after they are added to the sketch. Once you create the object as a 3D solid feature, as you will in the next step, you must use ADEditfeat from the Designer Main toolbar to change the resulting feature. After changing dimensions or features, use ADUpdate to update the model's appearance to reflect the change. ■

Next, use the Designer Viewing toolbar and change back to the isometric view so that you will be able to select back surfaces easily.

Pick: **Isometric View icon**

Your model should look similar to Figure 12.22.

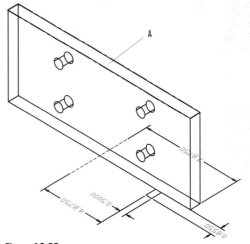

Figure 12.22

Now you are ready to extrude the profile to create the guide feature.

Pick: **Designer, Features, Extrude**

On your own, use the Designer Extrusion dialog box to select the Join operation and To Plane termination. Leave the draft angle set at 0.0. When you have made these selections, pick OK. You will make the remaining selections from your drawing. Refer to Figure 12.22 for your selections.

Xy/Yz/Zx/Ucs/<Select work plane or planar face>:
(pick surface A in Figure 12.22)

If the correct surface is highlighted on your screen as in Figure 12.22, then press ⏎ at the prompt *Next/<Accept>:*. If not, type *N* ⏎ to select the next surface until the correct one is highlighted, and then press ⏎ to accept.

The profile is extruded so that your drawing looks like Figure 12.23.

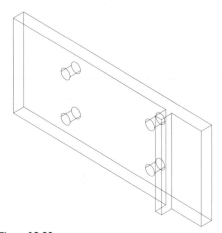

Figure 12.23

Next you will return to the sketch plane view and create the guide for the other side.

Pick: **Sketch View icon**

On your own, sketch the shape for the other guide. Make it into a profile and add the dimensions shown in Figure 12.24. Notice that you could use parametric dimensions for the width and height of the guide. You could also constrain the top of this guide to be colinear with

the top of the other guide. When you are designing a part, you must decide which methods will work best to carry out your design objectives. For the guides, you will type in the values rather than use global parameters, but you can edit any dimension and specify global parameters or equations at that point.

Figure 12.24

On your own, use the Extrude Features selection from the Designer menu. Use the Designer Extrusion dialog box to extrude the profile and join it, as you did for the previous guide. When you are finished, your screen should look like Figure 12.25.

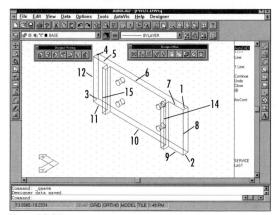

Figure 12.25

Editing the Sketch

Even though you have already made the sketch into a feature, you can use AutoCAD Designer to edit the dimensions of the sketch. In fact, you can even extrude profiles that are not fully constrained and add their dimensions later. Note, however, that you cannot go back and change the sketch geometry that you extruded or otherwise created as a feature; you can only change the dimensions and constraints. You will use the Edit Feature selection from the Designer menu bar.

Pick: **Designer, Edit Feature**

Sketch/<Select feature>: **S** ⏎

Select feature: *(pick on the right-hand guide)*

The original dimensions of the sketch return to your screen. You will pick the ADModdim icon from the Designer Main toolbar to change the dimensions. (You can also use the selection Designer, Change Dimension from the menu bar to select the same command.)

Pick: **ADModdim icon**

Select dimension to change: *(pick on the .855 dimension)*

New value for dimension <.855>: **.865** ⏎

Solved fully constrained sketch.

Select dimension to change: ⏎

To show the resulting part with the sketch dimension changed, you will use ADUpdate:

Pick: **ADUpdate icon**

Updating feature 6 of 6 (type: Extrude)...

The height of the guide should reflect the change. If you had difficulty creating any of the sketch features, use the commands you have learned to go back and make corrections to the sketch and then update your drawing.

Adding Fillets

Next you will add fillets to the edges of the pivot base. Designer has its own parametric fillet command, ADFillet. If the edges you fillet move when the model is updated, the fillets will also move. ADFillet can also be picked from the Designer Main toolbar.

Pick: **Designer, Features, Fillet**

Select edge: *(pick edges 1 through 12 in Figure 12.25)*

Select edge: ⏎

Fillet radius <0.25>: **.125** ⏎

Command: ⏎ *(to restart the ADFillet command)*

Select edge: *(pick 14 and 15 to fillet the outside edges of the guides)*

Select edge: ⏎

Fillet radius <0.125>: **.25** ⏎

The fillets added to the drawing are shown in Figure 12.26.

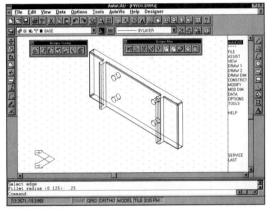

Figure 12.26

The pivot base is complete. Now that you have constructed the parametric model, you can always update it or refine it further by adding more features. Next, you will change the global parameter HOLEDIST2 and see how the model updates.

Pick: **Designer, Parameters, List**

An AutoCAD Text window appears on your screen, showing the list of global parameters, as shown in Figure 12.27.

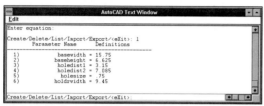

Figure 12.27

Next you will set HOLEDIST2 to 8. Then you will update the model and watch it change. You should still see the prompt for the Adparam command.

Create/Delete/List/Import/Export/<eXit>: **C**

Enter equation: **HOLEDIST2=8** ⏎

Parameter "holedist2" created: CURRENT VALUE == 8

Enter equation: ⏎

Parameter changes affect active part. Use ADUPDATE to see changes.

Create/Delete/List/Import/Export/<eXit>: ⏎

Pick: **ADUpdate icon**

 You will see a message saying *Updating feature...* and then the drawing will update! It may be hard to tell that the holes are still centered when you are looking at the isometric view, so you will double-check this in the front view. You can quickly switch to the front view by using the Front View icon from the Designer Viewing toolbar.

Pick: **Front View icon**

The front view appears on your screen, as shown in Figure 12.28.

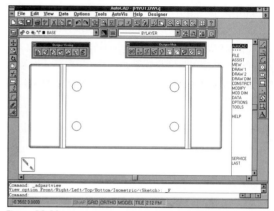

Figure 12.28

Pick: **Designer, Parameters, Create**

On your own, change the value for HOLEDIST2 back to 7.085 and then use the ADUpdate command to update the model again. Save your model.

In the next tutorial, you will create a new active part in this same drawing and model the pivot holder part. You will then import static parts for the bushing.

Save your drawing, close the toolbars, and exit AutoCAD; you have completed this tutorial.

clearance fit   global parameters   static base feature
datum surface

Add Constraints   Front View   Sketch Plane
ADFillet   List Parameters   Sketch View
Create Parameters

Create parametric models for the parts shown below using global parameters. The letter M after an exercise number means that the given dimensions are in millimeters. Use the method you learned in Tutorial 11 to produce a dimensioned detail drawing for the part.

12.1 Locator Block

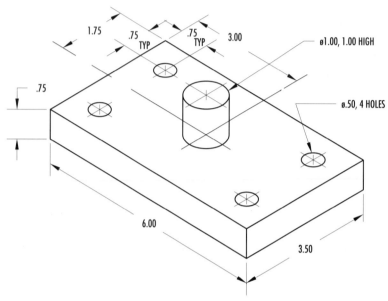

12.2M Guard

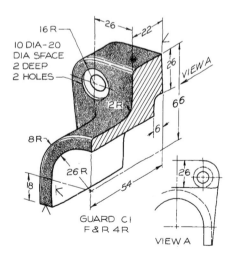

12.3 Shaft Arm

Model the Shaft Arm shown. Use parametric dimensions for the shaft sizes and distance between centers. Make the values for the shafts 1.00 and 2.00, and the distance between centers 6.00. Change the values for the parameters to create another drawing with shafts of 1.50 and 2.50 and with 6.50 between centers.

12.4M Radial Link

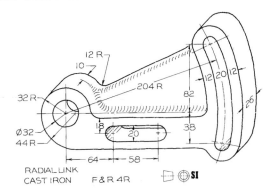

12.5M Yoke

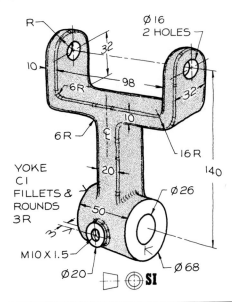

12.6M Clamp

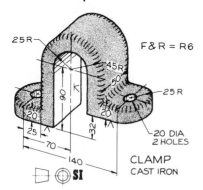

25R
F & R = R6
45R
60
25 R
90
18
20
25
32
70
140
20 DIA
2 HOLES
CLAMP
CAST IRON

12.9M Link

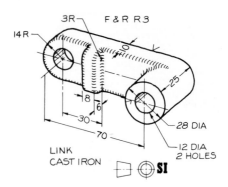

3R   F & R R3
14R
0
8
6
25
30
70
28 DIA
12 DIA
2 HOLES
LINK
CAST IRON

12.7 Shaft Base

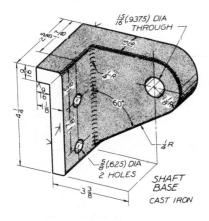

$\frac{15}{16}$ (.9375) DIA
THROUGH
60°
$\frac{1}{4}$R
$\frac{5}{8}$ (.625) DIA
2 HOLES
SHAFT
BASE
CAST IRON
$4\frac{1}{4}$
$3\frac{3}{8}$

12.10 Saddle Support

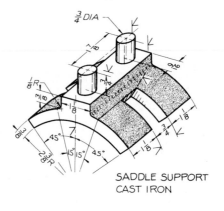

$\frac{3}{4}$ DIA
45°
15°   45°
SADDLE SUPPORT
CAST IRON

12.8M Bearing Cap

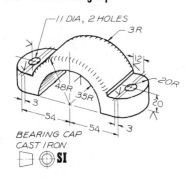

11 DIA, 2 HOLES
3R
12
20R
48R   35R
20
3
54
54
3
BEARING CAP
CAST IRON

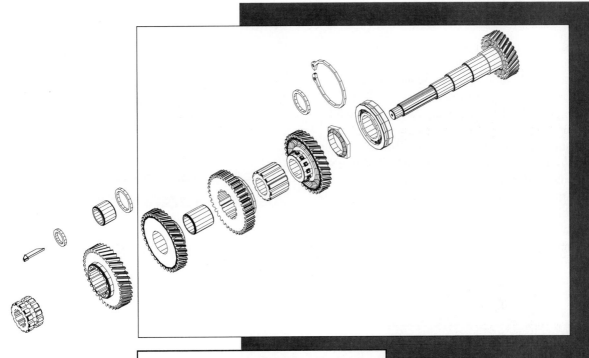

Creating Intelligent Assemblies Using AutoCAD Designer

Objectives

When you have completed this tutorial, you will be able to

1. Create assemblies with multiple AutoCAD Designer parts.

2. Make a new active part.

3. Switch between active parts.

4. Use work planes and work axes.

5. Create revolved features.

6. Export AutoCAD solid models for use as static parts.

7. Use point filters.

8. Add features to static parts.

Introduction

In this tutorial you will create the remaining parts for the pivot assembly you started in Tutorial 12 to finish creating an intelligent model of the entire assembly. You will also export a part created using AutoCAD's solid modeling for use as a static part in the AutoCAD Designer assembly. This will demonstrate how you can use AutoCAD Designer's powerful capabilities even with regular AutoCAD solid models. By importing your solid models as static parts, you can use AutoCAD Designer to automatically generate multiview drawings of solid models with hidden lines shown correctly. You can also add AutoCAD Designer features to static parts created from solid models.

Starting

Launch AutoCAD, making sure that AutoCAD Designer is loaded. You should see the AutoCAD drawing editor on your screen. The Designer pull-down menu should appear at the far right of the pull-down menu items. If you have not installed AutoCAD Designer and set up your AutoCAD program so that the AutoCAD Designer menu is automatically loaded, refer to the Getting Started section of this manual.

On your own, start a new drawing from *pivtbase.dwg*, which is provided with the data files. It is like the base part that you created in Tutorial 12. Use *pivtbase.dwg* as the prototype to start your new drawing. Name the new drawing *pivtasmb.dwg*. When you have completed these steps the drawing should appear on your screen, as shown in Figure 13.1.

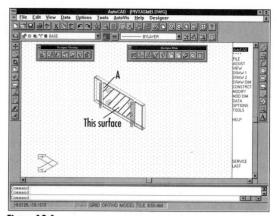

Figure 13.1

Next you will create a sketch plane parallel to the top surface (it faces you on the screen) of the pivot base. Then you will make a new active part so that you can create the pivot holder. Remember that AutoCAD Designer adds the features you create to the active part. Since the base and holder are separate parts, and the base is now active, you must make a new active part before you create features for the holder.

Pick: **Designer, Sketch, Sketch Plane**

Xy/Yz/Zx/Ucs/<Select work plane or planar face>: *(pick line A in Figure 13.1)*

Next/<Accept>: *(if the top surface of the base shown in Figure 13.1 is highlighted, press ⏎; if not, type N ⏎ until it is highlighted and then press ⏎)*

X/Y/Z/<Select work axis or straight edge>: *(pick A again)*

Rotate/<Accept>: **R** ⏎ *(repeat as necessary to align the UCS icon as shown in Figure 13.2.)*

Rotate/<Accept>: ⏎

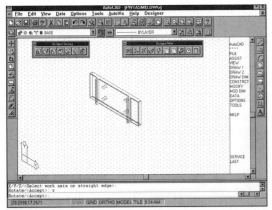

Figure 13.2

On your own, make a new layer named HOLDER with color blue and linetype CONTINUOUS. Set HOLDER as the current layer.

Making a New Active Part

Now you are ready to make a new active part. You can switch back and forth between active parts as necessary.

Pick: **Designer, Part, New**

You will see a message in the command window saying *Part 1 initialized.* The base you created previously is part 0. You can easily switch between active parts: use the selection Designer, Part, Make Active and follow the prompts to pick on the part you want to make active.

Next you will sketch the rectangular shape of the bottom surface of the pivot holder. After you create the rectangular feature as the base feature, you will add the bosses, the holes, the large cylindrical feature, and the fillets to it.

Aligning the View with the Sketch Plane

You will align your view with the sketch plane to make it easy to draw the new sketch. You will use the Designer Viewing toolbar to select the Sketch View icon.

Pick: **Sketch View icon**

Your view should be aligned so that you are looking directly on to the pivot base, as shown in Figure 13.3.

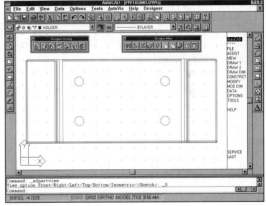

Figure 13.3

■ *TIP* If your view does not look like Figure 13.3, go back to the step where you created the sketch plane and try it again. Make sure to align the sketch plane with the surface shown in Figure 13.1. ■

Now you are ready to begin creating the pivot holder part. The completed pivot holder is shown in Figure 13.4.

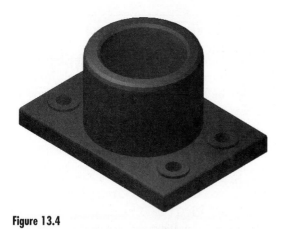

Figure 13.4

Sketching the Holder

You will do the next steps on your own, using the methods you learned in Tutorials 11 and 12. Sketch lines 1 through 4 to create the rectangular shape for the pivot holder, as shown in Figure 13.5. Now make the sketch into a profile. Make sure the fixed point for the sketch is the lower right-hand corner, as shown in Figure 13.5.

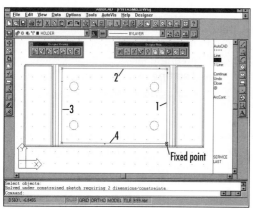

Figure 13.5

Show the constraints on your own. There should be horizontal constraints for each of the longer lines, and vertical constraints for each of the shorter lines. If there are not, add them on your own.

Now you are ready to add dimensions to fully constrain the profile. You will use a global parameter to dimension the width of the holder.

Listing the Parameters

You will use the List option of the Parameters command to list the global parameters as you did in Tutorial 12. You can use this to check the exact names and values of the parameters that already exist in your drawing. Remember that to use a global parameter in a dimension, you must specify its name exactly as it was created.

Pick: **Designer, Parameters, List**

A text window appears on your screen listing the global parameters, as shown in Figure 13.6.

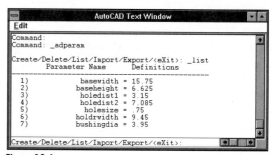

Figure 13.6

On your own, close the AutoCAD Text window and then press ⏎ to exit the Parameters command.

On your own, use AutoCAD Designer to add the dimensions you see in Figure 13.7. Refer back to Tutorial 12 if you need help adding the dimensions.

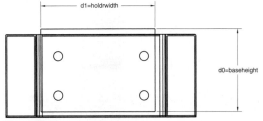

Figure 13.7

Don't worry if the profile for the holder is not lined up with the pivot base. You will move it into position later. Next you will change your viewing direction back to the isometric view, so that you can see the effect of the Extrude Feature command when you use it. On your own, select the Isometric View icon from the Designer Viewing toolbar.

Next you will extrude the profile to create a feature.

Extruding the Holder

On your own, use the Designer Extrusion dialog box to extrude the profile you created in the previous steps. This extrusion will act as the base feature for the holder part you are creating. Set the Designer Extrusion dialog box for a blind extrusion of distance .865. Leave the draft angle set to 0.00. When you have finished setting up the dialog box, pick OK. Your drawing will look like Figure 13.8.

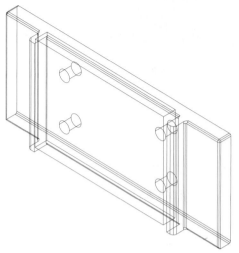

Figure 13.8

Now freeze layer BASE on your own. The pivot base, which was created on layer BASE, disappears from your screen. Save your drawing on your own.

Setting AutoCAD Designer's Isolines

Next you will set AutoCAD Designer's variable for Isolines. AutoCAD Designer's Display Isolines command is similar to the AutoCAD Isolines variable that you used in Tutorial 9 when creating solid models. Setting Display Isolines to a larger value causes rounded surfaces to be displayed with more tessellation lines. This can make it easier to recognize rounded surfaces on your screen. *Nurbs* (nonuniform rational B-splines) are used to model irregular curves. AutoCAD's Spline command uses nurbs to create curves. Since you do not have any irregularly curved surfaces on the pivot holder, you will not need to set the isoline display for nurbs.

Pick: **Designer, Display, Isolines**

Isolines for cones, cylinders and torii <2>: **8** ⏎

Isolines for nurbs <0>: ⏎

Next you will sketch a circle and construction centerline that you can extrude to form the boss (or raised surface) in the upper right corner of the holder.

Creating the Bosses

Using Figure 13.9 as a guide, do the next steps on your own. Make layer CONSTRUCTION current and draw the vertical and horizontal lines near the center of the pivot holder. Switch back to layer HOLDER and draw the circle in the upper right corner, as shown.

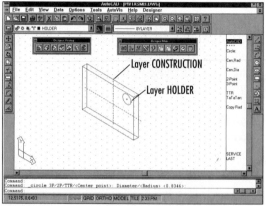

Figure 13.9

Continue on your own to make the two sketched lines and the circle into a profile. Then add projected constraints for all endpoints of the two lines to the rear surface of the holder (as it is currently facing). Add the global parametric dimensions shown in Figure 13.10. Remember to watch for the AutoCAD Designer message *Solved fully constrained sketch*.

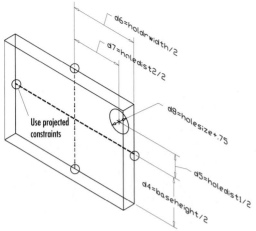

Figure 13.10

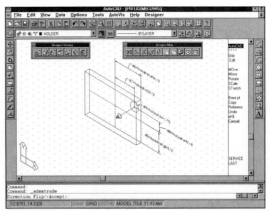

Figure 13.11

■ *TIP* When you are adding dimensions, pick near the endpoint closest to the side where you will place the dimension. This helps to create dimensions which do not have extension lines drawn over the top of the object. When the extension lines cross the object, it can make it difficult to select lines on the object to place additional dimensions. The parametric dimensions you create must be attached to the profile, not to other dimensions. ■

Redraw the screen on your own as needed. Now you are ready to extrude the circle to form the boss. Use the Designer Extrusion dialog box on your own to pick Blind, Join, a distance of 1.00, and 0.00 draft angle. When you have made these selections, pick OK. At the *Direction Flip/<Accept>:* prompt, press ↵ if the arrow shown on the screen is pointing so that the boss will go through the object and extend beyond the other side, as shown in Figure 13.11. If the arrow is pointing the wrong way, type *F* ↵ to flip the direction.

When you have finished creating the boss, your drawing should look like Figure 13.12.

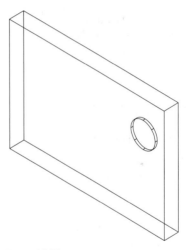

Figure 13.12

Next, sketch another circle in the lower right corner of the holder. You will constrain it so that its center has the same X value as the first boss, as you did in Tutorial 12 for the holes. On your own, draw the circle and make it into a profile. You should see the message, *Solved under constrained sketch requiring 3 dimensions/ constraints.* On your own, add the X value constraint. When you have successfully added the X value constraint, you will see the message, *Solved under constrained sketch requiring 2 dimensions/constraints.* Don't be concerned that the circle does not appear to line up exactly with the boss when you add the X value constraint. That is because the center of the extruded boss is on the bottom surface of the base feature, as that is where the sketch plane was aligned. When you extrude the circle, you should see that it is aligned correctly.

Next add the dimensions shown in Figure 13.13 on your own. Check for the message *Solved fully constrained sketch.* If your sketch is not fully constrained, check to see that you added the X value constraint correctly. If necessary, use AutoCAD's Erase command to erase the dimensions and then make the circle into a profile again. You should see a message specifying the number of constraints necessary for a fully constrained sketch. Add the constraints and dimensions until your sketch is fully constrained.

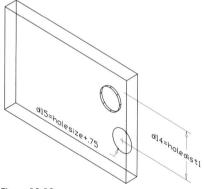

Figure 13.13

■ *TIP* If you see the message *Highlighted dimension cannot be attached and will be ignored,* continue trying to add the dimension until you are successful. It may help to use the Zoom In command and pick on the foremost circle of the boss. ■

Now use the Designer Extrusion dialog box to extrude the hole to create the lower right boss. Select the Join option to join it to the base feature. Pick Blind for the termination. Specify distance 1.00 and draft angle 0.00 on your own. When you have finished, pick OK. The lower right boss should appear below the upper boss.

Continue on your own to create the circle at the upper left for the upper left boss. Make it into a profile and use the Y value constraint so that the circle has the same Y value for its center as the upper right hole does. Review Tutorial 12 if you need help adding the Y value constraint. Add the dimensions shown in Figure 13.14.

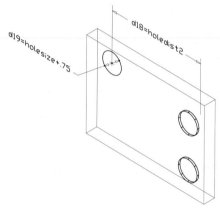

Figure 13.14

On your own, pick Blind and extrude the profile to a height of 1.00 using the Join option, with draft angle 0.00 as you did for the others. The new boss should appear similar to the previous two bosses you created.

Add the remaining boss on your own by drawing a circle and making it into a profile. Use the X value and Y value constraints so that its center is at the same X value as the upper left boss and its Y value is the same as the lower right boss. Specify a diameter dimension equal to HOLESIZE + .75.

Once your profile is fully constrained, extrude it on your own using blind as the method to a height of 1.00 with no draft angle so that it is joined to the base feature. You should now have four completed bosses in your drawing, as shown in Figure 13.15.

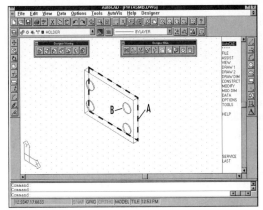

Figure 13.15

Adding the Holes

Next you will add holes concentric to the bosses, using AutoCAD Designer's Hole feature.

Pick: **Designer, Features, Hole**

On your own, use the Designer Hole dialog box to create a through hole using a drilled operation with a diameter of .75. Pick Concentric as the placement method. When you have finished making these selections, pick OK. You will return to your drawing to select the placement. Refer to Figure 13.15 for your selections.

Select work plane or planar face: *(pick surface A)*

Next/<Accept>: ⏎ *(if the surface shown in Figure 13.15 is highlighted; if not, type N ⏎ until it is and then press ⏎)*

Select concentric edge: *(pick circle B)*

Repeat this process on your own to create the remaining holes. When you have finished, your drawing will look like Figure 13.16.

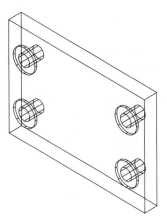

Figure 13.16

Save your drawing on your own. Next you will align the sketch plane with the front surface of the feature, identified as C in Figure 13.17.

Pick: **Designer, Sketch, Sketch Plane**

Xy/Yz/Zx/Ucs/<Select work plane or planar face>: *(pick surface C)*

Next/<Accept>: **N** ⏎

Next/<Accept>: ⏎ *(when the surface shown in Figure 13.17 is highlighted)*

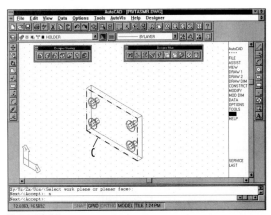

Figure 13.17

X/Y/Z/<Select work axis or straight edge>: *(pick surface C)*

Rotate/<Accept>: **R** ⏎

Rotate/<Accept>: **R** ⏎

Rotate/<Accept>: ⏎ *(when the UCS is shown as in Figure 13.17)*

Creating the Cylindrical Feature

Next, sketch the circular shape of the large center cylinder on the new sketch plane. On your own, use the Circle command to draw the circle near the center of the object. Make it into a profile. Add a diameter dimension of 5.50 and the parametric dimensions shown in Figure 13.18.

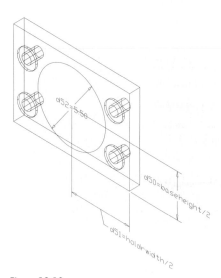

Figure 13.18

On your own, extrude the circle to create the cylindrical feature. Select Join as the operation and Blind for the termination in the Designer Extrusion dialog box. Use a distance of .50 and leave the draft angle set to 0.00. The Extrusion arrow that appears on your model should point toward you. When you are done, your drawing should look like Figure 13.19.

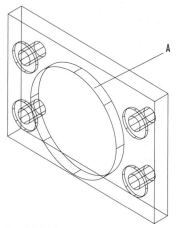

Figure 13.19

Notice that you only extruded the circle to a distance of .5 units, even though the cylindrical feature of the pivot holder is taller.

Next you will use the cylinder you created to define a work axis. You will use the work axis to define a work plane and a new sketch plane that you will use to revolve a sketch profile to create the cylindrical feature. Keep in mind that there are many ways to approach the design of each part. You could also create the same feature by extrusion and use the Cut option to remove material from the part. You could also use the Hole feature to create a series of different diameter holes. You will use revolution in this tutorial to demonstrate its use.

Using a Work Axis

A work axis is a parametrically defined line that you can use to locate sketch planes or parametrically defined work planes. You can define a work axis at the centerline of the existing cylindrical, conical, or toroidal surface that you select. You will create a work axis through the center of the cylindrical feature you just extruded. Refer to Figure 13.19 for your selection.

Pick: **Designer, Features, Work Axis**

Select cylindrical face: *(pick circular edge A)*

The work axis is created, as shown in Figure 13.20.

> ■ *TIP* If you do not see the work axis on your screen, select Designer, Display, Work Axis, On to display it. If you still cannot see the work axis, try creating it again. Remember, work planes, work axes, and work points are all located on the generated layer ADP_WORK. Do not edit the contents of this layer; doing so may corrupt or destroy your modeling database. ■

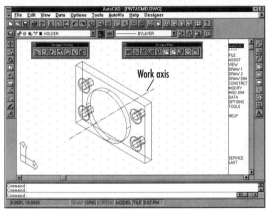

Figure 13.20

Creating a Work Plane

A work plane is a parametric plane that you can use to locate sketch planes or other features. When you create a work plane, you link it to the current feature geometry by an edge, axis, vertex, or surface of the part. This way, features located relative to the work plane will move and update when the parent feature to which the work plane is related moves. A sketch plane is different from a work plane.

Sketch planes are used for sketching the current profile only. Sketch planes are similar to UCSs. You define them in order to orient the current sketch. You do not need to create a work plane in order to create a sketch plane, but often it is useful to define a parametric work plane and then define the sketch plane relative to it.

Pick: **Designer, Features, Work Plane**

The Designer Work Plane dialog box appears on your screen. You will use it to make the selections shown in Figure 13.21. Picking the box to the left of Create Sketch Plane at the right side of the dialog box creates the sketch plane at the same time as the work plane. On your own, turn on Create Sketch Plane, then pick the box to the left of Planar Parallel to create a planar parallel work plane.

Figure 13.21

When you select Planar Parallel, the Planar Parallel dialog box appears on your screen, as shown in Figure 13.22, displaying the options available for creating parallel work planes.

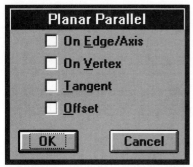

Figure 13.22

On your own, select the On Edge/Axis option and then pick OK. Pick OK again to return to your drawing for the final selections. You will select the work axis you created previously for the axis. Then you will select the Z X plane option so that the parallel work plane is created parallel to the Z X axis of the coordinate system. Refer to Figure 13.23 for your selections.

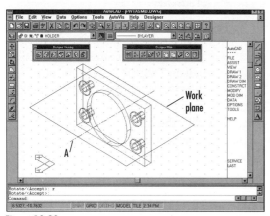

Figure 13.23

X/Y/Z/<Select work axis or straight edge>: *(pick work axis A)*

Xy/Yz/Zx/Ucs/<Select work plane or planar face>: **ZX** ⏎

X/Y/Z/<Select work axis or straight edge>: *(pick work axis A)*

Rotate/<Accept>: **R** ⏎
Rotate/<Accept>: **R** ⏎
Rotate/<Accept>: **R** ⏎
Rotate/<Accept>: ⏎ *(when the UCS appears as shown in Figure 13.23)*

The work plane appears in your drawing, as shown in Figure 13.23. The sketch plane is already aligned with the work plane because you chose the Create Sketch Plane option.

Pick: **Sketch View icon**

The view is aligned with the sketch plane, as shown in Figure 13.24.

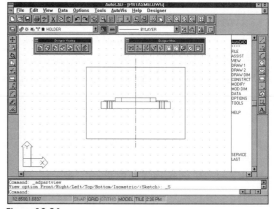

Figure 13.24

On your own, sketch the shape you see in Figure 13.25, using six lines. Make the lowest line above the existing feature so that you can select it later. You can use the Close option of the Line command to close the last line of the sketch to the first line of the sketch.

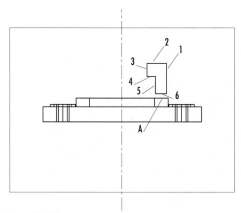

Figure 13.25

Pick: **Designer, Sketch, Profile**

Select objects for sketch: *(pick the six lines of the sketch)*

Select objects: ⏎

You will see a message similar to *Solved under constrained sketch requiring 6 dimensions/ constraints*. If there are more constraints required to solve your sketch, you probably need to add constraints for Join, Horizontal, or Vertical. If so, add these constraints on your own.

Now add the dimensions to further define the profile on your own. Add a vertical dimension with the value 0 between horizontal line 6 and the top surface of the existing feature, marked A in Figure 13.25. The profile should appear similar to Figure 13.26. Add the remaining dimensions on your own until you have a fully constrained sketch. Save your drawing on your own.

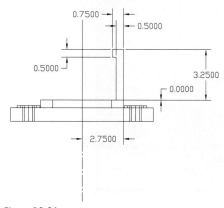

Figure 13.26

Creating a Revolved Feature

Now you are ready to revolve the profile to create the cylindrical feature.

Pick: **Designer, Features, Revolve**

The Designer Revolution dialog box appears on your screen, as shown in Figure 3.27. You use it to create revolved features similar to the ones you made in Tutorial 9 using AutoCAD's solid modeling. On your own, select the radio button to the left of Full to create a full 360° revolution. Pick the radio button to the left of Join so that the new feature is joined to the previous feature. The Size area is grayed out; it is only available when you select By Angle as the termination for the revolution. When you are finished selecting, pick OK to exit the dialog box. You will return to the drawing to select the axis of revolution.

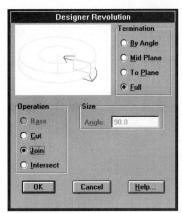

Figure 13.27

Select axis of revolution: *(pick the work axis)*

The profile revolves to create a cylindrical feature, but you cannot see its shape until you change the viewing direction. You will use the Designer Viewing toolbar to select the isometric view.

Pick: **Isometric View icon**

Your drawing should look like Figure 13.28.

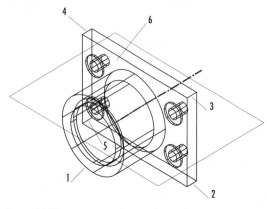

Figure 13.28

Adding the Fillets and Chamfer

Now you are ready to add the fillets, rounds, and chamfer to finish the pivot holder. (A *round* is the same as a fillet, but on an exterior corner.) On your own, use the Designer, Features, Fillet selection from the menu bar and follow the prompts to add a .25 round to the top edge of the cylinder identified as 1 in Figure 13.28. Add .125 radius fillets to the four corners (labeled 2 through 5) and then to the outer edge of the top horizontal surface (labeled 6). When you have finished the fillets and rounds, your drawing should appear similar to Figure 13.29.

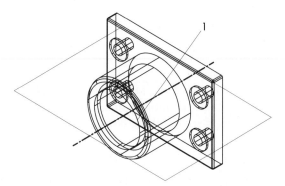

Figure 13.29

Next you will add a .125 × .125 chamfer to the top inside edge of the hole.

Pick: **Designer, Features, Chamfer**

The Designer Chamfer dialog box appears on your screen, as shown in Figure 13.30. You will use it to add the chamfer for the large hole. On your own, pick Equal Distance and set the distance .125 for the chamfer. When you are done setting up the dialog box, pick OK. You will return to your drawing for the remaining prompt.

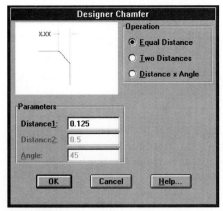

Figure 13.30

Select edge: *(pick edge 1 in Figure 13.29)*
Select edge: ⏎

The chamfer should appear in your drawing, as shown in Figure 13.31.

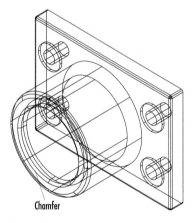

Chamfer

Figure 13.31

Now thaw layer BASE on your own to restore the pivot base to your screen. Use Zoom and Pan as needed so that you have a clear view of

the parts. In the next steps you will move the holder so that it is lined up to fit correctly with the base.

First you will turn off the display of the work plane and the work axis.

Pick: **Designer, Display, Work Plane, Off**
    Select/<All>: ⏎

The message *Work plane display mode turned off* appears in the command window and your screen is redrawn without the work plane. Next turn off the display of the work axis.

Pick: **Designer, Display, Work Axis, Off**

Your screen is redrawn without the work axis displayed. (If it does not redraw automatically, pick Redraw on your own.) It should look similar to Figure 13.32.

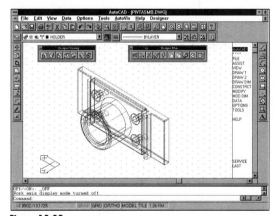

Figure 13.32

On your own, enlarge the upper left holes on your screen, using Zoom Window so that your screen looks like Figure 13.33. Type *REGEN* ⏎ at the command prompt to regenerate the circles so they appear round, if necessary.

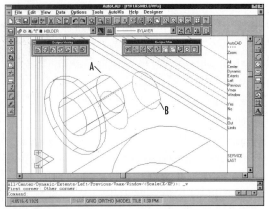

Figure 13.33

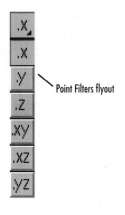

Point Filters flyout

Figure 13.34

Using Point Filters

You will use AutoCAD's *point filters* to select only certain coordinates during the following operations. Using point filters, you can choose to find the X, Y, Z, XY, YZ, or XZ coordinate(s) when you select. There are several ways to use point filters. To use point filters at the command prompt, type a period followed by the name of the coordinate you want to select and ⏎ when you are prompted to pick a point. Only the coordinate you identified is selected when you pick. You will then see another prompt stating *Needs Z* or whatever coordinate(s) remains unspecified. Point filters can be extremely useful when creating 3D drawing geometry. You can use them with any AutoCAD command that prompts for selection of a point or location.

You can also use the Point Filters flyout from the Standard toolbar shown in Figure 13.34, which is located on the Standard toolbar.

Third, you can show the Point Filters toolbar and position it anywhere on the screen.

You will use point filters while you move the holder so that it aligns with the base. You will move the holder to the X and Z coordinates of the center of the hole in the pivot base, but you will leave the Y coordinate for the move as it is by picking the center of the original circle once again. Refer to Figure 13.33 as you make selections.

Pick: **Move icon**

Select objects: *(pick the holder)*

Select objects: ⏎

Base point or displacement: *(pick Snap to Center icon)*

cen of *(pick circle A)*

Second point of displacement: *(pick .XZ icon)*

of *(pick Snap to Center icon)*

cen of *(pick circle B)*

(need Y): *(pick .Y icon)*

of *(pick Snap to Center)*

cen of *(pick circle A)*

The pivot holder moves so that the hole in the pivot holder lines up with the hole in the pivot base. Because the parts were both created so that they are symmetrical about their centers, and the holes are spaced at the same distances (HOLEDIST1 and HOLEDIST2), the two parts should now line up, as shown in Figure 13.35. Use Zoom Previous on your own to restore the original view.

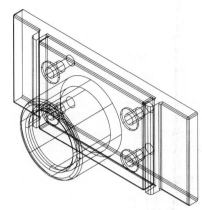

Figure 13.35

> ■ *TIP* If you are unsure whether the views line up exactly, first change to the sketch view. The holes should align. Next use the Left View or Right View selection to check that the parts align in the profile plane. ■

With AutoCAD Designer, you can also use multiple tiled viewports while in model space to show multiple views of your model on the screen. When you are creating complex parts this can be very helpful.

■ *Warning:* Do not control Tilemode, Pspace, and Mspace directly because AutoCAD Designer uses these commands and manipulates layers to create drawing views. Changing them at the command level can destroy AutoCAD Designer's part-to-drawing associativity. ■

Updating the Assembly

Next you will create a global parameter for HOLEDIST2 with a different value than you specified in Tutorial 12. Then you will update the parts in the assembly so that they reflect the change.

Pick: **Designer, Parameters, Create**

Enter equation: **HOLEDIST2=6** ⏎

Parameter "holedist2" created: current value == 6

Enter equation: ⏎

Parameter changes affect active part. Use ADUPDATE to see changes.

Create/Delete/List/Import/Export/<eXit>: ⏎

Pick: **Designer, Update**

Updating feature 38 of 38 (type: Chamfer)...

The active part in your drawing updates to reflect the new distance for the hole size. Only the active part updates. To update the other part, you must make it active and then update it.

Pick: **Designer, Part, Make Active**

Select part: *(pick on the pivot base)*

Pick: **ADUpdate icon**

Updating feature 24 of 24 (type: DatumOrientedPlane)...

Now the pivot base is updated to reflect the new distance between the holes, so that it matches the holes in the pivot holder.

Using Ad_updatepart

AutoCAD Designer provides an undocumented AutoLISP API function, Ad_updatepart, which you can use to verify the integrity of your model by forcing the part to update.

Ad_updatepart has four switch values that you can specify after the LISP command to cause various update conditions. They are:

0 - regenerates current active part only if required

1 - forces a bottom-up regeneration of the current active part

3 - forces a bottom-up regeneration of all parts in the drawing

4 - regenerates while checking for polyline lists

If you encounter problems when changing features or parameters, you may want to try typing *(AD_UPDATEPART 1)* ⏎ at the command prompt. This will force the current active part to regenerate from the bottom up, applying each constraint and dimension to determine the validity of the model. If you do this after each major step, you will know if you have made an error and can go back and fix it before continuing. If a certain feature is no longer valid, it will be identified and then you can go back and change it. (The regular AutoCAD Designer ADUpdate command only updates the part if it knows you have pending edits, i.e., you have made a change and the model does not yet reflect the change.)

You may be able to make corrections to a model that will not update by using Designer, Edit Feature. If you determine that a certain feature is causing a problem, try using Edit Feature with the Sketch option. Delete the dimensions or constraints that you suspect may be causing the problem, and then type *(AD_UPDATEPART 1)* ⏎ to force the part to regenerate. If the part regenerates successfully, go back and add new dimensions and constraints to define the profile. In AutoCAD Designer 1.2, the profile *does not* have to be fully constrained to create a feature. You can go back and add the constraints and dimensions later.

Keep in mind as you are changing parameters that if you specify values that are too extreme, you may unknowingly create a condition in which your model is no longer valid. For example, the model you created in this tutorial would be invalid if you changed the value for the HOLEDIST1 parameter so that it is larger than the BASEHEIGHT parameter (because the holes would be outside of the part).

> ■ *TIP* You can use Ad_updatepart to update all of the parts in your drawing at once by adding the undocumented switch 3 (for all parts) after the function. You must type it at the command prompt inside parentheses, like this: *(ad_updatepart 3)* ⏎. ■

Using Static Parts

Static parts in AutoCAD Designer are parts that do not have parametric constraints and dimensions. You cannot update them, but you *can* add features to them. Static parts work well for standard parts you are using in your design that will be purchased from a supplier, as well as for your company's standard parts, because they will not be changed. Copying a parametrically designed part with the AutoCAD Copy command creates a static part.

Save your *pivtasmb.dwg* drawing on your own.

Exporting Static Parts

On your own, open the *bushing.dwg* file that was provided with your data files. The bushing drawing appears on the screen, as shown in Figure 13.36. You will export the bushing as a *.sat* format file, and then import it into your pivot assembly drawing. The *.sat* format is used by the ACIS modeler, which is used by both AutoCAD's solid modeler and AutoCAD Designer. Many different CAD programs and third-party applications are able to read this file format. You will select AutoCAD's File, Export command from the menu bar.

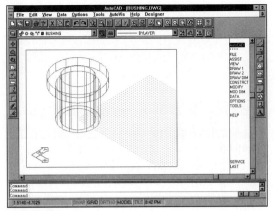

Figure 13.36

Pick: **File, Export**

The Export Data dialog box appears on your screen. You will select the ACIS (*.SAT) format to use for the export. Make the changes shown in Figure 13.37 on your own and then pick OK. Be sure to export the file to your working directory. You return to your drawing to select the objects to export.

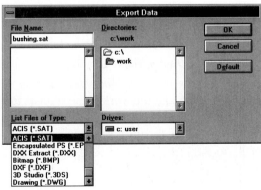

Figure 13.37

Select objects: *(pick on the bushing)*
Select objects: ⏎

You will see a message in the command window saying *File name <C:\WORK\BUSHING.SAT >:*.

Now you are ready to import *bushing.sat* into the assembly drawing.

Reopen *pivtasmb.dwg* on your own. Create a new layer named BUSHING with color green and linetype CONTINUOUS. Set it as the current layer on your own.

Transferring Models into AutoCAD Designer

You will import the *.sat* file using the commands provided in AutoCAD Designer.

Pick: **Designer, Utilities, Transfer, SAT In**

The SAT In dialog box appears on your screen. Use it to select the directory *c:\work* and the file *bushing.sat*. When you have the file selected, pick OK to exit the dialog box. On your own set layer HOLDER current.

The bushing appears in your drawing, as shown in Figure 13.38.

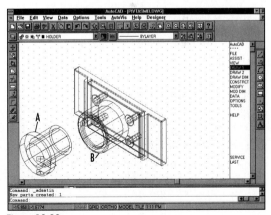

Figure 13.38

Next you will move the bushing so that it aligns with the pivot holder and pivot base.

Pick: **Move icon**

Select objects: *(pick the bushing)*

Select objects: ⏎

Base point or displacement: *(pick Snap to Center icon)*

cen of *(pick A)*

Second point of displacement: *(pick Snap to Center icon)*

cen of *(pick B)*

■ *TIP* Use the Zoom In command if necessary to pick the correct circle. ■

The bushing moves so that it fits in the holder, as shown in Figure 13.39. You may want to change to the sketch view to make sure that the parts line up.

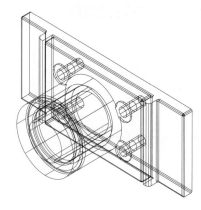

Figure 13.39

Next you will change the viewing direction, using the AutoCAD commands from the Standard toolbar. You will change to the SW isometric view.

Pick: **SW Isometric View icon**

The view changes so that your model appears in a more normal isometric orientation, as shown in Figure 13.40.

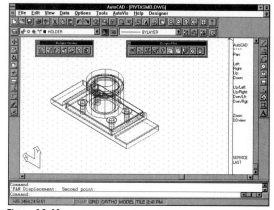

Figure 13.40

Now that you have the parts modeled and shown in their assembled positions, you are ready to create the assembly drawing directly from your model. (You will create an assembled section view in this tutorial, but isometric pictorial assembly views are also easily created.) To define where the section will pass through the model, you will make the pivot holder part active once again and turn on its work plane, which passes through the center of the part.

Pick: **Designer, Part, Make Active**

Select part: *(pick on the pivot holder)*

You will see the message *Part 1 is now active.* Next turn on the work plane and work axis so they are displayed on the screen again.

Pick: **Designer, Display, Work Plane, On**

You will see the message *Work plane display mode turned on* and the work plane returns to your screen.

Pick: **Designer, Display, Work Axis, On**

The work axis also returns to your screen, as shown in Figure 13.41.

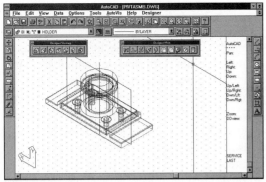

Figure 13.41

Now you are ready to create a section view through the assembly.

Pick: **Designer, Drawing, Create View**

The Designer Drawing View dialog box appears on your screen. On your own, use it to make the selections shown in Figure 13.42: Full Section, Hatching turned on, and All selected in the upper right in the Parts area. When you are done selecting, pick OK. You will return to your drawing for the following selections.

Figure 13.42

Work plane will be the cutting plane.

Select work plane: *(pick the work plane)*

X/Y/Z/<Select work axis or straight edge>: *(pick the work axis)*

Rotate/<Accept>: **R** ⏎

Rotate/<Accept>: **R** ⏎

Rotate/<Accept>: **R** ⏎

Rotate/<Accept>: ⏎

Rotate the UCS so that it aligns with the work plane and the X direction is to the right and the Y direction is up.

View center: *(pick near the center of the area inside the border)*

View center: ⏎

The section view of the assembled parts appears on your screen as shown in Figure 13.43.

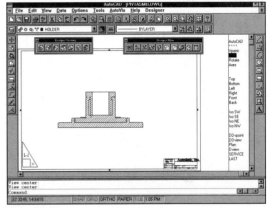

Figure 13.43

On your own, create the balltags shown in Figure 13.44, containing the numbers 1, 2, and 3. Add a leader from the edge of each part to the balltags. Create layer CENTERLINE and assign it color green and linetype CENTERX2. Set it as the current layer and use it to draw the centerline for the assembly. When you have finished your drawing should look like Figure 13.44.

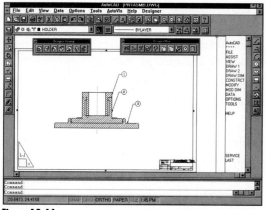

Figure 13.44

Create a parts list for the assembly drawing as you did in Tutorial 10, showing the following information.

Parts List

Item No.	Description	Quantity	Material
1	Bushing	1	Steel
2	Pivot Holder	1	Steel
3	Pivot Base	1	Steel

When you have finished adding these items on your own, your screen should appear similar to Figure 13.45.

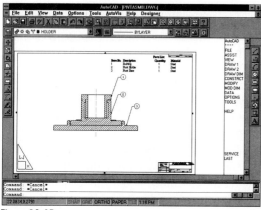

Figure 13.45

Creating Annotations

The AutoCAD drawing objects you added as callouts and the centerline you added are not tied to the drawing in the same way that AutoCAD Designer hole notes and dimensions are. You can make items you have drawn into annotations using the selection Designer, Drawing, Annotation, Create. Follow the prompts to select the previously drawn AutoCAD text or objects which you want to turn into annotations, then pick a location on the part where you want the item attached. AutoCAD Designer will create a leader from the attachment point to the object. Once the item is an annotation, if you move a drawing view, the annotation will move with it. Otherwise, regular AutoCAD items will not move when you move your views. You cannot attach annotation to the Section view you created because its lines are all generated silhouette edges calculated by AutoCAD Designer. To use annotation, you must be able to attach the item to a line which is part of a view. Experiment using annotation on your own.

Adding Features to a Static Part

Make the bushing the active part on your own and add a .25 fillet to the top outer edge.

■ *TIP* AutoCAD Designer files sometimes get very large. You can reduce their size using the *adpurge.lsp* bonus routine provided with the data files. To use *adpurge.lsp*, start the AutoCAD Applications command from the Tools pull-down menu and select *adpurge.lsp* as the file to load. Open or save the file to be used before running the command. At the command prompt, type *ADPURGE* ⏎ and then save your file again. You can also reduce file sizes by converting parametric parts to static parts using the command Ad_makebase, but this has the disadvantage that the part can no longer be updated. However, this can be a good way to reduce file size before sending a part to a colleague. ■

Close the Designer toolbars. Save your *pivtasmb.dwg* drawing and exit AutoCAD. You have completed the AutoCAD Designer portion of this tutorial manual.

nurb   point filter   round

Ad_makebase   Display Isolines   Point Filters
Ad_setvar   Display Work Axes   (.X, .Y, .Z, .XY, .XZ, .YZ)
Ad_updatepart   Export   Revolve Features
ADChamfer   Make Part Active   Transfer SAT In
Application Load   New Part

EXERCISES

Use AutoCAD Designer to model the assemblies shown below. Use global parameters for mating dimensions. Do not model thread, instead model the nominal diameter of the shaft or hole. A letter M after the exercise number indicates that the dimensions are in metric units.

13.1M U-bolt Pipe Strap

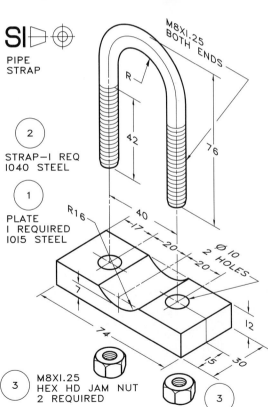

SI

PIPE
STRAP

2

STRAP—I REQ
1040 STEEL

1

PLATE
I REQUIRED
1015 STEEL

M8X1.25
BOTH ENDS

R

42

76

R16

40

17

20

20

Ø 10
2 HOLES

7

74

12

15

30

3

M8X1.25
HEX HD JAM NUT
2 REQUIRED

3

13.2M Fixture Guide

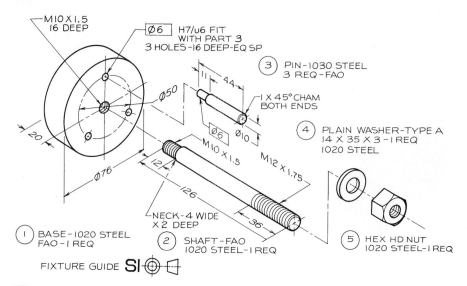

M10 X 1.5
16 DEEP

Ø6   H7/u6 FIT
WITH PART 3
3 HOLES–16 DEEP–EQ SP

③ PIN–1030 STEEL
3 REQ–FAO

I X 45° CHAM
BOTH ENDS

Ø50

Ø6

Ø10

M10 X 1.5

M12 X 1.75

④ PLAIN WASHER–TYPE A
14 X 35 X 3 –1 REQ
1020 STEEL

20

Ø76

11

44

12

126

NECK–4 WIDE
X 2 DEEP

36

① BASE–1020 STEEL
FAO–1 REQ

② SHAFT–FAO
1020 STEEL–1 REQ

⑤ HEX HD NUT
1020 STEEL–1 REQ

FIXTURE GUIDE SI⊕⊏

13.3M Centering Point

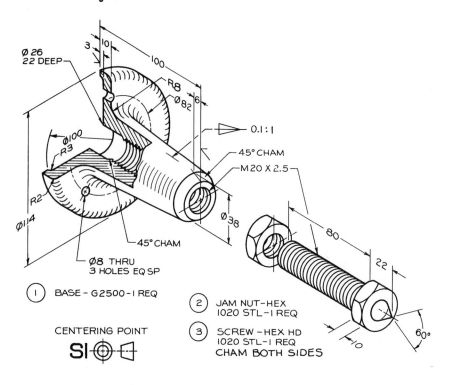

Ø 26
22 DEEP

3

10

100

R8

Ø82

6

Φ100

R3

0.1 : 1

45° CHAM

M20 X 2.5

R2

Ø114

Ø38

45° CHAM

Ø8 THRU
3 HOLES EQ SP

80

22

60°

10

① BASE – G2500 –1 REQ

② JAM NUT–HEX
1020 STL–1 REQ

③ SCREW –HEX HD
1020 STL–1 REQ
CHAM BOTH SIDES

CENTERING POINT

SI⊕⊏

13.4M Hanger

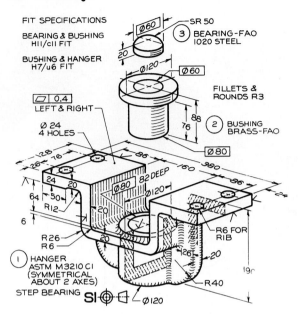

FIT SPECIFICATIONS

BEARING & BUSHING
H11/c11 FIT

BUSHING & HANGER
H7/u6 FIT

Ø60
SR 50
3 BEARING-FAO
1020 STEEL
20

Ø120 Ø60

FILLETS &
ROUNDS R3

0.4
LEFT & RIGHT

Ø 24
4 HOLES

88
76

2 BUSHING
BRASS-FAO

Ø80

128 76 96 160 390
26 24 20 Ø80 82 DEEP 96 2 φ
64 50 Ø120
R12 20
6
R26 20 R6 FOR
R6 RIB

1 HANGER
ASTM M3210 CI
(SYMMETRICAL
ABOUT 2 AXES)
STEP BEARING SI

126 20
190
R40
Ø120

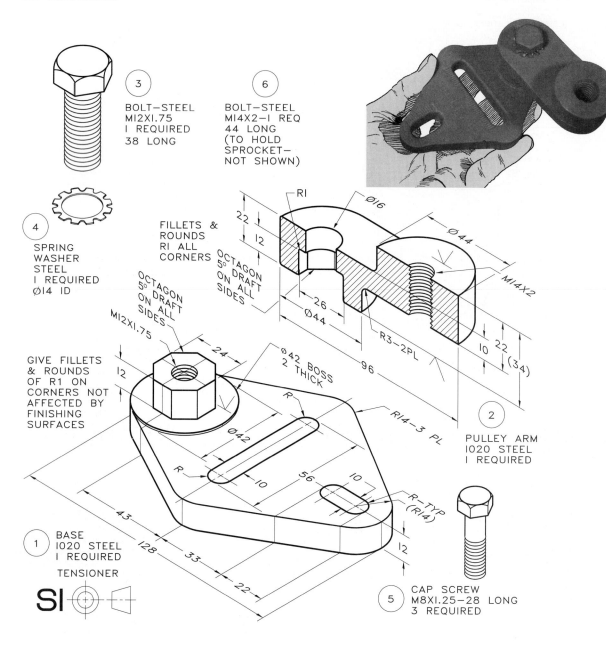

③ BOLT—STEEL
M12X1.75
1 REQUIRED
38 LONG

⑥ BOLT—STEEL
M14X2—1 REQ
44 LONG
(TO HOLD
SPROCKET—
NOT SHOWN)

④ SPRING
WASHER
STEEL
1 REQUIRED
Ø14 ID

FILLETS &
ROUNDS
R1 ALL
CORNERS

R1
Ø16
OCTAGON
5° DRAFT
ON ALL
SIDES
22
12
Ø44
26
Ø44
96
M14X2
R3—2PL
22
10
(34)

② PULLEY ARM
1020 STEEL
1 REQUIRED

OCTAGON
5° DRAFT
ON ALL
SIDES

M12X1.75

GIVE FILLETS
& ROUNDS
OF R1 ON
CORNERS NOT
AFFECTED BY
FINISHING
SURFACES

24
12
Ø42 BOSS
2 THICK
R
Ø42
R14—3 PL
R
10
56
10
R—TYP
(R14)
12

① BASE
1020 STEEL
1 REQUIRED

TENSIONER

SI ⊕ ⬡

43
128
33
22

⑤ CAP SCREW
M8X1.25—28 LONG
3 REQUIRED

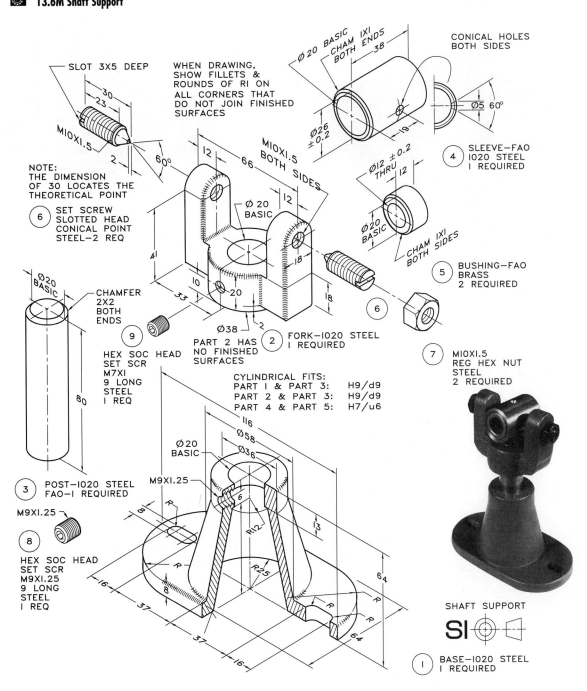

SLOT 3X5 DEEP

30
23

MIOXI.5
2

60°

NOTE:
THE DIMENSION
OF 30 LOCATES THE
THEORETICAL POINT

⑥ SET SCREW
SLOTTED HEAD
CONICAL POINT
STEEL—2 REQ

WHEN DRAWING,
SHOW FILLETS &
ROUNDS OF RI ON
ALL CORNERS THAT
DO NOT JOIN FINISHED
SURFACES

Ø20 BASIC
CHAM IXI
BOTH ENDS
38

Ø26 ±0.2

19

Ø5 60°

CONICAL HOLES
BOTH SIDES

④ SLEEVE—FAO
1020 STEEL
I REQUIRED

Ø12 ±0.2 THRU   12

Ø20 BASIC

CHAM IXI
BOTH SIDES

⑤ BUSHING—FAO
BRASS
2 REQUIRED

MIOXI.5
BOTH SIDES

12
66

12

Ø 20
BASIC

41

18

10   20

33

8

Ø38   2

PART 2 HAS
NO FINISHED
SURFACES

② FORK—1020 STEEL
I REQUIRED

⑥

⑦ MIOXI.5
REG HEX NUT
STEEL
2 REQUIRED

Ø20
BASIC

CHAMFER
2X2
BOTH
ENDS

⑨

HEX SOC HEAD
SET SCR
M7XI
9 LONG
STEEL
I REQ

80

③ POST—1020 STEEL
FAO—I REQUIRED

M9XI.25

⑧

HEX SOC HEAD
SET SCR
M9XI.25
9 LONG
STEEL
I REQ

CYLINDRICAL FITS:
PART I & PART 3:   H9/d9
PART 2 & PART 3:   H9/d9
PART 4 & PART 5:   H7/u6

116
Ø58
Ø36

Ø20
BASIC

M9XI.25

6

8

R

R12

13

R25

R

R

64

16

37

8

37

16

64

R

SHAFT SUPPORT

SI ⊕ ◁

① BASE—1020 STEEL
I REQUIRED

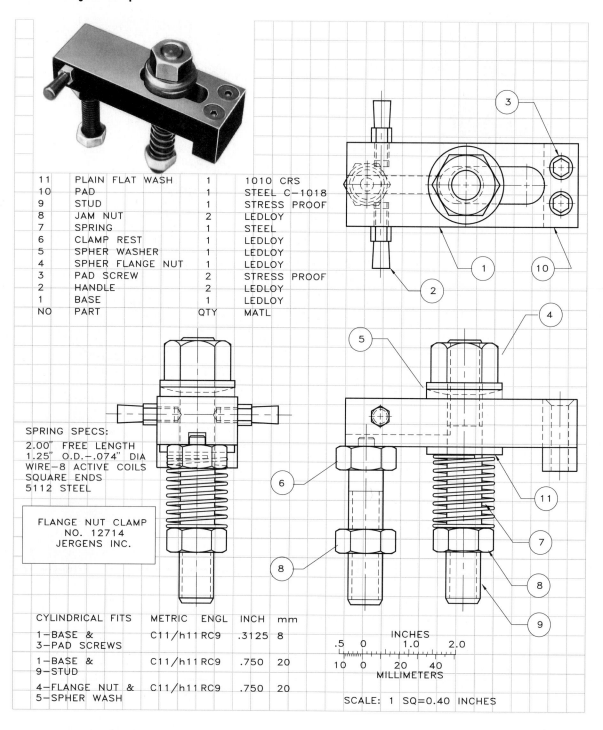

NO	PART	QTY	MATL
11	PLAIN FLAT WASH	1	1010 CRS
10	PAD	1	STEEL C−1018
9	STUD	1	STRESS PROOF
8	JAM NUT	2	LEDLOY
7	SPRING	1	STEEL
6	CLAMP REST	1	LEDLOY
5	SPHER WASHER	1	LEDLOY
4	SPHER FLANGE NUT	1	LEDLOY
3	PAD SCREW	2	STRESS PROOF
2	HANDLE	2	LEDLOY
1	BASE	1	LEDLOY

SPRING SPECS:
2.00" FREE LENGTH
1.25" O.D.−.074" DIA
WIRE−8 ACTIVE COILS
SQUARE ENDS
5112 STEEL

FLANGE NUT CLAMP
NO. 12714
JERGENS INC.

CYLINDRICAL FITS	METRIC	ENGL	INCH	mm
1−BASE & 3−PAD SCREWS	C11/h11	RC9	.3125	8
1−BASE & 9−STUD	C11/h11	RC9	.750	20
4−FLANGE NUT & 5−SPHER WASH	C11/h11	RC9	.750	20

INCHES
.5 0   1.0   2.0
10 0   20   40
MILLIMETERS

SCALE: 1 SQ=0.40 INCHES

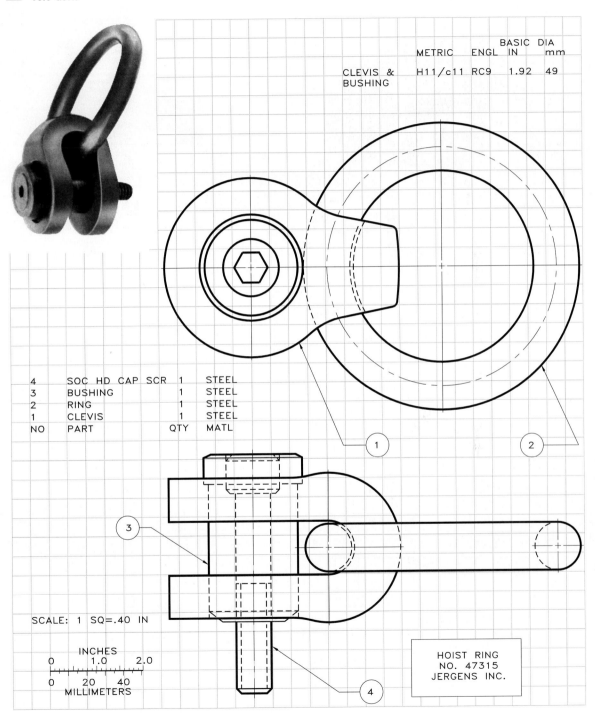

			BASIC DIA	
	METRIC	ENGL	IN	mm
CLEVIS & BUSHING	H11/c11	RC9	1.92	49

4	SOC HD CAP SCR	1	STEEL
3	BUSHING	1	STEEL
2	RING	1	STEEL
1	CLEVIS	1	STEEL
NO	PART	QTY	MATL

SCALE: 1 SQ=.40 IN

INCHES
0   1.0   2.0

0   20   40
MILLIMETERS

HOIST RING
NO. 47315
JERGENS INC.

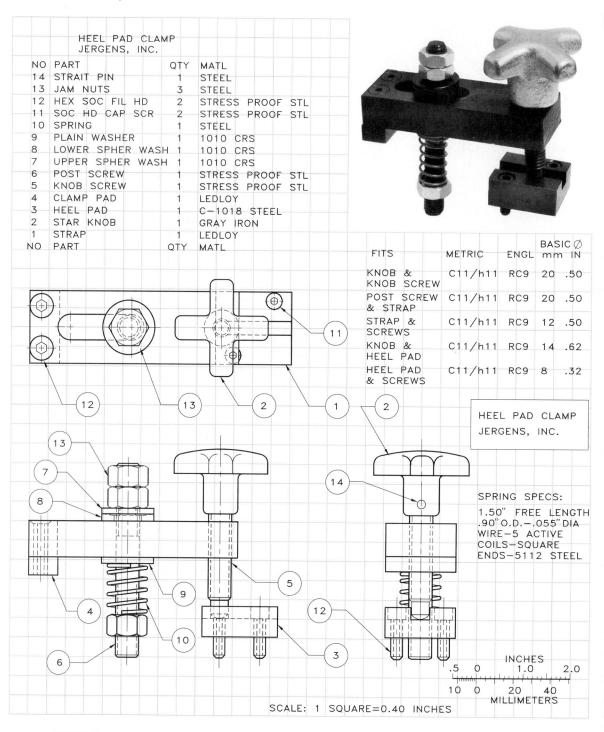

HEEL PAD CLAMP
JERGENS, INC.

NO	PART	QTY	MATL
14	STRAIT PIN	1	STEEL
13	JAM NUTS	3	STEEL
12	HEX SOC FIL HD	2	STRESS PROOF STL
11	SOC HD CAP SCR	2	STRESS PROOF STL
10	SPRING	1	STEEL
9	PLAIN WASHER	1	1010 CRS
8	LOWER SPHER WASH	1	1010 CRS
7	UPPER SPHER WASH	1	1010 CRS
6	POST SCREW	1	STRESS PROOF STL
5	KNOB SCREW	1	STRESS PROOF STL
4	CLAMP PAD	1	LEDLOY
3	HEEL PAD	1	C-1018 STEEL
2	STAR KNOB	1	GRAY IRON
1	STRAP	1	LEDLOY
NO	PART	QTY	MATL

FITS	METRIC	ENGL	BASIC ∅ mm	IN
KNOB & KNOB SCREW	C11/h11	RC9	20	.50
POST SCREW & STRAP	C11/h11	RC9	20	.50
STRAP & SCREWS	C11/h11	RC9	12	.50
KNOB & HEEL PAD	C11/h11	RC9	14	.62
HEEL PAD & SCREWS	C11/h11	RC9	8	.32

HEEL PAD CLAMP
JERGENS, INC.

SPRING SPECS:

1.50" FREE LENGTH
.90" O.D. –.055" DIA
WIRE–5 ACTIVE
COILS–SQUARE
ENDS–5112 STEEL

INCHES
.5   0   1.0   2.0

10   0   20   40
MILLIMETERS

SCALE: 1 SQUARE=0.40 INCHES

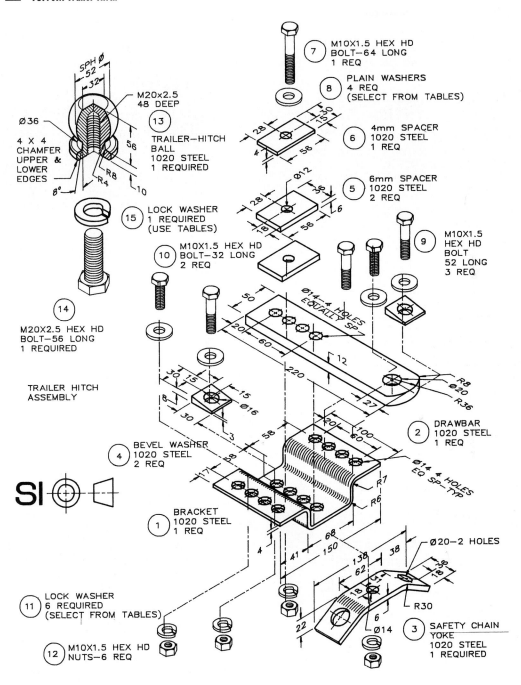

SPH Ø 52
32

Ø36

4 X 4 CHAMFER UPPER & LOWER EDGES

R8
R4
8°

M20x2.5 48 DEEP

13 TRAILER—HITCH BALL 1020 STEEL 1 REQUIRED

56
10

7 M10X1.5 HEX HD BOLT—64 LONG 1 REQ

8 PLAIN WASHERS 4 REQ (SELECT FROM TABLES)

30
15
28
4
56

6 4mm SPACER 1020 STEEL 1 REQ

Ø12
36
28
8
56
6

5 6mm SPACER 1020 STEEL 2 REQ

15 LOCK WASHER 1 REQUIRED (USE TABLES)

10 M10X1.5 HEX HD BOLT—32 LONG 2 REQ

9 M10X1.5 HEX HD BOLT 52 LONG 3 REQ

14 M20X2.5 HEX HD BOLT—56 LONG 1 REQUIRED

Ø14—4 HOLES EQUALLY SP

50
20
60
220
12
R8
Ø20
R36
27

2 DRAWBAR 1020 STEEL 1 REQ

TRAILER HITCH ASSEMBLY

30
15
15
Ø16
8
30
3
58

4 BEVEL WASHER 1020 STEEL 2 REQ

Ø14 4 HOLES EQ SP TYP

R7
R6

SI

1 BRACKET 1020 STEEL 1 REQ

4
41
150
68
138
62
18
22

Ø20—2 HOLES
38
36
18
R30
6
Ø14

11 LOCK WASHER 6 REQUIRED (SELECT FROM TABLES)

12 M10X1.5 HEX HD NUTS—6 REQ

3 SAFETY CHAIN YOKE 1020 STEEL 1 REQUIRED

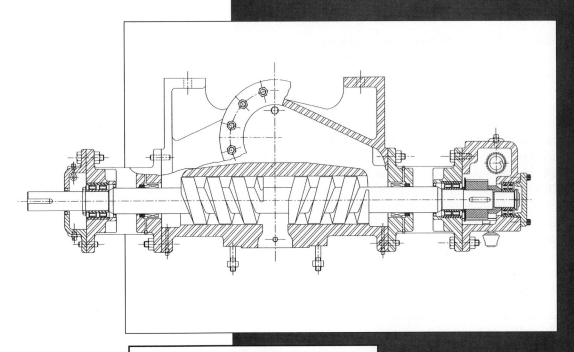

Creating Section Views Using 2D and Solid Modeling

Objectives

When you have completed this tutorial, you will be able to

1. Show the internal surfaces of an object, using 2D section views.

2. Locate and draw cutting plane lines and section lines on appropriate layers.

3. Hatch an area with a pattern.

4. Edit associative hatching.

5. Section and slice a solid model.

6. Control layer visibility within a viewport.

Introduction

In this tutorial you will learn how to draw section views. A *section view* is a special type of orthographic view used to show the internal structure of an object. It essentially shows what you would see if a portion of the object were cut away and you looked at the part that remains. Section views are often used when the normal orthographic views contain so many hidden lines that they are confusing and difficult to interpret. If you did Tutorial 13, you have used AutoCAD Designer to create a section view automatically. If you are creating your drawings with solid modeling, you can generate a section view directly from the solid model. In this tutorial you will learn to use both 2D and solid modeling to create section views.

Section View Conventions

Figure 14.1 shows the front and side views of a circular object. Notice that the side view contains many hidden lines and is somewhat difficult to interpret.

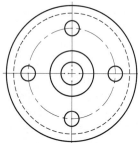

Figure 14.1

Figure 14.2 shows a pictorial drawing of the same object cut in half along its vertical centerline. This is called a *pictorial full section*.

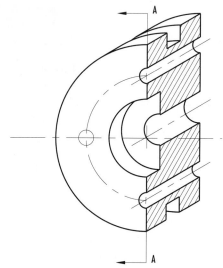

Figure 14.2

In many drafting applications, it is common practice to fill an area (such as the portion of an object that has been cut to show a section view) with a pattern to make the drawing easier to interpret. The pattern can help differentiate components of a three-dimensional object, or it can indicate the material composing the object. You can accomplish this *crosshatching* or pattern filling using AutoCAD's Boundary Hatch command.

In Figure 14.2, crosshatching shows the solid portions of the object, where material was cut to make the section view.

Figure 14.3 shows a front and section view of the same circular object. A *cutting plane line*, line A–A, defines where the sectional cut should be taken on the object and indicates the direction in which you should view the object to produce the section view.

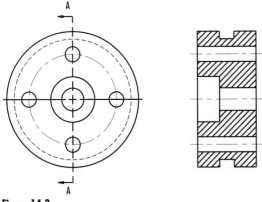

Figure 14.3

Compare the side view in Figure 14.1 with the section view in Figure 14.3 and the three-dimensional section view shown in Figure 14.2. Figures 14.2 and 14.3 are easier to interpret than the front and side views shown in Figure 14.1.

Figure 14.4 shows a different object with a cutting plane line and the appropriate section view.

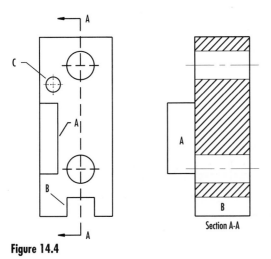

Figure 14.4

Crosshatching

Crosshatching is used to show where material has been cut. Note in Figure 14.4 that the surfaces labeled A and B are not crosshatched. Only material that has been cut by the cutting plane is crosshatched. A and B are shown in section A–A because they are visible surfaces once the cut is taken, but they are not hatched because they were not cut by the cutting plane.

A section view shows what you would see if you looked at the object in the direction the arrows on the cutting plane line point, with all of the material in back of the arrows removed. When you are creating a section view, ignore the portion of the object that will be cut off. Draw the remaining portion, toward which the arrows on the cutting plane line point. Remember that once the object is cut, some features that were previously hidden will be visible and should be shown. The hole labeled C in the side view of Figure 14.4 is in the remaining portion of the object, but is not visible, so it is not shown in the section view. The purpose of a section view is to show the internal structure without the confusion of hidden lines. You should only use hidden lines if the object would be misunderstood if they were not included. (You could use an offset cutting plane line—which bends 90° to pass through features that are not all in the same plane—to show hole C in the section view, if you wanted to. Refer to any standard engineering graphics text to review conventions for cutting plane lines and types of section views.)

Starting

Start AutoCAD and, on your own, use the data file *cast2.dwg* that came with your software as a prototype file to begin a new drawing. Name the drawing *bossect.dwg*.

Creating 2D Section Views

The top, front, and side orthographic views of a casting are displayed on your screen. The horizontal centerlines have been erased in the top view. Your screen should look like Figure 14.5.

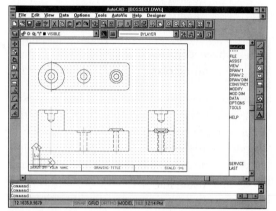

Figure 14.5

Often one or more of the orthographic views are replaced with section views. You will use the front orthographic view to draw a front section view, and the side view to draw a side section view through the boss.

Cutting Plane Lines

You can draw a cutting plane line using one of two different linetypes: either long dashes (DASHED), or long lines with two short dashes followed by another long line (PHANTOM). You should use only one type of cutting plane line in any one drawing. You will make dashed cutting plane lines by creating a new layer with the linetype DASHED. Linetype DASHED has longer dashes than linetype HIDDEN. You should use a different color for the cutting plane layer to help distinguish it from the other layers and linetypes. Also, cutting plane lines

are drawn with a thick pen when drawings are plotted. Since using different object colors allows for different pens to be selected when plotting, you should use a different color for thick cutting plane lines than for thin lines.

Creating Dashed Lines

Create a new layer for the cutting plane lines on your own. Name the new layer CUTTING_PLANE. Assign it color white (which will appear black on your screen if you are using a white background), and linetype DASHED. Set layer CUTTING_PLANE as the current layer. Make sure that Snap is turned on and set to .25.

Pick: **Line icon**

On your own, add a horizontal dashed cutting plane line in the top view of the boss that passes through the center of the holes in the top view, between points A–A in Figure 14.6. Then add a vertical dashed cutting plane line through the center of the top view of the boss between points B–B. The lines should extend beyond the edges of the object by at least .5 units.

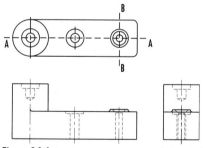

Figure 14.6

The cutting plane lines shown in Figure 14.6 should now appear in your drawing on layer CUTTING_PLANE. Save your drawing on your own.

Drawing the Leaders

As you saw in Tutorial 10, the Leader command lets you construct complex leaders. You will use the Leader command to add the arrowed line segments and the identifying letters for the cutting plane lines. Because the arrowed line segments are part of the cutting plane lines, they should be drawn on layer CUTTING_PLANE.

You will start the leader .5 units to the left of the top end of cutting plane line B–B.

> ■ *TIP* The Leader command will not draw the arrows if the first line segment of the leader is too short. For AutoCAD to draw an arrow, the first line segment must be longer than the current arrowhead size. You can set the arrowhead size using the dimensioning variables you learned in Tutorial 7. ■

On your own, turn on the Dimensioning toolbar that you used in Tutorial 7 before you continue. Use what you have learned about floating toolbars to position it on your screen so that it is out of the way of your drawing, but still handy to pick commands.

On your own, make sure that Snap is set to .25 before you draw the leader line. Use Ortho to assist you in drawing straight lines. Use the object snaps from the Standard toolbar.

Pick: **Leader icon**

From point: *(select a point .5 units to the left of, and even with, the top end of cutting plane line B–B)*

To point: *(pick Snap to Endpoint)*

endp of *(pick the top end of the cutting plane line B–B)*

To point (Format/Annotation/Undo)<Annotation>: ⏎
Annotation (or RETURN for options): **B** ⏎
MText: ⏎

> ■ *TIP* Remember that after typing text you must press ⏎ from the keyboard. The return button on your pointing device does not work to enter text. ■

Repeat these steps to create the leader for the bottom of cutting plane line B–B on your own.

> ■ *TIP* You can type *U* at the *To point* prompt to remove the last leader segment drawn if you make an error. ■

For cutting plane line A–A, you will end the command after the first segment of the line is drawn and use an option of the Leader command to have no annotation added. Then you will use the Dtext command to place the text labeling cutting plane line A–A.

Leader Command Options

The Leader command provides the following options once you have drawn the leader line: Format, Annotation, and Undo. The Format option has further options that allow you to select the shape for the leader line: Splined, Straight, Arrow, and None. The Splined and Straight options let you choose splined lines (so that you can make a curved leader) or straight lines. The Arrow and None options let you choose to draw an arrow at the end of the leader or draw a leader with no arrow.

The Annotation option also has further selections: Tolerance, Copy, Block, None, and Mtext. Mtext is the default option; you have used this option to create text at the end of the leader line, and you can also use it to create multiple lines of text. The Tolerance selection lets you add a geometric tolerance feature control frame to the end of the leader line. (You will use this option in Tutorial 16, when you learn about advanced dimensioning features.) The Copy option lets you pick on an existing feature control frame, block, or text object and copy it to the end of the new leader line. Choosing Block lets you select a block to be added to the end of the leader like you did in Tutorial 10. The None option lets you draw a leader line with no text added to the end. You will use the None option to end the Leader command when you draw the leaders for cutting plane line A–A.

Pick: **Leader icon**

From point: *(select a point .5 units above, and even with, the left endpoint of cutting plane line A–A)*

To point: *(pick Snap to Endpoint)*

of *(pick the left endpoint of cutting plane line A–A)*

To point (Format/Annotation/Undo)<Annotation>: ⏎

Annotation (or RETURN for options): ⏎

Tolerance/Copy/Block/None/<Mtext>: **N** ⏎

Command: ⏎ *(to restart the Leader command)*

From point: *(pick a point .5 units above, and even with, the right endpoint of cutting plane line A–A)*

To point: *(pick Snap to Endpoint)*

of *(pick the right endpoint of cutting plane line A–A)*

To point (Format/Annotation/Undo)<Annotation>: ⏎

Annotation (or RETURN for options): ⏎

Tolerance/Copy/Block/None/<Mtext>: **N** ⏎

Pick: **Dtext icon**

Justify/Style/<Start point>: **J** ⏎

Align/Fit/Center/Middle/Right/TL/TC/TR/ML/MC/MR/BL/BC/BR: **M** ⏎

Middle point: *(pick on the snap point below the left leader for cutting plane line A–A)*

Height: <0.2000>: ⏎

Rotation angle <0>: ⏎

Text: **A** ⏎

Text: *(pick a point below the end of the right leader for cutting plane line A–A)*

Text: **A** ⏎

Text: ⏎

When completed, your leader lines should look similar to the ones shown in Figure 14.7.

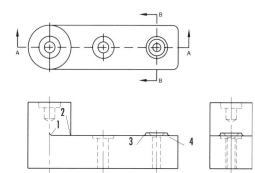

Figure 14.7

When you have finished creating the cutting plane lines as shown in Figure 14.7, turn off the Dimensioning toolbar on your own by clicking its Windows Control box.

Adjusting Linetype Scale

As you have seen in previous tutorials, you can change the general linetype scaling factor to improve the overall appearance of the cutting plane lines by adjusting their scale. You will type *LTSCALE* at the command prompt to activate the Linetype Scale command.

Command: **LTSCALE** ⏎

New scale factor <0.7500>: **.5** ⏎

Observe how the overall lines are affected. You can continue to try different values until the lines suit your particular needs.

Next, you will remove the unnecessary lines from the front view. When you draw a section view, you do not want to see the lines that represent intersections and surfaces that are on the outside of the object.

On your own, turn Snap off and use the Erase command to remove the *runout* (or curve which represents where two surfaces blend together) labeled 1 in Figure 14.7. Use Trim to remove the portion of the line to the left of point 2. Use the Erase command to remove the solid line from point 3 to point 4 that defines the surface in front of the boss. Use Zoom Window to help identify the points. Your drawing should look similar to Figure 14.8 when you are finished.

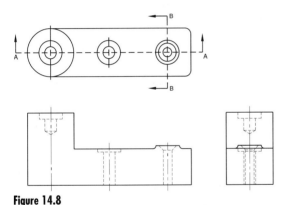

Figure 14.8

Changing Properties

When you section the drawing, the interior details become visible. You will use the Properties selection to change the layer of the hidden lines to layer VISIBLE. Pick the Properties icon from the Object Properties toolbar.

Pick: **Properties icon**

Select objects: *(pick all of the hidden lines in the front view)*

Select objects: ⏎

The Change Properties dialog box appears on your screen. You will use it to pick the Layer button and then use the Select Layer dialog box to set the new layer for the selected lines to layer VISIBLE.

Pick: **Layer**

Pick: **VISIBLE**

Pick: **OK *(to exit the Select Layer dialog box)***

Pick: **OK *(to exit the Change Properties dialog box)***

Now all of the hidden lines in the front view have been changed to layer VISIBLE. Your screen should now look similar to Figure 14.9.

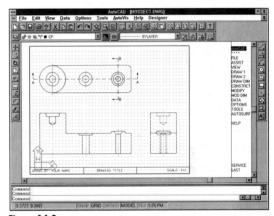

Figure 14.9

If you have zoomed your drawing, use Zoom Previous to return to the original size. You are almost ready to add the hatching to the drawing. The hatching will fill the areas of the drawing where the solid object was cut by the cutting plane.

Hatching should be on its own layer and should be plotted with a thin pen. Having the hatch on its own layer is useful so that you can freeze it if you do not want it displayed while you are working on the drawing. In the proto-type drawing, *cast2.dwg*, a separate layer for the hatching has already been created. Its color is red, and its linetype is CONTINUOUS.

Use Layer Control from the Object Properties toolbar to set HATCH as the current layer on your own.

Using Boundary Hatch

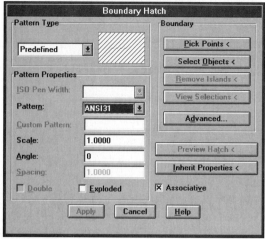

Hatches are composed of a series of lines, dashes, or dots that form a pattern. The Boundary Hatch command fills an area with a pattern. It creates *associative hatching*, which means that if the area selected as the hatch boundary is changed, the hatching is automatically updated to fill the new area. Associative hatching is the default for the Boundary Hatch command. If you update the selected boundary in a way that no longer forms a closed area, the hatching will no longer be associative. Boundary Hatch helps automatically select the boundary for the pattern through its various boundary selection options. To activate Boundary Hatch, select the Hatch icon from the Draw toolbar.

Pick: **Hatch icon**

The Boundary Hatch dialog box appears on your screen, as shown in Figure 14.10. Notice the checked box near the bottom right of the dialog box that indicates that associative hatch-ing is turned on.

Figure 14.10

Selecting the Hatch Pattern

You can select predefined hatch patterns by pulling down the list in the Pattern Properties area. ANSI31 is the default hatch pattern. It consists of continuous lines angled at 45°, spaced about 1/16" apart, and is typically used to show cast iron materials. It is also used as a general pattern when you do not want to specify a material by the hatch pat-tern. To select a hatch pattern, pick on the name ANSI31 shown to the right of the word Pattern in the Pattern Properties area of the dialog box.

Pick: **(on ANSI31 to the right of Pattern)**

The list of predefined hatch patterns pulls down, as shown in Figure 14.11.

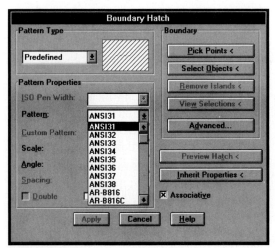

Figure 14.11

You can also select a hatch pattern by picking on the active image tile in the Pattern Type area of the dialog box to scroll through the available patterns. Refer to the discussion of image tiles in Tutorial 7 if you need help using them to select a pattern.

Pick: (on the image tile to the right of Predefined)

Notice that when you pick on the image tile, the hatch pattern changes. The name of the pattern appearing in the image tile is shown to the right of Pattern where the name ANSI31 used to appear. After you have looked at some of the hatch patterns, return the selection to ANSI31 on your own.

In addition to Predefined hatch patterns, you can also use *custom hatch patterns* (that you have previously created and stored in the file *acad.pat* or another *.pat* file of your own making) or *user-defined hatch patterns* (which you can create using the Boundary Hatch dialog box by changing the Pattern Type selection to Custom and specifying the angle and spacing for the hatching using the lower left portion of the dialog box). User-defined hatch patterns can be made more complex by setting the linetype for the layer on which you will create the hatch. Try these features on your own.

Remember to refer to AutoCAD's on-line help for more information about any of the commands described in these tutorials.

Scaling the Hatch Pattern

You can set the scale for the hatch pattern using the Boundary Hatch dialog box. This determines the spacing for the hatched lines in your drawing by scaling the entire hatch pattern. The hatch pattern is similar in some ways to a linetype. Hatch patterns are stored in an external file with the spacing set at some default value. Usually the default scale of 1.0000 is the correct size for drawings that will be plotted full-scale, resulting in hatched lines that are about 1/16" apart. If the views will be plotted at a smaller scale, half-size for instance, you will need to increase the spacing for the hatch by setting the scale to a larger value, so that the lines of the hatch are not as close together. For this drawing, you will not need to change the scale of the hatch pattern.

■ **Warning:** Be cautious when specifying a smaller scale for the hatch pattern, because if the hatch lines are very close together, AutoCAD will take a long time to calculate all of them and may even run out of space on your hard disk, causing your system to crash. ■

Angling the Hatch Pattern

The Angle area of the Boundary Hatch dialog box allows you to specify a rotation angle for the hatch pattern. The default is 0°, or no rotation of the pattern. In the hatch pattern you selected, ANSI31, the lines are already drawn at an angle of 45°. Any angle you choose in the Angle area of the dialog box will be added to the chosen pattern; for example, if you added a rotation angle of 10° to your current pattern, the lines would be drawn at an angle of 55°. Leave the default value of zero for now.

Selecting the Hatch Boundary

The Boundary area of the Boundary Hatch dialog box lets you select among the different methods you can use to specify the area to hatch. Pick Points allows you to pick inside a closed area to be hatched; Select Objects lets you select drawing objects forming a closed boundary for the hatching. Picking the Advanced button opens the Advanced Options dialog box, which lets you select more options for the creation of the hatch boundary.

Pick: **Advanced**

Advanced Options

You can create boundaries for hatching as one of two different types of objects: polylines or regions (as you recall from Tutorial 9, a region is a two-dimensional enclosed area). When determining the hatch boundary, AutoCAD analyzes all of the objects on the screen. This can be time-consuming if you have a large drawing database. To define a different area to be considered for the hatch boundary, you would pick the Make New Boundary Set button. You would then return to your drawing, where you can select the objects to form the new set. Using implied Window works well for this.

You can use the Style area of the Advanced Options dialog box to change the way islands inside the hatch area are treated. Observe the image tile as you explore the options. The selection Normal causes islands inside the hatched area to be alternately hatched and skipped, starting at the outer area. Choosing Outer causes only the outer area to be hatched; any islands inside are left unhatched. You can also choose Ignore; then all of the inside is hatched, regardless of the structure. When you are finished observing the effect of these selections on the image tile, make sure you leave this area of the dialog box set to Normal.

Turning off Island Detection (removing the X from this check box) allows you to select the boundary by *ray casting*. Using the defaults, ray casting finds the nearest closed boundary in any direction. For quicker results, you can change the direction from which the ray casting is started by picking on Nearest and using the list that appears to select a different direction. Leave Island Detection turned on for now.

Pick: **OK *(to exit the Advanced Options dialog box)***

Next you will specify Pick Points as the method for selecting the area to fill with the hatch pattern. You will return to your drawing to select points *inside* of the areas that you want to have hatched. (To use the Select Objects method, you would return to your drawing and pick *on the objects* that form the boundary.) AutoCAD will determine the boundary around the areas you picked. (If there are islands inside the area that you did not want hatched, you would use Remove Islands to pick them after you pick the area to be hatched.) When you are finished selecting, you will press ⏎ to tell AutoCAD you are finished selecting and want to continue with the Boundary Hatch command. You can use any standard AutoCAD selection method to select the objects. Refer to Figure 14.12 as you make selections.

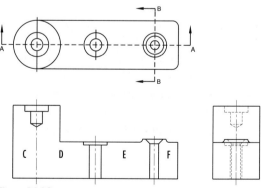

Figure 14.12

Pick: **Pick Points**

You are returned to your drawing screen and the following command prompt appears:

Select internal point: *(pick a point inside the area labeled C)*

The perimeter of the area you have selected is highlighted, and you again see the prompt:

Select internal point: *(pick a point inside the area labeled D)*

Select internal point: *(pick inside the area labeled E)*

Select internal point: *(pick inside the area labeled F)*

Select internal point: ⏎

You return to the Boundary Hatch dialog box. Some of the choices that were previously grayed out (so you couldn't select them) are now available.

■ *TIP* Turn Grid off before previewing, as it can make it difficult to read the hatch pattern. ■

Pick: **Preview Hatch**

You return to your drawing screen with the hatch showing, so that you can confirm that the areas you selected are correct. You will pick the Continue box when you are ready to return to the Boundary Hatch dialog box. If the area was not hatched correctly, make the necessary changes and then preview the hatch again. When what you see is correct, you will pick Apply to apply the hatch. (Your screen may not show the preview perfectly, but if the areas are generally correct, they will probably hatch correctly.) If the hatch appeared correctly,

Pick: **Continue** *(to return to the Boundary Hatch dialog box)*

Pick: **Apply** *(in the lower left of the dialog box)*

The area that you selected becomes hatched in your drawing. Your drawing should now look like Figure 14.13.

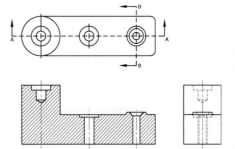

Figure 14.13

Adjusting Your Viewing Area

To draw section B–B, you will need to increase the space available to draw the side view on your screen. Use Zoom to zoom out so that more drawing area is shown in the viewport.

Command: **Z** ⏎

All/Center/Dynamic/Extents/Left/Previous/Vmax/Window/<Scale(X/XP)>: **.75XP** ⏎

Your drawing appears smaller, so more of it fits into the viewport.

Next you will use the Pan command to move the view to the left in the viewport to make more room to the right of the existing views.

Command: **P** ⏎

Displacement: *(pick a point to the left of the front view)*

Second point: *(pick a point inside the border, to the left of the first point you selected)*

Now you have space to create section B–B to the right of the top view.

You will use the Rotate icon from the Modify toolbar to re-orient the side view so that you can align it with the top view, as you see in Figure 14.14. On your own, make sure that Snap is on.

Pick: **Rotate icon**

Select objects: *(select all of the objects in the side view with implied Crossing)*

Select objects: ⏎

Base point: *(pick the top left corner of the side view)*

<Rotation angle>/Reference: **90** ⏎

> ■ *TIP* You can also use the hot grips to rotate. Use implied Crossing to select all of the objects you want to rotate. Once they are selected and you see all of the blue hot grips, pick a base grip that will be the base point in the Rotate command. Once you have selected the base grip, you will see the Stretch command echoed in the command prompt area. Press ⏎ twice to pass Stretch and Move; the next command is Rotate. Follow the prompts that appear on your screen to rotate the objects in the side view 90°. ■

Now you will use the Move command to align the rotated side view with the top view.

Pick: **Move icon**

Select objects: **P** *(for previous selection set)* ⏎

Select objects: ⏎

Base point or displacement: *(pick the upper left corner of the rotated object)*

Second point of displacement: *(pick a point on the snap that lines up with the top line in the top view)*

Your drawing should now look like Figure 14.14.

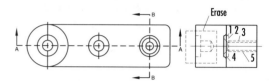

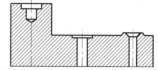

Figure 14.14

On your own, turn Snap off and use the Erase command to remove the unnecessary hidden lines (those for the counterbored hole and lines 1 through 5) from the rotated object. Refer to Figure 14.14 to help you make selections. Then change the hidden lines for the countersunk hole in the rotated side view to layer VISIBLE on your own, as this hole is the only one that will show in the section view.

> ■ *TIP* You may need to use Zoom Window to enlarge the area on the screen. When you are finished erasing and changing the lines, use Zoom Previous to return the area to its original size. ■

Your screen should look similar to Figure 14.15 when you are finished.

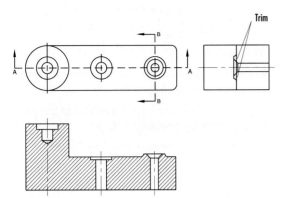

Figure 14.15

Next use the Trim command on your own to remove the visible line that crosses the outer edge of the boss, as shown in Figure 14.15. When you are finished trimming, your drawing should look like Figure 14.16.

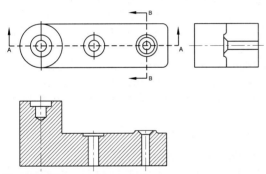

Figure 14.16

Now you are ready to add the hatching.

Pick: **Hatch icon**

In the Boundary Hatch dialog box, the hatch pattern ANSI31 should already be selected and shown at the top as the current pattern.

Pick: **Pick Points**

Select inside the areas you want to have hatched and press ⏎ to return to the Boundary Hatch dialog box when you have finished. To check and make sure the hatching shows correctly,

Pick: **Preview Hatch**

Pick: **Continue *(to return to the dialog box)***

If the hatching appears the way you want it to,

Pick: **Apply**

When you are done, the hatching in your drawing should appear similar to Figure 14.17.

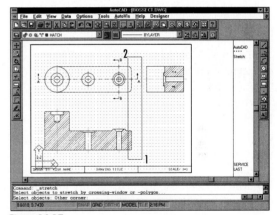

Figure 14.17

Plot your drawing on your own. Before you plot, be sure to switch to paper space. If you need help plotting, refer to Tutorial 3.

Changing Associative Hatches

The hatching created using the Boundary Hatch command is associative: it automatically updates so that it continues to fill the boundary when you edit the boundary by stretching, moving, rotating, arraying, scaling, or mirroring. You can also use grips to modify hatching that has been inserted into the drawing.

Next you will stretch the top and front views of the drawing. You will notice that the hatching automatically updates to fill the new boundary. Refer to Figure 14.17 for the points to select to create a crossing box. Make sure you are in model space and Snap is turned on.

Pick: **Stretch icon**

Select objects to stretch by crossing-window, or -polygon...

Select objects: **(pick point 1)**

Select objects: **(pick point 2)**

Select objects: ⏎

Base point or displacement: **(pick on the snap at the lower right corner of the front view)**

Second point of displacement: **(turn Snap off and pick to the right of the previous location)**

The front and top views of the drawing are stretched to the right. Your screen should appear similar to Figure 14.18.

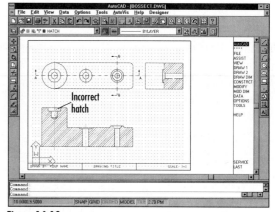

Figure 14.18

 The objects selected by the crossing box are stretched, as shown in Figure 14.18. The hatching updates to fill the new boundary. However, at the left of the drawing where the counterbored hole is located, the hatching does not show correctly. You will use the Edit Hatch icon on the Special Edit flyout on the Modify toolbar to correct this and to make other edits. You need pick on only one of the four hatch pattern sections in the front view, because they were all created in the same Boundary Hatch command sequence. Figure 14.19 shows the Special Edit flyout and the Edit Hatch icon.

Special
Edit flyout

Edit
Hatch

Figure 14.19

Pick: **Edit Hatch icon**

Select hatch object: *(pick anywhere on the hatching in the front view)*

The Hatchedit dialog box shown in Figure 14.20 appears on your screen. It looks identical to the Boundary Hatch dialog box. You can use this dialog box to edit hatching that has already been added to your drawing. You will try using it to fix the hatching that did not update correctly when you stretched your drawing. Don't make any changes to the dialog box; just opening it and clicking on Apply will update the hatch pattern.

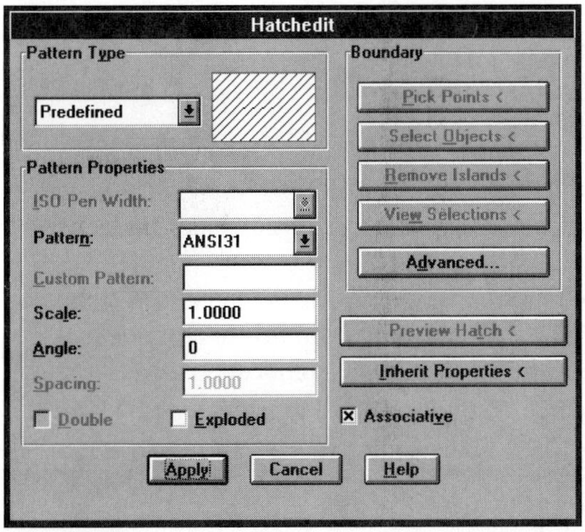

[Hatchedit dialog box]

Hatchedit

Pattern Type

Predefined

Pattern Properties

ISO Pen Width:

Pattern: ANSI31

Custom Pattern:

Scale: 1.0000

Angle: 0

Spacing: 1.0000

☐ Double   ☐ Exploded   ☒ Associative

Boundary

Pick Points <

Select Objects <

Remove Islands <

View Selections <

Advanced...

Preview Hatch <

Inherit Properties <

Apply   Cancel   Help

Figure 14.20

Pick: **Apply**

The hatching should update correctly, as shown in Figure 14.21.

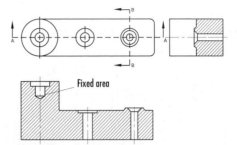

Fixed area

Figure 14.21

Next you will use the Hatchedit dialog box to change the angle of the hatching. Because the angled edge of the countersunk hole is drawn at 45°, you will change the angle of the hatch pattern so that the hatch is not parallel to features in the drawing. (Having the hatch run parallel or perpendicular to a major drawing feature is not good engineering drawing practice.) You will restart the Edit Hatch command by pressing ⏎ at the command prompt.

Command: ⏎

Select hatch object: *(pick the hatching in the front view)* ⏎

The Hatchedit dialog box returns to your screen. This time use the Angle area of the dialog box and type in *15* for the hatch angle on your own. When you have finished this,

Pick: **Apply**

The hatching updates so that it is angled an additional 15°. On your own, repeat this process for the hatching in the side view. All of the hatching on a single part should be at the same angle. Different angles are used on parts in assembly to help differentiate between them. The result of changing the hatch angle is shown in Figure 14.22.

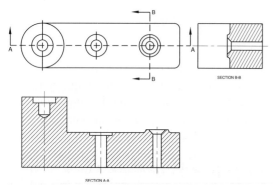

Figure 14.22

On your own, set layer TEXT current and use Dtext to label the section views SECTION A–A and SECTION B–B. Place the text directly below the section views, as shown in Figure 14.22. Save your *bossect.dwg* drawing.

■ *TIP* You can create open, irregular hatch boundaries by typing the Hatch command at the command prompt. Irregular hatch boundaries are useful when hatching large areas where you do not want to fill the entire area—for example, representing earth around a foundation on an architectural drawing—or in mechanical drawings to hatch only the perimeter of a large shape, instead of filling the entire area. Typing *HATCH* ⏎ at the command prompt will start the Hatch command, rather than the Boundary Hatch command. Select the default pattern and specify the default scale and angle for the pattern. Then at the *Select objects* prompt, press ⏎. This allows you to specify points to form a boundary for the hatch. You will see the question *Retain polyline? <N>*. Accept the default to indicate not to retain a polyline boundary formed by the points you will pick. You are then prompted to pick points. Type *C* ⏎ to use the close option to return to the first point when you are done selecting and end the command. The area you selected will be filled with the pattern. ■

You have completed the 2D portion of this tutorial. Next you will learn how to automatically create section views from a solid model, using the Section command.

Creating Sections from a Solid Model

This time you will start your drawing from a solid model prototype provided with the data files. It is similar to the object that you created in Tutorial 9.

Pick: **New**

In the Create New Drawing dialog box, use *solblk2.dwg* as the prototype for your new drawing. For the new file name, type *blksect.dwg* in the empty box at the bottom of the dialog box. Pick OK when you are done.

A drawing similar to the one you created in Tutorial 9 appears on your screen, as shown in Figure 14.23.

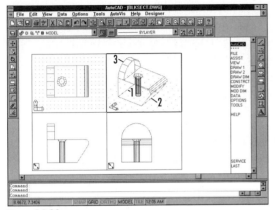

Figure 14.23

On your own, create layer HATCH. Assign it color red and linetype CONTINUOUS. Make it the current layer. AutoCAD places the region it generates during the Section command on the current layer. Turn on the Midpoint running object snap on your own, so that it is available for the next command.

You will use the Midpoint running object snap to create a UCS through the center of the part. You will be typing the UCS command at the command prompt. Refer to Figure 14.23 to make your selections.

Pick: *(the upper right viewport to make it active)*

Command: **UCS** ⏎

Origin/ZAxis/3point/OBject/View/X/Y/Z/Prev/Restore/Save/Del/?/<World>: **3** ⏎

Origin point <0,0,0>: *(pick near point 1)*

Point on positive portion of the X axis<3.0000,4.0000,0.0000>: *(pick near point 2)*

Point on the positive-Y portion of the UCS Xyplane<2.0000,5.0000,0.0000>: *(pick near point 3)*

You will type *UCSICON* to move the UCS icon to the origin of the newly created User Coordinate System. This helps you see whether the correct coordinate system has been selected.

Command: **UCSICON** ⏎

ON/OFF/All/Noorigin/ORigin <ON>: **OR** ⏎

The UCS icon should appear at the center of the part in the upper right view. Your screen should look like Figure 14.24. If it does not, repeat the preceding steps, being careful to select the midpoints indicated in Figure 14.23.

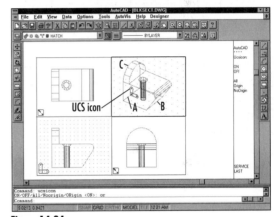

Figure 14.24

Use the Tools pull-down menu and select Toolbars on your own and choose to show the Solids toolbar. Then make sure that the upper right viewport is still active.

> ■ *TIP* If you want to select in a viewport that is not active, press Ctrl-R to toggle the active viewport during commands. ■

Using Section

 The Section command generates a region (a closed 2D area) from a solid and a cutting plane that you specify. The region is inserted on the current layer and at the location of the cross section you specified by selecting the cutting plane. You can pick the Section icon from the Solids toolbar.

Pick: **Section icon**

Select objects: *(pick the object displayed in the upper right viewport)*

Select objects: ⏎

Sectioning plane by Object/Zaxis/View/XY/YZ/ZX/<3points>: ⏎

You can specify the cutting plane by choosing

Object	aligns the cutting plane with a circle, ellipse, arc, 2D spline, or 2D polyline that you select.
Zaxis	prompts you for two points that determine the edge view of an XY plane. The cutting plane is the Z axis plane through the first point you select and is normal (perpendicular) to the line defined by the two points selected.
View	aligns the cutting plane parallel to the current viewing plane through the point you specify.
XY, YZ, or ZX	aligns the cutting plane parallel to the specified UCS plane through the point you specify.
3 points	prompts you for three points, through which the desired cutting plane passes.

You will specify three points to define the cutting plane, which will pass through the middle of the object. The Midpoint running object snap, which you turned on earlier, will select the midpoint of the lines you pick. Refer to Figure 14.24 for selections.

1st point on plane: *(pick line A)*

2nd point on plane: *(pick line B)*

3rd point on plane: *(pick line C)*

The Section command creates a region or a block in your drawing. You should see it appear in your drawing, as shown in Figure 14.25.

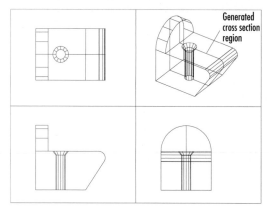

Generated cross section region

Figure 14.25

Next, use the Boundary Hatch command to hatch the section boundary that you created.

Pick: **Hatch icon**

The Boundary Hatch dialog box appears on your screen. On your own, choose the button for Select Objects to define the boundary. You will return to your drawing screen. Pick on the boundary that you created with the Section command and press ⏎ to return to the dialog box. Preview the hatching to make sure that the correct area will be hatched. When you are finished,

Pick: **Apply (to apply the hatch and exit the Boundary Hatch dialog box)**

The hatching appears in your drawing, as shown in Figure 14.26.

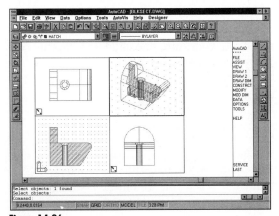

Figure 14.26

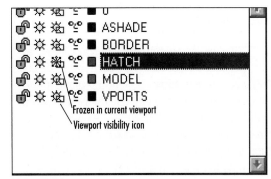

Frozen in current viewport

Viewport visibility icon

Figure 14.27

The hatching should only be drawn in the section view (in this case, the front view) according to good engineering drawing practice, but at this point it is still visible in every view. (It is on edge in the top and side views, so it just appears as a single line through the center of the part.) You will control the visibility for layer HATCH so that the hatching only shows in the front view.

Controlling Layer Visibility

You will use Layer Control from the Object Properties toolbar to freeze the hatching in all of the viewports except the lower left one. First you will freeze layer HATCH within individual viewports.

On your own, make the top left viewport current.

> *Pick: (on layer name HATCH on the Object Properties toolbar)*

The list of layers pulls down from the current layer name, as shown in Figure 14.27.

Pick on the viewport visibility icon to the left of layer name HATCH. It should change in appearance from a sun with a little viewport to its lower right to a snowflake with a little viewport. Pick anywhere in the graphics window for the selection to take effect. The line representing the edge view of the hatching should no longer be visible in the upper left viewport. Repeat this process on your own to freeze the hatching in the lower right viewport containing the side view. When you have finished, your screen should look like Figure 14.28.

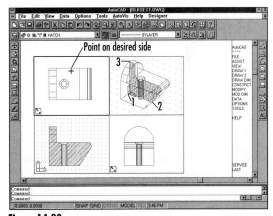

Figure 14.28

Using Slice

 The Slice command lets you remove a portion of a solid by specifying a slicing plane similar to the cutting plane in the Section command. You can use the Section command in conjunction with the Slice command to create pictorial section views from solid models. Next, you will slice the model and cut away the front half. Refer to Figure 14.28 for the points to select. (They will be the same points you selected last time to define the section.) Make sure your isometric view in the upper right viewport is active on your own before you continue. You can pick the Slice icon from the Solids toolbar.

Pick: **Slice icon**

Select objects: *(pick the object displayed in the upper right viewport)*

Select objects: ⏎

Slicing plane by Object/Zaxis/View/XY/YZ/ZX/ <3points>: ⏎

The methods by which you select the cutting plane for a section are very similar to the methods for using the Slice command. You will specify three points to define the cutting plane, which will pass through the middle of the object. The Midpoint running object snap will select the midpoint of the lines you pick. Refer to Figure 14.28 for the lines to select.

1st point on plane: *(pick line 1)*

2nd point on plane: *(pick line 2)*

3rd point on plane: *(pick line 3)*

Both sides/<Point on desired side of the plane>: *(pick once in the top view to activate it and then near the point shown)*

Using Hide

As you saw in Tutorial 9, the Hide command eliminates lines from the back surfaces of the object from the screen so that the object

appears correctly in the viewport. Next you will hide the back surface lines with the Hide command. On your own, make the upper right viewport active. Since you do not have the Render toolbar open, you will type *HIDE* at the prompt.

Command: **HIDE** ⏎

On your own, make the lower left viewport containing the front view active and use the Hide command to hide the back surfaces in that view. When you are finished, your screen should look like Figure 14.29.

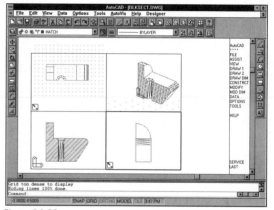

Figure 14.29

On your own, plot the upper right viewport from model space, if you want. Select the Hide Lines box in the Plot Configuration dialog box. To plot from paper space with hidden lines removed, use the Mview command and select the Hideplot option, and pick the viewports in which you want back surface lines removed when plotting.

Turn off the Midpoint running object snap on your own and close the Solids toolbar.

Save your drawing and exit AutoCAD.

You have completed Tutorial 14. Now you know how to create both 2D and 3D section views.

KEY TERMS

associative hatching	cutting plane line	runout
crosshatching	pictorial full section	section view
custom hatch pattern	ray casting	user-defined hatch pattern

KEY COMMANDS

Boundary Hatch	Hatch	Section
Edit Hatch	Savetime	Slice

Redraw the given front view and replace the given right side view with a section view. Use 2D methods or solid modeling. Assume that the vertical centerline of the front view is the cutting plane line for the section view. The letter M after an exercise means that the given dimensions are in millimeters.

14.1 High Pressure Seal

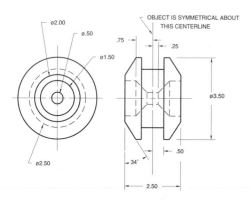

14.2M Pulley

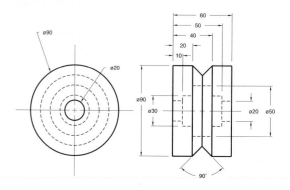

14.3 Valve Cover

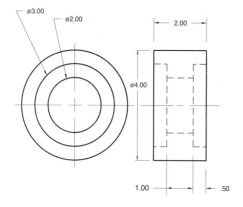

14.4 Double Cylinder

Use 2D techniques to draw this object; the grid shown is 0.25 inches. Show the front view as a half section and show the corresponding cutting plane in the top view. On the same views, completely dimension the object to the nearest 0.01 inches. Include any notes that are necessary.

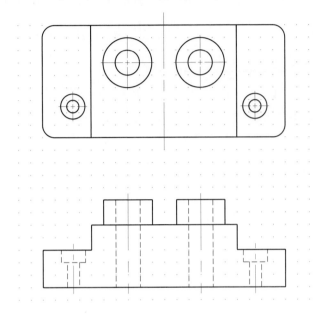

14.5 Plate

Use 2D to construct the top view with the cutting plane shown and the sectioned front view.

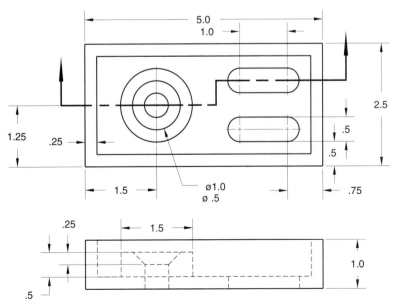

14.6 Wedge

Make a solid model based on the given views, then produce a plan view and a full-section front view. Make sure the sectioned front view is aligned with the plan view. Hidden lines and/or cutting planes are not needed in the plan view.

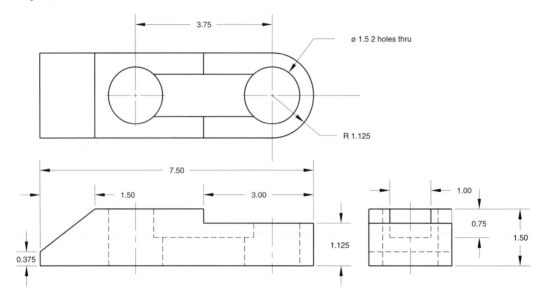

14.7 Support Bracket

Create a solid model of the object shown, then produce a plan view and full section front view. Note the alternate units in brackets; these are the dimensions in millimeters for the part.

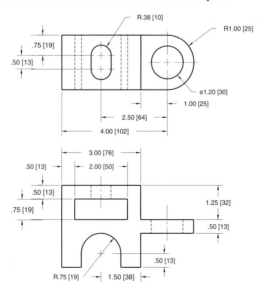

14.8 Turned Down Slab Detail

Use 2D methods to create the drawings shown. Use the Boundary Hatch command to fill the area with the pattern for concrete. Type the Hatch command and create the irregular areas to fill with patterns for sand and earth.

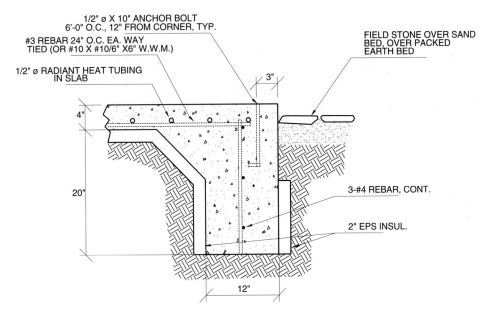

14.9 Shaft Support

Redraw the given front view and create a section view. Assume that the vertical centerline of the shaft in the front view is the cutting plane line for the section view.

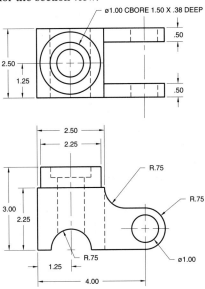

14.10M Pulley Arm

Use solid modeling to create a pictorial view like the one shown.

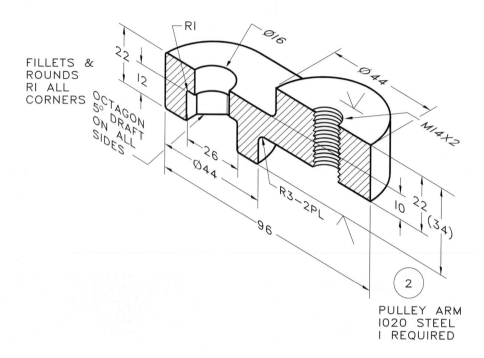

FILLETS &
ROUNDS
RI ALL
CORNERS

OCTAGON
5° DRAFT
ON ALL
SIDES

R1

Ø16

Ø44

M14X2

22

12

26

Ø44

R3—2PL

96

10

22

(34)

2

PULLEY ARM
1020 STEEL
I REQUIRED

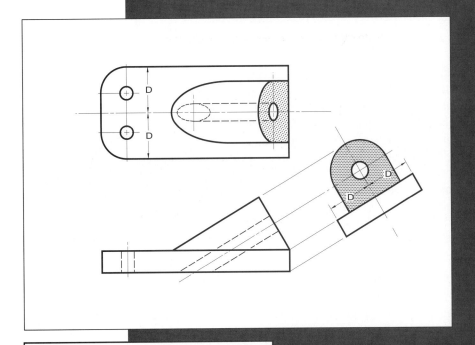

Creating Auxiliary Views with 2D and Solid Models

Objectives

When you have completed this tutorial, you will be able to

1. Draw auxiliary views, using 2D projection.

2. Set up a UCS to help create a 2D auxiliary view.

3. Rotate the snap to help create a 2D auxiliary view.

4. Draw auxiliary views of curved surfaces.

5. Create an auxiliary view of a 3D solid model.

Introduction

An *auxiliary view* is an orthographic view of the object that has a different line of sight from the six *basic views* (front, top, right-side, rear, bottom, and left-side). Auxiliary views are most commonly used to show the true size of a slanted or oblique surface. Slanted and oblique surfaces are *foreshortened* in the basic views because they are tipped away from the viewing plane, causing their projected size in the view to be smaller than their actual size. An auxiliary view is drawn with its line of sight perpendicular to the surface showing the true size and shape of a slanted or oblique surface.

In this tutorial, you will learn to create auxiliary views using 2D methods and 3D solid modeling. You can project 2D auxiliary views from the basic orthographic views by rotating the snap or by aligning a new User Coordinate System. You can generate auxiliary views directly from a 3D solid model by changing the viewpoint to show the true size of the surface.

Auxiliary Views

Figure 15.1 shows three views of an object and an auxiliary view.

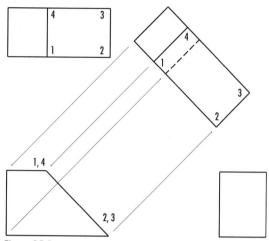

Figure 15.1

The surface defined by vertices 1–2–3–4 is inclined. It is perpendicular to the front viewing plane and shows on edge in the front view. Both the top and side views show foreshortened views of the surface; that is, neither shows the true shape and size of that surface of the object.

To project an auxiliary view showing the true size of surface 1–2–3–4, you create a new view by drawing projection lines perpendicular to the edge view of the surface, in this case the angled line in the front view. You transfer the width of the surface, measured from a *reference surface*, to the auxiliary view to complete the projection. Note that as with the front, top, and side views, the auxiliary view shows the entire object as viewed with the line of sight perpendicular to that surface, causing the base of the object to show as a hidden line.

Starting

Start AutoCAD and, on your own, start a new drawing from the prototype drawing, *adapt4.dwg*, provided with the data files. Name your drawing *auxil1.dwg*.

AutoCAD's drawing editor shows the drawing of the orthographic views for the adapter on your screen, as shown in Figure 15.2.

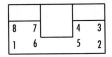

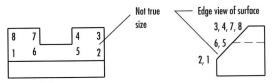

Figure 15.2

You will create an auxiliary view showing the true size of the inclined surface. The object is shown in Figure 15.2 with the surface 1–2–3–4–5–6–7–8 labeled.

Drawing Auxiliary Views Using 2D

You can create the auxiliary view using information from any two of the views shown, since any two views with a 90° relationship provide all of the principal dimensions. For this tutorial, you will erase the top view and use the front and side views to project the auxiliary view.

Pick: **Erase icon**

Select objects: *(use implied Window to select the top view)*

Select objects: ⏎

On your own, use Redraw as necessary. Your screen will look like Figure 15.3 once you have erased the top view.

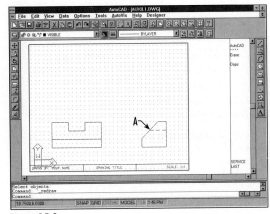

Figure 15.3

Next you will draw projection lines perpendicular to surface 1–2–3–4–5–6–7–8 in the side view. You will create a new UCS aligned with the edge view to help you project these lines easily.

Aligning the UCS with the Angled Surface

 You will set the UCS so that it is aligned with the angled surface. You will use the Object option to align the new coordinate system. The Object option prompts for the existing drawing object with which you want to align the UCS. You will align the UCS with the angled line in the side view. You will pick the Object UCS icon from the UCS flyout on the Standard toolbar.

Pick: **Object UCS icon**

Select object to align UCS: *(pick the angled line in the side view, identified as A in Figure 15.3)*

When you have finished aligning the UCS, your screen should look like Figure 15.4. Notice that the crosshairs and grid, as well as the UCS icon, now line up with the angled surface.

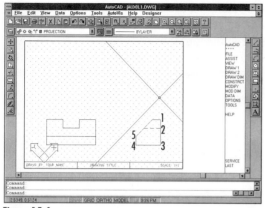

Figure 15.4

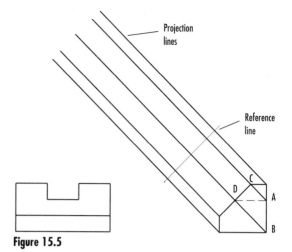

Figure 15.5

On your own, set layer PROJECTION as current and turn on the Intersection running object snap. Make sure that Ortho is on and Snap is off.

Now you will draw rays from each intersection of the object in the side view into the open area of the drawing above the front view. You will use the Ray icon on the Line flyout of the Draw toolbar; if you need help creating rays, refer to Tutorial 4.

Pick: **Ray icon**

From point: *(target intersection 1 in Figure 15.4)*

Through point: *(with Ortho on, pick a point above and to the left of the side view)*

Through point: ⏎

On your own, restart the Ray command and draw projection lines from points 2 through 5 in the left-side view. Then use the Line command to draw a *reference line* perpendicular to the projection lines. The location of this line can be anywhere along the projection lines, but it should extend about .5 units beyond the top and bottom projection lines.

Your drawing should look similar to Figure 15.5.

Note that the back edge of the slot (labeled A) and the lower right corner (labeled B) in the side view align with the projection lines through the corners of the angled surface (labeled C and D) to produce a single projection line. If points C and D did not line up on the projection lines you created, you would need to project them.

The width of the object is known to be 3. Next you will use the Offset command to draw a line 3 units away from and parallel to the reference line.

> ■ *TIP* If you did not know the dimensions, you could use the Distance command (available from the Inquiry flyout on the Object Properties toolbar) with the Intersection object snap in the front view to determine the distance. ■

Pick: **Offset icon**

Offset distance or Through <Through>: **3** ⏎

Select object to offset: *(pick the reference line, as shown in Figure 15.5)*

Side to offset? *(pick a point anywhere to the left of the reference line)*

Select object to offset: ⏎

Your drawing should look similar to Figure 15.6.

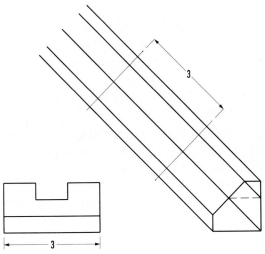

Figure 15.6

Add the width of slot on your own by using Offset to create two more parallel lines 1 unit apart. Your drawing should look similar to Figure 15.7.

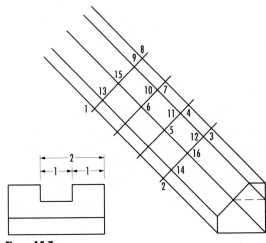

Figure 15.7

On your own, set layer VISIBLE as the current layer and make sure that the Intersection running object snap is on before you continue.

Next, you will draw the visible lines in the drawing over the top of the projection lines. Be sure you draw the correct shape of the object as it will appear in the auxiliary view. Refer to Figure 15.7 for your selections.

> *Pick:* **Line icon**
>
> From point: *(pick the intersection of the projection lines at point 1)*
>
> To point: *(pick point 2)*
>
> To point: *(pick point 3)*
>
> To point: *(pick point 4)*
>
> To point: *(pick point 5)*
>
> To point: *(pick point 6)*
>
> To point: *(pick point 7)*
>
> To point: *(pick point 8)*
>
> To point: **C** ⏎

Now you will restart the Line command and draw the short line segments where the two short top surfaces of the object intersect the slanted front surface:

> Command: ⏎
>
> From point: *(pick point 9)*
>
> To point: *(pick point 10)*
>
> To point: ⏎
>
> Command: ⏎
>
> From point: *(pick point 11)*
>
> To point: *(pick point 12)*
>
> To point: ⏎

Next, you will draw the line representing the back edge of the slot in the auxiliary view.

> Command: ⏎
>
> From point: *(pick point 10)*
>
> To point: *(pick point 11)*
>
> To point: ⏎

Now you will add the line showing the intersection of the vertical front surface with the angled surface.

Command: ⏎

From point: *(pick point 13)*

To point: *(pick point 14)*

To point: ⏎

On your own, set layer HIDDEN_LINES as the current layer.

Next you will draw the hidden line showing the intersection of the bottom surface and the back surface. It will be hidden from the line of sight.

Command: **LINE** ⏎

From point: *(pick point 15)*

To point: *(pick point 16)*

To point: ⏎

Freeze layer PROJECTION on your own. Use the Trim command to remove the portion of the hidden line that coincides with the visible line from point 5 to point 6 before you continue.

When you are finished, your screen should look like Figure 15.8.

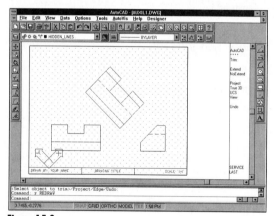

Figure 15.8

Adding Visual Reference Lines

Next, you will add two visual reference lines to your drawing, from the extreme outside edges of the surface shown true size in the auxiliary view to the inclined surface in the side view.

The lines will provide a visual reference for the angles in your drawing and help show how the auxiliary view aligns with the standard views. You can extend a centerline from the primary view to the auxiliary view, or use one or two projection lines for the reference lines. In this case, since there are no centerlines, you will create two reference lines. Leave a gap of about 1/16" between the reference lines and the views.

On your own, thaw layer PROJECTION. Then pick the Properties icon from the Object Properties toolbar and select the two rays that project from the ends of the angled surface to use as visual reference lines, and change them to layer THIN. Refer to Figure 15.9 if you are unsure which lines to select.

The rays should change to layer THIN. You can verify this visually because the lines will turn red, the color for layer THIN.

Linetype Conventions

Two lines of different linetypes should not meet to make one line. You must always leave a small gap on the linetype that has lesser precedence. In the case of the visible and the reference line, the reference line should have a small gap (about 1/16" on the plotted drawing) where it meets the visible line. In the case of the visible and the hidden line, the hidden line should have a small gap (about 1/16" on the plotted drawing) where it meets the visible line. You can use the Break command to do this.

On your own, freeze layer PROJECTION again. Then use the Break command to create a gap of about 1/16" between each view and the reference lines. Create a similar gap where the hidden line extends from the visible line, forming the front of the slot in the auxiliary view. Erase the extra portions of the rays where they extend past the auxiliary views.

Your drawing should look like Figure 15.9.

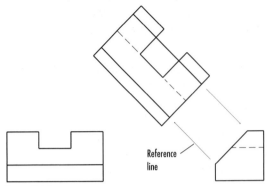

Figure 15.9

On your own, save your drawing.

Curved Surfaces

Now you will create an auxiliary view of the slanted surface in a cylinder drawing like the one you created at the end of Tutorial 6. The drawing has been provided for you as *cyl3.dwg* on the data files that came with your software.

On your own, begin a new drawing. Use *cyl3.dwg* as a prototype. Name your new drawing *auxil3.dwg*.

You return to AutoCAD's drawing editor. Figure 15.10 shows the drawing of the orthographic views of the cylinder with the slanted surface that should be on your screen.

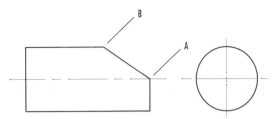

Figure 15.10

The drawing shows the front and side view of the slanted surface, but not the true shape of the surface. You will draw a *partial auxiliary view* to show the true shape of the surface. Partial auxiliary views are often used because in the auxiliary view, the inclined or oblique surface chosen is shown true size and shape, but the other surfaces are all foreshortened. Since these other surfaces have already been defined in the basic orthographic views, there is really no need to show them in the auxiliary view. A partial auxiliary view shows only the inclined surfaces (leaving out the normal surfaces), thus saving time in projecting the view and giving a clearer appearance to your drawing.

Figure 15.11 shows a front view of an object, plus two partial auxiliary views.

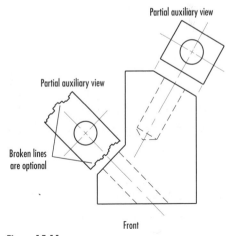

Figure 15.11

Rotating the Snap

This time you will use the Rotate option of the Snap command to rotate the grid and snap so that they align with the inclined surface, to make it easy to project the curved surface to the auxiliary view you will create.

On your own, make sure that the Intersection running object snap is on. Make sure that the grid is on and Ortho is off, and that layer PROJECTION is the current layer.

Now, you will rotate the snap angle with the Snap command. You will pick two points (identified in Figure 15.10) to define the angle.

> Command: **SNAP** ⏎
>
> Snap spacing or ON/OFF/Aspect/Rotate/Style <0.2500>: **R** ⏎
>
> Base point <0.0000,0.0000>: **(target the intersection labeled A)**
>
> Rotation angle<0>: **(target the intersection labeled B)**

You will see the message *Angle adjusted to 326.* Your screen should look similar to Figure 15.12. Notice that the crosshairs and grid have aligned with the inclined surface.

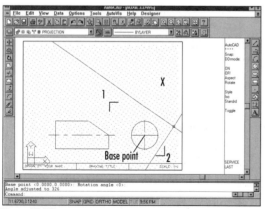

Figure 15.12

Next, you will copy the object in the side view and then rotate it so that it aligns with the rotated grid and snap. This makes it easy to

project the depth of the object from the side view into the auxiliary view, so that you do not have to use a reference surface to make depth measurements, as you did in the previous drawing. The shapes in this drawing are simple, but these methods will also work for much more complex shapes.

For the next steps you will use Ortho and Snap. Make sure that they are turned on. Refer to Figure 15.12 for the points to select.

> *Pick:* **Copy Object icon**
>
> Select objects: **(pick point 1)**
>
> Other corner: **(pick point 2 to form a window around the view)**
>
> Select objects: ⏎
>
> <Base point or displacement>/Multiple: **(target the intersection labeled Base Point)**
>
> Second point of displacement: **(pick a point on the snap above and to the right of the side view)**

A copy of the side view should appear on your screen, so that it looks like Figure 15.13.

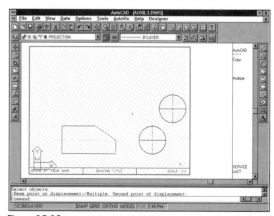

Figure 15.13

Now you will change the copy onto layer PROJECTION, because it is part of the construction of the drawing, not something you will leave in once you are finished. You will freeze layer PROJECTION when you have finished the drawing.

Pick: **Properties icon**

Select objects: *(window the new copy you created)*

Select objects: ⏎

Use the dialog box to change the layer to PROJECTION on your own.

The objects you selected are changed onto layer PROJECTION. You will notice that their color is now magenta, which is the color of layer PROJECTION.

Now you are ready to rotate the copied objects so that they align with the snap and can be used to project the auxiliary view. You will use the Rotate icon from the Modify toolbar and type *P* for the previous selection set to accomplish this.

Pick: **Rotate icon**

Select objects: **P** ⏎

> ■ **TIP** Typing *P* gives you the previous selection set, which you just defined when changing properties. ■

Select objects: ⏎

Base point: *(pick the intersection of the centerlines of the copied objects)*

<Rotation angle>/Reference: *(with Ortho turned on, select a point above and to the right of the base point you picked, as shown in Figure 15.14)*

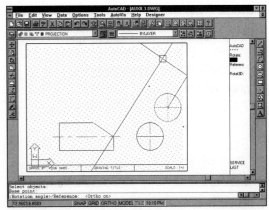

Figure 15.14

The objects you copied previously should now be rotated into position so that they align with the snap and grid. Your drawing should look like Figure 15.15.

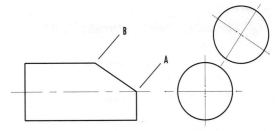

Figure 15.15

Next, you will draw projection lines from the front view out into the empty area where the auxiliary view will be located. Start the Ray command by picking its icon from the Line flyout on the Draw toolbar. (Ortho and the Intersection running object snap should be on.) Refer to Figures 15.15 for the points to select and 15.16 for an example of the lines.

Pick: **Ray icon**

From point: *(target the intersection labeled A in Figure 15.15)*

Through point: *(pick a point above the front view, as in Figure 15.16)*

Through point: ⏎

Command: ⏎ **(to restart the Ray command)**

From point: **(target the intersection labeled B in Figure 15.15)**

Through point: **(pick a point above the front view)**

Through point: ⏎

Your drawing should look like Figure 15.16.

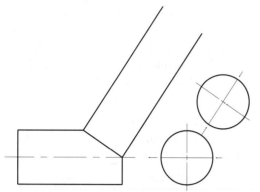

Figure 15.16

On your own, project rays from the intersections in the copied side view, so that your drawing looks like Figure 15.17. Then set layer VISIBLE as the current layer.

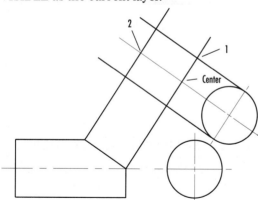

Figure 15.17

Because the object appears circular in the side view, and is tilted away in the front view, the true shape of the surface must be an ellipse. You will select the

Ellipse Center icon from the Draw toolbar to create the elliptical surface in the auxiliary view. Refer to Figure 15.17 for your selections.

Pick: **Ellipse Center icon**

Center of ellipse: **(target the intersection labeled Center)**

Axis endpoint: **(target point 1)**

<Other axis distance>/Rotation: **(target point 2)**

> ■ **TIP** If the surface in the side view were an irregular curve, you could project a number of points out into the auxiliary view and then connect them with a polyline. Once they were connected, you could connect them in a smooth curve, using the Fit Curve or Spline option of the Edit Polyline command. This would also work to draw an ellipse; however, the Ellipse command is quicker. ■

The ellipse in your drawing should appear as shown in Figure 15.18. You will need to trim off the lower portion of it to create the final surface.

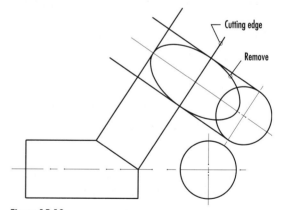

Figure 15.18

You will use the Trim command to eliminate the unnecessary portion of the ellipse in the auxiliary view:

Pick: **Trim icon**

Select cutting edges: (Projmode = UCS, Edgemode = No extend)

Select objects: *(pick the cutting edge in Figure 15.18)*
Select objects: ⏎

<Select object to trim>/Project/Edge/Undo: *(pick on the ellipse to the right of the cutting edge)*

Select objects: ⏎

The lower portion of the ellipse is now trimmed off.

Change the two projection lines from the front view onto layer THIN on your own. Use the Break command to break them so that they do not touch the views. Draw in the final line across the bottom of the remaining portion of the ellipse on layer VISIBLE. Freeze layer PROJECTION to remove the construction and projection lines from your display.

When you are finished, your screen should look like Figure 15.19. Save your drawing.

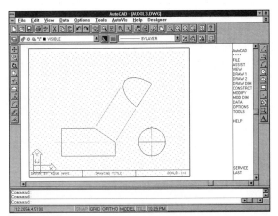

Figure 15.19

Creating an Auxiliary View of a 3D Solid Model

You can easily create auxiliary views of 3D solid models by changing the viewpoint in any viewport to show the desired view. You can view the model from any direction inside any viewport. However, good engineering drawing practice requires that adjacent views align; otherwise it

can be difficult or impossible to interpret the drawing. In general, you should show at least two principal orthographic views of the object. Additional views are sometimes placed in other locations on the drawing sheet, or on a separate sheet. If you do this, you should clearly label the view that is located elsewhere. When placing auxiliary views in other locations, a common practice is to show a viewing plane line, similar to a cutting plane line, that indicates the direction of sight for the auxiliary view. The auxiliary view should be clearly labeled and should preserve its correct orientation, rather than be rotated to a different alignment.

On your own, create a new drawing called *aux3d.dwg*. Use the drawing *anglblok.dwg* that is provided with the data files as the prototype.

The solid model shown in Figure 15.20 appears on your screen. You will continue working with this drawing, creating an auxiliary view showing the true size of the inclined surface.

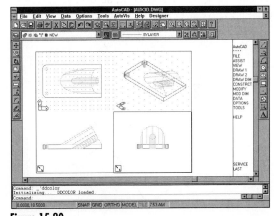

Figure 15.20

You will use hot grips to stretch the viewports to different sizes in order to make room for the auxiliary view. To do this, you will need to thaw the layer containing the viewports. It is turned off in the data file. On your own, thaw layer VPORT, but do not make it current. Leave layer

NEW set as the current layer for now. When layer VPORT is thawed, you should see the outlines of each viewport in your drawing.

Now you are ready to stretch the viewports to different sizes to make room for a new viewport that will show the auxiliary view.

On your own, switch to paper space.

Next activate the grips for each of the viewports. (Remember, viewports are paper space objects and you can modify them using Scale, Rotate, Stretch, and Copy, among other commands, but only when you are in paper space.)

■ *TIP* To activate the grip for each of the four viewports, make an implied Crossing box where the viewports come together at the center of the screen. ■

Pick: **(to activate the grips for the viewport borders)**

The grips become activated, as shown in Figure 15.21.

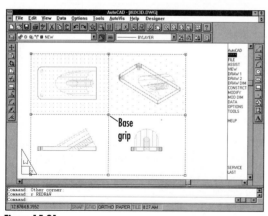

Figure 15.21

Next, pick the grip in the center of the four viewports as the base grip. You will stretch the viewports to the right, making the viewports

containing the pictorial view and side view smaller. On your own, pick the base grip, as shown in Figure 15.21, and stretch the viewports to the right so that the final result looks like Figure 15.22.

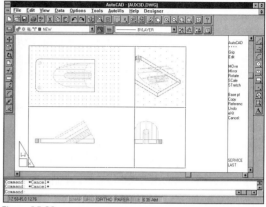

Figure 15.22

Now switch to model space and do the following steps on your own. Make the top right view active and zoom the pictorial view to .75X (so the view is shown 3/4ths its present size). Then make the lower right viewport active and use the Pan command with Ortho turned on to pan to the right in the side view to make room for a new viewport in the center of the drawing. (Make sure that the appropriate viewport is active before you select the Pan command.) Now you are ready to create the new viewport, in which you will show the auxiliary view.

On your own, switch to paper space. Set layer VPORT as current. The paper space icon is displayed in the lower left corner of your screen. The upper right viewport containing the pictorial view should be the active viewport. You will now create your new viewport, using the View menu.

Pick: **View, Floating Viewports, 1 Viewport**
ON/OFF/Hideplot/Fit/2/3/4/Restore/<First point>:
    (pick point A in Figure 15.23)
Other corner: *(pick point B)*

The new viewport appears, as shown in Figure 15.23. It doesn't matter that the viewport borders overlap because you will turn their layer off before plotting your drawing.

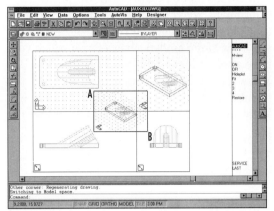

Figure 15.23

Enlarging a Viewport

You will use Zoom Window to enlarge the upper right viewport so that it fills the entire screen. Then you can switch back to model space to work in the viewport. This is especially useful if you do not have a large monitor. Sometimes several small viewports on the screen at once show the object too small for you to work effectively. By switching to paper space and zooming in on one viewport, you can enlarge just that viewport on the screen. (Zooming in model space will only enlarge the view within the viewport, while the size of the viewport on the screen remains the same.)

On your own, switch back to paper space.

Pick: **Zoom Window icon**

First corner: *(pick a point above and to the left of the upper right viewport)*

Other corner: *(pick a point below and to the right of the upper right viewport)*

The upper right viewport fills the screen. You are still in paper space. To work on the model, you must change to model space. Switch to model space on your own. The UCS icon shows inside the enlarged viewport.

Now you are ready to change the viewpoint in the new viewport so that it shows the auxiliary view. You will do this by aligning the UCS with the angled surface and then using the Plan command to show the view looking straight towards the UCS.

Using 3 Point UCS

 The 3 Point option of the UCS command creates a User Coordinate System aligned with the three points you specify. You are prompted to pick the origin, a point in the positive X direction, and a point in the positive Y direction. Refer to Figure 15.24 for the points to select.

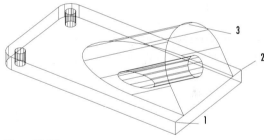

Figure 15.24

■ *TIP* Using the Endpoint object snap helps you pick the exact points you want on the solid model. The Intersection object snap can be slower or sometimes does not find locations on solid models at all. ■

Pick: **3 Point UCS icon**

Origin point <0,0,0>: *(pick Snap to Endpoint icon)*

of *(pick point 1)*

Point on positive portion of the X-axis
  <14.000,4.5000,0.5000>: *(pick Snap to Endpoint icon)*

of *(pick point 2)*

Point on positive-Y portion of the UCS XY plane
  <12.0000,4.5000,0.5000>: *(pick Snap to Endpoint icon)*

of *(pick on the upper tessellation line labeled as point 3)*

Switch back to paper space and use Zoom Previous on your own to restore the original zoom factor, showing the five viewports. When you have restored the previous zoom factor, switch back to model space. Your screen should look like Figure 15.25. Notice that the UCS icon now lines up with the angled surface. It is shown as a broken pencil in the lower left viewport. This is important to notice because it tells you that the User Coordinate System is correctly aligned with the angled surface. The broken pencil shows when the coordinate system is perpendicular to the viewing direction. Since the UCS is aligned with the angled surface, it is perpendicular to the front view (although along an angled line).

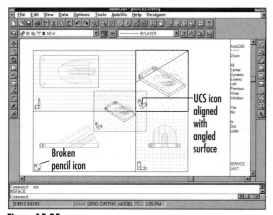

Figure 15.25

Using Plan

The Plan command aligns the viewpoint in the active viewport so that you are viewing the current UCS, a named UCS, or the World Coordinate System straight on. You will accept the default of the current UCS to show the true size view of the angled surface, with which you have already aligned the coordinate system. You will type the Plan command at the command prompt, but it is also available from the View menu, 3D Viewport Presets, Plan View.

On your own, make the center viewport active. The crosshairs appear in the center viewport, indicating that it is the active viewport.

> ■ *TIP* If you have trouble picking inside a viewport because it overlaps other viewports, type ⌃-*R* to toggle the viewport. ■

Command: **PLAN** ⏎

  <Current UCS>/Ucs/World: ⏎

The view in the center viewport changes so that you are looking straight on to the angled surface. On your own, use the Zoom command with a .5XP scaling factor to size the view. (The original views were zoomed to .5XP in the data file.) Next, use the Pan command to center the view inside the viewport. When you have finished these steps, your screen should look like Figure 15.26.

> ■ *TIP* If your viewport is not large enough to show the entire view zoomed to .5XP, switch back to paper space and use hot grips to stretch the viewport so that it is slightly larger than before. Switch back to model space when you are finished and use Pan once again to position the view. ■

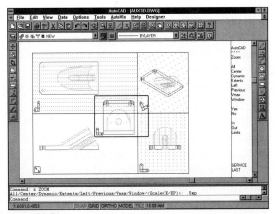

Figure 15.26

■ **TIP** You can set the Ucsfollow variable to automatically generate the plan view of the current UCS. Pick in a viewport to make it active, then set Ucsfollow to 1. AutoCAD automatically generates a plan view of the current UCS in that viewport whenever you change the UCS. ■

Now the view shows the true size of the angled surface, but it is not aligned correctly with the front view. Remember that to show a true-size view of the surface, you use a line of sight that is perpendicular to where the surface shows on edge, as in the front view.

Using 3D Dynamic View

The 3D Dynamic View command uses the metaphor of a camera and a target, where your viewing direction is the line between the camera and the target. It is called 3D Dynamic View because while you are in the command options, you are able to see the view of the objects you select change dynamically on the screen when you move your pointing device. If you do not select any objects for use with the command, you see a special block, shaped like

a 3D house, instead. Use AutoCAD's Help command to discover more information about the options of the 3D Dynamic View command.

In the next steps you will use the 3D Dynamic View command with the Twist option to twist the viewing angle. You will type *DVIEW* at the command prompt, but the 3D Dynamic View command is also available from the View menu. The angled surface is at 60° in the front view. You will type *30* for the twist angle. This will produce a view that is aligned 90° to the front view. On your own, make sure the center viewport is active.

Command: **DVIEW** ⏎

Select objects: *(pick on the solid model)* ⏎

CAmera/TArget/ Distance/POints/PAn/Zoom/TWist/ CLip/Hide/Off/Undo/<eXit>: **TW** ⏎

New view twist <0.00>: **30** ⏎

CAmera/TArget/ Distance/POints/PAn/Zoom/TWist/ CLip/Hide/Off/Undo/<eXit>: ⏎

The view is twisted 30° to line up with the edge view of the angled surface. Your screen should look like Figure 15.27.

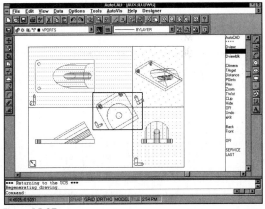

Figure 15.27

On your own, return to paper space and use hot grips to stretch the center viewport so that it is large enough to show the entire object. When you have it sized the way you want, set layer NEW as the current layer and then freeze layer VPORT. Make a new layer named REFLINES and assign it color blue and linetype CONTINUOUS. Set it as the current layer and draw reference lines between the front view and the auxiliary view in paper space. When you have finished, your screen should appear similar to Figure 15.28. Plot your drawing from paper space.

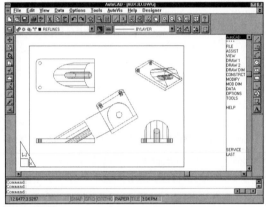

Figure 15.28

■ *TIP* To generate a drawing of a solid model with the hidden lines shown correctly, you can import any AutoCAD solid as a static part into Designer and use the powerful features of Designer to create the drawings. To export your solid as a static part, use pick File, Export and change the export filter to file type *ACIS *.sat*. Follow the prompts to export your model. Then use the Designer command Adsatin to import the part in Designer, as you learned in Tutorial 13. A static part does not contain parametric dimension information, so you cannot update the model using Designer, but you can still generate the drawing views. ■

Save your drawing, close any toolbars leaving only the Draw and Modify toolbars open, and exit AutoCAD. You have completed Tutorial 15.

KEY TERMS

auxiliary view
basic view
foreshortened
partial auxiliary view
reference line
reference surface

KEY COMMANDS

3 Point UCS
3D Dynamic View
Distance
Ellipse Center
Object UCS

Use the 2D or solid modeling techniques you have learned in the tutorial to create an auxiliary view of the slanted surfaces for the objects below. The letter M after an exercise number means that the given dimensions are in millimeters.

15.1 Stop Block

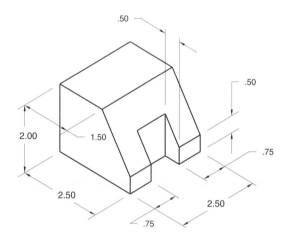

15.2M Lever

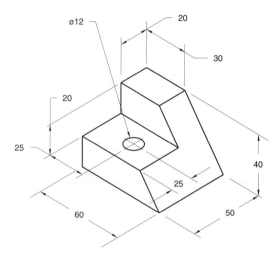

15.3 Bearing

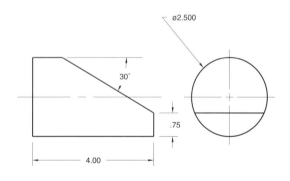

15.4M Router Guide

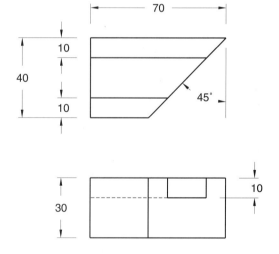

15.5 Tooling Support

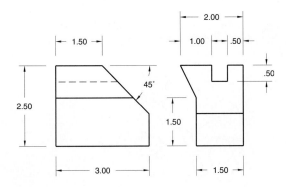

15.6 Incline Block

Use solid modeling to construct the object shown. Create an auxiliary view showing the true size of the slanted surface. Use the VPOINT (3, –3,1) for the pictorial view.

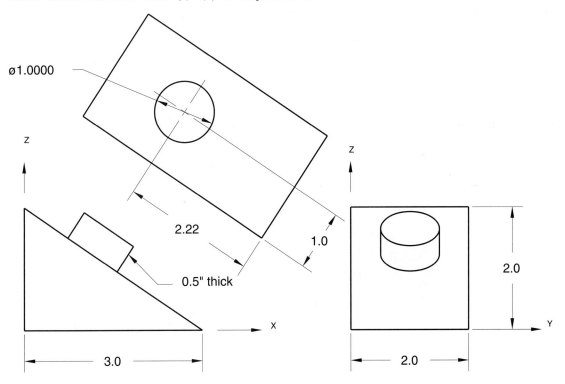

Use solid modeling to create the objects below. Produce the necessary primary and auxiliary views to describe each part.

15.7 Column Base

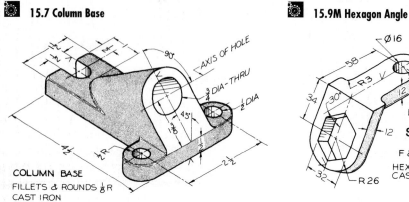

COLUMN BASE
FILLETS & ROUNDS ⅛ R
CAST IRON

15.9M Hexagon Angle

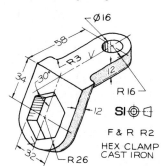

F & R R2
HEX CLAMP
CAST IRON

15.8M Hanger

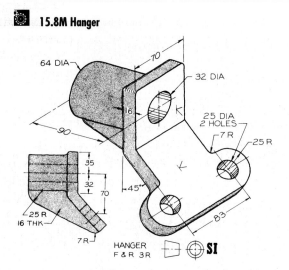

HANGER
F & R 3R

15.10M Dovetail Bracket

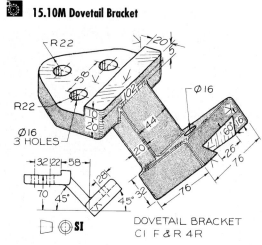

DOVETAIL BRACKET
CI F & R 4R

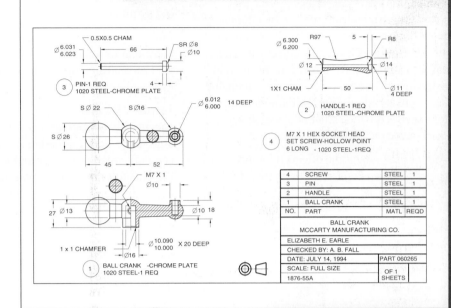

Advanced Dimensioning

Objectives

When you have completed this tutorial, you will be able to

1. Use tolerances in a drawing.

2. Set up the dimension variables to use limit tolerances and variance (plus/minus) tolerances.

3. Add geometric tolerances to your drawing.

4. Use dimension overrides.

Tolerance

No part can be manufactured to exact dimensions. There is always some amount of variation between the size of the actual object when it is measured and an exact dimension specified in the drawing. To take this into consideration, tolerances are specified along with the dimensions. A tolerance is the total amount of variation that an acceptable part is allowed to have.

To better understand the effects tolerances have on a part's dimensions, refer to Figure 16.1.

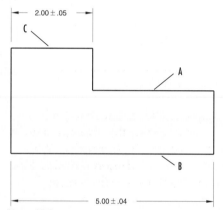

Figure 16.1

The dimensions given in the drawing are used to make and inspect the part. Included with the dimension is the allowable amount of variation (tolerance) that a part can have and still be acceptable. In order to determine whether a part is acceptable, you compare the measurements of the actual part to the toleranced dimensions specified in the drawing. If the part falls within the tolerance range specified, it is an acceptable part.

In Tutorial 7, you saw how you can add a general tolerance note to your drawing to specify this allowable variation. AutoCAD provides three ways that you can give a tolerance with

the dimension for the feature: variance tolerances, limit tolerances, and geometric tolerancing symbols.

Limit tolerances specify the upper and lower allowable measurements for the part. An actual part measuring anywhere between the two limits is acceptable.

Variance tolerances or *plus/minus tolerances* specify the *nominal dimension* and the allowable range that is added to it. From this you can determine the upper and lower limits. Add the plus tolerance to the nominal size to get the upper limit; subtract the minus tolerance from the nominal size to get the lower limit. (Or you can think of it as always adding the tolerance, but when the sign of the tolerance is negative, it has the effect of subtracting the value.) The plus and minus values do not always have to be the same. There are two types of variance tolerances: bilateral and unilateral. *Bilateral tolerances* specify a nominal size and both a plus and minus tolerance. *Unilateral tolerances* are a special case where either the plus or the minus value specified for the tolerance is zero. You can create both bilateral and unilateral tolerances using AutoCAD.

Geometric tolerances use special symbols inside *feature control frames* to describe tolerance zones that relate to the type of feature being controlled. AutoCAD Release 13's new Tolerance command allows you to quickly create feature control frames and select geometric tolerance symbols for them. You will learn to create these later in this tutorial.

To more clearly understand tolerances, consider surface A in Figure 16.1. Surface A on the actual object is longest when the 5.00 overall dimension shown for surface B is at its largest acceptable value, and when the 2.00 dimension shown for surface C is at its shortest acceptable value.

$$A_{max} = 5.04 - 1.95 = 3.09$$

Surface A is shortest when surface B is at its smallest acceptable value and surface C is at its largest acceptable value.

$$A_{min} = 4.96 - 2.05 = 2.91$$

In other words, the given dimensions with the added tolerances permit surface A to vary between 3.09 and 2.91 units.

Figure 16.2 shows the drawing you dimensioned in Tutorial 7 with tolerances added to the dimensions. Tolerances A, B, F, and G are examples of bilateral tolerances. Tolerances C, D, and E are limit tolerances.

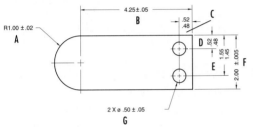

Figure 16.2

Next you will change the dimensions for the drawing you created in Tutorial 7 to add tolerances.

Starting

Start AutoCAD. On your own, use the drawing *obj-dim.dwg* as the prototype for your new drawing. Name your new drawing *tolernc.dwg*.

Your screen should appear similar to Figure 16.3. Be sure that layer DIM is set as the current layer.

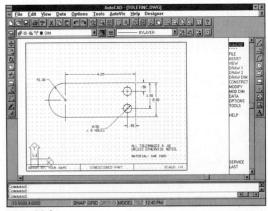

Figure 16.3

In the next steps, you will set up the dimension variables to use tolerances.

Automatic Bilateral Tolerances

You can have AutoCAD automatically add bilateral tolerances by setting the dimensioning variables. You will use the Dimension Styles dialog box to set the dimension variables. First you will show the Dimensioning toolbar.

Pick: **Tools, Toolbars, Dimensioning**

The Dimensioning toolbar appears on your screen. On your own, position it in a location where it will be easily available, but not in the way of your drawing. You will use the Dimensioning toolbar to select the Dimension Styles dialog box to set the variables.

Pick: **Dimension Styles icon**

The Dimension Styles dialog box appears on your screen, as shown in Figure 16.4. You will use it to set up the appearance for the tolerances to be added to the dimensions. The dimension style MECHANICAL is current. It is like the style you created in Tutorial 7. You will make changes to it so that the dimensions in the drawing that use that style will show a variance or deviation tolerance.

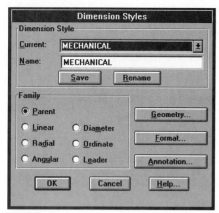

Figure 16.4

As you recall from Tutorial 7, you use the Name text box to select a name for a new dimension style. You *could* create a new style, but this time, you will change style MECHANICAL to include deviation tolerances. Changes that you make to the dimension style affect only the future dimensions you create, or dimensions you update, unless you pick the Save button in the Dimension Style area of the dialog box. Picking the Save button causes AutoCAD to make the changes to the existing dimensions that use that style when you exit the Dimension Styles dialog box. When you are done making the following changes to the dimension style, you will save the changes to style MECHANICAL so that the dimensions automatically reflect the changes.

Pick: **Annotation**

The Annotation dialog box appears. On your own, pick in the box to the right of Method in the Tolerance area of the dialog box, or pick on the downward-pointing arrow to pull down the list of tolerance methods. The list of tolerance methods available to use with dimensions pulls down, as shown in Figure 16.5. The choices for tolerance methods are: None, Symmetrical, Deviation, Limits, and Basic. None, of course, uses just the dimension and does not add a tolerance, as your current dimensions reflect. *Symmetrical tolerances* are lateral tolerances

where the upper and lower values are the same; these generally specify the dimension plus or minus a single value. *Deviation tolerances* are lateral tolerances that have a different upper and lower deviation. Limit tolerances show the maximum value for the dimension preceded by a plus sign, and the minimum value preceded by a minus sign. *Basic dimension* is a term used in geometric dimensioning and tolerancing to specify a theoretically exact dimension that does not have a tolerance applied to it. Basic dimensions appear in your drawing with a box drawn around them. If you choose any of the tolerance methods, you can see what your dimensions will look like reflected in the image tiles.

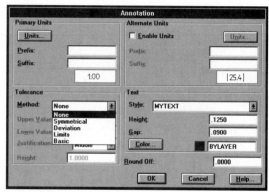

Figure 16.5

Pick: **Deviation**

Notice the image tile just above the Tolerance area; it shows the appearance of a dimension with a deviation style tolerance added. You can also make selections in the dialog box by picking on the image tile until it has the correct appearance; try this, as follows:

Pick: **(on the image tile to cycle its appearance to that of limits tolerances)**

Pick: **(again on the image tile to change it to a basic dimension)**

Continue picking the image tile on your own until you cycle through all of the options and return to deviation tolerance, as shown in Figure 16.6.

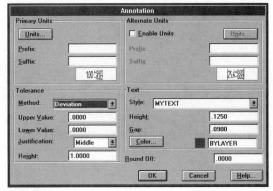

Figure 16.6

Specifying the Tolerance Values

The text boxes for setting the tolerance values were formerly grayed out. Now that you have selected a tolerance method, they are available for input. You will use them to specify the upper and lower values you want to use with the dimension to specify the allowable deviation. These values will both be positive. The upper value will be added to the dimension value; the lower value will be subtracted from the dimension value. AutoCAD will automatically add the plus or minus sign in front of the value when showing the tolerance. On your own, pick in the text box to the right of Upper Value.

Type: **.05**

On your own, pick in the text box to the right of Lower Value, or press (TAB).

Type: **.03**

Notice that you can also control the justification for the tolerance. For now you will leave this set to Middle, the default.

Setting the Tolerance Text Height

In the Height text box at the bottom of the Tolerance area, you will see the default setting of 1.0000. The value set for the tolerance height is a scaling factor. A setting of 1.0000 makes the tolerance values the same height as the standard dimension text. For this tutorial you will set tolerance height to .8 (so that the height of the tolerance values will be 8/10ths the height of the dimension values). Highlight the value in the box.

Type: **.8**

Setting the Tolerance Precision

The tolerance values can have a different precision than the dimension values themselves. You control this by selecting the Units button in the Primary Units area at the upper left of the Annotation dialog box.

Pick: **Units**

The Primary Units dialog box appears on your screen, as shown in Figure 16.7. You used it in Tutorial 7 to set the number of decimal places for the dimension. Now you will use it to set the precision for the tolerance values. You can also control whether leading and trailing zeros are used. On your own, pick on the Precision value in the Tolerance area (not the Dimension area). A list of precisions pulls down. Pick the two-place decimal (0.00) from the list. Pick OK to exit the Primary Units dialog box.

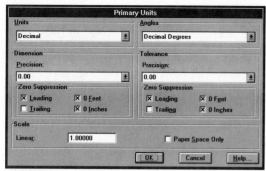

Figure 16.7

The same text style used for the dimension text, MYTEXT, will also be used for the tolerance text. You can control the text style and the dimension text height, color, and gap in the Text area of the Annotation dialog box. You will leave these features set as they are.

Pick: **OK** *(to exit the Annotation dialog box)*

Pick: **Save** *(to save your changes to the dimension style MECHANICAL)*

Pick: **OK** *(to exit the Dimension Styles dialog box)*

You return to your drawing. The dimensions that used style MECHANICAL automatically update to show the deviation tolerance. Your drawing should look similar to Figure 16.8.

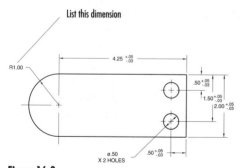

Figure 16.8

Why did the radial dimension not update? You can find out information about the dimension by using the List command. List is located on the Inquiry flyout on the Object Properties toolbar.

Pick: **List icon**

Select objects: **(pick the radial dimension at the left end of the object)** ⏎

The List command displays the information shown in Figure 16.9 in an AutoCAD Text window.

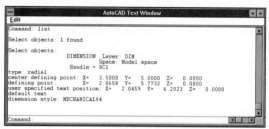

Figure 16.9

Notice that the style name for this dimension is MECHANICAL$4. When you list a dimension, the $# code tells you that the dimension uses a child style that is different from the parent style. Here it tells you that the dimension was created using a child style that has different settings than the parent style, MECHANICAL. If you need to review parent and child styles, refer to Tutorial 7. The codes for the child types are:

$0	linear
$2	angular
$3	diameter
$4	radial
$6	ordinate
$7	leader (also used for tolerance objects)

You must change the particular child dimension style (for example, radial) in order to change the appearance of the dimension. The reason the dimension did not update to show the tolerance is because no tolerance was created for the radial dimension child style MECHANICAL$4. To change the child style, you will use the Dimension Styles dialog box and pick the child style for radial and set its properties.

On your own, close the AutoCAD Text window.

Pick: **Dimension Styles icon**

The Dimensions Styles dialog box appears on your screen. You will use it to change the radial child style for MECHANICAL. In the Family area of the dialog box,

Pick: **Radial**

Pick: **Annotation**

The Annotation dialog box appears on your screen. On your own, use it to select Deviation for the tolerance method and set the upper value to .05 and the lower value to .03, as you did earlier in the tutorial for the MECHANICAL parent dimension style. Change the scale factor for the height in the Tolerance area to .8. Use the Units selection to set two decimal places for the tolerance precision. When you have finished making these selections, pick OK to return to the Dimension Styles dialog box. Save the changes you have made to the style so that they will be applied to the existing dimensions.

Pick: **Save**

Pick: **OK** *(to exit the dialog box)*

Notice that now the radial dimension has a tolerance value applied to it, as shown in Figure 16.10.

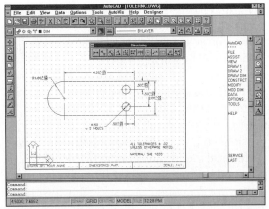

Figure 16.10

Using Dimension Overrides

Dimension overrides let you change the dimension variables controlling a dimension and leave its style as is. A good practice is to change the general appearance of the dimensions by changing the dimension style, and to change individual dimensions that need to vary from the overall group of dimensions, for instance to suppress an extension line, by using overrides.

You can change dimension variables by typing *DIMOVERRIDE* at the command prompt. Dimoverride has the alias Dimover. When using Dimoverride at the command prompt, you are prompted for the exact name of the dimension variable you want to control and its new value. Then you select the dimensions to which you will apply the override. The Clear option of the Dimoverride command lets you remove overrides from the dimensions you select.

You can also use the Properties selection from the Object Properties toolbar to accomplish the same thing as the Dimoverride command.

Pick: **Properties icon**

Select objects: *(pick on the radial dimension for the rounded end)* ⏎

The Modify Dimension dialog box appears on your screen. You will use it to override the dimension style for the radial dimension. Because the height of the part is controlled by the 2.00 vertical dimension at the right side of the part, providing the radial dimension is an example of over-dimensioning. In instances where you want to call attention to the full rounded end, which is tangent to the horizontal surfaces at the top and bottom of the object, you would either show the 1.00 radius value as a *reference dimension*, or just identify it with the letter R indicating the radius, but leaving its value to be determined by the other surfaces. You will turn off the tolerance and add REF to the end of the radial dimension value to indicate that it is provided as a reference dimension only. You can also note reference dimensions by including the dimension value in square brackets.

Pick: **Annotation**

The Annotation dialog box appears on your screen. On your own, use it to select None as the tolerance method. In the Primary Units area of the dialog box, type *REF* in the Suffix text box. When you have finished making these selections, pick OK. You return to the Modify Dimension dialog box.

You can also use the Modify Dimension dialog box to override the dimension text by picking the Edit button and using the Edit MText dialog box that appears to change the text. The angle brackets you see in the Mtext editor are AutoCAD's way of storing the default lengths of the dimensions, which allows them to update. Once you have overridden the text, your dimension values no longer update if you stretch or change the object. As usual with the Properties selection, you can change the layer, color, linetype, and linetype scale and thickness. Exit the Modify Dimension dialog box by picking OK.

Your drawing should look like Figure 16.11.

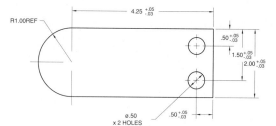

Figure 16.11

Next, use the List command to list the radial dimension.

Command: **LIST** ⏎

Select objects: **(pick the radial dimension for the rounded end)** ⏎

An AutoCAD Text window appears on your screen, as shown in Figure 6.12, listing the information about the radial dimension. Notice that now overrides for the dimension variables Dimpost and Dimtol are listed. Dimpost displays the dimension suffix, which is set to REF. Dimtol is set to Off, so that there is no deviation tolerance applied to the dimension.

Figure 6.12

When you are dimensioning, keep your dimension styles organized by creating new dimension styles and controlling the child styles as needed. Use dimension overrides when certain dimensions need a different appearance than the general case for that type of dimension.

Using Limit Tolerances

Next you will create a new style for dimensions, using limit tolerances, and then you will change the object's overall dimension to use that style.

Pick: **Dimension Styles icon**

The Dimension Styles dialog box appears on your screen once again. Next you will type in a new style name. On your own, highlight the name MECHANICAL in the Name text box.

Type: **LIMITTOL**

Make sure the Parent radio button is selected.

Pick: **Annotation**

On your own, use the image tile in the Annotation dialog box to select the Limits tolerance method. Once it is selected, the dialog box appears as shown in Figure 16.13.

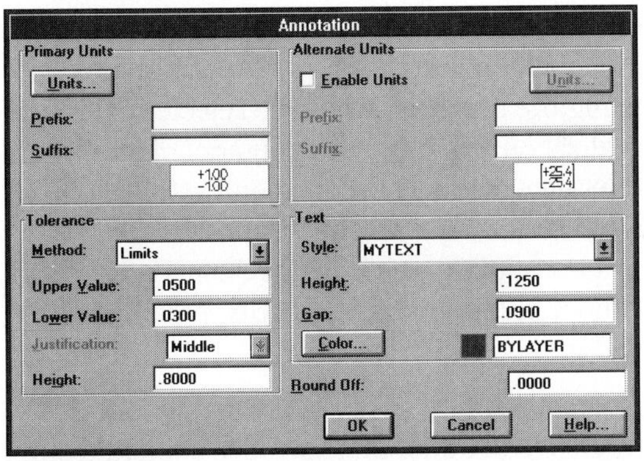

Figure 16.13

You can leave the upper and lower values and text height set as you did for deviation tolerances.

Pick: **OK** *(to exit the Annotation dialog box)*

Pick: **Save**

Pick: **OK** *(to exit the Dimension Styles dialog box)*

Next you will change the overall dimension of the object to use the new style, LIMITTOL. To do this you will use the Properties selection from the Object Properties toolbar.

Pick: **Properties icon**

Select objects: *(pick on the longest horizontal dimension)* ⏎

Use the Modify Dimension dialog box that appears to change the dimension style to LIMITTOL. To change the dimension's style, pick on the name MECHANICAL in the Style area near the center of the dialog box to pull down the list of available styles. On your own, pick LIMITTOL from the list of styles. (If you do not see the selections for changing the dimension style, perhaps you selected more than one dimension. The Properties selection is sensitive to the types of objects selected and displays a different dialog box depending on what you select.) Pick OK to exit the dialog box.

The limit tolerance appears, as shown in Figure 16.14.

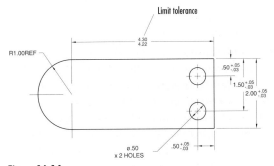

Figure 16.14

Next you will use the Tolerance command to add a feature control frame specifying a geometric tolerance to the drawing.

Creating Feature Control Frames

Geometric tolerancing allows you to define the shapes of *tolerance zones* in order to control the manufacture and inspection of parts accurately. Consider the location dimensions for the two holes shown in Figure 16.15. The location for the center of hole A is allowed to vary in a square zone that is .1 units wide, as defined by the location dimensions 1 and 2 and the +/−.05 tolerance applied to them.

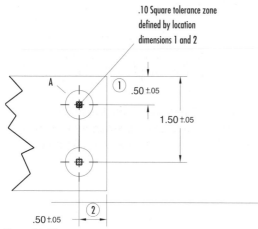

Figure 16.15

This tolerance zone is not the same length measured in all directions from the true center of the hole. The stated tolerances allow the location for the center of the actual hole to be off by a larger amount if it is off in a diagonal direction, as shown in Figure 16.16.

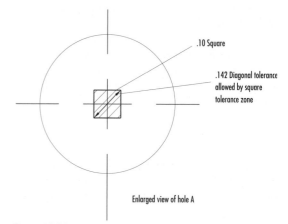

Figure 16.16

This means that either some parts that should be rejected would be accepted, because the tolerance stated for the location of the hole was considered as only being in the horizontal or vertical direction, or, if the diagonal was considered when the tolerance was stated, then some parts that are off in a horizontal or vertical direction, but could have been accepted using the diagonal, would be rejected. This realization led to the development of specific tolerance symbols used to control the geometric tolerance characteristics for position, flatness, straightness, circularity, concentricity, runout, total runout, angularity, parallelism, and perpendicularity.

Geometric tolerances allow you to define a feature control frame that specifies a tolerance zone that is shaped appropriately for the feature being controlled (sometimes this is called feature-based tolerancing). A feature control frame is shown in Figure 16.17.

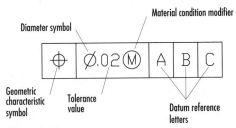

Diameter symbol

Material condition modifier

Geometric
characteristic
symbol

Tolerance
value

Datum reference
letters

Figure 16.17

The feature control frame begins with the *geometric characteristic symbol*, which tells the type of geometry being controlled. Figure 16.17 shows a geometric characteristic symbol for positional tolerance. The next section of the feature control frame specifies the total allowable tolerance zone; in this case a diametral-shaped zone .02 wide around the perfect position for the feature. These two symbols are followed by any modifiers for material condition or projected tolerance zone. The circled M shown in Figure 16.17 is the modifier for maximum material condition, meaning that at maximum material, in this case the smallest hole, the tolerance must apply, but when the hole is larger a greater tolerance zone may be calculated based on the size of the actual hole measured. The remaining boxes indicate that the position is measured from datum surfaces A, B, and C. Not every geometric tolerance feature control frame requires a datum; for example, flatness can just point to the feature and control it, without respect to another surface. Features like perpendicularity that are between two different surfaces require a datum.

Among other things, good application of geometric tolerancing symbols requires that you understand:

- The design intent for the part
- The shapes of the tolerance zones that are specified using particular geometric characteristic symbols
- The selection and indication of datum surfaces on the object
- The placement of feature control frames in the drawing

In this tutorial, you will learn how to create feature control frames and add them to your drawing. In order to fully understand the topic of geometric tolerancing, refer to your engineering design graphics textbook or a specific text on geometric tolerancing. Merely adding feature control frames to the drawing will not result in a drawing that clearly communicates your design intent for the part to the manufacturer.

 Next you will create the feature control frame shown in Figure 16.17. To create a feature control frame, you will pick the Tolerance icon from the Dimensioning toolbar.

Pick: **Tolerance icon**

The Symbol dialog box shown in Figure 16.18 appears on your screen. You will use it to pick the symbol for the geometric characteristic of the geometric tolerance feature control frame you will create.

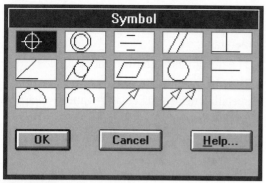

Figure 16.18

Pick: (the positional tolerance symbol in the upper left corner)

Pick: **OK**

■ *TIP* It is a feature of many Windows dialog boxes that you can double-click on the desired selection to select it directly and exit the dialog box in one action, without having to pick OK. You can double-click the picture of the positional tolerance symbol to select it directly and exit the dialog box. ■

The Geometric Tolerance dialog box appears on your screen. You will make selections so the dialog box appears as shown in Figure 16.19. The positional tolerance symbol should already be shown at the left.

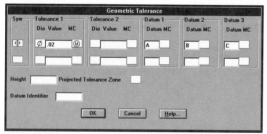

Figure 16.19

The box to the right of the positional tolerance symbol, below the heading Dia, is a toggle.

Pick in the empty box to turn on the diameter symbol. Once you pick in the empty box, the symbol should appear. Next, click in the banner below the heading Value and type in *.02*. To add a modifier, pick on the empty box below the heading MC (for material condition). The Material Condition dialog box appears on your screen, as shown in Figure 16.20.

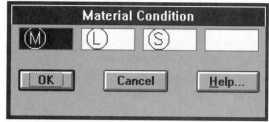

Figure 16.20

Double-click on the circled M to select the modifier for maximum material condition, or pick once and then pick OK. You return to the Geometric Tolerance dialog box. The modifier appears for the feature control frame. Next you will specify datum surfaces A, B, and C. To do this, pick in the empty text box below Datum 1 and type *A*; repeat this process for datums B and C on your own. Notice that datums can also have modifiers. If you wanted to, you could pick on the empty box below MC to add a modifier for the datum.

Now you are ready to add the feature control frame to your drawing. On your own, pick OK to exit the Geometric Tolerance dialog box. You will locate the feature control frame below the .50 diameter dimension for the two holes.

Enter tolerance location: *(pick the location for the tolerance symbol)*

Your drawing should look like Figure 16.21.

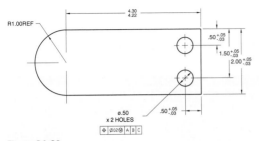

Figure 16.21

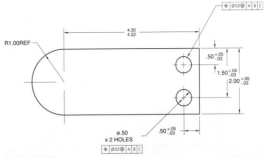

Figure 16.22

Using Leader with Tolerances

You can also use the Leader command to add a leader with an attached geometric tolerance symbol to your drawing. You will create the next feature control frame starting with the Leader command. You will pick the Leader icon from the Dimensioning toolbar.

Pick: **Leader icon**

From point: *(pick the Snap to Nearest icon)*

nea to *(pick at about 2 o'clock on the upper hole)*

To point: *(pick above and to the right of the drawing)*

To point (Format/Annotation/Undo)<Annotation>: ⏎

Annotation (or RETURN for options): ⏎

Tolerance/Copy/Block/None/<Mtext>: **T** ⏎

The Symbol dialog box appears; from there you can start creating the feature control frame. On your own, create another feature control frame like the first one. After you finish making selections in the Geometric Tolerance dialog box, pick OK. The feature control frame is added to the end of the leader line, as shown in Figure 16.22.

Next you will override the location dimensions for the holes so that they appear as basic dimensions. When you specify a positional tolerance, the true position is usually located with basic dimensions. Basic dimensions are theoretically exact dimensions. The feature control frame for positional tolerance controls the tolerance zone from which the true position can vary. To create basic dimensions by overriding the dimension style,

Pick: **Properties icon**

Select objects: *(pick the .50 vertical dimension at the upper right of the object)* ⏎

The Modify Dimension dialog box appears on your screen. On your own, use it to select Annotation, and then set the method in the Tolerance area of the dialog box to Basic. When you have finished, pick OK to exit the Annotation dialog box, and then pick OK to exit the Modify Dimension dialog box.

Figure 16.23 shows the drawing with the .50 basic dimension.

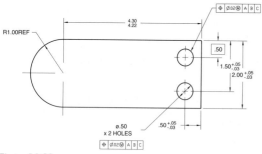

Figure 16.23

On your own, override the dimension style for the .50 horizontal dimension below the object, and then for the 1.50 vertical dimension. You must pick only one dimension each time you use the Properties dialog box, or you will get the more generic Change Properties dialog box. When you have finished, save your drawing on your own. Next, you will add datum flags to identify datum surfaces A and B, from which the position feature can vary.

Creating Datum Flags

Datum flags are boxed letters identifying the feature on the object that is being used as a datum. Often they are attached to the extension line that is parallel to the surface being identified as the datum surface. To create a datum flag,

Pick: **Tolerance icon**

From the lower right of the Symbol dialog box,

Pick: **(the blank box)**

The Geometric Tolerance dialog box appears on your screen. You will leave the selections blank, except for the datum identifier. On your own, type -A- in the Datum Identifier text box. When you have finished, the dialog box should

look like Figure 16.24. Pick OK. You will return to your drawing to select the location for the datum flag you are creating.

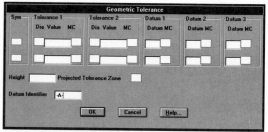

Figure 16.24

Enter tolerance location: **(pick on the extension line)**

The datum flag appears in your drawing, as shown in Figure 16.25. Add datum flag B to the drawing on your own.

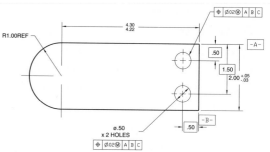

Figure 16.25

Close the Dimensioning toolbar, save your drawing, and exit AutoCAD. You have completed Tutorial 16.

KEY TERMS

basic dimension
bilateral tolerances
datum flags
deviation tolerances
feature control frames
geometric characteristic symbol
geometric tolerances
limit tolerances
nominal dimension
plus/minus tolerances
reference dimension
symmetrical tolerances
tolerance zones
unilateral tolerances
variance tolerances

KEY COMMANDS

Dimoverride
Dimpost
Dimtol
Leader
Tolerance

Draw the necessary views and dimension the following shapes. Include tolerances if indicated. Use either solid modeling techniques or 2D orthographic views.

 16.1 Stop Base

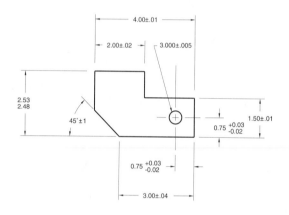

 16.2 Tolerance Problem

Calculate the maximum and minimum clearance between Parts A and B and Part C as shown.

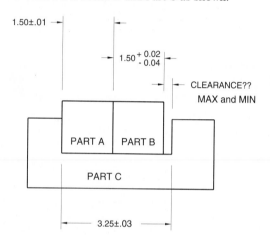

16.3M Lathe Stop

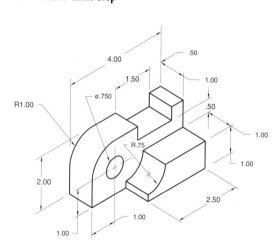

16.4M Mill Tool

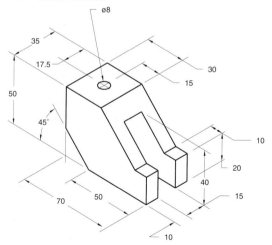

16.5M Shaft Bearing

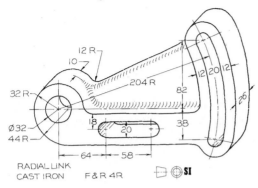

12 R

10

204 R

32 R

Ø32
44R

18
20

64 — 58

12 20 12

82

38

26

RADIAL LINK
CAST IRON   F & R 4R   SI

16.6 Positional Tolerance

Locate the two holes with an allowable size tolerance of 1.00 mm and a position tolerance of 0.50 mm DIA. Use the proper symbols and dimensions.

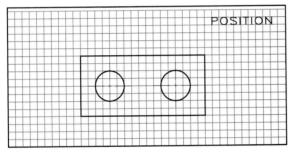

POSITION

16.7 Angular Tolerance

Using a feature control frame and the necessary dimensions, indicate that the allowable variation from the true angle specified for the inclined plane is 0.7mm, using the bottom surface of the object as the datum.

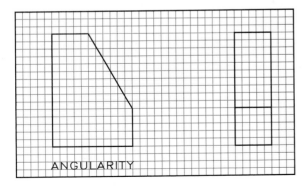

ANGULARITY

16.8 Hub

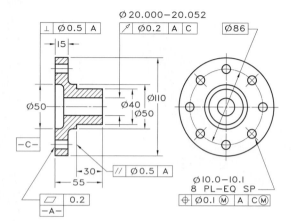

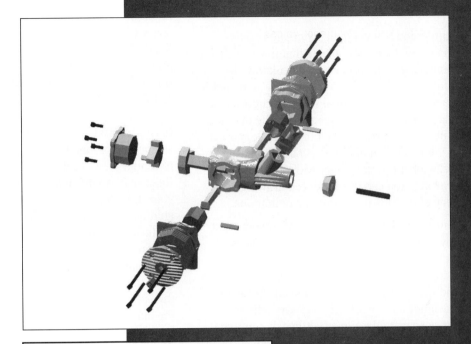

Rendering with AutoVision

Objectives

When you have completed this tutorial, you will be able to

1. Realistically shade a solid model or surface modeled drawing.

2. Set up and use lighting effects.

3. Apply materials and finishes.

4. Save rendered files for use with other programs.

5. Export to 3D Studio format.

Introduction

AutoVision lets you quickly create rendered images of AutoCAD objects that have surfaces. Rendering is the process of calculating the lighting effects, color, and shape of an object in a single two-dimensional view. A rendered view adds greatly to the appearance of the drawing. It makes the object appear more real, and aids in interpretation of the object's shape, especially for those unfamiliar with engineering drawing. You can use rendered drawings very effectively in creating presentations.

AutoVision can render AutoCAD objects that have surfaces. Objects that have surfaces include solid models, regions, objects created using AutoCAD's surface modeling commands, and models that you created using AutoCAD Designer and for which you have defined surfaces using Designer's Admesh command (as well as AutoSurf objects that have surfaces defined using AutoSurf's Dispsf command).

Rendering is a process that is mathematically complex. When you are using AutoVision or other programs to do rendering, the capabilities of your hardware, such as speed, amount of memory, and the *resolution* of the display, are particularly noticeable. The complex rendering calculations demand more of your hardware than just running the AutoCAD program, so you may notice that your system seems slower. Display resolution affects not only the speed with which objects are rendered, but the quality of the rendered appearance as well. The rendered screens shown in this tutorial were created using Super VGA resolution (800 × 600) and 256 colors. Better appearance can be gained by rendering to a display using higher resolution and more colors; however, increased resolution and colors take a longer time to render.

■ *Warning:* The AutoCAD minimum requirement is 16 MB RAM. AutoVision is 3 MB in size when loaded, so if you have only 16 MB of RAM, AutoVision may push your system to the limit. Refer to the Getting Started section of this manual for information on configuring AutoVision. ■

In this tutorial you will learn to apply materials, surface maps, and lighting to your drawing and to render views of the object using AutoVision Release 2. Applying a *material* to a surface in your drawing gives the surface its color and shininess; it is similar to painting the surface. Materials can also reflect light like a mirror, or be transparent like glass. Adding a *surface map* is similar to adding wallpaper; surface maps are patterns or pictures that you can apply to the surface of an object. Lighting illuminates the surfaces. Lighting is a very important factor in creating the desired effect in your rendered drawing.

Starting

Launch AutoCAD. You should see AutoVis as a selection on the menu bar at the top of your screen if you configured AutoVision as suggested in the Getting Started section of this manual. If not, refer to that section before continuing.

In this tutorial you will use AutoVision to shade the solid model of the assembly you created in Tutorial 10. On your own, start a new drawing from the prototype drawing *assemb1.dwg*, included in the data files for this manual. Use the name *asmshade.dwg* for the new drawing.

The drawing appears on your screen, as shown in Figure 17.1. Notice the AutoVis menu selection on the menu bar.

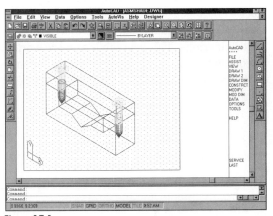

Figure 17.1

AutoVision also has its own toolbar. Next you will turn on the AutoVision toolbar so that you can use it to select commands.

Pick: **AutoVis**

The AutoVision menu selections pull down from the menu bar, as shown in Figure 17.2.

Figure 17.2

Pick: **AutoVis, AutoVis Toolbar**

The AutoVision toolbar appears on your screen, as shown in Figure 17.3. Pick on its title bar and, keeping the pick button held down, drag it to a location where it is handy, but not in the way of your drawing. (There is also a Render toolbar, which you can turn on using the Tools, Toolbars, Render selections. This Render toolbar contains many of the same icons, but if AutoVision is loaded, its commands will be used in place of the default rendering available in AutoCAD. The native AutoCAD rendering is similar to AutoVision, but does not contain as many options.)

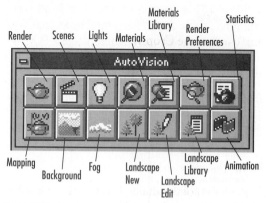

Render Scenes Lights Materials Materials Library Render Preferences Statistics

Mapping Background Fog Landscape New Landscape Edit Landscape Library Animation

Figure 17.3

You will use AutoVision to render the solid model drawing so that it appears realistically shaded on your screen.

The AutoVision Render Dialog Box

The Render option at the top of the AutoVision pull-down menu opens a dialog box to allow you to select rendering options. (You can also select the Render icon from the AutoVision toolbar.)

Pick: **AutoVis, Render**

AutoVision's Render dialog box appears on your screen, as shown in Figure 17.4.

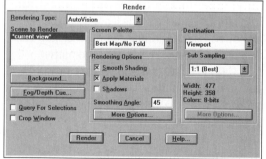

Figure 17.4

Rendering Algorithms

You should see AutoVision selected as the rendering type. The other rendering types available are AutoCAD Render and AutoVis Raytrace. The selection AutoCAD Render uses either *Gourard* or *Phong* shading to add smooth shading and materials, but not surface maps or shadows. Gourard shading calculates the shading value for each vertex of a surface and then averages the color across the surface. This method is faster, but less accurate than Phong shading, which calculates the shading value for each pixel on the surface, taking longer, but producing more realistic highlights. (When you select AutoCAD Render as the rendering type, you can use the More Options selection to choose between the Gourard and Phong rendering algorithms.) The selection AutoVis Raytrace uses a *ray tracing* algorithm to produce rendered images that have *reflections*, *refractions*, and detailed shadows. This can produce a more realistic result, but because it is more complex, it takes more time to render. The AutoVision rendering algorithm falls somewhere between AutoCAD Render and AutoVis Raytrace in complexity. It can render objects with surface maps and shadows, but not transparent materials, or reflections and refractions. It is often a good strategy when rendering to start with the least complex method and a lower resolution, and progress to more complex renderings. This way, while you are working out details of how to set up the lighting and surfaces for a realistic appearance, you are not waiting a long time for a complex rendering. For now, you will leave AutoVision selected.

Scene to Render

A *scene* establishes a direction from which you will view the model and the lights you will use for that scene. You make and name scenes with AutoVision's Scenes command. (You can select the Scenes command by picking the Scenes icon from the AutoVision toolbar.) To select a view for use in the Scenes command, you must first use AutoCAD's View command and save a named view. Then select the Scenes command and associate that named view with the new scene name. The default selection, current view, always appears and can be used to render the current view on your screen even if you have not created any named scenes.

Background

Selecting the Background button displays the Background dialog box, which lets you select colors, bitmapped images, gradient backgrounds that blend from one color to the next, and merge the rendered image with the current AutoCAD image as the background.

Fog/Depth Cue

Selecting the Fog/Depth Cue button displays the Fog dialog box, which you can use to turn on and adjust this effect. Fog and depth cue provide visual cues about the distance that the object is from the camera, as if you were looking at the object through a fog. As the distance to the object increases, there is more fog between the observer and the object. Fog uses a white color to obscure more of the object as it gets further from the camera. Depth cue has the same effect as Fog, but uses black to achieve the visual cue.

Query For Selections

If you check this box, when you choose Render you are returned to your drawing screen to select the objects you want to render. When you press ⏎ to end the selection, the objects you have selected are rendered. You do not return to the Render dialog box.

Crop Window

If you select the Crop Window box, when you choose Render you are returned to your drawing to pick a window defining the area you want to render. Once the window is defined, the area is rendered. You do not return to the Render dialog box.

Screen Palette

This section of the dialog box is only available when you have selected Viewport as the destination and are using at least 256 colors for the rendering. It controls the appearance of colors on the display. The selection Best Map/No Fold gives the truest appearance of colors on the rendering, but can affect the appearance of colors on the rest of the AutoCAD screen. The selection Best Map/Fold uses the screen colors along with the colors for the rendering so that colors for the remaining screen area do not change, but the colors for the rendering may be affected so that they are not the truest colors possible. The final selection, Fixed ACAD Map, uses only the AutoCAD color palette. This way the rendering colors always match the screen colors.

Destination

The Destination selection allows you to send the rendered output to a viewport, to the Render Window, or to a file. For now you will have the drawing rendered to your viewport. Choosing to send the output to the rendering window uses AutoVision's Render Window, which has its own menu bar selections for copying to the Windows Clipboard, saving the rendering as a bitmap file, and setting the resolution for the rendering window. You can also print the image shown in the rendering window. (This option is not available when you run the DOS version of AutoCAD Release 13.) Selecting rendering to a file allows you to select more options, which let you set different file types, number of colors, and resolutions for the rendered image saved in a file. Check to see that Viewport is selected under the heading Destination.

> ■ *TIP* Note that once you pick any AutoVision menu selection, AutoVision's Render Window is open, but not active, on your system. This may take up some of your available memory. To close the Render Window, click on the Windows Control box to pull down its menu. Select Switch To. The Task List dialog box opens on your screen. Select AutoVision from the list and pick End Task. Then return to your AutoCAD drawing session. ■

Rendering Options

Smooth Shading shades rounded surfaces so that they appear round. If it is off, rounded surfaces will appear faceted (made up from flat surfaces). Apply Materials tells AutoVision to apply materials to objects. Check to see that Smooth Shading and Apply Materials are turned on under Rendering Options. In the scene you will render, since you have not selected any finishes yet, the AutoCAD Color Index (ACI) will be used, giving your shaded objects a color similar to their original AutoCAD color. Later, when you apply materials to the surfaces of the objects in your drawing, selecting Apply Materials will cause the surfaces to be rendered with the material appearance, not the AutoCAD color.

Selecting Shadows causes shadows to be generated when using AutoVision or AutoVis Raytrace as the rendering algorithm. For now, leave Shadows turned off. Using shadows can create drawings that have a more realistic appearance, but take a longer time to render. Smoothing Angle controls which surfaces are shaded as smoothed rounded surfaces and which are shaded as flat surfaces. The value 45 smoothes all surfaces that meet with a change in angle of less than 45° and assumes that surfaces with a greater change in angle are flat intersecting surfaces.

The More Options button displays additional options you can use in the AutoVision Render Options dialog box. Pick it to see the additional options. A different dialog box showing additional options appears, depending on which rendering type you have selected: AutoVision, AutoCAD Render, or AutoVis Raytrace.

One of the additional options you can set is anti-aliasing, which produces a smoother appearance on edges of surfaces by blending additional pixels along the edge with a color in between the edge color and the surface color. You can set the level of anti-aliasing used, which is helpful because rendering takes longer when high levels of anti-aliasing are used. Because additional pixels are blended along edge lines, they appear thicker when anti-aliased.

Options for Texture Map Sampling control how bitmaps are projected onto objects that are smaller than the bitmap.

Another of the additional selections lets you control how the back surfaces of the object, called *back faces*, are shaded. Picking Discard Back Faces causes the back surfaces to be ignored. This speeds up rendering time when there are no surfaces in back of the object that you can see. You would not want to select Discard Back Faces when you are looking into an open box and can see the back surface. Back Face Normal is Negative controls the direction for the back surface of the objects. A *normal* is a vector perpendicular to a surface, telling AutoCAD which is the top side of the surface. Selecting Back Face Normal is Negative tells AutoCAD to shade the back surface inside a shape like an open box, not the outside back surface.

On your own, pick Cancel to close the AutoVision Render Options dialog box and return to the Render dialog box.

Help

You can get help for the items in the Render dialog box by picking the Help button at the lower right of the dialog box.

Pick: **Help**

The AutoVision Help window appears on your screen, as shown in Figure 17.5. Use it like you use the AutoCAD Help window. You can pick an item that appears highlighted in the Help window to get help for that topic. Use the buttons below the menu bar to select the contents, search for a particular topic, or move back to the previous help screen. Experiment with AutoVision Help on your own. When you are finished, double-click the Windows Control box to close the Help window on your own. You return to the Render dialog box. If you have experimented with any of the settings, return the Render dialog box to the selections shown in Figure 17.4.

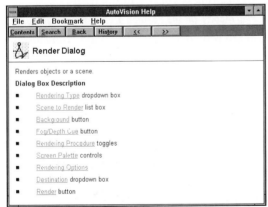

Figure 17.5

Pick: **Render**

Your drawing is rendered to the display. It should appear similar to Figure 17.6.

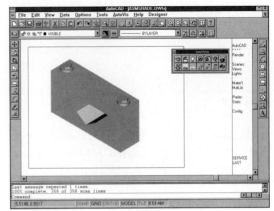

Figure 17.6

Adding Lights

 Notice that some of the surfaces in your rendered drawing blend into one another and are hard to see. This is because you have not set up any lighting in the drawing. The default lighting is *ambient* lighting with an intensity of 0.30, which is not very much illumination. Ambient lighting illuminates every surface equally, so it is difficult to distinguish one surface on the object from the next. The best use for ambient lighting is to add some light to back surfaces of the object, so that they do not appear entirely dark. You can improve the look of your rendered drawing by using AutoVision to add additional lights. Lighting is very important in order to achieve good-looking rendering results.

Pick: **Lights icon**

The Lights dialog box appears on your screen, as shown in Figure 17.7. You will use it to create lighting for your drawing.

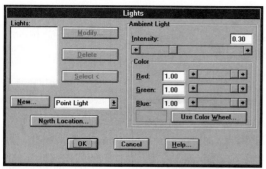

Figure 17.7

In addition to ambient light, AutoVision allows three different types of lights. A *point light* is like a bare light bulb that casts light in every direction (i.e., it is omnidirectional). A *distant light* is like sunlight. Distant lights illuminate all of the surfaces on which they shine with an equal brightness. A *spotlight* highlights certain key areas and casts light only toward the selected target. Point lights and spotlights create indoor lighting effects.

AutoVision provides a sun angle calculator to help you establish the location of the sun, or distant light. This is particularly useful for architectural applications. You will learn how to use the sun angle calculator later in this tutorial. To see the other Lights options,

Pick: (on Point Light)

The list of light types appears. Next, you will add a point light to your drawing. Because it is the default, it is already selected in the box to the right of the New button at the left of the dialog box. Leave this selection set to Point Light.

To create a new light for the drawing, you will use the New button.

Pick: **New**

The New Point Light dialog box appears, as shown in Figure 17.8. You will use this dialog box to name the new light and select the features that determine its appearance, such as

intensity (brightness), location, color, attenuation (strength), and shadows, in the drawing. You will name the light FILL because you are using it to fill the area where your object is with more lighting.

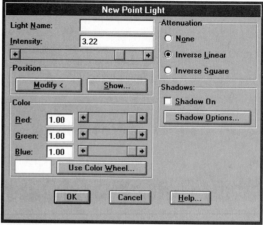

Figure 17.8

Type: **FILL** *(in the text box to the right of Light Name)*

Next you will position the light in your drawing. AutoVision places a block as a symbol in your drawing to represent the light.

Pick: **Modify** *(in the Position area of the dialog box)*

You are returned to your drawing. You can position the light by picking a location from the screen; however, for this example you will type in the coordinates for the light. It is difficult to position the light by picking from the screen because you are selecting only the X and Y coordinates. (You can use point filters and then set the Z location, if you want. Use Help to find out more about point filters or refer to T13.) When specifying the coordinates for the light position, do not position it too close to the object.

Enter light location <current>: **3,5,10** ⏎

You return to the New Point Light dialog box.

Pick: **Show**

The Show selection opens the Show Light Position dialog box and shows the coordinates of the light's location. This allows you to check to see whether you have made an error, or to identify the coordinates for lights that were placed previously or located by picking from the screen.

Pick: **OK *(to exit the Show Light Position dialog box)***

Next you will adjust the *intensity* or brightness of the light. You can think of this as the wattage of the bare light bulb (a 100-watt bulb in the location is brighter than a 60-watt bulb). You will type the value in the box provided. You can also use the slider bar to adjust the intensity of the light.

Select: **(the value in the text box to the right of Intensity)**

Type: **7**

You can use the Color area of the dialog box to select a different color for the light. Setting the light to a yellowish color will give a warmer effect, like indoor lamps. For this example you will leave it white.

Pick: **OK *(to exit the New Point Light dialog box)***

Pick: **OK *(to exit the Lights dialog box)***

Whenever you make changes that will affect the rendered appearance of your object — adding lights or materials, for instance — you need to re-render your drawing. You are now ready to use AutoVision's Render command to see the effect of adding the point light to your drawing. This time you will pick the icon from the AutoVision toolbar.

Pick: **Render icon**

The Render dialog box that you used earlier appears on your screen. Make sure that the options you set earlier are still current: Rendering Type should be AutoVision, Smooth Shading and Apply Materials should be on, and Destination should be Viewport. When you have made sure that these selections are chosen,

Pick: **Render**

The added lighting should improve the appearance of your rendered drawing so that it looks like Figure 17.9.

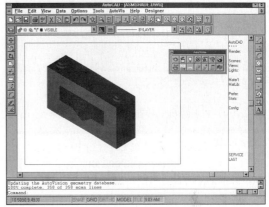

Figure 17.9

The right surface of the object is still not lit very well; not much light is striking that surface. Next you will add a spotlight behind the object to light that surface and provide highlights in the drawing. You will use the Lights dialog box once again to add a new light.

Pick: **Lights icon**

Pick on the name Point Light and from the list of light types that appears,

Pick: **Spotlight**

Pick: **New**

The New Spotlight dialog box appears on your screen, as shown in Figure 17.10. You will use it to name and select the attributes for the spotlight.

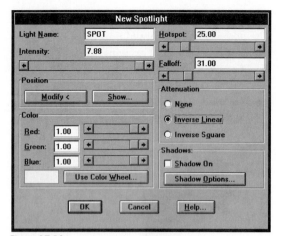

Figure 17.10

Type: SPOT (in the text box to the right of Light Name)

Set the intensity for the spotlight by moving the scroll bar to the highest value possible, or type a value near 8.00 in the box to the right of Intensity.

Now you are ready to position the light in your drawing. When positioning a spotlight, you first specify the target toward which it will point. Then you give the location for the light. In the Position area of the dialog box,

Pick: Modify

You will press ↵ to accept the current location for the target (near the center of the object), then type in the location for the light at the next prompt:

Enter light target <current>: ↵

Enter light location <current>: **14,4,–6** ↵

The New Spotlight dialog box returns to your screen. Next, you will set the *falloff* and *hotspot* for the spotlight. Falloff is the angle of the cone of light from the spotlight. Hotspot is the angle of the cone of light for the brightest area of the beam. The angles for the falloff and hotspot must be between 0 and 160°. AutoVision will not allow you to set Hotspot to a greater value than Falloff.

On your own, use the slider bar to set the value for Hotspot to 25.00 and the value for Falloff to 31.00. When you are finished, your dialog box should have settings similar to those in Figure 17.10. If so, pick OK to exit the New Spotlight dialog box and then pick OK once again to exit the Lights dialog box on your own.

You return to the AutoCAD drawing editor. You will use the Render dialog box to render your drawing again to see the effect of adding the spotlight to the drawing.

Pick: Render icon

Check the settings in the dialog box to make sure that they are the same ones that appear in Figure 17.4, then,

Pick: Render

The drawing is rendered to your screen and should appear similar to Figure 17.11. Notice the bright area of the spotlight beam, which you set using Hotspot. The size of the entire circle of light from the spotlight is determined by the setting you entered for Falloff.

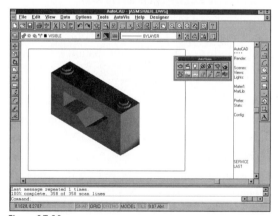

Figure 17.11

Choosing Materials

 You can also improve your rendered drawing by selecting or creating realistic-looking materials for the objects in it. You can either use the materials that are provided in the Materials Library or create your own materials using AutoVision. As with lights, a block is added to the drawing to represent the material. To add materials,

Pick: **Materials icon**

The Materials dialog box appears on your screen, as shown in Figure 17.12.

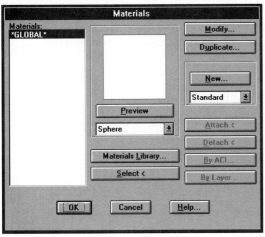

Figure 17.12

You will use the Materials dialog box to select materials from the Materials Library for the objects in your drawing. A single block or solid object in the drawing can have only one material assigned. This is usually not a problem in engineering drawings, because you typically make each part in an assembly a separate object or block in the drawing. Do not use Boolean operators to union together parts that you want to assign different materials.

To select from the library of pre-made materials,

Pick: **Materials Library** *(from the lower center of the dialog box)*

The Materials Library dialog box shown in Figure 17.13 appears on your screen. You will choose materials for your assembly drawing from the list of materials shown at the right-hand side of the dialog box.

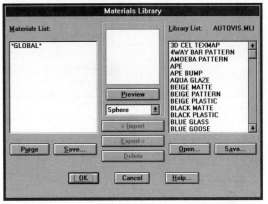

Figure 17.13

Use the scroll bar or pick on the arrows to scroll through the list of materials until you see the selection BRASS VALLEY, which you will use for the washers in the assembly drawing. Position the cursor over the name BRASS VALLEY and press the pick button to select it. The material becomes highlighted in the list.

Many of the materials dialog boxes allow you to preview the selection to see approximately what the material will look like. This saves time in case you do not want to use the particular material once you see it, or your system is not capable of displaying it. Remember that rendering a complex drawing can be time-consuming. Use the Preview function to see what different materials would look like.

Pick: **Preview**

The material you selected is shown applied to a spherical shape in the Preview box, as shown in Figure 17.14.

Figure 17.14

■ *Warning:* If you do not see it, you may not have set the variable for Avemaps before you started Windows, or perhaps your system cannot display the colors being used. If this happens, open your Preferences dialog box. Confirm that the Environment card Support path includes the directories *c:\av\avwin* and *c:\av\avis_sup*, and that the Render card Map Files path includes the directories *c:\av\maps* and *c:\av\tutorials*. ■

On your own, pick on the Sphere selection that is shown below the Preview button and change it to Cube. Preview the material once again. Return the selection to Sphere. Once you have previewed the materials, you will import BRASS VALLEY so that it is available for use in your drawing. Note that you can preview only one material at a time, but you can select several materials to import at once.

Pick: **Import**

The selected material is shown in the Materials List at the left of the dialog box. Now you are ready to select another material.

Now highlight the materials GRAY MATTE and GRAY SEMIGLOSS on your own from the list of materials at the right of the dialog box. Pick Import to make the materials available for use in the current drawing. The names are

added to the list at the left. When you have finished importing the materials,

Pick: **OK** *(to exit the Materials Library dialog box)*

You return to the Materials dialog box, where you will attach the materials to the objects in your drawing.

Select: **BRASS VALLEY** *(from the Materials list at the left of the dialog box)*

Pick: **Attach** *(at the right of the dialog box)*

You are returned to your drawing to select the objects, in this case the two washers. If you accidentally select something else, use the Select Remove option to take it out of the selection set, or use Detach later in the dialog box to detach the wrong material. If you have trouble selecting, try to pick on an edge, not in the center of an object, or use implied Windowing.

Pick: **(on the edge of each of the two washers)**

Press: ⏎

The Materials dialog box returns to your screen so that you can continue attaching materials.

Select: **GRAY SEMIGLOSS**

Pick: **Attach**

Pick: **(on the edge of the two screws)**

Press: ⏎

Next, attach the material GRAY MATTE to the base and cover of the assembly.

Select: **GRAY MATTE**

Pick: **Attach**

Pick: **(on the edge of the base and the cover)**

Press: ⏎

Pick: **OK** *(to exit the Materials dialog box)*

Now that you are finished selecting the materials, you will render the drawing once again in order to see the results of the changes you have made.

Pick: **Render icon**

Accept the same defaults that you have been using previously.

Pick: **Render**

Your drawing should now show the new materials. Your screen should look similar to Figure 17.15.

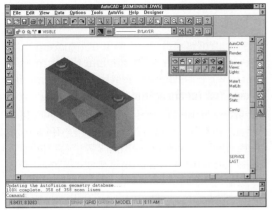

Figure 17.15

You can also use AutoVision to create your own materials. Next you will create a material that looks like steel to use for the base and cover of the assembly.

Pick: **Materials icon**

When the Materials dialog box appears,

Pick: **New** *(from the right-hand side of the dialog box)*

The New Standard Material dialog box shown in Figure 17.16 appears on your screen.

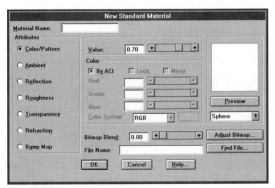

Figure 17.16

Type: **STEEL** *(in the text box to the right of Material Name)*

At the left of the New Standard Material dialog box are the various qualities, or material attributes. They are Color/Pattern, Ambient, Reflection, Roughness, Transparency, Refraction, and Bump Map. These are the attributes that you can control to create materials with different appearances. In general, these qualities tell how the surface or material responds when light strikes it. To change these attributes, pick in the circle to the left of the attribute for which you want to set the values.

Color/Pattern

This selection determines the color for the material. The default selection for color is By ACI (AutoCAD Color Index), which results in the object being rendered in the color assigned to its lines in the AutoCAD model. You saw this when you rendered the assembly for the first time and the objects were shown in the original layer colors. If you unselect By ACI in the Color area of the dialog box, you can select the other Color options. Near the bottom of the Color area the selection RGB appears. RGB stands for Red, Green, Blue, the primary colors of light. You can use the slider bars in that section of the dialog box to set how much of each color of light is used to make up the material color. As you move the slider bars, the resulting color is shown in the empty box at the bottom of the Color area. This is called the color *swatch*.

Keep in mind that colors on your computer screen are made of the primary colors of light (Red, Green, and Blue), not the primary colors of pigment (Red, Blue, and Yellow) that you may be used to using. Changing the selection at the bottom of the Color area from RGB to HLS (Hue, Lightness, Saturation) lets you use the same slider bars to define the color in terms of these qualities instead of RGB. When you have changed the color slightly, pick Preview.

In addition, you can use a bitmap (a pattern or picture), such as a scanned image or drawing created with a paint-style program, instead of just solid color. To do this, in the lower center area of the dialog box, type the name of a bitmap file in the text box to the right of File Name. You can use bitmaps with the selections Color/Pattern, Reflectivity, Transparency, and Bump Map. The Bitmap Blend value determines how much of the underlying object color shows through the bitmap. Values less than 1.00 allow some of the underlying object color to show through. The lower the value, the more of the underlying color shows through the bitmap.

Ambient

This selection allows you to set the color for the object's shadow. You can use the same color controls as discussed above.

Reflection

This selection allows you to adjust the color of the material's reflective highlight. When you select ray tracing, you can specify the value for the material's reflectivity from 0, where it does not reflect any light, to 1, where it is perfectly reflective, like a mirror.

Roughness

Roughness lets you specify the surface *roughness* of the material. Roughness is also related to reflectivity. The less rough the surface is, the more reflective it will be. Change the value and use Preview to see how roughness changes the size of the material's highlight.

Transparency

The settings for *transparency* allow you to make objects that are clear, like glass, or somewhat clear, like colored liquids. Keep in mind that objects that are completely transparent, so that you do not even see them in the drawing when they are rendered, still take time to render.

Refraction

This control is only used with transparent objects. Refraction is the amount that light is bent when entering and leaving a transparent object. You must use the AutoVis Raytrace rendering algorithm in order to see the effect of this setting. Ray tracing uses a more sophisticated rendering algorithm to generate reflections, refractions, and detailed shadows. In ray tracing, the result is calculated for each "ray" of light striking the object. This produces very realistic and detailed effects, but increases the complexity and rendering time.

Bump Map

The Bump Map selection allows you to enter the file name of a *bump map*. Bump maps convert the color intensity or grayscale information to heights, to give the appearance that features are raised above the surface, like embossed letters.

Next you will set the attributes for the material STEEL that you are going to create. On your own, make sure that Color/Pattern is selected and By ACI is turned off.

Then use the slider bars to adjust the colors for Red, Green, and Blue so that Red and Green are each set to a value of .88 and Blue is set to .92. Notice how the color swatch changes as you move the slider bars. The resulting color should be gray. Preview the result on your own.

Next you will make the surface somewhat reflective and assign a small value for surface roughness, so that the material will look like steel.

Pick: **Reflection**

Use the slider bar to set the value for Reflection to .65. Then,

Pick: **Roughness**

Set the value for Roughness to .30.

To see the effects of your selections on the material being created,

Pick: **Preview**

Your screen should appear similar to Figure 17.17.

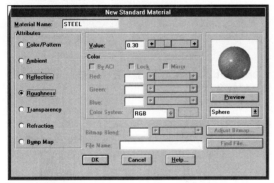

Figure 17.17

Pick: **OK *(to exit the New Standard Material dialog box)***

You will return to the Materials dialog box. Next you will attach this material to the base and cover in the assembly drawing in place of GRAY MATTE. To do this, use the dialog box on your own to select the material STEEL that you created and attach it to the base and cover. When you have attached the material, pick OK to exit the dialog box. Then select the AutoVision Render command and render the drawing to the display on your own.

Adjusting Mapping

 When bitmaps, or materials that use bitmapped patterns, are applied to surfaces in your drawing, you can control the type of mapping, coordinates, scale, repeat, and other aspects of the maps'

appearance. Next you will choose to apply a bitmapped material to the base part of the assembly to see these effects.

Pick: **Materials icon**

Pick: **Materials Library**

Pick: **V Pattern *(from the list of materials at the right)***

Pick: **Import**

Pick: **OK *(to exit the Materials Library dialog box)***

Pick: **Attach**

Select objects to attach "V PATTERN" to: *(pick the base)* ⏎

Pick: **OK *(to exit the Materials dialog box)***

The V Pattern material is attached to the base part of the drawing. On your own, render the current scene to the viewport. Your screen should look like Figure 17.18.

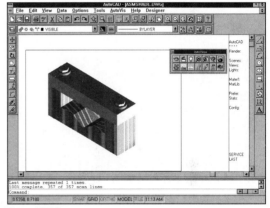

Figure 17.18

Pick: **Mapping icon**

The Setuv command is echoed at the command prompt, followed by the prompt,

Select objects: *(pick the base)*

Select objects: ⏎

The Mapping dialog box appears on your screen, as shown in Figure 17.19. To preview the mapping for the Planar method of projection, which is currently selected,

Pick: **Preview**

The preview with the bitmapped material shown using planar projection shows in the upper right of the dialog box, as you see in Figure 17.19.

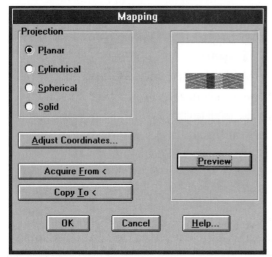

Figure 17.19

To adjust the mapping for a cylindrical or spherical object, you can pick the Cylindrical or Spherical radio buttons from the left of the dialog box. Try each one on your own, and use Preview to see the effect of that projection method. In general, you would use cylindrical mapping for cylindrical objects, spherical mapping for spherical objects, and so forth. Since the base is composed of plane surfaces, on your own, return the selection to Planar. Solid mapping is a special type that you should use when you use granite, marble, and wood solid materials. Figure 17.20 shows the Materials dialog box with the list of these materials pulled down. Picking one of these materials and then the New button from the Materials dialog box displays a dialog box you can use to create custom materials of that type.

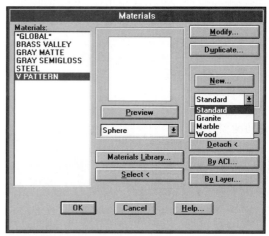

Figure 17.20

If you experimented with the solid materials, return to the Mapping dialog box on your own. Next you will adjust the mapping coordinates to change the direction for the pattern.

Pick: **Adjust Coordinates**

The Adjust Planar Coordinates dialog box appears on your screen.

Pick: **WCS XZ Plane**

Pick: **Preview**

The preview of the object with the pattern applied in the XZ direction appears, as shown in Figure 17.21.

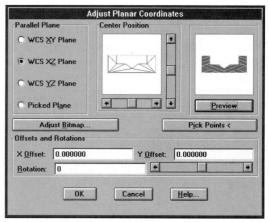

Figure 17.21

Next you will change the scale of the pattern.

Pick: **Adjust Bitmap**

The Adjust Object Bitmap Placement dialog box appears on your screen. On your own, use it to change the scale to 2.5 and preview the result, as shown in Figure 17.22.

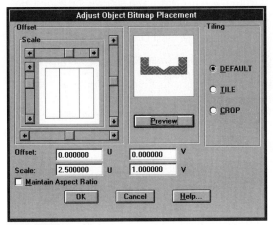

Figure 17.22

In addition to changing the scale for the bitmapped image, you can select the radio buttons at the right of the dialog box to tile or crop the bitmapped image. Tiling the bitmapped image causes it to repeat its pattern across the object. Crop uses the bitmapped image like a decal: it only appears at the single location specified by the offsets. When you create materials that use bitmaps, they can be either tiled or cropped. The DEFAULT selection uses whatever was set for the material as a whole when it was created. Mapped coordinates are given relative to the U and V directions. The U and V coordinates are equivalent to X and Y coordinates for the map. The letters U and V are used to indicate that they are part of a local coordinate system for the map that is independent of the X and Y coordinate system of the drawing. You specify offset and scale using these U and V directions. When you select Maintain Aspect Ratio, the U and V sliders work as a unit to maintain the proportion of the bitmapped

pattern. The scale and offsets set in this dialog box are in addition to the ones set for the material as a whole. The offset is added to the materials offset and the scale is multiplied times the scale for the material. To exit the dialog box,

Pick: **OK**

Pick: **OK (to exit the Adjust Planar Coordinates dialog box)**

Pick: **OK (to exit the Mapping dialog box)**

Render the drawing on your own.

Using mapped materials is a good way to add detail to your rendered drawing without creating very complex models. For example, you could model every brick in a wall as a separate solid and create a different solid for the cement between them. This would create a very complex solid model with a lot of vertices. Instead, you can create the entire wall as a single flat surface and then apply the appearance of brick to it as a bitmapped pattern. This will create a drawing that looks very detailed, but has far fewer vertices and takes less file space than the previous example. You can also combine bitmaps and bump maps in the same drawing to give a raised appearance to parts of the material.

Rendering to a File

Rendering your drawing to a file makes it possible to use it in other programs. For example, you can render your drawing to a Windows metafile format and paste it into a wide variety of Windows applications.

Pick: **Render icon**

Use the Destination area of the Render dialog box to select File on your own. (Pick on the word Viewport and use the list that appears to make your selection.)

Pick: **File**

Pick: **More Options (below Destination)**

The File Output Configuration dialog box appears on your screen, as shown in

Figure 17.23. You can use the selection at the upper left of the dialog box to save in many different file formats, among them TGA (TARGA), TIFF, GIF, and RND. If you have a paint-style program that uses one of these formats, you may be able to edit and print the saved image file. You can often also use these formats to insert the drawing into a word processing program. You can set the resolution for the output file using the selection below File Type, where you see 640 × 480 (VGA) and the number of colors for the resulting file. You do not need to change any of the settings here. To save the image file using the defaults,

Pick: **OK**

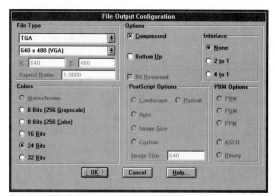

Figure 17.23

The file will be saved with the extension TGA, since you accepted the defaults. To render to the file,

Pick: **Render**

Use the Rendering File dialog box that appears to specify the name and directory in which to save the file. On your own, try importing your rendered drawing as a graphic into your Windows word processor, then return to your AutoCAD drawing session.

Using the Sun Angle Calculator

The sun angle calculator is particularly useful in architectural drawings for showing the different

effects that sunlight has on the rendered scene during different times of the day and year. You can use the sun angle calculator to easily figure these effects, instead of the laborious process of looking the information up in tables and adding the correct lighting effects to your drawing. You can even set the geographic location by picking it from the map. Since the sun will be added as a distant light in your rendered drawing,

Pick: **Lights icon**

Pick: **Distant Light** *(from the list of light types that pulls down to the right of New)*

Pick: **New**

The New Distant Light dialog box shown in Figure 17.24 appears on your screen.

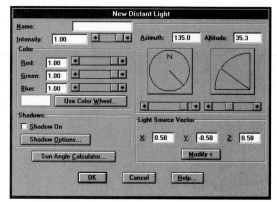

Figure 17.24

Type: **SUN** *(in the text box to the right of Name)*

You can use the slider bars or the color wheel to set the color for the distant light. Leave it white for now, as the sun usually casts white light, which is composed of the full spectrum of light colors. The area at the top right of the dialog box allows you to specify the azimuth and altitude for the sun if you want. You can use the Sun Angle Calculator button at the bottom left of the dialog box and let it determine these values for you. You will do that now.

Pick: **Sun Angle Calculator**

The Sun Angle Calculator dialog box appears on your screen. Using the Sun Angle Calculator dialog box, you can specify the date, time of day, time zone, latitude, and longitude to determine the angle of the sun.

Set the date to March 20 (3/20) and the clock time to 6:00 hours (6 o'clock a.m.) MST on your own. Turn Daylight Savings on by picking in the box to its left so that an X is displayed in the box. Notice the azimuth and altitude settings change as you make selections. When you are finished the dialog box should appear as shown in Figure 17.25.

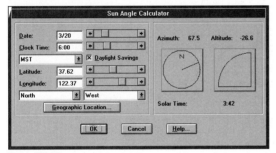

Figure 17.25

Next you will specify the geographic location, because it works with date and time of day to determine the angle of the sun.

Pick: **Geographic Location**

The Geographic Location dialog box shown in Figure 17.26 appears on your screen.

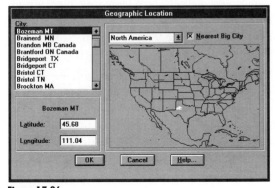

Figure 17.26

Move the cursor over the map until it is positioned as you see in Figure 17.26 and press the pick button. Notice that Bozeman, MT (cultural Mecca and center of the universe) appears as the Nearest Big City in the list at the left. Below, the latitude and longitude for Bozeman are displayed. You can also scroll down the list of cities to make a selection. Notice the words North America at the upper left of the map area.

Pick: **(on the words North America)**

A list of the other continents pulls down, as shown in Figure 17.27. You can use the scroll bar to select from the other continents. The major cities of other continents are stored in the database and you can select them by picking from the map or by scrolling down the list of cities.

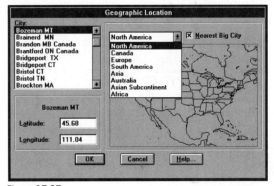

Figure 17.27

Do not change the continent selection at this time. Leave it set to North America.

Pick: **OK (to exit the Geographic Location dialog box)**
Pick: **OK (to exit the Sun Angle Calculator dialog box)**
Pick: **OK (to exit the New Distant Light dialog box)**
Pick: **OK (to exit the Lights dialog box)**

Your drawing returns to the screen. Now you will render the drawing again to see the part as though the sun were shining on it at 6 o'clock a.m. on March 20 in Bozeman, Montana.

On your own, use the Render dialog box and set the Destination back to Viewport. Render the drawing and notice the effect of the sunlight.

Then you will modify the light, SUN, in the Lights dialog box. On your own, select the Lights command and pick SUN as the light you want to modify. Change the time of day to 4 o'clock p.m. (The time-of-day clock is a 24-hour clock.) Render the drawing again and notice the change. Although this assembly is not an architectural drawing, you can see what a great advantage the sun angle calculator is for architectural uses. You can also use it to evaluate the suitability of a site for solar power, and other heat transfer analysis.

Adding Backgrounds

 You can add backgrounds to your rendering, using the Background icon on the AutoVision toolbar. Backgrounds can be solid colors, gradient colors (which blend one to three colors), or images (bitmaps), or you can use Merge to select the current AutoCAD image as the background.

Pick: **Background icon**

You will set up the Background dialog box so that it looks like Figure 17.28. You will use it to select an image as the background. Using the radio buttons across the top of the dialog box,

Pick: **Image**

Pick: **Find File**

Select the file *c:\av\maps\cloud.tga* from the dialog box that appears. When it is selected, pick OK on your own to return to the Background dialog box. You can adjust the appearance of the bitmap, as you did earlier in the tutorial, by picking Adjust Bitmap. Environment lets you use an image that AutoVision maps onto a sphere surrounding your drawing. This environment is then reflected onto other objects when you use reflective and refractive materials in your drawing. When you use the AutoVision renderer, reflective objects mirror the background image. When you use the Raytrace renderer, the environment is raytraced.

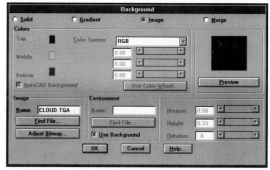

Figure 17.28

Pick: **OK** *(to exit the Background dialog box)*

On your own, render your drawing to show the cloud background.

Next you will restore the world coordinate system so that the landscape objects you add will have the correct orientation.

Command: UCS ⏎

Origin/ZAxis/3point/OBject/View/X/Y/Z/Prev/Restore/ Save/Del/?/<World>: ⏎

The UCS icon changes back to the WCS.

Using Landscape Objects

 Landscape objects are bitmapped images that are applied to *extended entity* objects and added to your drawing. Landscape objects have grips at the base, top, and corners that you can use to move, scale, rotate, and otherwise edit them. You can create your own landscape objects and add them to the landscape library. To add a landscape object to the drawing,

Pick: **Landscape New icon**

The Landscape New dialog box appears on your screen, as shown in Figure 17.29. On your own, select People #2. Change Height to 6.0 and pick Preview to preview the person you will add to your drawing. To locate the landscape object in your drawing,

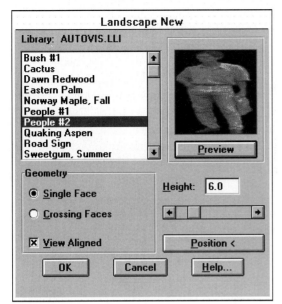

Figure 17.29

Pick: **Position**

Choose the location of the base of the landscape object:
   5,5,0 ⏎

You return to the Landscape New dialog box.

Pick: **OK** *(to add the landscape object and exit the dialog box)*

The landscape object is added to your drawing. It appears as a triangular shape. On your own, use Zoom Out to zoom out so that you can see the entire landscape object. If necessary, use Zoom Window afterwards to size your view of the landscape object and the assembly. Render the drawing on your own. While the effect of the background and person are not particularly applicable to this engineering assembly drawing, you can see that you can create realistic effects fairly easily, using backgrounds and landscape objects to add to the realism of the rendering.

Exporting to 3D Studio

You can use the materials and lights that you have created using AutoVision with Autodesk's 3D Studio animation program. Once your drawing is exported for use with 3D Studio, you can import your drawing into 3D Studio and add 3D animation effects. AutoVision also has its own animation commands, which you can use if you do not have 3D Studio available.

> ■ *TIP* Solid materials like granite, marble, and wood do not export to 3D Studio. ■

To export to 3D Studio, use the File selection at the left of the menu bar to pick the Export command.

Pick: **File, Export**

The Export Data dialog box shown in Figure 17.30 appears on your screen.

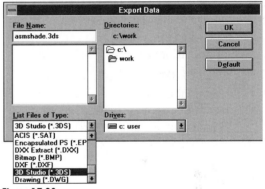

Figure 17.30

Use the selection under the heading List Files of Type to select 3D Studio (*.3DS) as the type of file to export. Use the dialog box to select the *c:\work* directory and name your drawing file *asmshade.3ds* on your own. The file extension .3DS will automatically be assigned to your file once you pick the 3D Studio export type. When you are finished naming the file, pick OK on your own to exit the dialog box.

   Select objects: **ALL** ⏎

   Select objects: ⏎

When you exit the file naming dialog box, the 3D Studio File Export Options dialog shown in Figure 17.31 appears on your screen. You will use it to set the method that 3D Studio should use to determine individual objects, and the smoothing and welding to be done.

Figure 17.31

3D Studio can determine which objects are individual items and can therefore be moved and animated separately from one another by three methods. If you select the Layer option, each object that you want to have as a separate 3D Studio object must be on its own layer in the drawing. This method works quite well if the objects are on separate layers. (It is easy to change the layer of an object, using AutoCAD's Properties icon, if they are not on separate layers.) Choosing AutoCAD Color Index creates the object from objects that have the same AutoCAD color. Choosing Object Type creates individual objects from the same type of object in the drawing. Arcs, lines, and polylines must have a non-zero thickness in order to be exported as 3D Studio objects.

> Pick: **AutoCAD Object Type** *(from the selections available in the dialog box)*

> Pick: **Override** *(so that each block is one object)*

Auto-smoothing makes faceted surfaces appear smooth in the drawing. The smooth appearance of contoured objects, such as spheres, is controlled by the Facetres variable in AutoCAD. Setting Facetres to a higher value causes the object to appear to be made of more facets approximating the surface. Even when a lower value for Facetres is used, AutoVision and 3D Studio have the ability to make such an object appear smooth and round, like a ball. When Auto-Smoothing is turned on, the setting in the Degrees area specifies the maximum number of degrees for smoothing; in other words, surfaces that meet with an angle greater than the number of degrees specified will not be smoothed. 30° is the default.

Leave Auto-Smoothing turned on and the angle set to 30°.

Auto-Welding joins vertices that are no farther apart than the specified *threshold value* into one vertex. This simplifies complex geometry so that it is quicker to animate and render in 3D Studio. As in AutoVision, the complexity of the objects and the demands of the lighting and shading can make rendering and animating slow. Simplifying the drawing in this way can improve the speed of rendering and animating in 3D Studio. For the assembly drawing, multiple vertices are not close together, so this will not have a large effect.

Leave Auto-Welding turned on and Threshold set to .001.

> Pick: **OK** *(to export the assembly drawing to 3D Studio format)*

Your computer will take a moment to process the file and export it into 3D Studio format. When it is finished, you are ready to exit AutoCAD. Close the AutoVision toolbar, save your drawing, and exit AutoCAD.

ambient	intensity	resolution
back faces	material	roughness
bump map	normal	scene
distant light	Phong	spotlight
extended entity	point light	surface map
falloff	ray tracing	swatch
Gourard	reflection	threshold value
hotspot	refraction	transparency

KEY COMMANDS

Background	Lights	Materials
Landscape New	Mapping	Scenes

Use solid modeling to create the object shown below or retrieve your file from the exercises you did for Tutorial 10. Produce three different renderings of the object, varying the type, location, intensity, and color of the light. Experiment with different materials and backgrounds.

 17.1 Clamping Block

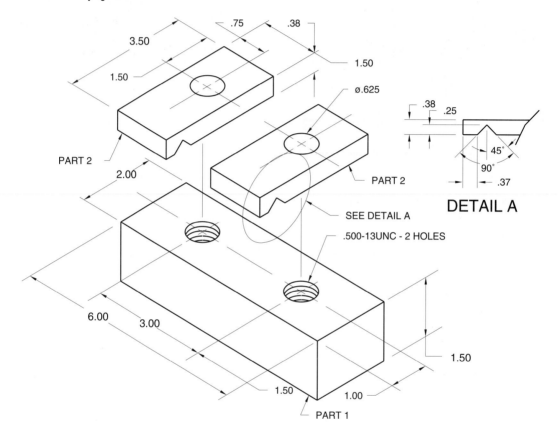

17.2 Pressure Assembly

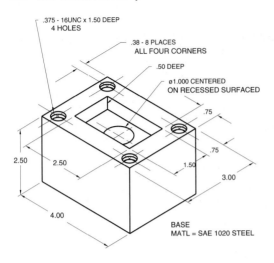

.375 - 16UNC x 1.50 DEEP
4 HOLES

.38 - 8 PLACES
ALL FOUR CORNERS

.50 DEEP

ø1.000 CENTERED
ON RECESSED SURFACED

.75

.75

2.50

2.50

1.50

3.00

4.00

BASE
MATL = SAE 1020 STEEL

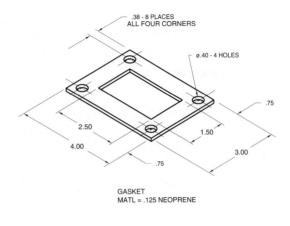

.38 - 8 PLACES
ALL FOUR CORNERS

ø.40 - 4 HOLES

.75

2.50

1.50

4.00

.75

3.00

GASKET
MATL = .125 NEOPRENE

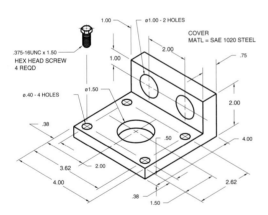

1.00

ø1.00 - 2 HOLES

COVER
MATL = SAE 1020 STEEL

.375-16UNC x 1.50
HEX HEAD SCREW
4 REQD

2.00

1.00

.75

ø.40 - 4 HOLES

ø1.50

2.00

.38

.50

4.00

3.62

2.00

2.62

4.00

.38

1.50

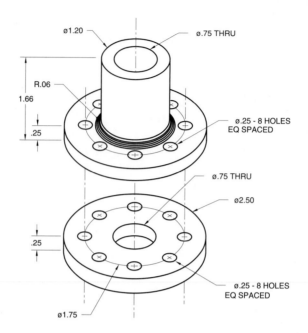

ø1.20

ø.75 THRU

R.06

1.66

.25

ø.25 - 8 HOLES
EQ SPACED

ø.75 THRU

ø2.50

.25

ø.25 - 8 HOLES
EQ SPACED

ø1.75

17.4 Turbine Housing Assembly

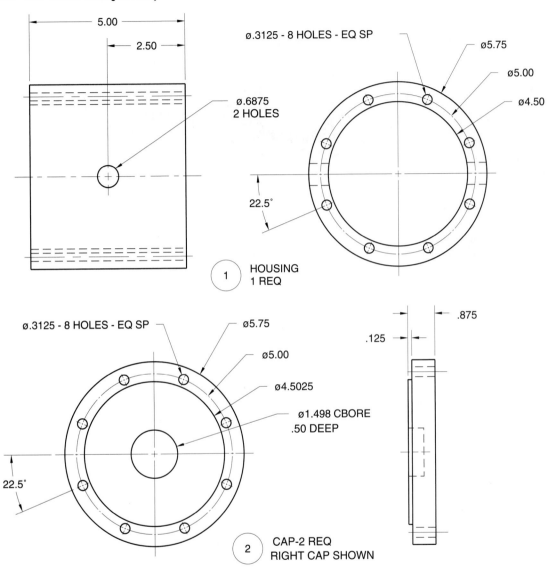

5.00

2.50

ø.6875
2 HOLES

ø.3125 - 8 HOLES - EQ SP

ø5.75

ø5.00

ø4.50

22.5°

1 HOUSING
1 REQ

ø.3125 - 8 HOLES - EQ SP

ø5.75

ø5.00

ø4.5025

ø1.498 CBORE
.50 DEEP

22.5°

.875

.125

2 CAP-2 REQ
RIGHT CAP SHOWN

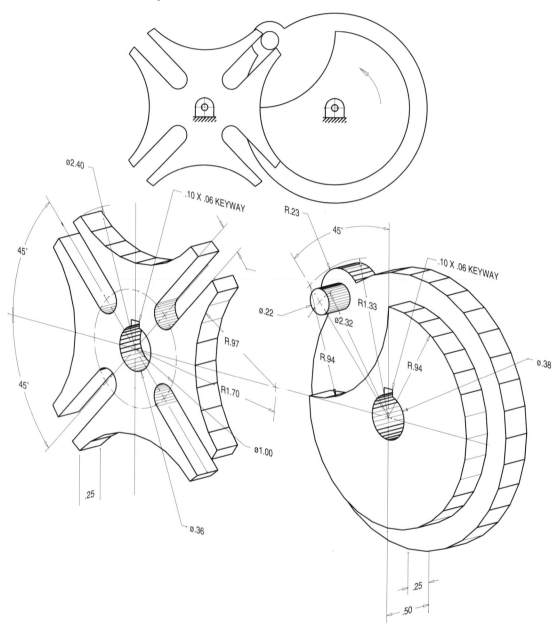

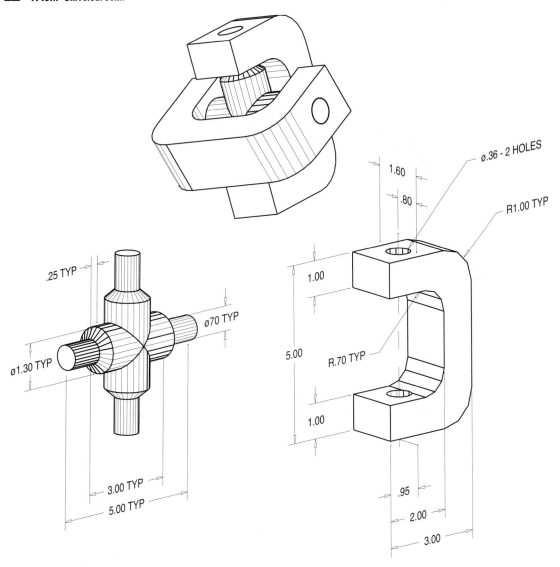

17.6M Universal Joint

.25 TYP

ø70 TYP

ø1.30 TYP

3.00 TYP

5.00 TYP

1.60

.80

ø.36 - 2 HOLES

R1.00 TYP

1.00

5.00

R.70 TYP

1.00

.95

2.00

3.00

17.7 Staircase

Retrieve your file from Exercise 10.6 and render the stairs using wood materials and a distant light source.

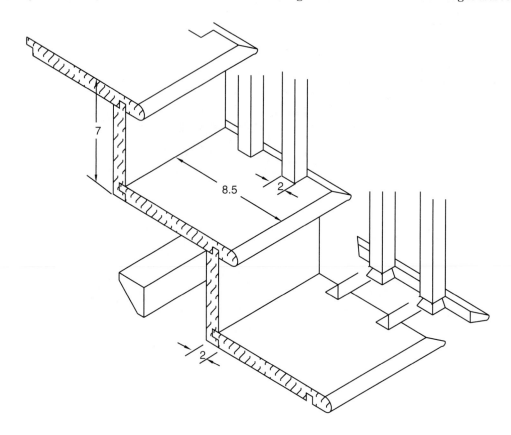

17.8M Pulley Arm

Use solid modeling to create a pictorial view like the one shown. Render the model using AutoVision to create a pictorial section.

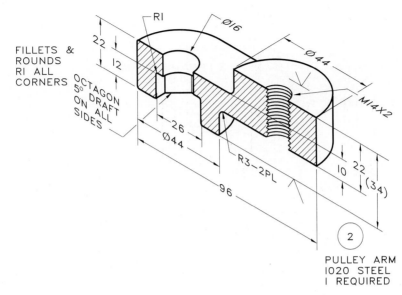

FILLETS &
ROUNDS
RI ALL
CORNERS

OCTAGON
5° DRAFT
ON ALL
SIDES

RI

Ø16

Ø44

M14X2

22

12

26

Ø44

R3-2PL

96

22

10

(34)

2

PULLEY ARM
1020 STEEL
1 REQUIRED

17.9M Step Bearing

Retrieve your file from Exercise 10.8M or 10.9M. Assign materials, create lighting and render your drawing to a file. Import the rendered drawing into your Windows word processor and add ball tags and a parts list to create a rendered isometric assembly drawing.

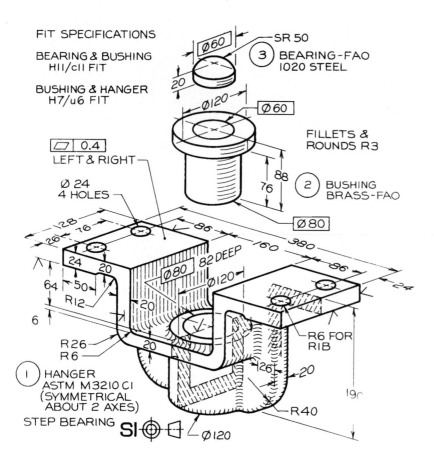

FIT SPECIFICATIONS

BEARING & BUSHING
  HII/cII FIT

BUSHING & HANGER
  H7/u6 FIT

SR 50

Ø60

3) BEARING-FAO
     1020 STEEL

20

Ø120   Ø60

FILLETS &
ROUNDS R3

0.4
LEFT & RIGHT

Ø 24
4 HOLES

88
76

2) BUSHING
     BRASS-FAO

Ø80

128
76
20
24 20
64 50
R12
6

86
160   380
86
82 DEEP
Ø80
Ø120
20
24

R26
R6

20

R6 FOR
RIB

1) HANGER
     ASTM M3210 CI
     (SYMMETRICAL
     ABOUT 2 AXES)

126   20
19

STEP BEARING SI⊕∏ Ø120

R40

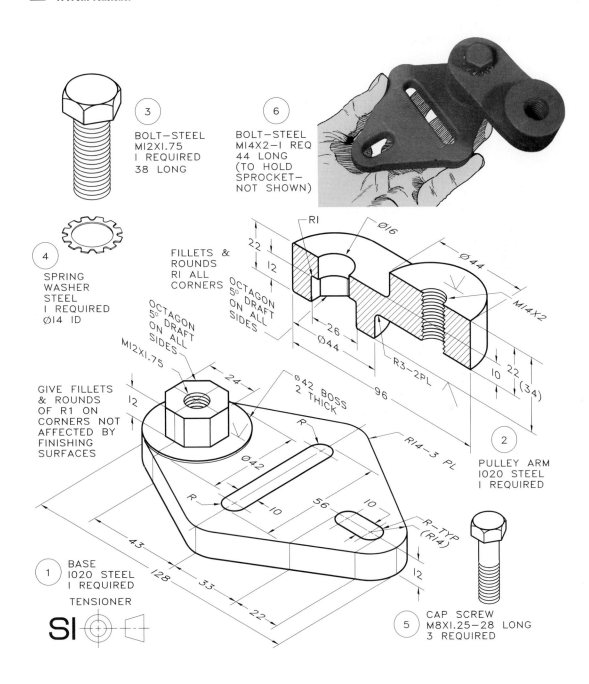

③
BOLT—STEEL
M12X1.75
1 REQUIRED
38 LONG

⑥
BOLT—STEEL
M14X2—1 REQ
44 LONG
(TO HOLD
SPROCKET—
NOT SHOWN)

④
SPRING
WASHER
STEEL
1 REQUIRED
∅14 ID

FILLETS &
ROUNDS
R1 ALL
CORNERS

OCTAGON
5° DRAFT
ON ALL
SIDES

R1

∅16

22

12

26

∅44

∅44

96

R3—2PL

∅44

M14X2

10

22

(34)

②
PULLEY ARM
1020 STEEL
1 REQUIRED

OCTAGON
5° DRAFT
ON ALL
SIDES

M12X1.75

24

12

∅42 BOSS
2 THICK

∅42

R

R14—3 PL

GIVE FILLETS
& ROUNDS
OF R1 ON
CORNERS NOT
AFFECTED BY
FINISHING
SURFACES

R

10

56

10

R—TYP
(R14)

43

128

33

22

12

①
BASE
1020 STEEL
1 REQUIRED

TENSIONER

SI ⊕ ◁

⑤
CAP SCREW
M8X1.25—28 LONG
3 REQUIRED

This grid includes the commands available for use with AutoCAD Release 13 for Windows. You should use AutoCAD's on-line help feature for complete information about the commands and their options.

Command or Icon Name	Description	Toolbar	Icon	Menu Bar	Command:
Acadprefix	System variable that stores the directory path, if any, specified by the ACAD environment variable.	N/A		N/A	ACADPREFIX
Acadver	System variable that stores the AutoCAD version number.	N/A		N/A	ACADVER
Aerial View	Opens the Aerial View window.	Standard (Aerial View flyout)		N/A	DSVIEWER
Align	Moves and rotates objects to align with other objects.	Modify (Rotate flyout)		N/A	ALIGN
Align Dimension Text Center	Centers dimension text.	Dimensioning (Align Dimension Text flyout)		N/A	DIMTEDIT
Align Dimension Text Home	Moves dimension text to Home position.	Dimensioning (Align Dimension Text flyout)		N/A	DIMTEDIT
Align Dimension Text Left	Aligns dimension text left.	Dimensioning (Align Dimension Text flyout)		N/A	DIMTEDIT
Align Dimension Text Right	Aligns dimension text right.	Dimensioning (Align Dimension Text flyout)		N/A	DIMTEDIT
Align Dimension Text Rotate	Rotates dimension text.	Dimensioning (Align Dimension Text flyout)		N/A	DIMTEDIT

Command or Icon Name	Description	Toolbar	Icon	Menu Bar	Command:
Aligned Dimension	Creates an aligned linear dimension.	Dimensioning		N/A	DIMALIGNED
AME Convert	Converts AME solid models to AutoCAD solid objects.	Solids		N/A	AMECONVERT
Angular Dimension	Dimensions an angle.	Dimensioning		N/A	DIMANGULAR
Aperture	Controls the size of the object snap target box.	N/A		N/A	APERTURE
Application Load	Loads AutoLISP, .ads, and .arx files.	N/A		Tools, Applications	'APPLOAD
Arc 3 Points	Draws an arc based on three points.	Draw		N/A	ARC or A
Arc Center Start Angle	Draws an arc of a specified angle based on the center and start points.	Draw		N/A	ARC or A
Arc Center Start End	Draws an arc based on the center start and endpoints.	Draw		N/A	ARC or A
Arc Center Start Length	Draws an arc of a specified length based on the center and start points.	Draw		N/A	ARC or A
Arc Continue	Draws an arc from the last point of the previous line or arc drawn.	Draw (Arc flyout)		N/A	ARC or A
Arc Start Center Angle	Draws an arc of a specified angle based on the start and center points.	Draw (Arc flyout)		N/A	ARC or A
Arc Start Center End	Draws an arc based on the start, center and endpoints.	Draw (Arc flyout)		N/A	ARC or A
Arc Start Center Length	Draws an arc of a specified length based on the start and center points.	Draw (Arc flyout)		N/A	ARC or A
Arc Start End Angle	Draws an arc of a specified angle based on the start and endpoints.	Draw (Arc flyout)		N/A	ARC or A

Command or Icon Name	Description	Toolbar	Icon	Menu Bar	Command:
Arc Start End Direction	Draws an arc in a specified direction based on the start and endpoints.	Draw (Arc flyout)		N/A	ARC or A
Arc Start End Radius	Draws an arc of a specified radius based on the start and endpoints.	Draw (Arc flyout)		N/A	ARC or A
Area	Finds the area and perimeter of objects or defined areas.	Object Properties (Inquiry flyout)		N/A	AREA
Attach	Uses another drawing as an external reference, without really adding it to your current drawing.	External Reference		N/A	XREF
Attdia	System variable that controls whether the Insert command uses a dialog box for attribute value entry.	N/A		N/A	ATTDIA
Attribute Display	Globally controls attribute visibility.	N/A		Options, Display, Attribute Display	ATTDISP
Audit	Evaluates the integrity of a drawing.	N/A		File, Management, Audit	AUDIT
AutoCAD	Refreshes the screen menu and clears any active commands.	N/A		N/A	N/A
Back View	Sets the viewing direction for a three-dimensional visualization of the drawing to the back.	Standard (View flyout)		View, 3D Viewpoint Presets, Back	VPOINT OR DDVPOINT
Base	Specifies the insertion point for a block or drawing.	N/A		N/A	BASE
Baseline Dimension	Continues a linear, angular, or ordinate dimension from the baseline of the previous or selected dimension.	Dimensioning		N/A	DIMBASELINE

Command or Icon Name	Description	Toolbar	Icon	Menu Bar	Command:
Bind All	Binds all dependent symbols of an external reference to a drawing.	External Reference (Bind flyout)		N/A	XREF
Bind Block	Binds blocks of an external reference to a drawing.	External Reference (Bind flyout)		N/A	XBIND
Bind Dimension Style	Binds dimension styles of an external reference to a drawing.	External Reference (Bind flyout)		N/A	XBIND
Bind Layer	Binds layers of an external reference to a drawing.	External Reference (Bind flyout)		N/A	XBIND
Bind Linetype	Binds linetypes of an external reference to a drawing.	External Reference (Bind flyout)		N/A	XBIND
Bind Text Style	Binds text styles of an external reference to a drawing.	External Reference (Bind flyout)		N/A	XBIND
Block	Groups a set of objects together so that they are treated as one object.	Draw (Block flyout)		N/A	BLOCK
Bottom View	Sets the viewing direction for a three-dimensional visualization of the drawing to the bottom.	Standard (View flyout)		View, 3D Viewpoint Presets, Bottom	VPOINT or DDVPOINT
Boundary	Creates a region or poly-line of a closed boundary.	Draw (Polygon flyout)		N/A	BOUNDARY
Boundary Hatch	Automatically fills the area you select with a pattern.	Draw (Hatch flyout)		N/A	BHATCH
Box Center	Creates a three-dimensional rectangular-shaped solid from a specified center point.	Solids (Box flyout)		N/A	BOX
Box Corner	Creates a three-dimensional rectangular-shaped solid specifying a point for the first corner.	Solids (Box flyout)		N/A	BOX

Command or Icon Name	Description	Toolbar	Icon	Menu Bar	Command:
Break 1 Point	Removes part of an object or splits it in two.	Modify (Break flyout)		N/A	BREAK
Break 1 Point Select	Removes part of an object or splits it in two.	Modify (Break flyout)		N/A	BREAK
Break 2 Points	Removes part of an object or splits it in two.	Modify (Break flyout)		N/A	BREAK
Break 2 Points Select	Removes part of an object or splits it in two.	Modify (Break flyout)		N/A	BREAK
Calculator	Calculates mathematical and geometrical expressions.	Standard (Object Snap flyout)		Tools, Calculator	'CAL or CAL
Cancel	Exits current command without performing operation.	N/A		N/A	(ESC)
Celtype	System variable that sets the linetype of new objects.	N/A		N/A	CELTYPE
Center Mark	Draws center marks and centerlines of arcs and circles.	Dimensioning		N/A	DIMCENTER
Chamfer	Draws a straight line segment (called a chamfer) between two given lines.	Modify (Chamfer flyout)		N/A	CHAMFER
Circle Center Diameter	Draws circles of any size.	Draw (Circle flyout)		N/A	CIRCLE or C
Circle Center Radius	Draws circles of any size.	Draw (Circle flyout)		N/A	CIRCLE or C
Circle Tan Tan Radius	Draws circles of any size.	Draw (Circle flyout)		N/A	CIRCLE or C
Circle 3 Point	Draws circles of any size.	Draw (Circle flyout)		N/A	CIRCLE or C
Circle 2 Point	Draws circles of any size.	Draw (Circle flyout)		N/A	CIRCLE or C
Clip	Inserts and clips an external reference to a drawing.	External Reference		N/A	XREFCLIP

Command or Icon Name	Description	Toolbar	Icon	Menu Bar	Command:
Cmddia	System variable controls whether dialog boxes are enabled for more than just Plot and external database commands.	N/A		N/A	CMDDIA
Color Control	Sets the color for new objects.	Object Properties		Data, Color	'COLOR or COLOR or 'DDCOLOR or DDCOLOR
Cone Center	Creates a three-dimensional solid primitive with a circular base tapering to a point perpendicular to its base.	Solids (Cone flyout)		N/A	CONE
Cone Elliptical	Creates a three-dimensional solid primitive with an elliptical base tapering to a point perpendicular to its base.	Solids (Cone flyout)		N/A	CONE
Configure	Reconfigures AutoCAD.	N/A		Options, Configure	CONFIG
Construction Line	Creates an infinite line.	Draw (Line flyout)		N/A	XLINE
Continue Dimension	Adds the next chained dimension, measured from the last dimension.	Dimensioning		N/A	DIMCONTINUE
Copy	Copies an existing shape from one area on the drawing to another area.	Modify (Copy flyout)		Edit, Copy	COPY or CP
Copy View	Copies the current view to the Windows Clipboard for linking to other OLE applications.	N/A		Edit, Copy View	COPYLINK
Copyclip	Copies objects to the Windows Clipboard.	Standard		Edit, Copy	COPYCLIP
Cylinder Center	Creates a three-dimensional solid primitive similar to an extruded circle, but without a taper.	Solids (Cylinder flyout)		N/A	CYLINDER

Command or Icon Name	Description	Toolbar	Icon	Menu Bar	Command:
Cylinder Elliptical	Creates a three-dimensional solid primitive similar to an extruded ellipse, but without a taper.	Solids (Cylinder flyout)		N/A	CYLINDER
Define Attribute	Creates attribute text.	Attributes		N/A	ATTDEF or DDATTDEF
Delobj	Controls whether objects used to create other objects are retained or deleted from the drawing database	N/A		N/A	DELOBJ
Detach	Removes an externally referenced drawing.	External Reference		N/A	XREF
Diameter Dimension	Adds center marks to a circle when providing its diameter dimension.	Dimensioning (Radial Dimension flyout)		N/A	DIMDIAMETER
Diastat	System variable that stores the exit method of the most recently used dialog box.	N/A		N/A	DIASTAT
Dimalt	System variable that enables alternate units.	N/A		N/A	DIMALT
Dimaltd	System variable that sets alternate unit precision.	N/A		N/A	DIMALTD
Dimaltf	System variable that sets alternate unit scale.	N/A		N/A	DIMALTF
Dimalttd	System variable that stores the number of decimal places for the tolerance values of an alternate units dimension.	N/A		N/A	DIMALTTD
Dimalttz	System variable that toggles the suppression of zeros for tolerance values.	N/A		N/A	DIMALTTZ
Dimaltu	System variable that sets the units format for alternate units of all dimension style family members, except angular.	N/A		N/A	DIMALTU

Command or Icon Name	Description	Toolbar	Icon	Menu Bar	Command:
Dimaltz	System variable that toggles suppression of zeros for alternate units dimension values.	N/A		N/A	DIMALTZ
Dimapost	System variable that specifies the text prefix and/or suffix to the alternate dimension measurement for alternate types of dimensions except angular.	N/A		N/A	DIMAPOST
Dimaso	System variable that controls the creation of associative dimension objects.	N/A		N/A	DIMASO
Dimasz	System variable that controls the size of dimension line and leader line arrowheads.	N/A		N/A	DIMASZ
Dimaunit	System variable that sets the angle format for angular dimensions.	N/A		N/A	DIMAUNIT
Dimblk	System variable that sets the name of a block to be drawn instead of the normal arrowhead at the end of the dimension line or leader line.	N/A		N/A	DIMBLK
Dimblk1	System variable that, if Dimsah is on, specifies the user-defined arrowhead block for the first end of the dimension line.	N/A		N/A	DIMBLK1
Dimblk2	System variable that, if Dimsah is on, specifies the user-defined arrowhead block for the second end of the dimension line.	N/A		N/A	DIMBLK2
Dimcen	System variable that controls the drawing of circle or arc centermarks or centerlines.	N/A		N/A	DIMCEN

Command or Icon Name	Description	Toolbar	Icon	Menu Bar	Command:
Dimclrd	System variable that assigns colors to dimension lines, arrowheads, and dimension leader lines.	N/A		N/A	DIMCLRD
Dimclre	System variable that assigns colors to dimension extension lines.	N/A		N/A	DIMCLRE
Dimclrt	System variable that assigns color to dimension text.	N/A		N/A	DIMCLRT
Dimdec	System variable that sets the number of decimal places for the tolerance values of a primary units dimension.	N/A		N/A	DIMDEC
Dimdle	System variable that extends the dimension line beyond the extension line when oblique strokes are drawn instead of arrowheads.	N/A		N/A	DIMDLE
Dimdli	System variable that controls the dimension line spacing for baseline dimensions.	N/A		N/A	DIMDLI
Dimension Status	Shows the dimension variables and their current settings.	N/A		N/A	DIM
Dimension Styles	Controls dimension styles.	Dimensioning (Dimension Styles flyout)		N/A	DDIM or DIMSTYLE
Dimexe	System variable that determines how far to extend the extension line beyond the dimension line.	N/A		N/A	DIMEXE
Dimexo	System variable that determines how far extension lines are offset from origin points.	N/A		N/A	DIMEXO

Command or Icon Name	Description	Toolbar	Icon	Menu Bar	Command:
Dimfit	System variable that controls the placement of text and attributes inside or outside extension lines.	N/A		N/A	DIMFIT
Dimgap	System variable that sets the distance around the dimension text when you break the dimension line to accommodate dimension text.	N/A		N/A	DIMGAP
Dimjust	System variable that controls the horizontal dimension text position.	N/A		N/A	DIMJUST
Dimlfac	System variable that sets the global scale factor for linear dimensioning measurements.	N/A		N/A	DIMLFAC
Dimlim	System variable that, when on, generates dimension limits as the default text.	N/A		N/A	DIMLIM
Dimoverride	Allows you to override dimensioning system variable settings associated with a dimension object, but will not affect the current dimension style.	N/A		N/A	DIMOVERRIDE or DIMOVER
Dimpost	System variable that specifies a text prefix and/or suffix to the dimension measurement.	N/A		N/A	DIMPOST
Dimrnd	System variable that rounds all dimensioning distances to the specified value.	N/A		N/A	DIMRND
Dimsah	System variable that controls the use of user-defined attribute blocks at the ends of the dimension lines.	N/A		N/A	DIMSAH

Command or Icon Name	Description	Toolbar	Icon	Menu Bar	Command:
Dimscale	System variable that sets overall scale factor applied to dimensioning variables that specify sizes, distances, or offsets.	N/A		N/A	DIMSCALE
Dimsd1	System variable that, when on, suppresses drawing of the first dimension line.	N/A		N/A	DIMSD1
Dimsd2	System variable that, when on, suppresses drawing of the second dimension line.	N/A		N/A	DIMSD2
Dimse1	System variable that, when on, suppresses drawing of the first extension line.	N/A		N/A	DIMSE1
Dimse2	System variable that, when on, suppresses drawing of the second extension line.	N/A		N/A	DIMSE2
Dimsho	System variable that, when on, controls redefinition of dimension objects while dragging.	N/A		N/A	DIMSHO
Dimsoxd	System variable that, when on, suppresses drawing of dimension lines outside the extension lines.	N/A		N/A	DIMSOXD
Dimtad	System variable that controls the vertical position of text in relation to the dimension line.	N/A		N/A	DIMTAD
Dimtdec	System variable that sets the number of decimal places to display the tolerance values for a dimension.	N/A		N/A	DIMTDEC

Command or Icon Name	Description	Toolbar	Icon	Menu Bar	Command:
Dimtfac	System variable that specifies a scale factor for the text height of tolerance values relative to the dimension text height.	N/A		N/A	DIMTFAC
Dimtih	System variable that controls the position of dimension text inside the extension lines for all dimension types except ordinate dimensions.	N/A		N/A	DIMTIH
Dimtix	System variable that draws text between extension lines.	N/A		N/A	DIMTIX
Dimtm	System variable that, when Dimtol or Dimlim is on, sets the minimum (or lower) tolerance limit for dimension text.	N/A		N/A	DIMTM
Dimtofl	System variable that, when on, draws a dimension line between the extension lines even when the text is placed outside the extension lines.	N/A		N/A	DIMTOFL
Dimtoh	System variable that, when turned on, controls the position of dimension text outside the extension lines.	N/A		N/A	DIMTOH
Dimtol	System variable that generates plus/minus tolerances.	N/A		N/A	DIMTOL
Dimtolj	System variable that sets the vertical justification for tolerance values relative to the nominal dimension text.	N/A		N/A	DIMTOLJ

Command or Icon Name	Description	Toolbar	Icon	Menu Bar	Command:
Dimtp	System variable that, when Dimtol or Dimlim is on, sets the maximum (or upper) tolerance limit for dimension text.	N/A		N/A	DIMTP
Dimtsz	System variable that specifies the size of oblique strokes drawn instead of arrowheads for linear, radius, and diameter dimensioning.	N/A		N/A	DIMTSZ
Dimtvp	System variable that adjusts text placement relative to text height.	N/A		N/A	DIMTVP
Dimtxsty	System variable that specifies the text style of a dimension.	N/A		N/A	DIMTXSTY
Dimtxt	System variable that specifies the height of the dimension text, unless the current text style has a fixed height.	N/A		N/A	DIMTXT
Dimtzin	System variable that controls the suppression of zeros for tolerance values.	N/A		N/A	DIMTZIN
Dimunit	System variable that sets the units format for all dimension family members, except angular.	N/A		N/A	DIMUNIT
Dimupt	System variable that controls the cursor functionality for User Positioned Text.	N/A		N/A	DIMUPT
Dimzin	System variable that controls the suppression of the inch portion of a feet-and-inches dimension when the distance is an integral number of feet, or the feet portion when the distance is less than one foot.	N/A		N/A	DIMZIN

Command or Icon Name	Description	Toolbar	Icon	Menu Bar	Command:
Dispsilh	System variable that controls the display of silhouette curves of body objects in wire-frame mode.	N/A		N/A	DISPSILH
Distance	Reads out the value for the distance between two selected points.	Object Properties (Inquiry flyout)		N/A	'DIST or DIST
Divide	Places points or blocks along an object to create segments.	Draw (Point flyout)		N/A	DIVIDE
Donut	Draws concentric filled circles or filled circles.	Draw (Circle flyout)		N/A	DONUT
Dragmode	Controls the way dragged objects are displayed	N/A		N/A	'DRAGMODE or DRAGMODE
Drawing Aids	Sets the drawing aids for the drawing.	N/A		Options, Drawing Aids	'DDRMODES or DDRMODES
Dynamic Text	Adds text to a drawing.	Draw (Text flyout)		N/A	DTEXT
Edge	Changes the visibility of three-dimensional face edges.	Surfaces		N/A	EDGE
Edge Surface	Creates a three-dimensional polygon mesh.	Surfaces		N/A	EDGESURF
Edit Attribute	Edits attribute values.	Attribute		N/A	ATTEDIT
Edit Attribute Globally	Edits attribute values.	Attribute		N/A	DDATTE
Edit Hatch	Modifies an existing associative hatch block.	Modify (Special Edit flyout)		N/A	HATCHEDIT
Edit Links	Updates, changes, and cancels existing links.	N/A		Edit, Links	OLELINKS
Edit Multiline	Edits multiple parallel lines.	Modify (Special Edit flyout)		N/A	MLEDIT

Command or Icon Name	Description	Toolbar	Icon	Menu Bar	Command:
Edit Polyline	Changes various features of polylines.	Modify (Special Edit flyout)		N/A	PEDIT
Edit Spline	Edits a spline object.	Modify (Special Edit)		N/A	SPLINEDIT
Edit Text	Allows you to edit text inside a dialogue box.	Modify (Special Edit flyout)		N/A	DDEDIT
Elev	Sets elevation and extrusion thickness of new objects.	N/A		N/A	ELEV
Elevation	System variable that stores the current three-dimensional elevation relative to the current UCS for the current space.	N/A		N/A	ELEVATION
Ellipse Arc	Used to draw ellipses.	Draw (Ellipse flyout)		N/A	ELLIPSE
Ellipse Axis End	Used to draw ellipses.	Draw (Ellipse flyout)		N/A	ELLIPSE
Ellipse Center	Used to draw ellipses.	Draw (Ellipse flyout)		N/A	ELLIPSE
End	Saves current drawing file under name specified at beginning and returns you to the operating system.	N/A		N/A	END
Erase	Erases objects from drawing.	Modify		N/A	ERASE or E
Explode	Separates blocks or dimensions into component objects so that they are no longer one group and can be edited.	Modify (Explode flyout)		N/A	EXPLODE
Export	Exports a drawing to a file of type DXF so that another application can use it.	N/A		File, Import	EXPORT

Command or Icon Name	Description	Toolbar	Icon	Menu Bar	Command:
Export Links	Exports link information for selected objects.	External Database		N/A	ASEEXPORT
Extend	Extends the length of an existing object to meet a boundary.	Modify (Trim/Extend flyout)		N/A	EXTEND
Extract Attributes	Extracts attribute data from a drawing.	N/A		N/A	ATTEXT or DDATTEXT
Extrude	Extrudes a two-dimensional shape into a three-dimensional object.	Solids		N/A	EXTRUDE
Extruded Surface	Creates a tabulated surface from a path curve and direction vector.	Surfaces		N/A	TABSURF
Facetres	System variable that adjusts the smoothness of shaded and hidden line-removed objects.	N/A		N/A	FACETRES
Filedia	System variable that suppresses the display of the file dialog boxes.	N/A		N/A	FILEDIA
Files	Manages files.	N/A		Files, Management, Utilities	FILES
Fill	Controls the filling of multilines, traces, solids, and wide polylines.	N/A		N/A	'FILL or FILL
Fillet	Connects lines, arcs, or circles with a smoothly fitted arc.	Modify (Chamfer flyout)		N/A	FILLET
Front View	Sets the viewing direction for a three-dimensional visualization of the drawing to the front.	Standard (View flyout)		View, 3D Viewpoint Presets, Front	VPOINT or DDVPOINT
Global Linetype	System variable that sets the current global linetype scale factor for objects.	N/A		N/A	CELTSCALE

Command or Icon Name	Description	Toolbar	Icon	Menu Bar	Command:
Grid	Displays a grid of dots at desired spacing on the screen.	Status bar		N/A	'GRID or GRID, or toggle off and on with (F7)
Grips	Enables grips and sets their color.	N/A		Options, Grips	'GRIPS or GRIPS or 'DDGRIPS or DDGRIPS
Handles	System variable that controls the use of handles in solid models.	N/A		N/A	HANDLES
Hatch	Draws an unassociative hatch pattern.	N/A		N/A	HATCH
Help	Displays information explaining commands and procedures.	Standard		Help	'HELP or '?, or HELP or ?
Hide	Hides an object.	N/A		N/A	HIDE
Highlight	System variable that controls object highlighting.	N/A		N/A	HIGHLIGHT
Insert Block	Places blocks or other drawings into a drawing via the command line (Insert) or via a dialog box (Ddinsert).	Draw (Block flyout)		N/A	INSERT or DDINSERT
Insert Multiple Blocks	Inserts multiple instances of a block in a rectangular array.	Miscellaneous		N/A	MINSERT
Insert Objects	Inserts a linked or embedded object.	N/A		Edit, Insert Objects	INSERTOBJ
Interfere	Shows where two solid objects overlap.	Solids		N/A	INTERFERE
Intersection	Forms a new solid from the area common to two or more solids or regions.	Modify (Explode flyout)		N/A	INTERSECT
Isolines	System variable that specifies the number of isolines per surface on objects.	N/A		N/A	ISOLINES

Command or Icon Name	Description	Toolbar	Icon	Menu Bar	Command:
Isometric Plane	Controls which isometric drawing plane is active: left, top, or right.	N/A		N/A	'ISOPLANE or ISOPLANE, or toggle with Ctrl E
Layers	Creates layers and controls layer color, linetype, and visibility.	Object Properties		Data, Layers	'LAYER or LAYER or 'LA or LA or 'DDLMODES or DDLMODES
Leader	Creates leader lines for identifying lines and dimensions.	Dimensioning		N/A	LEADER
Left View	Sets the viewing direction for a three-dimensional visualization of the drawing to the left side.	Standard (View flyout)		View, 3D Viewpoint Presets, Left	VPOINT or DDVPOINT
Limits	Sets up the size of the drawing.	N/A		Data, Drawing Limits	'LIMITS or LIMITS
Line	Draws straight lines of any length.	Draw (line flyout)		N/A	LINE or L
Linear Dimension	Creates linear dimensions.	Dimensioning		N/A	DIMLINEAR
Linetype	Changes the pattern used for new lines in the drawing.	Object Properties		Data, Select Linetype	'LINETYPE or LINETYPE or 'DDLTYPE or DDLTYPE
Linetype Scale	System variable that changes the scale of the linetypes in the drawing.	N/A		Options, Linetypes, Global Linetype Scale	'LTSCALE or LTSCALE
Links	Manipulates links between objects and an external database.	External Database		N/A	ASELINKS
List	Provides information on an object, such as the length of a line.	Object Properties (Inquiry flyout)		N/A	LIST

Command or Icon Name	Description	Toolbar	Icon	Menu Bar	Command:
List External References	Adds one drawing inside another, linking them so that changes appear in both drawings.	External Reference		N/A	XREF
List Variables	Lists the values of system variables.	N/A		Options, System Variables, Lists	'SETVAR or SETVAR
Locate	Specifies an area on the drawing to be displayed in the Aerial View window.	Standard (Aerial View flyout, pick Aerial View, then Locate icon from Aerial View window)		Tools, Aerial View	DSVIEWER
Locate Point	Finds the coordinates of any point on the screen.	Object Properties (Inquiry flyout)		N/A	'ID or ID
Mass Properties	Calculates and displays the mass properties of regions and solids.	Object Properties (Inquiry flyout)		N/A	MASSPROP
Measure	Marks segments of a chosen length on an object by inserting points or blocks.	Draw (Point flyout)		N/A	MEASURE
Menu	Loads a menu file.	N/A		N/A	MENU
Menuload	Loads partial menu files.	N/A		Tools, Menu Customization	MENULOAD
Menuunload	Unloads partial menu files.	N/A		Tools, Menu Customization	MENUUN-LOAD
Mirror	Creates mirror images of shapes.	Modify (Copy flyout)		N/A	MIRROR
Model Space	Enters model space.	Status bar		View, Tiled Model Space	MSPACE or MS
Move	Moves an existing shape from one area on the drawing to another area.	Modify		N/A	MOVE or M

Command or Icon Name	Description	Toolbar	Icon	Menu Bar	Command:
Mtprop	Changes paragraph text properties.	N/A		N/A	MTPROP
Multiline	Creates multiple, parallel lines.	Draw (Polyline flyout)		N/A	MLINE
Multiline Style	Defines a style for multiple, parallel lines.	Object Properties		Data, Multiline Style	MLSTYLE
Multiline Text	Creates a paragraph that fits within a nonprinting text boundary.	Draw (Text flyout)		N/A	MTEXT
MV Setup	Helps set up drawing views using a LISP program.	N/A		View, Floating Viewports, MV Setup	MVSETUP
Mview	Creates viewports.	N/A		View, Floating Viewports (all options, except MV Setup)	MVIEW
Named UCS	Selects a named and saved user coordinate system.	Standard (UCS flyout)		View, Named UCS	DDUCS
Named Views	Sets the viewing direction to a named view.	Standard (View flyout)		View, Named Views	'VIEW or VIEW or 'DDVIEW or DDVIEW
NE Isometric View	Sets the viewing direction for a three-dimensional visualization of the drawing.	Standard (View flyout)		View, 3D Viewpoint Presets, NE Isometric	VPOINT or DDVPOINT
New Drawing	Sets up the screen to create a new drawing.	Standard		File, New	NEW
NW Isometric View	Sets the viewing direction for a three-dimensional visualization of the drawing.	Standard (View flyout)		View, 3D Viewpoint Presets, NW Isometric	VPOINT or DDVPOINT

Command or Icon Name	Description	Toolbar	Icon	Menu Bar	Command:
Object Group	Creates a named selection set of objects.	Standard		N/A	GROUP
Object UCS	Selects a user coordinate system based on the object.	Standard (UCS flyout)		View, Set UCS, Object	UCS
Oblique Dimensions	Edits dimensions.	Dimensioning (Dimension Styles flyout)		N/A	DIMEDIT
Offset	Draws objects parallel to a given object.	Modify (Copy flyout)		N/A	OFFSET
Oops!	Restores erased objects.	Miscellaneous		N/A	OOPS
Open Drawing	Loads a previously saved drawing.	Standard		File, Open	OPEN
Ordinate Dimension	Creates ordinate point dimensions.	Dimensioning (Ordinate Dimension flyout)		N/A	DIMORDINATE
Origin UCS	Creates a user-defined coordinate system at the origin of the view.	Standard (UCS flyout)		View, Set UCS, Origin	UCS
Ortho	Restricts movement to only horizontal and vertical.	Status bar		N/A	'ORTHO or ORTHO, or toggle on and off with F8
Overlay	Places one drawing over another, linking them so that changes appear in both drawings.	External Reference		N/A	XREF
Pan Down	Moves the drawing around on the screen without changing the zoom factor.	Standard (Pan flyout)		View, Pan, Down	'PAN or PAN or 'P or P
Pan Down Left	Moves the drawing around on the screen without changing the zoom factor.	Standard (Pan flyout)		View, Pan, Down-Left	'PAN or PAN or 'P or P
Pan Down Right	Moves the drawing around on the screen without changing the zoom factor.	Standard (Pan flyout)		View, Pan, Down-Right	'PAN or PAN or 'P or P

Command or Icon Name	Description	Toolbar	Icon	Menu Bar	Command:
Pan Left	Moves the drawing around on the screen without changing the zoom factor.	Standard (Pan flyout)		View, Pan, Left	'PAN or PAN or 'P or P
Pan Point	Moves the drawing around on the screen without changing the zoom factor.	Standard (Pan flyout)		View, Pan, Point	'PAN or PAN or 'P or P
Pan Right	Moves the drawing around on the screen without changing the zoom factor.	Standard (Pan flyout)		View, Pan, Right	'PAN or PAN or 'P or P
Pan Up	Moves the drawing around on the screen without changing the zoom factor.	Standard (Pan flyout)		View, Pan, Up	'PAN or PAN or 'P or P
Pan Up Left	Moves the drawing around on the screen without changing the zoom factor.	Standard (Pan flyout)		View, Pan, Up-Left	'PAN or PAN or 'P or P
Pan Up Right	Moves the drawing around on the screen without changing the zoom factor.	Standard (Pan flyout)		View, Pan, Up-Right	'PAN or PAN or 'P or P
Paper Space	Enters paper space when TILEMODE is set to 0.	Status bar		View, Paper Space	PSPACE or PS
Paper Space Linetype Scale	System variable that causes the paper space LTSCALE factor to be used for every viewport.	N/A		Options, Linetypes, Paper Space Linetype Scale	PSLTSCALE
Paste	Places objects from the Windows Clipboard into the drawing using the Insert command.	Standard		Edit, Paste	PASTECLIP
Path	Displays and edits the path associated with a particular externally referenced drawing.	External Reference		N/A	XREF

Command or Icon Name	Description	Toolbar	Icon	Menu Bar	Command:
Perimeter	System variable that stores the last perimeter value computed by the Area or List commands.	N/A		N/A	PERIMETER
Plan	Creates a view looking straight down along the Z axis to see the XY coordinates.	N/A		View, 3D Viewpoint Presets, Plan View	PLAN
Point	Changes the properties of existing objects.	Modify (Resize flyout)		N/A	CHANGE
Point	Creates a point object.	Draw (Point flyout)		N/A	POINT
Point Filters (.X, .Y, .Z, .XY, .XZ, .YZ)	Lets you selectively find single coordinates out of a point's three-dimensional coordinate set.	Standard (Point Filters flyout)		N/A	.X or .Y or .Z or .XY or .XZ or .YZ
Point Style	Specifies the display mode and size of point objects.	N/A		Objects, Display, Point Style	DDPTYPE
Polar Array	Creates multiple copies of objects in a regularly spaced circular pattern.	Modify (Copy flyout)		N/A	ARRAY
Polygon	Draws regular polygons with 3 to 1024 sides.	Draw (Polygon flyout)		N/A	POLYGON
Polyline	Draws a series of connected objects (lines or arcs) that are treated as a single object called a polyline.	Draw (Polyline flyout)		N/A	PLINE
Preferences	Customizes the AutoCAD settings.	N/A		Options, Preferences	PREFERENCES
Preset UCS	Selects a preset UCS.	Standard (UCS flyout)		View, Preset UCS	DDUCSP
Previous UCS	Returns you to the previously used user coordinate system.	Standard (UCS flyout)		View, Set UCS, Previous	UCS
Print/Plot	Prints or plots a drawing	Standard		File, Print	PLOT

Command or Icon Name	Description	Toolbar	Icon	Menu Bar	Command:
Projmode	System variable that sets the current projection mode for trim or extend operations.	N/A		N/A	PROJMODE
Properties	Used to change the properties of an object.	Object Properties		N/A	CHPROP or DDCHPROP
Purge	Removes unused items from the drawing database.	N/A		Data, Purge	PURGE
Quit	Exits AutoCAD with the option not to save your changes, and returns you to the operating system.	N/A		File, Exit	QUIT
Radius Dimension	Adds centermarks to an arc when providing its radius dimension.	Dimensioning (Radial Dimension flyout)		N/A	DIMRADIUS
Ray	Creates a semi-infinite line.	Draw (Line flyout)		N/A	RAY
Recover	Repairs a damaged drawing.	N/A		File, Management, Recover	RECOVER
Rectangle	Creates rectangular shapes.	Draw (Polygon flyout)		N/A	RECTANG
Rectangular Array	Creates multiple copies of objects in a regularly spaced rectangular pattern.	Modify (Copy flyout)		N/A	ARRAY
Redefine Attribute	Redefines a block and updates associated attributes.	Attribute		N/A	ATTREDEF
Redo	Reverses the effect of the most recent UNDO command.	Standard		Edit, Redo	REDO
Redraw All	Redraws all viewports at once.	Standard (Redraw flyout)		View, Redraw All	'REDRAWALL or REDRAWALL

Command or Icon Name	Description	Toolbar	Icon	Menu Bar	Command:
Redraw View	Removes excess blipmarks added to the drawing screen and restores objects partially erased while you are editing other objects.	Standard (Redraw flyout)		View, Redraw View	'REDRAW or REDRAW or 'R or R
Regenerate	Recalculates the display file in order to show changes.	N/A		N/A	REGEN
Regenerate All	Regenerates all viewports at once.	N/A		N/A	REGENALL
Region	Creates a two-dimensional area object from a selection set of existing objects.	Draw (Polygon flyout)		N/A	REGION
Reload	Reattaches the externally referenced drawing.	External Reference		N/A	XREF
Rename	Changes the names of objects.	N/A		Data, Rename	RENAME or DDRENAME
Render Configure	Reconfigures the rendering setup.	N/A		Options, Render Configure	RCONFIG
Replay	Displays a GIF, TGA, or TIFF image	N/A		Tools, Image, View	REPLAY
Restore UCS	Restores a saved user coordinate system.	Standard (UCS flyout)		View, Set UCS, Restore	UCS
Revolve	Creates a three-dimensional object by sweeping a two-dimensional polyline, circle or region about a circular path to create a symmetrical solid.	Solids		N/A	REVOLVE
Revolved Surface	Creates a rotated surface about a selected axis.	Surfaces		N/A	REVSURF
Right View	Sets the viewing direction for a three-dimensional visualization of the drawing to the right side.	Standard (View flyout)		View, 3D Viewpoint Presets, Right	VPOINT or DDVPOINT
Rotate	Rotates all or part of an object.	Modify (Rotate flyout)		N/A	ROTATE

Command or Icon Name	Description	Toolbar	Icon	Menu Bar	Command:
Ruled Surface	Creates a polygon mesh between two curves.	Surfaces		N/A	RULESURF
Running Object Snap	Sets running objects snap modes and changes target box size.	Standard (Object Snap flyout)		Options, Running Object Snap	'OSNAP or OSNAP or 'DDOSNAP or DDOSNAP
Save	Saves a drawing using Quick Save (Qsave).	Standard		File, Save	QSAVE
SAVE	Saves a drawing to a new file name, without changing the name of the drawing on your screen.	N/A		N/A	SAVE
Save As	Saves a drawing, allowing you to change its name or location.	N/A		File, Save As	SAVEAS
Save As R12	Saves the current drawing in AutoCAD Release 12 format.	N/A		File, Save R12 DWG	SAVEASR12
Save Image	Saves a rendered image to a file.	N/A		Tools, Image, Save	SAVEIMG
Save Slide	Creates a slide file of the current viewport.	N/A		Tools, Slide, Save	MSLIDE
Save Time	System variable that controls how often a file is automatically saved.	N/A		N/A	SAVETIME
Save UCS	Saves a user coordinate system.	Standard (UCS flyout)		View, Set UCS, Save	UCS
Scale	Changes the size of an object in the drawing database.	Modify (Resize flyout)		N/A	SCALE
Script	Executes a sequence of commands from a script.	N/A		Tools, Run Script	SCRIPT
SE Isometric View	Sets the viewing direction for a three-dimensional visualization of the drawing.	Standard (View flyout)		View, 3D Viewpoint Presets, SE Isometric	VPOINT or DDVPOINT
Section	Uses the intersection of a plane and solids to create a region.	Solids		N/A	SECTION

Command or Icon Name	Description	Toolbar	Icon	Menu Bar	Command:
Select Add	Places selected objects in the previous selection set.	Standard (Select Objects flyout)		N/A	'SELECT or SELECT or 'DDSELECT or DDSELECT
Select All	Places all objects in the selection set.	Standard (Select Objects flyout)		N/A	'SELECT or SELECT or 'DDSELECT or DDSELECT
Select Crossing	Defines selection mode.	Standard (Select Objects flyout)		N/A	'SELECT or SELECT or 'DDSELECT or DDSELECT
Select Crossing Polygon	Defines selection mode.	Standard (Select Objects flyout)		N/A	'SELECT or SELECT or 'DDSELECT or DDSELECT
Select Fence	Defines selection mode.	Standard (Select Objects flyout)		N/A	'SELECT or SELECT or 'DDSELECT or DDSELECT
Select Last	Re-highlights last selection set.	Standard (Select Objects flyout)		N/A	'SELECT or SELECT or 'DDSELECT or DDSELECT
Select Previous	Places previously selected objects in the selection set.	Standard (Select Objects flyout)		N/A	'SELECT or SELECT or 'DDSELECT or DDSELECT
Select Remove	Removes selected objects in the previous selection set.	Standard (Select Objects flyout)		N/A	'SELECT or SELECT or 'DDSELECT or DDSELECT
Select Window	Defines selection mode.	Standard (Select Objects flyout)		N/A	'SELECT or SELECT or 'DDSELECT or DDSELECT
Select Window Polygon	Defines selection mode.	Standard (Select Objects flyout)		N/A	'SELECT or SELECT or 'DDSELECT or DDSELECT

Command or Icon Name	Description	Toolbar	Icon	Menu Bar	Command:
Selection Filters	Creates lists to select objects based on properties.	Standard (Select Objects flyout)		N/A	'FILTER or FILTER
Set Variables	Changes the values of system variables.	N/A		Options, System Variables, Set	'SETVAR or SETVAR
Shade	Creates a shaded view of a model.	N/A		N/A	SHADE
Shape	Inserts a shape into the current drawing.	Miscellaneous		N/A	SHAPE
Shell	Accesses operating system commands.	N/A		N/A	SHELL
Single-Line Text	Creates a single line of user-entered text.	Draw (Text flyout)		N/A	TEXT
Sketch	Creates a series of free-hand line segments.	Miscellaneous		N/A	SKETCH
Slice	Slices a set of solids with a plane.	Solids		N/A	SLICE
Snap	Limits cursor movement on the screen to set interval so objects can be placed at precise locations easily.	Status bar		N/A	SNAP, or toggle off and on with (F9)
Snap From	Establishes a temporary reference point as a basis for specifying subsequent points.	Standard (Object Snap flyout)		N/A	'OSNAP or OSNAP or 'DDOSNAP or DDOSNAP
Snap to Apparent Intersection	Snaps to a real of visual three-dimensional intersection formed by objects you select or by an extension of those objects.	Standard (Object Snap flyout)		N/A	'APP or APP or 'APPINT or APPINT
Snap to Center	Snaps to the center of an arc, circle, or ellipse.	Standard (Object Snap flyout)		N/A	'CEN or CEN
Snap to Endpoint	Snaps to the closest endpoint of an arc, elliptical arc, ray, multiline, or line, or to the closest corner of a trace, solid, or three-dimensional face.	Standard (Object Snap flyout)		N/A	'ENDP or ENDP

Command or Icon Name	Description	Toolbar	Icon	Menu Bar	Command:
Snap to Insertion	Snaps to the insertion point of text, a block, a shape, or an attribute.	Standard (Object Snap flyout)		N/A	'INS or INS
Snap to Intersection	Snaps to the intersection of a line, arc, spline, elliptical arc, ellipse, ray, construction line, multiline, or circle.	Standard (Object Snap flyout)		N/A	'INT or INT
Snap to Midpoint	Snaps to the midpoint of an arc, elliptical arc spline, ray, solid, construction line, multiline, or line.	Standard (Object Snap flyout)		N/A	'MID or MID
Snap to Nearest	Snaps to the nearest point of an arc, elliptical arc, ellipse, spline, ray, multi-line, line, circle, or point.	Standard (Object Snap flyout)		N/A	'NEA or NEA
Snap to Node	Snaps to a point object.	Standard (Object Snap flyout)		N/A	'NOD or NOD
Snap to None	Turns object snap mode off.	Standard (Object Snap flyout)		N/A	'NONE or NONE
Snap to Perpendicular	Snaps to a point perpendicular to an arc, elliptical arc, ellipse, spline, ray, construction line, multi-line, line, solid, or circle.	Standard (Object Snap flyout)		N/A	'PER or PER
Snap to Quadrant	Snaps to a quadrant point of an arc, elliptical arc, ellipse, solid, or circle.	Standard (Object Snap flyout)		N/A	'QUA or QUA
Snap to Quick	Snaps to the first object snap point found.	Standard (Object Snap flyout)		N/A	'QUI or QUI
Snap to Tangent	Snaps to the tangent of an arc, elliptical arc, ellipse, or circle.	Standard (Object Snap flyout)		N/A	'TAN or TAN
Spelling	Checks spelling in a drawing.	Standard		Tools, Spelling	SPELL
Sphere	Creates a three-dimensional solid sphere.	Solids		N/A	SPHERE

Command or Icon Name	Description	Toolbar	Icon	Menu Bar	Command:
Spline	Fits a smooth curve to a sequence of points within a specified tolerance.	Draw (Polyline flyout)		N/A	SPLINE
Status	Displays drawing statistics, modes, and extents.	N/A		N/A	STATUS
Stretch	Stretches the objects selected.	Modify (Resize flyout)		N/A	STRETCH
Style	Lets you create new text styles and modify existing ones.	N/A		Data, Text Style	STYLE
Subtract	Removes the set of a second solid or region from the first set.	Modify (Explode flyout)		N/A	SUBTRACT
Surftab1	System variable that sets the number of tabulations to be generated for Rulesurf and Tabsurf.	N/A		N/A	SURFTAB1
Surftab2	System variable that sets mesh density in the N direction for Revsur and Edgesurf.	N/A		N/A	SURFTAB2
SW Isometric View	Sets the viewing direction for a three-dimensional visualization of the drawing.	Standard (View flyout)		View, 3D Viewpoint Presets, SW Isometric	VPOINT or DDVPOINT
Tablet	Calibrates the tablet with a paper drawing's coordinate system.	N/A		Options, Tablet	TABLET
Tabmode	System variable that controls the use of tablet mode.	N/A		N/A	TABMODE
Target	System variable that stores location (in UCS coordinates) of the target point for the current viewport.	N/A		N/A	TARGET
Text Screen	Switches from the graphics screen to the text screen.	N/A		Edit, Text Window	'TEXTSCR or TEXTSCR, or toggle back and forth with (F2)

Command or Icon Name	Description	Toolbar	Icon	Menu Bar	Command:
3 Point UCS	Selects a user coordinate system based on any three points in the graphics window.	Standard (UCS flyout)		View, Set UCS, 3 Point	UCS
3D Dynamic View	Changes the 3D view of a point	N/A		View, 3D Dynamic View	DVIEW
3D Face	Creates a three-dimensional surface.	Surfaces		N/A	3DFACE
3D Mesh	Creates a free-form polygon mesh.	Surfaces		N/A	3DMESH
3D Mirror	Creates a mirror image of objects about a plane.	Modify (Copy flyout)		N/A	MIRROR3D
3D Polar Array	Creates a three-dimensional polar array.	Modify (Copy flyout)	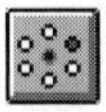	N/A	3DARRAY
3D Polyline	Creates a polyline of straight line segments.	Draw (Polyline flyout)		N/A	3DPOLY
3D Rectangular Array	Creates a three-dimensional rectangular array.	Modify (Copy flyout)		N/A	3DARRAY
3D Rotate	Rotates an object in a 3D coordinate system.	Modify (Rotate flyout)		N/A	ROTATE3D
Tiled Viewports	Divides the graphics window into multiple tiled viewports and manages them.	N/A		View, Tiled Viewports	VPORTS
Tilemode	When set to 1, allows creation of tiled viewports in model space; when set to 0, enables use of paper space.	Status bar		View, Tiled Model Space	TILEMODE
Tolerance	Creates geometric tolerances and adds them to a drawing in feature control frames.	Dimensioning		N/A	TOLERANCE
Toolbar	Displays, hides, and positions all toolbars.	Standard (Aerial View flyout)		Tools, Toolbars or Toolbar Customization	TOOLBAR
Tooltips	System variable that controls the display of ToolTips.	N/A		N/A	TOOLTIPS

Command or Icon Name	Description	Toolbar	Icon	Menu Bar	Command:
Top View	Sets the viewing direction for a three-dimensional visualization of the drawing to the top.	Standard (View flyout)		View, 3D Viewpoint Presets, Top	VPOINT or DDVPOINT
Torus	Creates a donut-shaped solid.	Solids		N/A	TORUS
Trace	Creates solid lines.	Miscellaneous		N/A	TRACE
Treestat	Displays information on the drawing's current spatial index.	N/A		N/A	TREESTAT
Trim	Removes part of an object at its intersection with another object.	Modify (Trim/Extend flyout)		N/A	TRIM
2D Solid	Creates solid-filled polygons.	Draw (Polygon flyout)		N/A	SOLID
UCS	Aligns a user coordinate system with the edge or face of a solid.	N/A		N/A	UCS or DDUCS
UCS Follow	System variable that allows you to generate a plan view of the current UCS in a viewport whenever you change the UCS.	N/A		Options, UCS, Follow	UCSFOLLOW
UCS Icon	System variable that repositions and turns on and off the display of the UCS icon.	N/A		Options, UCS, Icon	UCSICON
Undo	Reverses the effect of previous commands.	Standard		Edit, Undo	UNDO or U
Union	Adds two separate sets of solids or regions together to create one solid model.	Modify (Explode flyout)		N/A	UNION, or SOLUNION
Units	Controls the type and precision of the values used in dimensions.	N/A		Data, Units	'UNITS or UNITS or 'DDUNITS or DDUNITS
Update	Updates the dimensions based on changes to the dimension styles and variables.	N/A		N/A	DIM

Command or Icon Name	Description	Toolbar	Icon	Menu Bar	Command:
View	Saves views of an object to be plotted, displayed, or printed.	Standard (View flyout)		View, Named Views	'VIEW or VIEW or 'DDVIEW or DDVIEW
View Resolution	Sets the resolution for object generation in the current viewport	N/A		N/A	VIEWRES
View Slide	Displays a raster-image slide file in the current viewport.	N/A		Tools, Slide, View	VSLIDE
View UCS	Selects a user coordinate system aligned to a specific view.	Standard (UCS flyout)		View, Set UCS, View	UCS
Viewpoint	Establishes different directions from which to view the x,y,z coordinate system.	Standard (View flyout)		View, 3D Viewpoint	VPOINT or DDVPOINT
Viewport Layer	Controls the visibility of layers within specific viewports.	Object Properties		N/A	VPLAYER
Wedge Center	Creates a three-dimensional solid with a closed face tapering along the X axis by specifying the center.	Solids (Wedge flyout)		N/A	WEDGE
Wedge Corner	Creates a three-dimensional solid with a closed face tapering along the X axis by specifying the first corner.	Solids (Wedge flyout)		N/A	WEDGE
World UCS	Selects the world coordinate system as the user coordinate system.	Standard (UCS flyout)		View, Set UCS, World	UCS
Write Block	Saves a block as a separate file so that it may be used in other drawings.	N/A		N/A	WBLOCK
X Axis Rotate UCS	Selects a user coordinate system by rotating the X axis.	Standard (UCS flyout)		View, Set UCS, X Axis Rotate	UCS
Xplode	Breaks a component object into its component objects; see also Explode.	N/A		N/A	XPLODE

Command or Icon Name	Description	Toolbar	Icon	Menu Bar	Command:
Y Axis Rotate UCS	Selects a user coordinate system by rotating the Y axis.	Standard (UCS flyout)		View, Set UCS, Y Axis Rotate	UCS
Z Axis Rotate UCS	Selects a user coordinate system by rotating the Z axis.	Standard (UCS flyout)		View, Set UCS, Z Axis Rotate	UCS
Z Axis Vector UCS	Selects a user coordinate system defined by a vector in the Z direction.	Standard (UCS flyout)		View, Set UCS, Z Axis Vector	UCS
Zoom All	Resizes to display the entire drawing in the current viewport.	Standard (Zoom flyout)		View, Zoom, All	ZOOM or Z
Zoom Center	Resizes to display a window by entering a center point, then a magnification value or height.	Standard (Zoom flyout)		View, Zoom, Center	'ZOOM or ZOOM or 'Z or Z
Zoom Dynamic	Resizes to display the generated portion of the drawing with a view box.	Standard (Zoom flyout)		View, Zoom, Dynamic	'ZOOM or ZOOM or 'Z or Z
Zoom Extents	Resizes to display the drawing extents.	Standard (Zoom flyout)		View, Zoom, Extents	'ZOOM or ZOOM or 'Z or Z
Zoom In	Resizes areas of the drawing in the screen to make it appear closer or larger.	Standard		View, Zoom, In	'ZOOM or ZOOM or 'Z or Z
Zoom Left	Resizes to display a window by using the lower-left corner of the display window.	Standard (Zoom flyout)		View, Zoom, Left	'ZOOM or ZOOM or 'Z or Z
Zoom Limits	Resizes to display the drawing limits.	Standard (Zoom flyout)		View, Zoom, Limits	'ZOOM or ZOOM or 'Z or Z
Zoom Out	Resizes areas of the drawing in the screen so that more of the drawing fits in the screen.	Standard		View, Zoom, Out	'ZOOM or ZOOM or 'Z or Z
Zoom Previous	Resizes to the previous view.	Standard (Zoom flyout)		View, Zoom, Previous	'ZOOM or ZOOM or 'Z or Z
Zoom Scale	Resizes to display at a specified scale factor.	Standard (Zoom flyout)		View, Zoom, Scale	'ZOOM or ZOOM or 'Z or Z

Command or Icon Name	Description	Toolbar	Icon	Menu Bar	Command:
Zoom Vmax	Resizes so that the drawing is viewed from as far out as possible on the current viewports virtual screen without forcing a complete regeneration of the drawing.	Standard (Zoom flyout)		View, Zoom, Vmax	'ZOOM or ZOOM or 'Z or Z
Zoom Window	Resizes to display an area specified by two opposite corner points of a rectangular window.	Standard		View, Zoom, Window	'ZOOM or ZOOM or 'Z or Z

AUTOVISION COMMAND SUMMARY

This grid includes the commands available for use with AutoVision. You should use AutoVision's on-line help feature for complete information about the commands and their options.

Command or Icon Name	Description	Toolbar	Icon	Menu Bar	Command:
Animation	Puts your AutoVision drawing into motion.	AutoVision		AutoVis, Animation	ANIMATE
Background	Sets a background for your AutoVision scene.	AutoVision		AutoVis, Background	BACKGROUND
Convert Visual Link Data	Converts Visual Link rendering data into objects for use with the AutoVision rendering software.	N/A		N/A	VLCONV
Fog	Provides visual information about the distance of objects from the camera.	AutoVision		AutoVis, Fog	FOG
Landscape Edit	Allows you to edit a landscape object.	AutoVision		AutoVis, Landscape Edit	LSEDIT
Landscape Library	Maintains libraries of landscape objects.	AutoVision		AutoVis, Landscape Library	LSLIB

Command or Icon Name	Description	Toolbar	Icon	Menu Bar	Command:
Landscape New	Lets you add realistic landscape items, such as trees and bushes, to your drawings.	AutoVision		AutoVis, Landscape New	LSNEW
Lights	Controls the light sources in rendered scenes.	AutoVision		AutoVis, Lights	LIGHT
Mapping	Lets you map materials onto selected geometry.	AutoVision		AutoVis, Mapping	SETUV
Materials	Controls the material a rendered object appears to be made of, and allows the creation of new materials.	AutoVision		AutoVis, Materials	RMAT
Materials Library	Manages all the materials available for your drawing.	AutoVision		AutoVis, Materials Library	MATLIB
Render	Produces a realistically shaded 2D view of a 3D model.	AutoVision		AutoVis, Render	RENDER
Render File Options	Sets the render to file options for rendering.	N/A		N/A	RFILEOPT
Render Preferences	Lets you control how images are rendered and how you use the AutoVision commands.	AutoVision		N/A	RPREF
Render Screen	Renders to a full-screen display only.	N/A		N/A	RENDSCR
Render Unload	Unloads the AutoVision program from memory.	N/A		N/A	RENDERUN-LOAD
Replay	Displays a GIF, TGA, or TIFF image	N/A		Tools, Image, View	REPLAY
Save Image	Saves a rendered image to a file.	N/A		Tools, Image, Save	SAVEIMG
Scenes	Saves and recalls named scenes, making it easier to quickly change to a different viewpoint and lighting.	AutoVision		AutoVis, Scenes	SCENE

Command or Icon Name	Description	Toolbar	Icon	Menu Bar	Command:
Show Materials	Shows you the material and attachment method for the selected object.	N/A		N/A	SHOWMAT
Statistics	Displays information about your last rendering.	AutoVision		N/A	STATS
3D Studio In	Imports a 3d Studio (.3ds) file into an AutoCAD drawing.	N/A		N/A	3DSIN
3D Studio Out	Saves a file formatted for use in 3D Studio (.3ds).	N/A		N/A	3DSOUT

AUTOCAD DESIGNER COMMAND SUMMARY

This grid includes the commands available for use with AutoCAD Designer. You should use AutoCAD Designer's on-line help feature for complete information about the commands and their options.

Command or Icon Name	Description	Toolbar	Icon	Menu Bar	Command:
About Designer	Displays information about AutoCAD Designer.	N/A		Designer, About Designer	ADABOUT
Adborder	System variable that controls the display of view borders.	N/A		N/A	ADBORDER
ADChamfer	Creates a chamfer on the selected edge(s) of the active part.	Designer Main		Designer, Features, Chamfer	ADCHAMFER
Adcondspsz	System variable that controls the height of displayed constraints on the active sketch.	N/A		N/A	ADCONDSPSZ
Add Annotation	Adds items belonging to an annotation on an AutoCAD Designer drawing.	N/A		Designer, Drawing, Annotation, Add	ADANNOTE
Add Constraints	Adds parametric constraints to paths and profiles for features in the active sketch.	N/A		Designer, Sketch, Constraints, Add	ADADDCON

Command or Icon Name	Description	Toolbar	Icon	Menu Bar	Command:
ADEditfeat	Displays and modifies the dimension values of an active part's features.	Designer Main		Designer, Edit Feature	ADEDITFEAT
ADFillet	Creates a rolling ball fillet on the selected edge(s) of the active part.	Designer Main		Designer, Features, Fillet	ADFILLET
Adhidltype	System variable that determines the linetype of hidden lines.	N/A		N/A	ADHIDLTYPE
Adisocyl	System variable that controls the display of isolines on cylinders or spheres.	N/A		N/A	ADISOCYL
Adisonurb	System variable that controls the number of isolines in NURBS surfaces.	N/A		N/A	ADISONURB
ADList	Provides information on parts, features, and drawing views.	Designer Main		Designer, Utilities, List	ADLIST
ADModdim	Changes dimension values on the active sketch or on the drawing.	Designer Main		Designer, Change Dimension	ADMODDIM
ADPardim	Dimensions the active sketch interactively, creating AutoCAD dimension objects that control the feature creating parametrically.	Designer Main		Designer, Sketch, Add Dimension	ADPARDIM
ADProfile	Applies constraints to turn a sketch into a shape that can be extruded, revolved, or swept.	Designer Main		Designer, Sketch, Profile	ADPROFILE
Adprojtype	System variable that sets the method used when you unfold orthographic or ancillary views.	N/A		N/A	ADPROJTYPE
Adreusedim	System variable that toggles the automatic display of parametric dimensions.	N/A		N/A	ADREUSEDIM

Command or Icon Name	Description	Toolbar	Icon	Menu Bar	Command:
Adrulemode	System variable that controls whether constraints are applied automatically.	N/A		Designer, Settings	ADRULE-MODE
Adsecltype	System variable that sets the linetype for section lines in the parent view of a cross-section.	N/A		N/A	ADSECLTYPE
ADSettings	Sets the AutoCAD Designer system variables	Designer Main	 ⊘	Designer, Settings	ADSETTINGS
Adskangtol	System variable that controls the tolerance angle for constraints.	N/A		Designer, Settings	ADSKANGTOL
Adskmode	System variable that controls whether a sketch is interpreted as precise or rough.	N/A		Designer, Settings	ADSKMODE
Adskstyle	System variable that lists AutoCAD linetypes used for sketch borders.	N/A		N/A	ADSKSTYLE
ADUpdate	Regenerates the active part or drawing using any new dimension values or changed sketches.	Designer Main	 ⚡	Designer, Update	ADUPDATE
Bottom View	Changes the view orientation in Part mode.	Designer View	 ▦	Designer, Part Viewing, Bottom	ADPARTVIEW
Create Annotation	Creates annotations on an AutoCAD Designer drawing.	N/A		Designer, Drawing, Annotation, Create	ADANNOTE
Create Drawing View	Creates any of the following drawing view types from an original AutoCAD Designer solid model: base view, auxiliary view, orthogonal view, isometric view, or detail view.	N/A		Designer, Drawing, Create View	ADVIEW
Create Parameters	Lets you create global parameters.	N/A		Designer, Parameters, Create	ADPARAM

Command or Icon Name	Description	Toolbar	Icon	Menu Bar	Command:
Create Reference Dimensions	Creates a reference dimension on geometry in a drawing view.	N/A		Designer, Drawing, Dimension, Ref Dim	ADREFDIM
Create Work Axis	Creates a line at the center line of a cylindrical, conical, or toroidal surface that can be used to define other features.	N/A		Designer, Features, Work Axis	ADWORKAXIS
Create Work Plane	Creates a construction plane on the active part that can be used to define other features.	N/A		Designer, Features, Work Plane	ADWORKPLN
Create Work Point	Creates work points used for locating holes.	N/A		Designer, Features, Work Point	ADWORKPT
Delete Annotation	Deletes annotations on an AutoCAD Designer drawing.	N/A		Designer, Drawing, Annotation, Delete	ADANNOTE
Delete Constraints	Deletes constraints from the active sketch and changes the geometric relationships between sketched objects.	N/A		Designer, Sketch, Constraints, Delete	ADDELCON
Delete Features	Deletes features from the active part.	N/A		Designer, Features, Delete	ADDELFEAT
Delete Parameters	Lets you delete global parameters.	N/A		Designer, Parameters, Delete	ADPARAM
Delete Reference Dimensions	Deletes reference dimensions from drawing views.	N/A		Designer, Drawing, Dimension, Delete Ref Dim	ADDELREF
Delete View	Deletes specified drawing view and all of its dependent views.	N/A		Designer, Drawing, Edit View, Delete	ADDELVIEW

Command or Icon Name	Description	Toolbar	Icon	Menu Bar	Command:
Designer Help	Accesses the AutoCAD Designer on-line Help facility.	N/A		Designer, Designer Help (or Search for Help on or How to Use Help)	N/A
Designer Mode	Controls whether Part or Drawing mode is in effect.	Designer Main		Designer, Mode	ADMODE
Dimension Attributes	Modifies the appearance, precision, and tolerance of drawing view dimensions.	N/A		Designer, Drawing, Dimension, Attributes	ADDIMATT
Dimension Display	Changes the display mode for dimensions of all parts and sketches without affecting the drawing so that you can see the dimension parameters to use in equations for other dimensions.	N/A		Designer, Display, Dim Display	ADDIMDSP
Display Isolines	Controls the display of the wire representation of a part, which uses iso-lines to help you visualize the curved surfaces.	N/A		Designer, Display, Isolines	ADISOLINES
Display Work Axes	Toggles the display of all work axes on the active part.	N/A		Designer, Display, Work Axis	ADAXISDSP
Display Work Plane	Toggles the display of work planes on the active part.	N/A		Designer, Display, Work Plane	ADPLNDSP
Display Work Point	Toggles the display of the work points on the active part.	N/A		Designer, Display, Work Point	ADPTDSP
Edit View	Modifies the scale, associated text, and hidden line display of the selected drawing view.	N/A		Designer, Drawing, Edit View, Attributes	ADEDITVIEW

Command or Icon Name	Description	Toolbar	Icon	Menu Bar	Command:
Export Parameters	Exports global parameters to another drawing.	N/A		Designer, Parameters, Export	ADPARAM
Extrude Features	Creates an extruded solid feature from the active sketch.	N/A		Designer, Features, Extrude	ADEXTRUDE
Fix Point	Fixes a point on the active sketch so it is immovable in x,y,z space relative to all other sketch objects.	N/A		Designer, Sketch, Fix Point	ADFIXPT
Freeze Dimensions	Hides the specified dimension on a drawing.	N/A		Designer, Drawing, Dimension, Freeze	ADFRZDIM
Front View	Changes the view orientation in Part mode.	Designer View		Designer, Part Viewing, Front	ADPARTVIEW
Hole	Creates a drilled, counterbored, or countersunk hole in the active part.	N/A		Designer, Features, Hole	ADHOLE
Hole Note	Creates a standard hole note with diameter depth and angle information for the selected hole.	N/A		Designer, Drawing, Annotation, Hole Note	ADHOLENOTE
Import Parameters	Lets you import global parameters from another drawing.	N/A		Designer, Parameters, Import	ADPARAM
Isometric View	Changes the view orientation in Part mode.	Designer View		Designer, Part Viewing, Isometric	ADPARTVIEW
Left View	Changes the view orientation in Part mode.	Designer View		Designer, Part Viewing, Left	ADPARTVIEW
Linked File Parameters	System variable that defines the name of the linked parameter file for sharing global parameters between .dwg files.	N/A		Designer, Parameters, Linked File	ADPARFILE

Command or Icon Name	Description	Toolbar	Icon	Menu Bar	Command:
List Parameters	Lets you list global parameters.	N/A		Designer, Parameters, List	ADPARAM
Load Designer	Loads AutoCAD Designer for your current drawing session.	N/A		Designer, Utilities, Load Designer	N/A
Make Base	Converts the active part into a static part and compresses the part information to take up less disk space.	N/A		Designer, Utilities, Make Base	ADMAKEBASE
Make Part Active	Lets you switch between parts in a multiple part drawing.	N/A		Designer, Part, Make Active	ADACTPART
Mass Properties	Lists the mass properties for the specified parts.	N/A		Designer, Utilities, Mass Properties	ADMASSPROP
Mesh	Displays AutoCAD Designer wireframe surfaces as AutoCAD mesh surfaces.	N/A		Designer, Display, Mesh	ADMESH
Move Annotation	Moves annotations on an AutoCAD Designer drawing.	N/A		Designer, Drawing, Annotation, Move	ADANNOTE
Move Dimensions	Moves dimensions on a drawing while maintaining their association to the drawing view geometry.	N/A		Designer, Drawing, Dimension, Move	ADMOVEDIM
Move Leader	Moves an annotation leader arrowhead on the drawing.	N/A		Designer, Drawing, Annotation, Move Leader	ADMOVELDR
Move View	Moves a drawing view, within its restrictions, anywhere on the drawing.	N/A		Designer, Drawing, Edit View, Move	ADMOVEVIEW

Command or Icon Name	Description	Toolbar	Icon	Menu Bar	Command:
New Part	Lets you create a new solid part definition while in Part mode.	N/A		Designer, Part, New	ADNEWPART
Remove Annotation	Removes items belonging to an annotation on an AutoCAD Designer drawing.	N/A		Designer, Drawing, Annotation, Remove	ADANNOTE
Revolve Features	Creates a revolved solid feature from the active profile.	N/A		Designer, Features, Revolve	ADREVOLVE
Right View	Changes the view orientation in Part mode.	Designer View		Designer, Part Viewing, Right	ADPARTVIEW
Show Active Part	Highlights the active part, sketch, or sketch plane.	N/A		Designer, Utilities, Show Active	ADSHOWACT
Show Constraints	Displays the constraint symbols on the active sketch.	N/A		Designer, Sketch, Constraints, Show	ADSHOWCON
Sketch Path	Creates an active sketch by solving the AutoCAD two-dimensional geometry and dimensions on the active sketch plane.	N/A		Designer, Sketch, Path	ADPATH
Sketch Plane	Sets the sketch plane location in its XY axis orientation as you specify.	N/A		Designer, Sketch, Sketch Plane	ADSKPLN
Sketch View	Changes the view orientation in Part mode.	Designer View		Designer, Part Viewing, Sketch	ADPARTVIEW
Sweep Features	Creates a solid feature defined by a planar cross-section (profile) swept along a planar trajectory (path).	N/A		Designer, Features, Sweep	ADSWEEP

Command or Icon Name	Description	Toolbar	Icon	Menu Bar	Command:
Thaw Dimensions	Displays frozen dimensions.	N/A		Designer, Drawing, Dimension, Thaw	ADTHAWDIM
Toolbar	Displays the AutoCAD Designer toolbars.	N/A		Designer, Utilities, Toolbars	TOOLBAR
Top View	Changes the view orientation in Part mode.	Designer View		Designer, Part Viewing, Top	ADPARTVIEW
Transfer AutoSurf Out	Converts AutoCAD Designer parts into a collection of AutoSurf surfaces.	N/A		Designer, Utilities, Transfer, AutoSurf Out	ADASFCONV
Transfer Part In	Reads parts from another file into the current one.	N/A		Designer, Utilities, Transfer, Part In	ADPARTIN
Transfer Part Out	Saves selected parts in a separate file.	N/A		Designer, Utilities, Transfer, Part Out	ADPARTOUT
Transfer SAT In	Reads a *.sat* format file (ACIS) into AutoCAD Designer.	N/A		Designer, Utilities, Transfer, SAT In	ADSATIN
Transfer SAT Out	Writes an AutoCAD Designer file in standard *.sat* format (ACIS).	N/A		Designer, Utilities, Transfer, SAT Out	ADSATOUT

Command or Icon Name	Description	Toolbar	Icon	Menu Bar	Command:
Unload Designer	Unloads AutoCAD Designer from your current drawing session.	N/A		Designer, Utilities, Unload Designer	N/A
Version	Displays the release number of the AutoCAD Designer software.	N/A		N/A	ADVER

absolute coordinates The exact location of a specific point in terms of x, y, and z from the fixed point of origin.

absolute value The numerical value or magnitude of the quantity, without regard to its positive or negative sign.

alias A short name that can be used to activate a command; you can customize command aliases by editing the file *acad.pgp*.

ambient The overall amount of light that exists in the environment of a rendered scene.

And gate An electronic logic symbol; if either input is zero, the output is zero.

angle brackets A value that appears in angle brackets < > is the default option for that command, which will be executed unless it is changed.

aperture A type of cursor resembling a small box placed on top of the crosshairs; used to select in the object snap mode.

architectural units Drawings made with these units are drawn in feet and fractional inches.

aspect ratio The relationship of two dimensions to each other.

associative dimensioning Dimensioning where each dimension is inserted as a group of drawing objects relative to the points selected in the drawing. If the drawing is scaled or stretched, the dimension values automatically update.

associative hatching The practice of filling an area with a pattern which automatically updates when the boundary is modified.

attribute Text information associated with a block.

attribute prompt The prompt, which you define, that appears in the command area when you insert the block into a drawing.

attribute tag A variable name that is replaced with the value that you type when prompted as you insert the attribute block.

auxiliary view An orthographic view of an object using a direction of sight other than one of the six basic views (front, top, right-side, rear, bottom, left-side).

ball tag A circled number identifying each part shown in an assembly drawing; also called a balloon number.

base feature In AutoCAD Designer, the main feature of a drawing, from which other features are defined based on the constraints put on the model.

base grip The selected grip, used as the base point for hot grip commands.

baseline dimensioning A dimensioning method in which each successive dimension is measured from one extension line or baseline.

basic view One of the six standard views of an object: front, top, right side, rear, bottom, or left side.

bearing The angle to turn from the first direction stated toward the second direction stated.

bicubic surface A sculptured plane, mathematically describing a sculptured surface between three-dimensional curves.

bidirectional associativity Describes the link between the drawing and the model in AutoCAD Designer. If a change is made to the model, the drawing is automatically updated to reflect the change; if a change is made to the parametric dimensions in the drawing, both the drawing and the model update automatically.

bilateral tolerances Tolerances specified by defining a nominal dimension and the allowable range of deviation from that dimension, both plus and minus.

blipmarks Little crosses that appear on the screen, indicating where a location was selected.

block A set of objects that have been grouped together to act as one, and can be saved and used in the current drawing and in other drawings.

block name Identifies a particular named group of objects.

block reference A particular insertion of a block into a drawing (blocks can be inserted more than once).

Boolean operators Find the union (addition), difference (subtraction), and intersection (common area) of two or more sets.

Buffer A electronic logic symbol.

bump map Converts color intensity or grayscale information to heights to give the appearance that features are raised above the surface, like embossed letters.

buttons A method of selecting options by picking in a defined area of the screen resembling a box or push button.

Cartesian coordinate system A rectangular coordinate system created by three mutually perpendicular coordinate axes, commonly labeled x, y, and z.

chained (continued) dimensioning A dimensioning method in which each

successive dimension is measured from the last extension line of the previous dimension.

chamfer A straight line segment connecting two otherwise intersecting surfaces.

chord length The straight line distance between the start point and the endpoint of an arc.

circular view A view of a cylinder in which it appears as a circle (looking into the hole).

circumscribed Drawn around the outside of a base circle.

clearance fit The space available between two mating parts, where the greatest shaft size always is smaller than the smallest hole size, thus producing an open space.

colinear constraint Lying on the same straight line.

command aliasing The creation and use of alternative short names for commands, such as LA for Layer.

command prompt The word or words in the command window that ask for the next piece of information.

command window The lines of text below the graphics window that indicate the status of commands and prompt for user input.

context sensitive Recognizes when you are in a command, and displays information for that command.

Coons patch A bicubic surface interpolated between four edges.

coordinate system locator An icon to help you visually refer to the current coordinate system.

coordinate values Used to identify the location of a point, using the Cartesian coordinate system; x represents the horizontal position on the X axis, and y represents the vertical position on the Y axis. In a three-dimensional drawing, z represents the depth position on the Z axis.

crosshatching The practice of filling an area with a pattern to differentiate it from other components of a drawing.

current layer The layer you are working on. New drawing objects are always created in the layer that is current.

cursor (crosshairs) A mark that shows the location of the pointing device in the graphics window of the screen; used to draw, select, or pick icons, menu items, or objects. The appearance of the cursor may change, depending on the command or option selected.

custom hatch pattern A design that you have previously created and stored in the file *acad.pat* or another *.pat* file of your own making to use to fill an area.

customize To change the toolbars, menus, and other aspects of the program to show those commands and functions that you want to use.

cutting edges Objects used to define the portions to be removed when trimming an object.

cutting plane line Defines the location on the object where the sectional view is taken.

datum surface A theoretically exact geometric reference used to establish the tolerance zone for a feature.

default The value that AutoCAD will use unless you specify otherwise; appears in angle brackets < > after a prompt.

default directory The DOS directory to which AutoCAD will save all drawing files unless instructed otherwise.

delta angle The included angular value from the start point to the endpoint.

diameter symbol ø Indicates that the value is a diameter.

difference The area formed by subtracting one region or solid from another.

dimension line Drawn between extension lines with an arrowhead at each end; indicates how the stated dimension relates to the feature on the object.

dimension style A group of dimension features saved as a set.

dimension value The value of the dimension being described (how long, how far across, etc.); placed near the midpoint of the dimension line.

dimension variables (dim vars) Features of dimensions can be altered by the user; you control the features by setting the variables.

dimensions Describe the sizes and locations of a part or object so that it can be manufactured.

distance across the flats A measurement of the size of a hexagon from one flat side to the side opposite it.

docked Refers to the toolbar's ability to be attached to any edge of the graphics window.

draft angle The taper on a molded part that makes it possible to easily remove the part from the mold.

drag To move an object on the screen and see it at the same time, in order to specify the new size or location.

edge view A line representing a plane surface shown on its end.

elements Multilines comprising up to 16 lines.

engineering units Drawings made with these units are drawn in feet and decimal inches.

export To save a file from one application as a different file type for use by another application.

extension line Relates a dimension to the feature it refers to.

extension line offset Specifies a distance for the gap between the end of the extension line and the point that defines the dimension.

extrusion Creates a long three-dimensional strip with the shape of a closed two-dimensional shape, as if material had been forced through a shaped opening.

falloff The angle of the cone of light from a spotlight.

feature Any definable aspect of an object—a hole, a surface, etc.

file extension The part of a file name that is composed of a period, followed by one to three characters, and that helps you to identify the file.

fillet An arc of specified radius that connects two lines, arcs, or circles, or a rounded interior corner on a machined part.

floating Refers to the toolbar's or command window's ability to be moved to any location on the screen.

floating viewports A window, created in paper space, through which you can see your model space drawing. They are very useful for plotting the drawing, adding drawing details, showing an enlarged view of the object, or showing multiple views of the object.

flyout A sub-toolbar that becomes visible when its representative icon on the main tool-bar is chosen.

font A character pattern defined by the text style.

foreshortened Appears smaller than actual size, due to being tipped away from the viewing plane.

fractional units Drawings made with these units express lengths less than 1 as fractions (e.g., 15 1/4).

freeze The practice of making a layer invisible and excluding it from regeneration and plotting.

global linetype scaling factor The factor of the original size at which all lines in the drawing are displayed.

global parameters The variable names that act as dimensions to control your AutoCAD Designer model.

globe Used to select the direction for viewing a three-dimensional model.

graphics window The central part of the screen, which is used to create and display drawings.

grid Regularly spaced dots covering an area of the graphics window to aid in drawing.

group A named selection set of objects.

heads-up You can enter the command while looking at the screen; by picking an icon, for example, you do not need to look down to the keyboard to enter the command.

hidden line Represents an edge that is not directly visible because it is behind or beneath another surface.

highlight A change of color around a particular command or object, indicating that it has been selected and is ready to be executed or worked on in some way.

hot grips A method for editing an already drawn object using only the pointing device, without needing to use the menus, toolbars, or keyboard.

hotspot The angle of the cone of light for the brightest area of the beam of a spotlight.

image tile An active picture which displays dialog box choices as graphic images rather than words.

implied Crossing mode When this mode is activated, the first corner of a box is started when you select a point on the screen that is not on an object in the drawing. If the box is drawn from right to left, everything that partially crosses as well as items fully enclosed by the box are selected.

implied Windowing mode When this mode

is activated, the first corner of a box is started when you select a point on the screen that is not on an object in the drawing. If the box is drawn from the left to right, a window is formed that selects everything that is entirely enclosed in the box.

import To open and use a file created by a different application than the one being used.

inclined surface Slanted at an angle, a surface that is perpendicular to one of the three principal views and tipped away from the other principal views.

included angle The angular measurement along a circular path.

inscribed Drawn inside a base circle.

inspection The act of examining the part against its specifications.

instance Each single action of a command.

intensity The brightness level of a light source.

intersection The point where two lines or surfaces meet, or the area shared by overlapping regions or solid models.

Inverter gate A symbol used to draw an electronic logic circuit; for inputs of 1, the output is 0, for inputs of 0, the output is 1.

layer A method of separating drawing objects so that they can be viewed individually or stacked like transparent acetates, allowing all layers to show. Used to set color and linetype properties for groups of objects.

leader line A line from a note or radial dimension that ends in an arrowhead pointing at the feature.

limiting element The outer edge of a curved surface.

limiting tolerances Tolerances specified by defining an upper and lower allowable measurement.

lock To prohibit changes to a layer, although the layer is still visible on the screen.

major arc An arc which comprises more than 180° of a circle.

major axis The long axis of symmetry across an ellipse.

manufacture To create a part according to its specifications.

mass properties Data about the real-world object being drawn, such as its mass, volume, and moments of inertia.

matrix An array of vertices used to generate a surface.

menu bar The strip across the top of the screen showing the names of the pull-down menus, such as File, Edit, etc.

minor axis The short axis of symmetry across an ellipse.

mirror line A line that defines the angle and distance at which a reversed image of a selected object will be created.

miter line A line drawn above the side view and to the right of the top view, often drawn from the top right corner of the front view, used to project features from the side view onto the top view of an object.

model space The AutoCAD drawing database, when an object exists as a three-dimensional object.

Nand gate A symbol used to draw an electronic logic circuit; when both inputs are 1, output is 0, when any input is 0, output is 1.

Nor gate A symbol used to draw an electronic logic circuit; when both inputs are 0, output is 1, any input of 1, output is 0.

normal surface A surface that is perpendicular to two of the three principal orthographic views and appears at the correct size and shape in a basic view.

noun/verb selection A method for selecting an object first and then the command to be used on it.

nurb A cubic spline curve.

Object Properties toolbar Contains the icons whose commands control the appearance of the objects in your drawing.

offset distance Controls the distance from an existing object at which a new object will be created.

options The choices associated with a particular command or instruction.

Or gate A symbol used to draw an electronic logic circuit; any input of 1 results in output of 1, when both inputs are 0, output is 0.

orthographic view A two-dimensional drawing used in representing a three-dimensional object. A minimum of two orthographic views are necessary to show the shape of a three-dimensional object, because each view shows only two dimensions.

output file The file to which attribute data is extracted.

overall dimensions The widest measurements of a part; needed by manufacturers to determine how much material to start with.

override mode When the desired object snap mode is selected during each command it is to be used for.

paper space A mode that allows you to arrange, annotate, and plot different views of a model in a single drawing, as you would arrange views on a piece of paper.

parametric A modeling method that allows you to input sizes for and relationships between the features of parts. Changing a size or relationship will result in the model being updated to the new appearance.

partial auxiliary view An auxiliary view that shows only the desired surfaces.

parts list Provides information about the parts in an assembly drawing, including the item number, description, quantity, material, and part number.

perspective icon A cube drawn in perspective that replaces the UCS icon when perspective viewing, activated by the 3D Dynamic View command, is in effect.

pictorial full section A section view showing an object cut in half along its vertical center line.

pin registry A process in which a series of transparent sheets are punched with a special hole pattern along one edge, allowing the sheets to be fitted onto a metal pin bar. The metal pin bar keeps the drawings aligned from one sheet to the next. Each sheet is used to show different map information.

plan view The top view or view looking straight down the Z axis toward the XY plane.

plus/minus tolerances Another name for variance tolerances.

point filter Option that lets you selectively find single coordinates out of a point's three-dimensional coordinate set.

point light A light source similar to a bare light bulb that casts light in every direction.

pointer An indicator established from the current drawing to the original drawing, or external reference.

polar array A pattern created by repeating objects over and over in a circular fashion.

polar coordinates In AutoCAD, the location of a point, defined as a distance and an angle from another specified point, using the input format @DISTANCE<ANGLE

polygon A closed shape with sides of equal length.

polyline A series of connected objects (lines or arcs) that are treated as a single object.

precedence When two lines occupy the same space, precedence determines which one is drawn. Continuous lines take precedence over hidden lines, and hidden lines take precedence over center lines.

primitives Basic shapes that can be joined together or subtracted from each other to form more complex shapes.

profile A two-dimensional view of a three-dimensional model; or a closed 2D shape that can be extruded, revolved, or swept.

project To transfer information from one view of an object to another by aligning them and using projection lines.

projected constraint The selected point on the first object joins to the unbounded definition of the second object.

projection lines Horizontal lines that stretch from one view of an object to another to show which line or surface is which.

prototype drawing A drawing saved with certain settings that can be used repeatedly as the basis for starting new drawings.

quadrant point Any of the four points that mark the division of a circle or arc into four segments: 0°, 90°, 180°, and 270°.

ray tracing The method of calculating shading by tracking reflected light rays hitting the object. Both AutoVision and 3D Studio use the Phong algorithm to perform ray tracing to produce realistic shading.

real-world units Units in which drawings in the database are drawn; objects should always be drawn at the size that the real object would be.

rectangular array A pattern created by repeating objects over and over in columns and rows.

rectangular view A view of a cylinder in which it appears as a rectangle (looking from the side).

reference surface A surface from which measurements are made when creating another view of the object.

reflection The degree to which a surface bounces back light.

refraction The degree to which an object changes the angle of light passing through it.

regenerate To recalculate from the drawing database a drawing display that has just been zoomed or had changes made to it.

region A two-dimensional area created from a closed shape or a loop.

relative coordinates The location of a point in terms of the X and Y distance, or distance and angle, from a previously specified point @X,Y.

rendered Lighting effects, color, and shape applied to an object in a single two-dimensional view.

resolution Refers to how sharp and clear an image appears to be and how much detail can be seen; the higher the resolution, the better the quality of the image. This is determined by the number of colors and pixels the computer monitor can display.

revolution Creating a three-dimensional object by sweeping a two-dimensional polyline, circle, or region about a circular path to create a symmetrical solid that is basically circular in cross-section.

romans A font.

root page The original standard screen menu column along the right side of the screen, accessed by picking the word AutoCAD at the top of the screen.

roughness The apparent texture of a surface. The lower the surface roughness, the more reflective the surface will be.

round A convex arc, or a rounded external corner on a machined part.

running mode When the current object snap mode is automatically used any time a command calls for the input of a point or selection.

scale factor The multiple or fraction of the original size of a drawing object or linetype at which it is displayed.

section lines Show surfaces that were cut in a section view.

section view A type of orthographic view that shows the internal structure of an object; also called a cutaway view.

selectable group A named selection set in which picking on one of its members selects all members that are in the current space, and not on locked or frozen layers. An object can belong to more than one group.

selection filters A list of properties required of an object for it to be selected.

selection set All of the objects chosen to be affected by a particular command.

session Each use, from loading to exiting, of AutoCAD.

shape-compiled font Character shapes drawn in AutoCAD using vectors.

sign Positive or negative indicator that precedes a centermark value.

solid modeling A type of three-dimensional modeling that represents the volume of an object, not just its lines and surfaces; this allows for analysis of the object's mass properties.

spotlight A light source used to highlight certain key areas and cast light only toward the selected target.

Standard toolbar Contains the icons which control file operation commands.

static base feature A model that does not include the constraint and parametric information necessary for it to automatically update.

status bar The rectangular area at bottom of screen that displays time and some command and file operation information.

style Controls the name, font file, height, width factor, obliquing angle, and generation of the text.

submenu A list of available options or commands relating to an item chosen from a menu.

surface modeling A type of three-dimensional modeling that defines only surfaces and edges so that the model can be shaded and have hidden lines removed; resembles an empty shell in the shape of the object.

swatch An area that shows a sample of the color or pattern currently selected.

sweeping Extrusion along a non-linear path rather than a straight line.

target area The square area on an object snap aperture, in which at least part of the object to be selected must fit.

template file Specifies the file format for the Extract Attribute command as *.cdf* of *.sdf*.

tessellation lines To cover a surface with a grid or lines, like a mosaic. Tessellation lines are displayed on a curved surface to help you visualize it; the number of tessellation lines determines the accuracy of surface area calculations.

threshold value Determines how close vertices must be in order to be automatically welded during the rendering process.

through point The point through which the offset object is to be drawn.

tiled viewports If more than one viewport is created, they cannot overlap. They must line up next to each other, like tiles on a floor.

tilemode An AutoCAD system variable that controls the use of tiled or floating viewports or paper space.

title bar The rectangular area at the top of the screen which displays the application and file names.

toggle A switch that turns certain settings on and off.

tolerance The amount that a manufactured part can vary from the specifications laid out in the plans and still be acceptable.

tool tip The name of the icon that appears when you hold the cursor over the icon.

toolbar A strip containing icons for certain commands.

transparency A quality that makes objects appear clear, like glass, or somewhat clear, like colored liquids.

transparent command A command that can be selected while another command is operating.

typing cursor A special cursor used in dialog boxes for entering text from the keyboard.

UCS icon Indicates the current coordinate system in use and the direction in which the coordinates are being viewed in 3D drawings.

unidirectional dimensioning The standard alignment for dimension text, which is to orient the text horizontally.

union The area formed by adding two regions or solid models together.

units AutoCAD can draw figures using metric, decimal, scientific, engineering, or architectural measurement scales. These are set with the Units command.

update To regenerate the active part or drawing using any new dimension values or changed sketches.

User Coordinate System (UCS) A set of x, y, and z coordinates whose origin, rotation, and tilt are defined by the user. You can create and name any number of User Coordinate Systems.

user-defined hatch pattern A predefined hatch pattern for which you have specified the angle, spacing, and/or linetype.

variance tolerances Tolerances specified by defining a nominal dimension and the allowable range of deviation from that dimension; includes bilateral and unilateral tolerances.

vector A directional line. Also, a way of storing a graphic image as a set of mathematical formulas.

Venn diagram Pictorial description named after John Venn, an English logician, where circles are used to represent set operations.

viewpoint The direction from which you are viewing a three-dimensional object.

viewport A "window" showing a particular view of a three-dimensional object.

virtual screen A file containing only the information displayed on the screen. AutoCAD uses this to allow fast zooming without having to regenerate the original drawing file.

Windows Control box A box that appears in the upper left corner of a window that controls closing and resizing the window.

wireframe modeling A type of three-dimensional modeling that uses lines, arcs, circles, and other objects to represent the edges and other features of an object; so called because it looks like a sculpture made from wires.

work axis In AutoCAD Designer, a parametrically defined line at the center line of a cylindrical, conical, or toroidal surface that can be used to define other features.

work plane In AutoCAD Designer, a parametrically defined plane attached to a feature of the model that you can use to define other features.

work point In AutoCAD Designer, a parametrically located point that can be placed to define the location of holes.

World Coordinate System (WCS)
AutoCAD's system for defining three-dimesional model geometry using x, y, and z coordinate values; the default orientation is a horizontal X axis with positive values to the right, a vertical Y axis with positive values above the X axis, and a Z axis that is perpendicular to the screen and has positive values in front of the screen.

World Coordinates The basis for all user coordinate systems.

Cover drawing courtesy of Shawn Murphy

Chapter 1 and 2 openers, and Tutorial 1, 2, 3, 4, 5, 11, and 14 openers courtesy of Autodesk, Inc.

Exercises 1.2, 1.3, 1.6, 1.7, 2.2, 2.3, 2.5, 3.4, 4.1, 4.2, 4.4, 5.1, 6.4, 6.5, 6.6, 6.7, 7.1, 7.2, 7.3, 7.6, 7.7, 8.1, 8.3, 8.4, 8.8, 9.5, 9.6, 10.3, 10.4, 10.5, 10.7, 17.3, 17.4, 17.5, and 17.6 courtesy of the Spocad Centers of the Gonzaga University School of Engineering

Exercises 1.9, 3.5, 3.6, 8.6, 8.7, 14.4, 14.5, 14.6, and 15.6 courtesy of Karen L. Coen-Brown, Engineering Mechanics Department, University of Nebraska-Lincoln

Exercises 1.10, 2.7, 2.8, 2.9, 2.10, 3.8, 3.9, 3.10, 4.7, 4.8, 4.9, 4.10, 5.2M, 5.3, 5.4, 5.5M, 5.10M, 6.8M, 6.9, 6.10, 7.9, 7.10, 8.9, 8.10, 9.9, 10.8M, 10.9M, 10.10, 11.1, 11.2, 11.3, 11.4, 11.5, 11.6, 11.7, 11.8, 11.9, 12.2, 12.3, 12.4, 12.5, 12.6, 12.7, 12.8, 12.9, 12.10, 13.1, 13.2, 13.4, 13.5, 13.6, 13.7, 13.8, 13.9, 13.10, 14.10, 15.7, 15.8, 15.9, 15.10, 16.5, 16.6, 16.7, 16.8, 17.8, 17.9, 17.10, and Tutorial 7, 15, and 16 openers courtesy of James H. Earle

Exercise 4.3 courtesy of D. Krall, Norfolk State University

Exercises 4.6, 7.4, 7.5, 9.1, 9.2, 9.3, 14.7, and 14.9 courtesy of Tom Bryson, University of Missouri-Rolla

Exercise 5.8 courtesy of Kyle Tage

Tutorial 6 opener courtesy of Kim Manner, Department of Mechanical Engineering, University of Wisconsin-Madison

Tutorial 8 opener courtesy of Torian Roesch

Tutorial 9 opener courtesy of Craig Bradley

Exercise 9.4 courtesy of Mary Ann Koen, University of Missouri-Rolla

Tutorial 10 opener courtesy of Wendy Warren

Exercise 12.1 courtesy of John S. Walker, Program Coordinator-CADD, Essex Community College, Baltimore, MD

Tutorial 13 opener courtesy of Doug Baese

Tutorial 17 opener courtesy of Randy Harris

BUILDING VICTORIA

Corner of Government Street and Port Streets, Victoria, B.C.

In the early 1900s, telegraph poles marched along Government Street sidewalks, and streetcars vied with left-hand-drive jitneys. Today only the buildings—some much altered and most with different occupants—remind us of those bygone times.

BUILDING VICTORIA

MEN, MYTHS, AND MORTAR

DANDA HUMPHREYS

HERITAGE HOUSE

Copyright © 2004 by Danda Humphreys
First edition

Library and Archives Canada Cataloguing in Publication

Humphreys, Danda
    Building Victoria: men, myths and mortar / Danda Humphreys.

    Includes bibliographical references and index.
    ISBN 1-894384-68-7

    1. Historic buildings—British Columbia—Victoria. 2.Victoria (B.C.)—Buildings,
structures, etc. 3. Victoria (B.C.)—History. I. Title.

FC3846.7.H84 2004   971.1'28   C2004-905047-8

Cover photos: Front cover—View of Government Street (centre), postcard from Heritage House
collection; City Hall, Craigdarroch Castle, Victoria Public Library (top, left to right), postcards from
collection of John and Glenda Cheramy; detail of Rithet Building (bottom right), Christine Toller.
Back cover—Chinese Public School (upper right), Heritage House collection; CPR Steamship Ter-
minal (lower left), postcard from collection of John and Glenda Cheramy
Cover and book design and layout: Christine Toller.
Map: Nancy St. Gelais.
Editor: Audrey McClellan.
This book is set in Adobe Jenson.

Heritage House acknowledges the financial support for its publishing program from the Government
of Canada through the Book Publishing Industry Development Program (BPIDP), The Canada
Council for the Arts, and the Province of British Columbia through the British Columbia Arts Council.

Heritage House Publishing Co. Ltd.
#108-17665 66A Ave.
Surrey, BC, Canada
V3S 2A7
greatbooks@heritagehouse.ca
www.heritagehouse.ca

Printed in Canada

The Canada Council | Le Conseil des Arts
for the Arts | du Canada

BRITISH COLUMBIA
ARTS COUNCIL
We acknowledge the support of the Province of British Columbia
through the British Columbia Arts Council.

Contents

For J.

DH

In the early 1900s Victorians viewed their new Parliament Buildings with pride. The provincial legislature is still the most striking feature on Victoria's Inner Harbour.

Acknowledgements

One of the best things about finishing a book is reaching the point where I can express gratitude to all the people who helped make it happen. It's been seven years since my first articles appeared in the Victoria *Times Colonist*, and some of those "helpers" have been with me every step of the way.

First, Carey Pallister and Trevor Livelton at the City of Victoria Archives and Michael Carter and colleagues at the B.C. Archives and Retrieval Service—you do us all a wonderful service by having so much information on-line, but nothing can replace the time you spend and your patience with hapless authors facing imminent deadlines. I deeply appreciate your input and your help.

I also obtained information from the helpful people at City Hall's Planning Department, the Old Cemeteries Society, Stuart Stark, and Jennifer Nell Barr, whose own work in "putting heritage on paper" has been such an inspiration.

Don't let me forget the *Times Colonist* staff, particularly "Islander" editor Peter Salmon, with whom I enjoyed a happy and rewarding relationship over more than five years as a weekly historical columnist.

Thanks to the following people who spent valuable time with me in connection with specific stories and passed on information and photographs: Judith Hudson Beattie (Hudson's Bay Company), Jim Ralph (Rogers Block), Eden Jaycock (E.A. Morris Building), Jim Munro (Royal Bank of Canada), Islay Avren (Windsor Hotel), Sister Margaret Cantwell (Sisters of St. Ann), Jim McGrath and Tony Heeley of the Provincial Capital Commission (CPR Steamship Terminal and Crystal Garden), Ken O'Connor (Queen's Printer), Kathy Summers (Craigdarroch Castle), Mike Hamilton (Union Club), Greg Evans (Old Court House), and Rick Allen (James Bay Hotel).

And special thanks this time to John and Glenda Cheramy, tireless and energetic postcard collectors, whose patience and enthusiasm have been a boon. It's a pleasure to be able to feature so many items from your priceless collection in my book.

Hats off to Rodger Touchie at Heritage House for sharing my enthusiasm for Victoria's early history, kudos to Christine Toller for her design, and two thumbs up to editor Audrey McClellan for steering me through four books and making working with an editor such a treat.

Finally, for the *Times Colonist* readers whose interest and encouragement inspired me to research and write … the families who contacted me with information, then helped bring the stories to life … the visitors who follow me along sidewalks, through alleyways, into the fascinating areas of Victoria's historic downtown … this one's for you.

DANDA HUMPHREYS   9

Government Street once boasted several financial institutions, including the 1909 Neoclassical-style Royal Bank of Canada (second from left), the 1897 Chateau-style Bank of Montreal (centre), and the 1912 white-glazed, terra cotta-adorned Union Bank (right, flying the flag).

Introduction

The 1931 Causeway Tower epitomized the Art Deco style of the 1920s and '30s. The building at 812 Wharf Street, once an Imperial Oil gas station, now houses Victoria's downtown Information Centre.

No doubt about it. For a relatively young city, Victoria has a surprisingly rich and colourful history. Maybe that's because it was settled not by a handful of hardy pioneers, but by representatives of a British fur-trading company. Maybe it's because the people who settled here happened to be larger-than-life characters. Maybe it's because the buildings left behind by those characters tell so many stories.

A few years ago, when I was researching the historic origin of Victoria's street names, I walked along pioneer pathways, followed remnants of early railway lines, and visited hundreds of older buildings. It was like taking a step back in time. Victoria's downtown heritage structures, in particular, tell a fascinating tale. And fortunately for us all, groups of people dedicated to preserving this heritage have lobbied tirelessly over the years to make sure that these structures survived.

Many are much altered or have assumed a different guise over the years, but some still hold clues to what went on before. The 1861 Rithet Building on Wharf Street, for example, contains one of the few remaining remnants of Fort Victoria—its water well. The basement of the 1909 Royal Bank building on Government Street still houses the bank vault, and space for a "shooting gallery," where bank employees learned to use firearms in case of a robbery. Under the black-and-white, mock-Tudor façade of the old Windsor Hotel at the corner of Government and Courtney, the sturdy walls of Victoria's first brick-built hostelry stand as solid as the day they were built in 1858.

This was perhaps the most significant year in Victoria's history, for it was the 1858 discovery of gold in the mainland's Fraser River that changed the city's fortunes forever.

Twenty years earlier, realizing that the proposed British-U.S. border would place them in American territory, Columbia River-based officers of the Hudson's Bay Company (HBC) had explored Vancouver Island as a potential future base of operations. In 1842

James Douglas, the assistant of Fort Vancouver's chief trader, investigated all the nooks and crannies on the south Island. He declared the Port of Camosack the most suitable spot, and by mid-1843 Fort Camosun—quickly renamed Fort Victoria after the young British queen of the day—had been established on the Inner Harbour's east shore. Fort Street, originally the pathway from the front gate of the fort to the back, marks its location.

The fort occupied the area bordered by Wharf Street, Bastion Square, Government Street, and Broughton Street. Today bricks in Bastion Square and along the Government Street sidewalk outline the fort's boundaries. Those bricks bear the names of early Victoria pioneers and businesses and show where 18-foot-high palisades and gun bastions once stood, guarding the fort's occupants.

Ideally located for fur and salmon trading, Fort Victoria flourished. But in the spring of 1858 the fort changed almost overnight from a remote outpost of the Hudson's Bay Company to a way station for gold miners. On April 25, when the first shipload of gold seekers arrived, the settlement was home to some 300 souls. The ship's passengers doubled that number, and in the ensuing months close to 20,000 prospectors stepped ashore. Americans, Canadians, Europeans, Chinese … no one, it seemed, could resist the glitter of gold. Obtaining licences at the fort and gathering gear for the goldfields, they moved on as soon as possible, but their presence and their sheer numbers guaranteed their place in Victoria's history.

For most prospectors, the mother lode proved elusive. Desolate, disappointed, or delighted, they deserted the goldfields and came back … to a very different settlement than the one they had left behind. The HBC, realizing its fur-trading days on the west coast were numbered, had suspended operations. James Douglas, chief factor at the fort and governor of Vancouver Island, oversaw demolition of the fort buildings and sale of the land.

By 1865 almost all visible traces of the HBC's northern headquarters had disappeared. In its place were warehouses, stores, businesses, and banks. North of the fort, where the first settlers had been laid to rest on the side of a steep ravine, Johnson Street filled the gap. The remains of those early pioneers were moved by prison chain gangs to a dedicated burying ground (now Pioneer Square) near where Christ Church Cathedral stands today.

On the other side of Johnson Street, Chinatown—once reached by wooden walkways that crossed the ravine—was accessible, at least to its Chinese inhabitants, via Pandora Avenue and the closely guarded Fan Tan Alley. Residents of the close-knit community lived in wooden shanties long after downtown builders began constructing with brick. Eventually the shanties too were gone, replaced by the structures we see in Chinatown today.

Incorporated as a city in August 1862, Victoria grew and prospered, but in the post-gold rush era its fortunes declined. Although some buildings sprang up on Wharf Street's Commercial Row, on Store Street, and along Government Street, most of the current structures downtown date from a later era.

Tall, eight-sided bastions with cannons—one near Broughton and Wharf streets, the other at Bastion and Government streets—protected the stockaded settlement called Fort Victoria.

Established in 1855 and the final resting place of more than 1,000 settlers, gold seekers, and fortune hunters, the Old Burying Ground has been the scene of many a spectral sighting. Most frequently spotted is the ghost of Adelaide Griffin, whose husband, Ben, ran the Boomerang Saloon in Bastion Square in the early 1860s. Now a public park called Pioneer Square, the Old Burying Ground nestles in the shadow of Christ Church Cathedral on Quadra Street.

Hooper & Watkins designed Victoria's first public library, built in 1904 on the corner of Yates and Blanshard streets with financial assistance from the Andrew Carnegie Foundation.

Victoria was an exciting place for architects, who came from far and wide. First to arrive were Hermann Otto Tiedemann, John Wright, John Teague, and Thomas Trounce. Of these, Teague had perhaps the most influence, particularly during the rapid-growth era of the 1870s–90s. Then came talented and imaginative individuals such as A. Maxwell Muir, Edward Mallandaine, William Ridgeway Wilson, Thomas Sorby, Thomas Hooper, J. Gerhard Tiarks, Charles Elwood Watkins, Warren Heywood Williams, Leonard B. Trimen, Percy L. James, Samuel Maclure, the ill-fated Francis Rattenbury, and others too numerous to mention.

By the 1880s criss-crossing downtown streets contained the commercial centre, while James Bay was still the preferred neighbourhood for Victoria's elite. Gradually, those worthies went east to the tree-lined, peaceful seclusion of the Rockland area, where Government House reigned supreme and coal baron Robert Dunsmuir planned a castle-home high on the hill. Victoria became a destination of choice for seasoned travellers, and soon the large, popular, but expensive downtown hostelries faced friendly and more reasonably priced competition from the red-tiled James Bay Hotel.

Chinatown grew by leaps and bounds with the influx of men who had laboured on the Canadian Pacific Railway's transcontinental line. This "carrot" proffered by the Dominion government to coax British Columbia into Confederation in 1871 had been a long time coming. Eventually, still barely started and under pressure to complete its part of the bargain, the CPR hired thousands of Chinese labourers from San Francisco and China to build a route through the Canadian Rockies. Laid off in 1885 when the line was completed, they came to Victoria and were integrated into the Chinese community. Interestingly, many of the major Chinatown buildings, such as the Chinese Consolidated Benevolent Association and the Chinese Public School, were designed for the Chinese community by non-Chinese architects using a mix of styles.

As the 1900s moved into high gear, Victoria seemed light years away from its humble beginnings. Rattenbury's legislature graced the Inner Harbour's south shore, soon to be joined by his equally elegant Empress Hotel, Crystal Garden, and the Greek temple-like CPR Steamship Terminal. The queen the city was named after was dead. Edwardian was "in." Art Deco was still to come.

But many of the stories in this book are about people who arrived in Victoria's first half-century—from the mid-1840s to mid-1890s. And the book isn't just about buildings. It's about the people who designed them, built them, conducted business in them. If you've read my *On the Street Where You Live* series, many names will be familiar. People like James Douglas, Roderick Finlayson, Judge Matthew Baillie Begbie, the Sisters of St. Ann, R.P. Rithet, George Richardson, Hannah and Richard Maynard, and more were major players in the early days. They surface again in this selection of stories, along with the architects who made their dreams come true.

A good way to get the most out of this book is to conduct your own self-guided tour, using the map that's included, and admire the products of their labours for yourself. These were interesting times and interesting people. Read their stories and you'll realize we will never see their like again!

Government House, Victoria, B. C.

Francis Rattenbury and Samuel Maclure worked together on the second (1903) Government House, traditionally the home of B.C.'s lieutenant-governor. When this Rockland Avenue mansion burned down in 1957, its only surviving section—the stone porte cochère—was incorporated into the current residence.

Map Legend

1.   Hudson's Bay Company Store
1A. The Bay Centre
2.   Southgate and Lascelles Building
3.   Dominion Customs House
4.   Rithet Building
5.   Yates Block
6.   Deluge Fire Company
7.   Old Court House
8.   Hibben Bone, Block
9.   E.A. Morris Building
10. Royal Bank Building
11. Windsor Hotel
12. Rogers Block
13. Galpin Block
14. Hotel Janion
15. Craigdarroch Castle
16. Chinese Public School
16A. Chinese Consolidated Benevolent Association
17. City Hall
18. Maynard Building
19. E. G. Prior Building
20. New England Hotel
21. Theatres McPherson Playhouse
21A. Royal Theatre
22. Union Club
23. Alexandra Club
24. Legislative Buildings
25. Empress Hotel
26. Crystal Garden
27. CPR Steamship Terminal
28. Queen's Printer
29. James Bay Hotel
29A. Carr House
30. St. Ann's Academy
30A. St. Joseph's Hospital

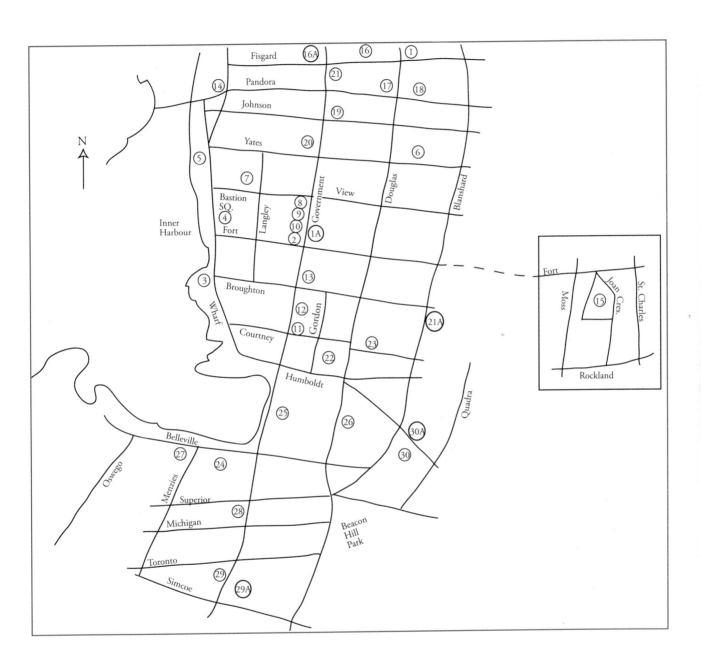

N

Fisgard
Pandora
Johnson
Yates

Inner
Harbour

Bastion
SQ.
Fort

Broughton

Wharf

Courtney

Humboldt

Belleville

Oswego

Menzies

Superior

Michigan

Toronto

Simcoe

Langley

Government

View

Douglas

Blanshard

Gordon

Quadra

Beacon
Hill
Park

Fort

Moss

Joan Cres.

St. Charles

Rockland

16A 16 1
21 17 18
14
19
20 6
5
7
8
9
10
4 1A
2
3 13
12
11 21A
23
22
25 26 30A
27 30
24
28
29 29A
15

17

The Hudson's Bay Company store, completed in 1921 and decorated with the company crest, proudly proclaimed its presence with huge letters on its roof.

Hudson's Bay Company Store

1701 Douglas Street

Detail of the HBC company crest today.

The building on the corner of Douglas and Fisgard streets may be less than 100 years old, but its history goes back to the late 1600s, when King Charles II granted a charter to "The Company of Adventurers of England Trading Into Hudson's Bay."

From its date of founding in 1670, and for about 100 years after, the Hudson's Bay Company possessed exclusive trading rights in Rupert's Land, the area that drained into Hudson's Bay. After that, it faced stiff competition from the North West Company, formed in Montreal by a group of Scots, and the American-owned Pacific Fur Company.

It was the North West Company that built fur-trading posts all through what Simon Fraser called New Caledonia (now northern B.C.). But by 1821 the HBC had bought out its rival, and in 1824 HBC governor George Simpson made the 80-day cross-country trek from Hudson's Bay to the Columbia River and supervised the building of Fort Vancouver on the river's north shore.

Seventeen years later, when news came that the U.S.–British North America border was to be extended along the 49th parallel to the Pacific, Simpson sent instructions to John McLoughlin, Fort Vancouver's chief factor, to relocate HBC's western headquarters farther north.

James Douglas, McLoughlin's assistant, chose the site, and Fort Victoria was built on the east side of the Inner Harbour in 1843. Fur stores and warehouses were on the fort's north and south sides. Now fur trading could continue as before.

Almost overnight the discovery of gold on the mainland in 1858 changed the settlement's focus from fur trading to commercial enterprise, and one by one Fort Victoria's structures were torn down. Long after the last one bit the dust, the red brick store and warehouse built at the foot of Bastion Street (now Bastion Square) in 1859 dominated the Wharf Street waterfront.

It was five storeys high, with a basement and a sub-basement stocked with food and ammunition against the unlikely possibility of attack. Twenty-two-inch-thick brick walls and iron-barred windows provided a measure of security, and a tunnel leading away from the building, under the city, provided a means of escape. Years later, when the building was demolished, a cache of cannonballs was discovered in one of its chimneys.

For more than six decades this building served as the HBC's chief commercial warehouse on the southern west coast. But early in the 1900s the HBC decided to replace it with a more modern structure, and in 1913 the chief stores commissioner, Richard Burbidge, announced that a grand new store was to be built in Victoria.

The company's focus had changed since the late 1850s, when fur trading was the main business and Company "saleshops" sold foodstuffs, prospecting licences, and gear for gold miners. By 1901 the HBC had declared itself "Western Canada's Great Headquarters for anything obtainable to Eat, Drink, Wear or Use."

The site chosen for the new store was a long way from the harbour—at the corner of Douglas and Fisgard streets, on a spot previously owned by the HBC and by this time occupied by the iron church of St. John. This prefabricated structure, shipped out from England in pieces and lovingly reconstructed on arrival, had in 1860 been dedicated to St. John the Evangelist. Its tall steeple is visible in many an archival photo. In 1912 the HBC bought back the property, paying enough—$140,000—to facilitate the building of a new, brick church that stands to this day on Quadra Street. The iron church was demolished to make way for a new Hudson's Bay Store.

Construction began on the store in 1913, but stopped a year later when the First World War began. In 1920, during the 250th anniversary of the company's Charter, the new HBC store was finally dedicated, and in September 1921 it opened its doors to customers for the first time.

The 192,000-square-foot structure at Douglas and Fisgard, designed in trademark terra cotta-columned style, took up the whole city block. It was the tenth in the HBC chain, with 50 departments and over 250 employees. Store services included an information bureau, a post office, and a library for customers' pleasure. Elevators were staffed by uniformed operators, who called out each floor's merchandise as the elevators ascended and descended.

Over the years the stores constantly adapted and updated to meet consumer demands, and in 1991 the HBC announced the closure of its most famous department. "The company," stated its owners, "will no longer deal in furs."

It was the end of an era. Today the Hudson's Bay Company is Canada's oldest corporation and largest department store retailer, with 100 locations coast to coast. But the reason for its existence—the beaver and sea otter fur trade on North America's west coast—is now just a part of its past.

Ironically, in Victoria, the Bay has recently moved to a new home on Government Street, across from where the company's fur-trading post once stood. On the fourth floor of the Bay Centre, an archives and gallery features visual and interactive displays that highlight the company's history and the creation of Vancouver Island's first European settlement, on Victoria's Inner Harbour.

Southgate and Lascelles Building

1102 Government Street

Southgate and Lascelles' 1869 building, complete with elaborate window surrounds on its 1887 second-floor addition, stands on the original site of HBC Chief Factor James Douglas's home.

The building on the northwest corner of Government and Fort streets, across from the Bay Centre, bears little resemblance to the first structure on this site—a Hudson's Bay Company fur-trading post.

In 1869, when J.J. Southgate proudly contemplated his newest commercial enterprise, he must have marvelled at how far he had come in just one decade. When he first stood near this spot in 1859, it was for a very different purpose—as one of a group of concerned residents, ready to petition for the organization of a fire department.

The group met with Governor James Douglas in the Fort Victoria yard.

Now, 10 years later, the Hudson's Bay headquarters was completely gone. Douglas, knighted by Queen Victoria in London in 1864, had retired. And Southgate was a force to be reckoned with on the local business and political scene.

James Johnson Southgate, native of England and a retired sea captain, was living in California in 1858 when news of the Fraser River gold rush sent him sailing up the coast to Victoria. He arrived in 1859, and while others journeyed on to the mainland goldfields, Southgate set up as a wholesale commission merchant on Wharf Street.

He wasted no time throwing his hat in the local political ring. By late 1859 he had managed to get himself elected as a member of the legislative assembly for Salt Spring Island. In March 1860 he took possession of the charter, lately arrived from England, that authorized formation of Victoria's first Masonic Lodge. The following year he was elected chair of the newly formed chamber of commerce. But in 1865, unbelievably, he threw it all up to return to England.

The *British Daily Colonist* reported that on May 6 of that year, a large party of friends accompanied Southgate on the tug *Diana* to Race Rocks. Champagne corks popped and two small cannon were fired as he boarded the mail steamer bound for San Francisco and thence to the country of his birth. No reason was given for his leaving … and no reason was given for his return in mid-1866. By the end of that year he was fully engaged in local politics once more, this time as MLA for Nanaimo.

The spring of 1867 saw him sailing for San Francisco and New York, en route to an exposition in France. Along the way he met up with an old friend in England—the Honourable Horace Douglas Lascelles, RN. Born the seventh son of the third Earl of Harewood in 1835, Lascelles had joined the navy at the age of 13 and was 25 years old when he arrived on the Pacific Station, based at Esquimalt, in 1860 as first lieutenant of HMS *Topaze*. In 1862 he took command of the gunboat *Forward*, then went back to England in 1865 and retired from the navy.

Lascelles had enjoyed his time in Victoria, and it wasn't difficult for Southgate to persuade his fun-loving friend to

return. The two sailed here together in the fall of 1868. The following year they invested jointly in a prime piece of real estate on Government Street. Soon the *Colonist* announced that Southgate had advertised for tenders for "the erection of two handsome brick stores with a basement. They are to front on Government Street, extending down Fort Street. The fronts, all glass, will be supported on iron pillars. The glass front will extend some distance down Fort Street. We cannot," gushed the newspaper, "cite a better proof of coming prosperity."

Within days, bids for the brick, woodwork, roofing, and ironwork had been received. The *Colonist* announced that contracts had been awarded to the firms of Kinsman & Styles and Spratt & Keimler, and that "the entire cost of the building will be about $7,000"—a veritable fortune at that time.

Construction took place as planned. Soon the splendid new structure stood on almost the exact spot as the Fort Victoria building that, 20 years earlier, had housed Chief Factor James Douglas and his family, the mess hall, and the junior clerks' quarters.

Southgate and Lascelles' single-storey brick building was designed for general merchandising. Street-level verandahs lined its longer, Government Street side. Cast iron, the traditional symbol of solidity and success, adorned its frontage. More than a decade later, a second floor was added. Detailing included pediments. Their distinctive, sharply arched shape lent added interest to the tall, narrow windows.

By 1900 streetcars were running past Southgate's building. Large awnings sheltered passersby, encouraging them to linger at the huge picture windows. Later still, the Canadian Pacific Steamship Company bought the building and used it as a ticket and administrative office. Then the Canadian Pacific Railway took it over. By that time, both Southgate and Lascelles were long gone. Lascelles died in Esquimalt in 1869 and was buried in the naval cemetery. Southgate returned to England one last time and died there in 1894.

Today there is an intriguing reminder of the simple, stockaded structure that once occupied this site. On the building's west side, a 140-foot microwave tower, designed to resemble a campanile, stands at the spot formerly taken by the Fort Victoria bell tower. Over 160 years ago the bell clanged greetings, warnings, and congratulations for the inhabitants of the Hudson's Bay Company's new northern headquarters on the Pacific Northwest coast. Now the tower stands as a mute reminder of Victoria's fur-trade beginnings and of J.J. Southgate, the man who conducted business on this corner more than a century ago.

This 1860 view of Fort Victoria from its Government Street entrance, sketched by Sarah Crease, shows the pathway through the fort (now Fort Street) that lead to its western gate at Wharf Street and the harbour beyond. The fort's bell tower is shown at right.

The building's microwave tower, viewed from the Fort Street side, marks the exact location of Fort Victoria's bell tower.

Dominion Customs House
1002 Wharf Street

The 1874 brick-and-stone Dominion Customs House, the city's oldest federal building, marked British Columbia's 1871 entry into the Dominion of Canada and signalled the end of Victoria's days as a free port.

With its deep-pink brick and cream cornices, the colourful Customs House building is quite the eye-catcher along Wharf Street. It has stood in this spot since the mid-1870s, once flanked on its north side by a Hudson's Bay Company warehouse, now standing in solitary splendour just north of Ship Point.

When the men of the HBC built Fort Victoria in 1843, the fort's western palisades parallelled the waterfront, close to what is now Wharf Street. At the foot of Fort Street—originally the wide pathway leading from one side of the stockade to the other—was a gate. Opposite the gate the ground sloped, then dropped, to the waters of the Inner Harbour. A small wharf afforded a landing place where boats could tie up safely, using one of two mooring rings set firmly into the rock.

In 1846, when the 49th parallel was declared the boundary between American and British land on the west coast, the HBC abandoned Fort Vancouver, its post on the Columbia River, and made Fort Victoria its northern headquarters. The stockade expanded to the south and now measured 300 feet by 400 feet. But the Wharf Street gate was still the main entry point.

There were several new arrivals over the years—HBC contract workers, a few independent settlers, and a handful of HBC retirees—but nothing equalled the influx of people that took place after gold was discovered along the Fraser River.

James Sangster, who served in turn as pilot, harbour master, and collector of customs, had just taken the position of postmaster when the hordes arrived. They came in their thousands, intent on purchasing licences and gathering gear for the goldfields. Entering the fort through the Wharf Street gate, the newcomers traipsed past postmaster James Sangster's log house, where he had thus far managed to keep his distance from the world by handing out mail through a small window. A melancholy man, fragile of health and overwhelmed by this unprecedented invasion, Sangster quickly retired to a cabin in Colwood, where shortly afterward he took his own life.

A dozen years earlier the HBC had abandoned Fort Vancouver. Now it all but abandoned Fort Victoria. The fur-trading post on the east shore of the Inner Harbour had outlived its usefulness. The land would be sold and developed as a business area to accommodate successful gold seekers and those who wished to service them.

The Wharf Street side of the fort—which afforded easy access to the waterfront for movement of goods—was the first to be demolished. By 1864 only a few remnants of the once-vital HBC stronghold remained. Unfortunately, by that time it was clear to almost everyone that the early promise of the gold-rush days was unlikely to be fulfilled. Newly incorporated as a city, and with the recent experience of a growth rate second to none, Victoria nevertheless watched its fortunes decline.

In 1867, when the Dominion of Canada was formed, its supporters in Victoria jumped for joy. British Columbia, they declared, should join Confederation. Others were skeptical. Anxious to secure the Pacific Northwest, Ottawa promised a railroad that would join Canada from coast to coast. It was enough to sway the naysayers. In 1871 British Columbia became the sixth province to join Confederation, and Victoria became the provincial capital.

It would take quite a while for the railroad to reach the west coast. Meanwhile, the federal government sought to impress its newest recruits with a federal building in the provincial capital that would show British Columbians they had made the right decision.

Designed by Thomas Seaton Scott, a federal works department architect, the Dominion Customs House was built in 1874–75. Three storeys high on the Wharf Street side, with an extra lower storey on the harbour side, its solid square structure and mansard roofline were reminiscent of other federal buildings. It was sited on a prominent piece of land just south of the spot where, 33 years earlier, the HBC had established Fort Victoria, and where, 15 years after that, thousands of prospectors had stepped ashore.

Twenty-five years after the Customs House opened for business, its employees were engaged in yet another gold rush. This time the men who lined up for licences were headed for the Klondike. George Fry was collector of customs, and A.R. Milne, brother of medical health officer Dr. George Milne, was customs appraiser. When George Fry died, A.R. Milne (whose name still graces his former Empire Hotel building on lower Johnson Street) assumed both his widow and his title, becoming collector of customs himself in 1890.

After a new Customs House was built in 1914 on the corner of Government and Wharf streets, the old Customs House took on many different guises over the years. Functioning as the headquarters for HMCS *Malahat* for a while, and known as the Malahat Building, it now houses lawyers' offices. Just below it and slightly to the north, the old HBC mooring rings, often hidden by brambles, serve as the only reminder of the trading post that once stood across the street. Where there were HBC warehouses, now there are public parking lots. Beautifully restored, the old Customs House stands sandwiched between them, as proud and straight as the day it was built, over a century ago.

Rithet Building
1117 - 1125 Wharf Street

Robert Paterson Rithet's 1861 building dominated the east side of Wharf Street, where the Hudson's Bay Company once reigned supreme.

The Rithet Building, on Wharf Street's east side, looks very different today compared to its beginnings almost 150 years ago.

The year was 1861. Three years of gold-rush traffic had transformed what was once a sleepy settlement surrounding a fur-trading post into a thriving commercial centre. Several astute businessmen hastened to make their mark. One, Robert Burnaby, bought a parcel of land from the Hudson's Bay Company on the western perimeter of the fort, facing the harbour.

Burnaby was a Leicestershire boy, 33 years old in 1858 when he arrived in British Columbia with a letter of introduction from colonial secretary Sir Edward Bulwer-Lytton. He was offered a job with Colonel Richard C. Moody, whose corps of Royal Engineers was busily surveying the land around New Westminster. Burnaby Lake and the municipality that surrounded it were named after Moody's new private secretary.

Leaving Moody's employ a year later, Burnaby became involved with many business ventures, mostly in Victoria. He built a one-storey brick warehouse on his parcel of Inner Harbour land. Over the next two decades the building was extended along Wharf Street, and two more storeys were added. By 1889 the building looked pretty much the way it looks today, but by that time Burnaby was long gone. He had returned to England in 1874 and died there a few years later.

The newest owner of the building, Robert Paterson Rithet, was pleased with his purchase. He had arrived in Victoria from Scotland in 1862, at the age of 18. Joining the firm of Sproat & Company as accountant, he had proved himself adept in business during Gilbert Sproat's many absences, and at the age of 25 he had been put in charge of the company's San Francisco affairs.

These were interesting times for Victoria. Despite the promise of major business gains when the Island and mainland colonies were combined in 1858, and despite the fact that Victoria was now the capital of British Columbia, many of the city's merchants were disappointed. There was talk of Confederation with the rest of Canada. Several local businessmen, anxious to re-establish free trade with their U.S. neighbours, lobbied for annexation.

Rithet was one of the few who believed that B.C. should remain Canadian.

Leaving Sproat & Company to work with J. Robertson Stewart, another well-established Victoria merchant, Rithet found himself managing the office when his employer became ill. Soon after, the business was put up for sale. It was bought by Andrew Welch and in 1870 was renamed Welch, Rithet & Company. When Andrew Welch died in 1888, Rithet bought out his interest. The wide-fronted building on Wharf Street now bore, in big letters, the words "R.P. Rithet & Company Limited … Wholesale Merchants, Shipping and Insurance Agents."

Rithet imported sugar from the Hawaiian Islands, imported groceries and liquor from all over the world, and soon expanded his business to include interests in lumber, sealing, whaling, canning, farming, shipping, insurance, mining, and railways. He built a wharf at Ogden Point (north of today's breakwater) that made Victoria accessible to travellers aboard Canadian Pacific's ocean-going Empress liners. He ploughed the money he made back into local business ventures and became a very wealthy man. He and his wife Lizzie lived with their family in a large home called "Hollybank" on Humboldt Street.

Rithet was an astute businessman, but when he entered politics for the first time in 1884, it was his campaign for adequate drainage that resulted in a successful bid for the city's top position. In 1884–85 he was mayor of Victoria, and from 1894 to 1898 he served as a member of the legislative assembly for Victoria City.

Rithet was 75 years old when he died in 1919. His company continued in business until 1948, when the building was taken over by one of its tenants, the Dowell Moving and Storage Company. Twenty years later the city bought the building and designated it a heritage site, then in 1974 sold it to the provincial government. The façade could not be touched because of its heritage status, so the original cast iron-decorated clay masonry remains. The interior, however, was extensively renovated.

When the old ground-level floor was removed, a well was discovered. It is believed to be the original Fort Victoria well, with pipes leading directly to the wharf across the street, probably designed to carry fresh water to the ships unloading cargo there in the 1840s and 1850s.

Today a walk through the Rithet Building's lobby, with its open brickwork, archival photographs, and original HBC well, is like taking a history lesson. It's a reminder of the days when Wharf Street was lined with wharves and warehouses and men like Robert Burnaby and R.P. Rithet featured prominently in Victoria's business scene.

In the main-floor lobby of this large building on Wharf Street, you will find a brick-lined water well—one of the few remnants of Fort Victoria.

Yates Block

1252 Wharf Street

Wholesale Section, Yates Street, Victoria, B.C.

The tall structure on Wharf Street at the foot of Yates (seen here in the distance, centre of photo) marks the site of Victoria's first liquor store and saloon. The 1882 brick structure that replaced it was enlarged in the 1890s.

The Yates Block on Wharf Street has had many occupants since it was built in 1882, but few were as colourful as the man who first set up a business on that spot and after whom Yates Street is named.

James Stuart Yates was a feisty Scot who travelled to Vancouver Island aboard the *Harpooner* with his wife, Mary, in 1849. He had been hired as a carpenter by the Hudson's Bay Company to work at Fort Victoria and might have done quite well had he not fallen afoul of his new boss. At that point, Chief Factor James Douglas was newly returned from Fort Vancouver, on the Columbia River, and was busily asserting his authority

The Yates Block still stands tall at the foot of the street bearing its original owner's name.

over the new arrivals. Douglas thrived on discipline; Yates did not. After 18 long months the latter could stand it no longer. Toward the end of 1850 he bought out his HBC contract and assumed independent status.

Yates purchased a town lot on the harbourside, north of the fort. To Douglas's chagrin, he became a wine and spirit merchant and proud owner of the settlement's first watering hole. It was a popular move. Business was brisk at the "Ship Inn" on Wharf Street, and before long Yates was able to expand his land interests. In 1852 he purchased almost 400 acres along the north side of the Gorge waterway for a family home.

By 1853 Douglas had found a way to profit from Yates's ventures by suggesting a licence on the sale of spirits. Yates professed indignation but didn't really mind—he realized that having the only licensed tavern assured his ability to sell liquor without fear of competition.

He continued to make money from the import and sales of good Scotch whisky and expanded his business endeavours.

In 1858, when prospectors swarmed into the settlement on their way to the Fraser River, Yates quickly lined his street with wooden shanties, which he rented out to other entrepreneurs. Supply stores stood cheek-by-jowl with saloons, where the prospectors could fortify themselves while they gathered their gear for the goldfields.

By 1860 Yates owned all the land between Langley and Wharf streets and was arguably the richest man in town. It was a far cry from his humble beginnings at the fort. In October of that year Yates took most of his family home to Scotland so that his two oldest children could have proper schooling. The two young boys—James and Harry—were left in Victoria with their Aunt Isabella, James's sister. In 1864 James returned briefly to attend to business and took his sons home to Edinburgh. They later came back to Victoria and settled here permanently.

It was James Jr. who built the interesting structure at the foot of Yates Street. Architect John Teague designed what is now the northern section of the building. It was four storeys high—two below the street level and two above. The lower levels were used for warehousing. The main floor contained offices and vaults, and the upper floor was designed for sales displays. In 1892 another storey was added. Four years later the building was extended southward, following designs by architect A.C. Ewart. This is the building we see today, its impressive arched entranceway giving way to fine panelling just inside.

Turner, Beeton & Co. Ltd. was the building's tenant for almost four decades. John H. Turner had arrived in 1862, just as celebrations of Victoria's incorporation as a city were getting underway. He went into partnership with another recent arrival, J. H. Todd, in the Victoria Produce Market on Langley Street, then started importing dry goods for his own enterprise—the London House, at Government and Fort streets. By 1879 Turner was mayor of Victoria and was elected premier of British Columbia in 1895. In 1901, the year Queen Victoria died, Turner returned to England as agent-general for the province of British Columbia, based in London.

In 1939 the W.H. Malkin Company took over the Yates Building for its grocery enterprise, and for the next 13 years the second floor was home to stored canned goods. These were so heavy that they strained the internal structure, causing the joists to drop and creating the need for additional under-floor support in the form of wooden posts. When Malkin moved out in 1953, the joists assumed their original placement, and the posts were removed.

McQuade's Ship Chandlers Ltd., established in 1858, was the building's next occupant, its endeavours advertised by a huge ship's wheel set into the frame of the arched window above the entrance. In recent times McQuade's activities were commemorated in the name of the restaurant, Chandlers, occupying part of the main floor of a building that bears little resemblance to the first small structure erected on James Yates's property all those years ago.

Deluge Fire Company

636 Yates Street

Home of the Deluge Fire Company, this 1877 structure once was topped by an ornate tower that contained the fire-alarm bell.

Look closely at 636 Yates Street, on the north side between Broad and Douglas, and you'll see no hint whatsoever of this building's former function—headquarters for one of Victoria's first firefighting companies.

In 1858, when the Fraser River gold rush focused attention on Fort Victoria, the streets that surrounded the fort sprouted shanties that served as saloons, stores, and temporary accommodations for the thousands of prospectors who journeyed to these shores. If one of these wooden buildings caught fire, bucket brigades carried water from wells and from the Inner Harbour. It was a primitive, and woefully inadequate, system.

Local businessmen petitioned Governor James Douglas for a fire service. The Hudson's Bay Company agreed to foot the bill. Two hand-operated pumping engines and 1,500 feet of leather hose were ordered from San Francisco. Volunteers were recruited. Two water cisterns were built, one at Store Street, the other on Government.

They were just in time—shortly afterward, a warehouse fire threatened the entire business district and would have destroyed it, but for the firefighters' efforts.

Today at 636 Yates Street, no hint remains of the Deluge Fire Company's former activities.

Still, it was a close call. Local businessmen J.J. Southgate and C.W. Wallace Jr. took it upon themselves to raise funds for more sophisticated equipment in the form of a hook-and-ladder rig and an alarm bell.

By November 1859, the Union Hook and Ladder Company was organized. Victoria's first fire hall was at Bastion and Wharf streets. By August 1862, when Victoria was incorporated as a city, the Union Hook and Ladder had been joined by the Deluge Engine Company No. 1 and Tiger Engine Company No. 2. Now the city boasted five water cisterns with a capacity of more than 100,000 gallons. Soon two additional cisterns brought the total to more than 150,000 gallons.

Deluge operated out of a rented building on Government Street, between Yates and Johnson, while the Tiger Company leased property on Johnson, between Government and Broad. The city was small—contained on its west and south

sides by the waters of the harbour, it was bordered by Johnson to the north and Government to the east. Thus the three fire companies could easily keep tabs on each other, and the fight to be first to the fire became fierce.

In 1867 the firefighting service received a boost in the form of a steam pumper, ordered especially from England. Hoses were still made of leather, which had to be constantly oiled to prevent cracking. It was 1871 before rubber and canvas hose was purchased from San Francisco.

Six years later the Deluge Company was operating out of a brand-new building constructed on a rented lot on Yates Street, just east of Broad. A fundraising drive for the new fire hall had been remarkably successful. The three main sources of donations were public subscription, money raised from performances by the fire company's own band, and, last but not least, citizens whose

Victoria's first fire department, the Union Hook and Ladder Company, was established in 1859 on the corner of Bastion and Wharf streets.

property had been saved through the volunteer firefighters' efforts.

The new fire hall's imposing entrance allowed for fast dispersal of men, horses, and equipment. An upper storey supported an ornate tower that did double duty as a hose-drying facility and as housing for a fire alarm bell.

Nothing matched the excitement when the huge bell in one or other of the fire towers clanged its insistent warning! All over the town, clerks dropped their pens and workmen downed their tools. Volunteer firemen from all walks of life raced to their respective fire halls, eager to be the first to haul out the pumpers. As the city grew and spread, however, it became clear that its firefighting operations left a lot to be desired.

On New Year's Day 1886, the three volunteer companies disbanded, and a paid department was organized. Wages were not high. The fire chief received $700 a year, more than twice as much as his assistant chief. Company foremen were paid $16.25 a month each, while the monthly stipend for hook-and-ladder men was $14.

By 1900 Victoria boasted five fire halls, 37 officers and men, 17 horses, two steam engines, an aerial ladder, a chemical engine, two combined chemical-and-hose wagons, a hose carriage, and a buggy for the chief.

In the century that has passed since then, firefighting has become more sophisticated than any of those early volunteers could ever have imagined. Sirens blaring, huge yellow engines packed with sophisticated equipment race through the city, down Yates Street, past the former headquarters of the Deluge Fire Company, which stands to this day as a silent monument to the brave men who fought Victoria's fires in days gone by.

Old Court House
28 Bastion Square

Over the years more than a dozen miscreants sentenced by Chief Justice Matthew Baillie Begbie met their demise in the jailyard once located behind the Old Court House, pictured in the early 1900s. Begbie—the "hanging judge" who hated hanging—may still frequent the premises, his ghostly presence having been noted near the third-floor courtroom in which he once presided.

In Bastion Square, the imposing structure we now call the Maritime Museum once housed the Victoria law courts.

Today's Bastion Square is a fascinating mix of building styles that tell us much about the area's former occupants. Back in the 1840s and 1850s it was short, stubby Bastion Street, which ran along the northern perimeter of Fort Victoria from the fort's northeast bastion, or gun tower (on Government Street), to the police barracks and city jail at Langley Street.

The jail was built in 1858 to help cope with the scores of lawbreakers who sailed in and out of Victoria

Replaced in the 1960s by a larger building in a different location, the Old Court House took out a new lease on life as the Maritime Museum. At the base of the building (centre left) is the original top of the Trial Island lighthouse.

in those early gold-rush days. Sentencing was swift, and the penalty for crime was harsh. The hangman had no mercy. Close to a dozen murderers, miscreants, and other miserable souls, who may or may not have been guilty, breathed their last on the jail yard's gallows. Rumour had it that some were buried under the exercise yard on the jail's northeast side.

By the mid-1880s the city needed a larger, more salubrious venue for its judicial activities, and in 1887 the chief commissioner of land and works authorized the building of a new courthouse on the site of the old jail. Architect of choice was Hermann Otto Tiedemann, who over the previous three decades had proved himself more than capable of designing sturdy structures.

Tiedemann had arrived in Victoria from Germany in 1858 to find the small settlement surrounding the Hudson's Bay Company fort transformed into a gath-

ering-place for gold miners. Tiedemann found work with the surveyor general's department, and two years later had celebrated two design "firsts"—a cluster of square, wood-frame legislative buildings on the Inner Harbour's south shore and a lighthouse on Fisgard Island.

Almost three decades later Tiedemann was called upon to draw up plans for the new law courts in Bastion Square. His design, which apparently resembled a similar structure in Munich, incorporated several styles ranging from renaissance revival to neo-baroque. Made of brick and supported by a stone foundation, it was reportedly the first structure in Victoria to make extensive use of reinforced concrete. Smith and Clark were the contractors, and S.G. Burris was the architect-in-charge. The contract for the interior finishing was awarded to Charles Hayward who was a master carpenter, as well as being the city's first funeral director.

The building was completed, at a total cost of just over $35,000, in 1889. People came from miles around to celebrate its official opening. Lieutenant-Governor Hugh Nelson and Chief Justice Sir Matthew Baillie Begbie led the parade of dignitaries to Bastion Square. In his speech at the opening ceremonies, Begbie recalled how, 40 years earlier, he had dispensed justice in the mainland goldfields from the saddle of his horse or seated on the stump of a fallen tree. Now he and his colleagues could preside over a very fine courtroom on the third floor of the new building, which loomed over others in the immediate vicinity.

Less than a dozen years after its completion, the Court House underwent $48,000 worth of extensive alterations, carried out by architect Francis Rattenbury. Most major of these was the installation of steel beams to support an ornate open-cage elevator that is, to this day, the oldest in use in B.C.

By this time—the start of the 1900s—the Court House was just one of several interesting buildings lining the square. Preceding it by seven years and one year respectively, Thomas Burnes' Beaver Building and high-class Burnes House Hotel (16 Bastion Square), designed by architect John Teague, served successful gold miners and others. Italianate in style, the $20,000 Burnes House featured tall, paired windows on the ground floor, with oriel windows on the second and third floors under a bracketed cornice.

Thomas Burnes, a tall, handsome, top-hatted man, always impeccably attired in a long, flowing cape and carrying a gold-headed cane, was justifiably proud of what he had accomplished barely 30 years after arriving in his adopted home. Sadly, his satisfaction was short-lived. In 1892 a smallpox epidemic swept the city. Hundreds died. Shipping was curtailed, visitors stayed away, and the days of the luxury hotel were numbered. His hotel closed later that same year.

On the opposite side of the square the 1892 British Columbia Board of Trade building (31 Bastion Square), which demonstrated architect A. Maxwell Muir's enthusiasm for decorative elements, rose steadily to its full four-floor height. And on the Langley Street corners, Rattenbury's 1899 Law Chambers (45 Bastion Square) were soon to be joined by his 1905 Chancery Chambers (1218 Langley Street).

The Court House fulfilled its ever-increasing obligations until the new, much larger courthouse on Burdett Avenue was completed toward the end of 1961. The last court case was heard in the Bastion Square building in February 1962. In 1963–64 it served as a temporary City Hall during the latter's renovations. Then in 1965 the old Court House took a new lease on life as the Maritime Museum.

Its brickwork long since stuccoed so that its surface resembles large granite blocks, the museum is a commanding presence in Bastion Square. With flags flying and heavy old door invitingly open, it welcomes the curious into its cool, old-world interior. But amidst all the hustle and bustle of summer activities, few visitors notice the plaque on the wall outside that proudly proclaims this building's earlier life as Victoria's centre for justice.

Hibben-Bone Block

1118 Government Street

Thomas Hibben's first stationery store, on the east side of Government Street, burned down in a spectacular 1910 fire.

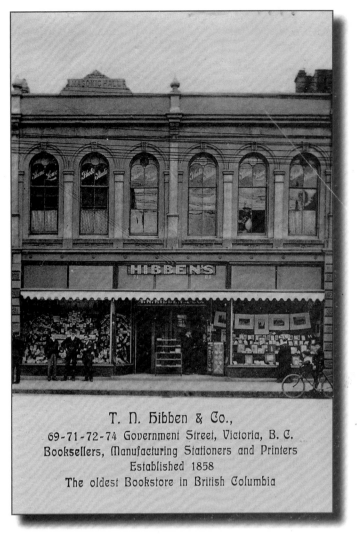

T. N. Hibben & Co.,
69-71-72-74 Government Street, Victoria, B. C.
Booksellers, Manufacturing Stationers and Printers
Established 1858
The oldest Bookstore in British Columbia

Once there were several hotels along Government Street between the Inner Harbour and Johnson Street. Now there is only one—and it was built to serve a very different function.

When gold was discovered on the Fraser River in 1858, thousands of California prospectors arrived in Victoria, eager to sail to the mainland and stake their claim. Suddenly fur trading took second place. Newcomers, and even some Hudson's Bay Company employees on the Island, were bitten by the gold bug.

HBC officials decided to cut their losses. They made a handy profit by demolishing Fort Victoria and selling the land. On the west and east sides of the fort compound—now called Wharf Street and Government Street—lots were quickly snapped up by enthusiastic entrepreneurs. On the fort's north side, shacks, shanties and saloons sprouted along Yates Street, selling supplies and sustenance to gold miners. And it was here that Thomas Napier Hibben started a business of another sort.

Hibben was an American, born and educated in Charleston, South Carolina. At 21 he travelled to California, intent on searching for gold there. Instead,

Established 1858.
T. N. HIBBEN & CO.,
Booksellers and Stationers,
HIBBEN-BONE BLOCK,
1122 Government Street,
VICTORIA, B. C.

A postcard like this was sent to inform customers that their stationery orders had arrived.

he set up a bookstore and stationery business and made enough money that when news of the Fraser River gold rush reached San Francisco, he was ready to move north.

Hibben sailed into Esquimalt in the summer of 1858, along with hundreds of other hopefuls. He didn't go on to the goldfields, however. Instead he bought a small bookstore on the south side of Yates Street, where he and partner James Carswell set up a combined printing and bookselling business. It was a smart move. Situated well inside the still-small settlement's boundaries, Hibben and Carswell's books were a welcome addition to the fledgling newspapers of the day.

By the end of 1858 Victoria had its first official street map, and the newly numbered bookstore and reading room at 37 Yates Street was a popular place for men to discuss the business of the day and local political shenanigans. Hudson's Bay Company influence was dwindling fast, and there was talk of uniting the colony of Vancouver Island with the mainland colony in order to protect British interests.

In 1864, as the last of the Fort Victoria buildings bit the dust, Thomas Hibben took a wife. He and Janet Parker Brown were married in January. They moved into a house on Pandora Avenue, and their first son was born later that same year. Not long afterwards, James

Today the much-renovated Hibben–Bone Block houses a downtown hotel.

Carswell set up a legal publishing firm. Hibben bought his partner's interest in the bookstore and continued the business on his own as T.N. Hibben & Co. He had moved his store to the east side of Government Street, near W.J. Wilson's clothing store at the corner of Thomas Trounce's alley.

Over the decades the store enjoyed continued growth. Advertised as "Importing Stationers and Booksellers," it declared itself "prepared to furnish nearly every variety of stationery in use," along with popular literature; printing, ruling and binding services; legal and office supplies; Admiralty coast charts; photographic albums; mathematical instruments; fine pocket cutlery; wrapping paper; music; and more. The store was a firm favourite with at least one local youngster. Emily Carr wrote of her fondness for Mr. Hibben's store, especially at Christmastime when picture books were left open invitingly at a perfect height for small children to see.

In 1910 Hibben's original building was burned to the ground by the fire that destroyed the nearby Spencer Arcade. From its ashes rose an impressive structure—on the opposite side of Government Street. The Hibben-Bone Block's five floors were topped by a large electric sign depicting St. George and the Dragon, emblem of another local company. At one time the roof sported a huge pencil—a clear indication of the kind of business being conducted below.

Thomas Hibben would have been proud of that building, but by then he was long gone, dead at the age of 62 and buried at Ross Bay Cemetery. His sons, Parker and T.N. Jr., represented their mother in the business, which was co-managed by long-time employees C.W. Kammerer and W.H. Bone.

The building's stationery store days eventually came to an end, and it went on to house the Churchill Hotel. Today, looking much the same as when it was built, all those decades ago, it is home to the Bedford Regency Hotel.

E.A. Morris Building
1116 Government Street

The store built in 1882 to house "E.A. Morris Tobacconist" (centre in this picture) retains its Edwardian-era opulence. It was designed in 1909 by Thomas Hooper.

Tucked away beside the Bedford Regency Hotel is a Government Street store that, unlike its neighbours on the block, is essentially unchanged from when Edward Arthur Morris had it redesigned to cater to his business needs almost 100 years ago.

Morris was born in London, England, in 1858, the year gold was discovered on British Columbia's Fraser River. Nineteen years later, Morris came to find some of that gold for himself. Sailing to Victo-ria in April 1877 via London, New York, and San Francisco, he spent the next five years working at several mainland mines and a couple of explosives factories before returning to the Island.

By 1892 he was ready to settle in Victoria. Buying a 10-year-old, two-storey brick structure that had been operated as a dry goods store, Morris soon made his mark on Government Street. He was the first to import choice cigars and tobacco

Entering the store, with its leaded-glass door surround and well-preserved interior, is like taking a step back in time.

from England, and his store was the largest of its kind between San Francisco and Alaska.

His efforts paid off. By 1899, when the Klondike gold rush was in full swing, Morris was able to move his headquarters and his warehouse to Vancouver, and before long he opened two stores there. Old store records show that his products were firm favourites with many Klondike-based companies catering to gold miners, the Royal Navy ships at Esquimalt, and individual purchasers all over British Columbia.

Business was brisk. Morris was now reportedly the largest distributor of smokers' supplies in the west, and in 1909 was ready to make made major renovations to his Government Street store. He hired Thomas Hooper to prepare the drawings and oversee the work. Since arriving in Victoria in 1889, Hooper had designed two churches, an orphanage, and several commercial buildings, including the Mahon Building at 1110 Government, next door to Morris's shop.

Hooper's alterations to 1116 Government created a store that would survive essentially unchanged to this day. Above the doorway, which was cut from Mexican onyx, a dome-shaped leaded window extended across the width of the store. Inside, leaded mirror domes and large wall mirrors created the illusion of extended space. Counters and cabinets were remodelled. The walls were panelled with polished mahogany, and the entrance to the humidor—a tile-lined, walk-in cabinet for dry storage of cigars—boasted classical carved columns. Pipes from all over the world were displayed in glassed-in pipe racks. A container for walking sticks was reminiscent of similar items in tobacco shops in England.

Serving staff stood behind long side counters fronted with Mexican onyx baseboards that reached the mosaic-tiled floor. The store's basement was blasted out of solid rock, with metallic paint on its walls reflecting the light that filtered through purple-coloured glass prisms in the sidewalk outside the entrance.

During the renovations, Morris ordered a unique item from the Keenan Company Marble Works of San Francisco. It was an electrolier—a Mexican onyx column on a marble pedestal, topped by a globe and with gas jets extending from either side of the column at "cigar level." A cigar cutter rested on a small shelf just below the jets. Now Morris's customers could light up their cigars before leaving. Apart from providing a convenience to those who had just made a purchase, the device would attract potential customers along Government Street, who would follow the tantalizing aroma right into the store.

Morris was actively engaged in the business until he died of bronchopneumonia in Vancouver in 1937, at the age of 60. He left the store to his wife, Elsie, and two daughters. Since they sold it in 1947, the store has had two other owners—Jack Delf and Don Taylor—who were eager to preserve the past. Thanks to them, the store is a remarkable example of what can be done when there is a clear commitment to maintaining the original rather than resorting to imitation.

The sidewalk prisms are gone—removed when the city extended the paving north along Government Street in 1975—but the interior and exterior remain intact. To enter "Old Morris Tobacconists" is like taking a step back in time. The sweet aroma of pipe tobacco assails your nostrils as you walk along the tiled floor between the side counter and the central display. The stock has been expanded to include non-tobacco-related items, but the onyx electrolier is still the store's central feature.

If you close your eyes, you can picture the parade of well-dressed gentlemen, indicating their intended purchase, then lighting it before taking their leave. Listen carefully and you can almost hear the mumble of conversation—friends greeting friends during the never-ending coming and going of customers at E.A. Morris's store.

Mahogany and mirrors surrounded Edward Morris's pride and joy—a Mexican onyx, gas-jet electrolier where cigars could be lit as soon as they were purchased.

Royal Bank Building
1108 Government Street

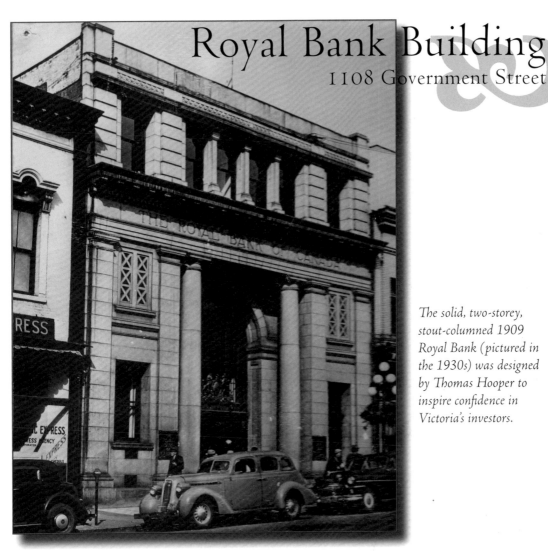

The solid, two-storey, stout-columned 1909 Royal Bank (pictured in the 1930s) was designed by Thomas Hooper to inspire confidence in Victoria's investors.

Victoria had no need of banks before prospectors returning from the gold rush on the Fraser River sought a safe haven for their spoils. By 1859 a smart young Scot named Alexander J. Macdonald had opened the town's first official public financial institution on lower Yates Street. It did well for five years. Then disaster struck. One night it was cleaned out during a mysterious, unsolved robbery. Depositors closed their accounts. The doors remained shut. Macdonald crossed the border and was never heard from again.

Fortunately for Victorians, other banks stood ready to help. Around the corner on Government Street and nearby, several financial institutions sprang up over the decades. At one time, five major banks operated within the few short blocks of Government between Johnson Street and James Bay.

The new $60,000 Royal Bank of

Looking the same, but minus its second storey, the former bank now houses a fine bookstore. Inside, a 1980s renovation revealed the original cast-plaster ceiling.

Canada was built on property that once formed a part of the Hudson's Bay Company fort. As the fort was demolished, piece by piece, the site became Victoria's business centre. The Royal Bank was located on the former fur-trading post's east side, partway between its main gate (at Fort Street) and its northern perimeter (Bastion Square).

The plum position for a bank was on a corner, where it could present two faces to the public, but the Royal Bank made up for its mid-block location by turning heads with its design. Architect Thomas

Hooper was no stranger to the area, having already been chosen to prepare drawings for several other buildings on the same block. He designed a classical façade with a central entrance flanked by solid granite pillars 31 inches in diameter and 24 feet tall. Cast-iron grillwork graced the recessed entrance doorway, which framed massive copper-clad doors. Granite, used exclusively for the front of the two-storey building, added to its air of solidity.

Once inside the doors, a terrazzo-tile floor led customers into the central area. Behind the counters, tellers moved quietly over oak flooring. A magnificent coloured-glass ceiling dome displayed the shields of the nine Canadian provinces (the tenth—Newfoundland—did not join Confederation until 1949). With its lofty ceilings and generous use of rich, dark wood, the bank's interior inspired trust and confidence. In the basement, a shooting gallery allowed employees to practise their skills with firearms against the ever-present threat of bank robbers.

In the 1950s the Royal Bank, bowing like so many other commercial businesses to changing styles and times, modernized the building. The original second storey and the glass dome were removed. The dome was filled in with concrete, and the ceiling was lowered. A Langley Street addition provided space for offices and storage. The terrazzo and oak floors were covered with cheap linoleum and shag carpeting. Plastic-topped plywood counters replaced the hardwood originals. The massive front entrance doors were removed in favour of a small aluminum door with "Royal Bank of Canada" in equally small letters above it.

Bit by bit, piece by piece, Hooper's handiwork was either obliterated or removed. It wasn't until the mid-1980s, when Yates Street bookstore owner Jim Munro bought the building, that glimpses of its former glory were seen.

Munro was intent on restoring the building and enhancing its original features. It was a mammoth task. Extensive refurbishing was carried out under the direction of Marshall, Goldsworthy & Associates. Wakefield Construction built a new, solid dome and installed a central marble-topped oak counter.

The ornate ceiling was restored, along with the beams and column capitals. The various layers of flooring were removed, and the old terrazzo tile and wood floors were refinished. In the process, a bullet was discovered lodged in the floor—the result, says a former employee, of a botched bank robbery in 1948.

Architect Alan Hodgson designed the interior colour scheme, which was brought to life by local artist Carole Sabiston's wall hangings depicting the four seasons from the land and from the sea. To complete the $155,000 restoration, the granite slabs above the entrance doors were refinished and engraved with the building's new name: Munro's Books of Victoria.

Today few people who enter the store to browse among its many tall shelves realize that customers once walked across the central area to undertake transactions with tellers standing on oak floors behind hardwood counters, and that today's fine bookstore is yesterday's Royal Bank.

Windsor Hotel
901 - 905 Government Street

In 1876, owner George Richardson almost demolished his hotel, reputedly Victoria's first all-brick hostelry, by accidentally causing an explosion.

Despite George Richardson's attempts to blow up his own building, the Windsor Hotel—built in 1859—has survived the ravages of time.

The British-born Richardson was just 23 years old when he boarded the *Norman Morison* at Gravesend in October 1849. He was in good company. More than 80 souls were aboard, most journeying to the southern tip of Vancouver's Island as contracted employees of the Hudson's Bay Company.

When their ship sailed up Juan de Fuca Strait in the early spring of 1850, most of the barque's passengers saw a fort and farms and cheap land in their future, once they were free of their HBC contracts. Richardson saw a small settlement with all kinds of potential for a smart young businessman like himself.

Five years later he had made enough money to buy 300 acres of land northeast of town, and he sailed back to England to find a bride. Mary Ann Parker

was 11 years younger than George, but just as adventurous. The newlyweds boarded the *Princess Royal* and arrived in Victoria in early 1858—a scant two months ahead of the first rush of Fraser River-bound gold miners. When the rush started, Richardson could almost hear the coins clinking into his cash box as these gold seekers sought to slake their thirst.

There were other saloons and hotels around, but Richardson decided to build his hotel toward the south end of the little town. Fort Victoria was being demolished piece by piece, and the rough, grass-lined dirt track along its eastern side—called Government Street—was fast becoming the most important north-south route.

The gold rush had brought more than 20,000 prospectors to these shores. Victoria was the jumping-off point for the Fraser River goldfields. Here the men bought licenses to dig for gold and gathered supplies—pickaxes, shovels, food, clothing—before sailing to the mainland.

Victoria's first directory, published in 1860, listed several hotels in the downtown core. These were mostly wooden buildings, hastily erected to provide for the newcomers' needs. But Richardson catered to the more discriminating traveller. His Victoria Hotel, completed in 1859 and later renamed the Windsor Hotel, was built almost entirely of brick.

Meanwhile, over on the Inner Harbour's south shore, a strange-looking set of administration buildings—the new home of the legislative assembly—was taking shape. It was reached via a wooden bridge that extended from the south end of Government Street and spanned the murky waters of James Bay. Richardson couldn't believe his good fortune. Anyone headed for those Legislative Buildings had little option but to walk right past his front door.

Business was brisk until 1864. Then the city experienced a post-goldrush slump. Richardson leased out the hotel and moved his growing family to his farm on North Park. However, by the mid-1870s he was back on the corner of Government and Rae (as Courtney was then called). And one night in 1876 he almost destroyed his precious building.

After retiring to the second floor for the night, the Richardsons were alarmed by a strange odour emanating from below. George went downstairs to investigate, armed with a lighted candle so he could see where he was going. The resulting explosion of gas, reported the *Colonist* the following day, flattened a couple of lamp standards along Government Street and was heard half a mile away.

The force of the blast blew down brick partitions, tore plaster from the walls, and wrenched doors from their frames. The parlour and dining room were wrecked. The stairs were partly destroyed. Ignited gas rushing up the stairwell blew out windows on the upper level. Shattered glass littered the street below. Amazingly, no one was seriously hurt. Richardson suffered severely singed hair and a burned hand, but lived to tell the tale … and to repair the damage to his hotel.

Around 1915, another building was constructed as an extension to the Windsor Hotel, running up Courtney to Gordon Street. It's believed that architects Percy Fox & Berrill designed the extension, but the existing structure may be a much-modified version of their original plans. Compared to the original, it lacked imagination, being a plain and unremarkable single-storey brick building.

By that time the Richardsons had moved to the other end of Government Street in James Bay. Mary Ann died there in 1911, aged 74. George died in 1922 at the ripe old age of 96. Today his original building, which still fronts onto Government Street, is home to retail stores. The Windsor Block stands exactly where Richardson left it—an enduring legacy to another era and to a man who was the proud proprietor of Victoria's first brick hotel.

Rogers Block
913 Government Street

Charles Rogers made a fortune selling the hand-filled chocolates he made each evening in the kitchen at the back of his store.

For well over 100 years, sweet-toothed Victorians have followed the tantalizing aroma of chocolate to this vegetable grower-turned-candy maker's store.

An institution in Victoria for more than a century, Rogers' Chocolates now opens its door to the public during normal shopping hours. But in the old days it was a very different story. Charles W. Rogers opened and closed his store whenever he pleased. Victorians and visitors alike lined up outside, willing to cater to his whims in return for the opportunity to purchase his delectable delicacies.

It's fortunate for Victorians that Rogers chose to start his legendary business in this city. Born in Petersham, Massachusetts, in 1854, he wandered west in his early teens and in 1885, at the age of 31, decided to try his luck in Victoria.

Selling produce on the west side of Government Street, he soon recognized a more lucrative opportunity. The demand for chocolates, first made in Switzerland a few years earlier, was steadily growing. Rogers decided to experiment. By trial and error he married his fruits with the finest, freshest ingredients and eventually found the magic combination that, to this day, no one has quite been able to match.

By this time, Rogers wasn't alone in his efforts. He had met and fallen in love with a local lass, Leah Morrison of James Bay. The two were married in May 1888 and set up home on Kingston Street. Shortly after, they started their candy-making business at 916 Government. And in 1890 their son Frederick was born.

Rogers started investing in real estate. One of his acquisitions was a new two-storey block across the street, designed and built for him in 1903 by Hooper and Watkins. Quite happy to stay where he was, Rogers rented his new property to Brown and Cooper, fish and fruit merchants. By 1905 jeweller W.B. Shakespeare had taken it over, and in 1909 another jeweller, W.B. Wilkerson, moved in.

With a thriving business and young Fred being groomed to join it, Charles and Leah seemed to be leading a charmed life. But all was not well in the Rogers household. Showing little interest from the start in his father's candy business, Fred had developed a morbid fascination with explosives. In 1905 tragedy struck twice. First, Fred lost three fingers in an explosives-related accident. A few months later he rented a room at the New England Hotel, wrote a note to his parents, then shot himself to death. He was just 15 years old.

Charles and Leah buried themselves in their business. They rose early in the morning and worked late into the night. Charles hung up a sign proclaiming the day's operating hours, and by the time they opened the door, people were lined up down Government Street. The entire inventory was sold, often within an hour, and the shop was closed again. In the afternoons they attended to their mail-order business. Rogers' chocolates were now being shipped to British royalty, U.S. presidents, movie stars, and chocolate lovers all over the world.

In 1917 Charles moved his business across the street and into his own

building. The couple lived a simple life. Never part of Victoria's social scene, they preferred to entertain themselves. When he was able, Charles had bought expensive gifts for Leah, and nothing pleased him more than to have her put on the furs and diamonds he had given her and sit quietly with him in the kitchen while he worked.

Eccentric? Yes. Devoted to each other? Absolutely. Hard workers? Without a doubt. Charles had looked after his money well. When he died of a heart attack in 1927 at the age of 73, he left an estate worth almost $300,000. A generous and trusting soul, with the best of intentions but bereft of her husband's business sense, Leah donated money to charities, made poor investments, and gradually lost all that Charles had left her. When she died in 1952 at the age of 88, she was living on an old-age pension.

A few years after Charles died, Leah sold the business to a wealthy American customer, who in turn bequeathed it to his heirs. Since 1968 the business has been owned and operated by Canadians, and the present, local owners have expanded it dramatically. Today Rogers' chocolates are sold at 600 retail outlets across Canada.

Victoria has grown and Government Street has changed, but if Charles Rogers were alive today, he would doubtless be delighted that his original recipes and methods are still used and that his store survives, exactly where he left it all those years ago.

Galpin Block
1017 - 1021 Government Street

Harris & Hargreaves designed this building for British investor Thomas Galpin in 1884. Two decades later, its second floor was home to the Alexandra Club for Ladies. In the building attached to the Galpin Block's south side, Charles Redfern advertised his jewellery store by hanging a large clock outside.

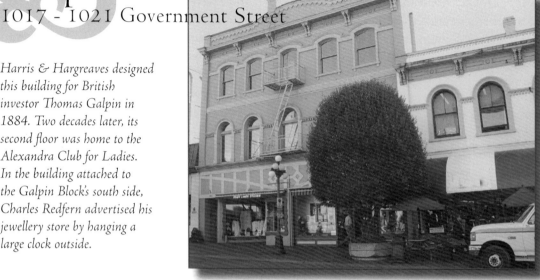

This is the tale of two men—one who owned the block that carries his name on Government Street, the other an enterprising jeweller who operated the business next door.

As you walk along Government Street today, the Galpin Block isn't hard to miss. Tucked away on the east side, between Fort and Broughton streets, it reminds us of a man who invested heavily in British Columbia just as the Canadian Pacific Railway joined the country from coast to coast.

London-based Thomas Dixon Galpin was a partner in Cassell, Petter and Galpin, at that time the largest publisher in the world. Well-known titles on their list included *Illustrated London News*, *Sphere*, *Tatler*, and *Punch*. Through Victoria realtor and financial agent Thomas Alsop, who visited him in England in the 1880s, Galpin learned of

British Columbia's potential and decided to make substantial mortgage and real estate investments here.

In 1886 Galpin arranged for Cuyler Armstrong Holland, the son of a friend, to take a post in Alsop's Victoria office. Holland, 22 years old, with a law degree from Cambridge and influential contacts, travelled on the Canadian Pacific Railway's first transcontinental journey with Sir John A. Macdonald, prime minister of Canada. Soon after the train reached its terminus at the townsite of Hastings (soon to be renamed Vancouver), Sir John A. returned to Ottawa; Holland sailed on to Victoria.

Two years later Holland returned to England and presented Galpin with a favourable report on the future of this fine land. Galpin responded by buying out Alsop's firm and reorganizing it into the B.C. Land and Investment Agency, with its head office in London. Holland married into the Galpin family, came back to Victoria with his wife and small son, and bought a large house in Rockland.

Galpin and various family members—he had 13 children of his own—visited the city many times. He acquired huge tracts of ranchland in the Interior and thousands of head of cattle. Well-known in Victoria business and social circles, he endeared himself to the locals by donating more than 300 of Cassell's best books to the Victoria Public Library.

Together, Galpin and Holland bought up land in Fairfield and Rockland, as well as huge chunks of the downtown core. At one point it was said that they owned every important corner in Victoria. The three-storey, Italianate Galpin Block on Government Street was built in 1884. Twenty years later its second floor was home to the organization formed by local ladies who were denied access to the gentlemen-only Union Club. Galpin's building served as the Alexandra Club's meeting place from 1900 to 1911.

Attached to the south side of the Galpin Block was the building where Charles Redfern operated his jewellery store. A clockmaker by trade, Redfern had travelled to Victoria in 1862 aboard the *Tynemouth*, one of the famous "bride ships" bringing young women from England who were destined to become the wives of gold miners and others who had gravitated here. Redfern was bound for the goldfields, but changed his mind when he saw the potential for business in the newly incorporated City of Victoria.

By 1875 he had opened his jewellery store on Government Street. Hitting on a unique advertising gimmick, he ordered a huge clock from England and installed it on the wall above his store. It could be seen from either end of Government Street, and—much to the annoyance of those trying to sleep close by—its chimes could be heard clear out to Oak Bay.

Redfern became the city's 15th mayor in 1883, and though he was defeated in the next election, he served a second term from 1896–99. In 1891 he won the contract to install Victoria's most famous clock—the one atop City Hall—which he ordered from England. He was also responsible for raising $4,950 toward the purchase of a new, state-of-the-art steam pumper for fighting fires. The nickel-domed "Charles E. Redfern" caused a great deal of excitement as it clattered through the downtown streets, the rumble of its wheels competing with the thundering hooves of the horses that pulled it.

Redfern's fortunes seemed set, but in the depression years leading up to the First World War, his business went bankrupt. The building that housed his jewellery store still stands, but you'll have to go a long way to find his magnificent clock. It was moved in 1938 to Cowichan Station's Fairbridge Farm, which became a residential school for children evacuated from England at the start of the Second World War.

By that time Galpin and Redfern were both long gone. But the buildings that housed their respective enterprises stand in the 1000 block of Government Street to this day.

Hotel Janion

1612 - 1614 Store Street

The staff of John Turner's 1891 high-class Hotel Janion had only two short years to enjoy their employment. By 1895 the hotel had closed, and the building was used by the Esquimalt & Nanaimo Railway as a business office.

Today it looms sullenly over the Esquimalt & Nanaimo Railway's tiny Victoria station house. But in happier times the Hotel Janion was its busy neighbour, ready to provide food and comfort to those who rode the rails.

Although the hotel was built next to the Janion warehouse and wharf, and borrowed its name, there was apparently no connection between the two or the men who owned them. Richard Janion arrived in Victoria via Hawaii, with a large supply of goods, in 1859. He was a commission merchant who, according to the *British Colonist*, "imported liquors and cleared cargoes from many parts of the world."

Before long Janion went into

Today, the once-grand Hotel Janion, now empty and forlorn, stands like a silent ghost on Store Street.

business with fellow Englishman Henry Rhodes, who had arrived from Hawaii the previous year. The ships arriving at their dock supplied many of Victoria's wholesale businesses. They opened an office in Portland, Oregon, and in April 1875 they became in-laws when Richard, Janion's eldest son, married Annie, Rhodes' eldest daughter, at St. John's Church on Douglas Street.

By that time the Janion–Rhodes business partnership had been dissolved. Henry Rhodes and Company took over the Victoria office; R.C. Janion and Company looked after the Portland operation. Rhodes eventually retired. Janion returned to England and died there in 1881.

A little over 10 years later an Esquimalt & Nanaimo train made its way

slowly across the Johnson Street bridge. The railway connecting Esquimalt to Nanaimo and places between had been in operation for some three years and was proving a boon to up-Island businesses. Malahat Drive had not yet been blasted through the rocky ridge north of Victoria, and water-based travel was efficient but expensive. The E&N was a fast, favourable alternative to the Cobble Hill–Goldstream trail or the long loop through Shawnigan and Sooke. And now, thanks to Nanaimo-based coal magnate Robert Dunsmuir, the line had been extended from Esquimalt and over the bridge to its new terminus on Store Street, near the former Janion-Rhodes location.

Before long local builder John Turner seized the opportunity to make his mark near the old Janion warehouse.

Turner couldn't have found a better spot for his latest enterprise. The E&N brought passengers right to his door. The area surrounding Johnson, Wharf, and Store streets was the heart and soul of downtown. What better place to build a major new hotel?

It was one of several such establishments in this part of town and was undoubtedly one of the finest. The 200-foot-deep, three-storey brick structure with imposing bay windows was impressive by anyone's standards. High above its 50-foot-wide Store Street entrance were the words "The Janion, 1891." Along the top of its south-facing wall—the side nearest the railway line—large white letters proclaimed "Hotel Janion."

The *Colonist* was impressed. In 1892 it noted that despite reports of "general dullness and stagnation from the mainland and all parts of Puget Sound … one look around this city can see no abatement in the constant building operations that are being carried on on so large a scale in nearly every business thoroughfare." The newspaper declared Victoria's newest hostelry to be "one of the best built and prettiest buildings in the city."

The Janion's 50 large bedrooms, it reported, were "all fitted up in the most attractive and substantial manner and filled with many comforts not generally seen in ordinary hotel rooms. Every room is thoroughly heated and well lighted with electricity, while large windows open into each room … giving from all directions a view almost unequalled in any part of the city. Fire escapes are numerous and handy on all sides, being practically useless, however, as the place is fireproof."

The hotel incorporated a novel concept: sleeping apartments ran along its whole length in parallel lines, with a wide corridor between them, and "several handsome bathrooms and lavatories" complemented the accommodations on each floor—considered a luxury at that time. Proprietors Leopold Rheinhart and William Walker were certain that, in the relatively uncertain business environment of the 1890s, their establishment would turn a profit. Apparently this wasn't the case. In the 1894 edition of the Victoria City Directory there is no mention of the Janion in the listing of commercial establishments.

Meanwhile, the railway it was built to serve was doing well, and in 1895 the E&N bought the Janion and used it as a business office until 1948. Then it was used as a warehouse. Close by, on the northwest corner of Johnson and Store streets, the old red-brick Hotel Janion still stands, a sad, silent reminder of olden days in Victoria.

Craigdarroch Castle
1050 Joan Crescent

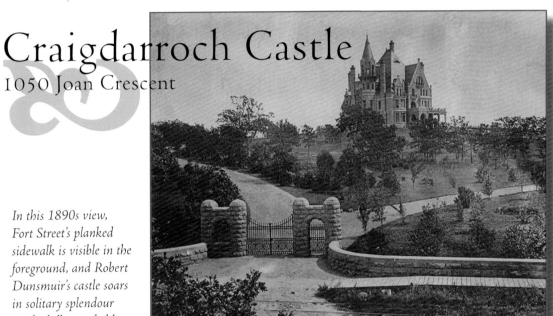

In this 1890s view, Fort Street's planked sidewalk is visible in the foreground, and Robert Dunsmuir's castle soars in solitary splendour on the hill, guarded by gates that once stood on today's Joan Crescent.

There's something magical about castles, even if their purpose seems a little obscure. Modern-day castles are really opulent homes, intended to defend nothing more dangerous than the reputation of the erstwhile "knights" who reside within. Toronto has Casa Loma. California has Hearst Castle. Victoria has Craigdarroch.

Growing up in Ayrshire, Scotland, the oldest of four in a middle-class family, Robert Dunsmuir could not have imagined that he would one day live in such a place. But he was young, a hard worker, and ambitious. When opportunity knocked in late 1850, he opened the door and let it in.

Dunsmuir was a coal-mine manager in a small Scottish town when the London-based Hudson's Bay Company offered him a job. With little ado he packed his family and belongings aboard the *Pekin* and set sail for Vancouver's Island. It was a smart move. Free to work for himself after his three-year contract expired, he managed a company mine independently, selling the coal back to the HBC. He was self-employed at last, but no better off financially.

A few years later two separate but related events helped to improve Dunsmuir's fortunes. The discovery of Fraser River gold in 1858 changed the British government's focus, and in 1859 the lease giving the Hudson's Bay Company control of Vancouver Island expired, along with its rights to Island coal. The Vancouver Coal Mining and Land Company bought the Nanaimo mines and hired Dunsmuir as superintendent. These were his last few years as anybody's employee. In October 1869 he discovered

a new seam of coal and established his own operation—the Wellington Mine.

With hard work, hundreds of labourers, and many willing investors, Dunsmuir's worth increased. By the mid-1880s his empire included coal mines, a sawmill, a railway, ironworks, ships, a theatre, and property. He was the wealthiest man in British Columbia, ready to celebrate his success with a long-overdue move to the province's capital.

Over the years he had managed to acquire 28 moss-and-wildflower-covered acres at the top of the Fort Street hill. He wanted to build a baronial mansion befitting his station in life, and he looked for a designer who could turn his dreams into reality.

At that time, the busiest and best-known architect in Victoria was John Teague. But Teague had never tackled anything like the dwelling Dunsmuir had in mind. Wanting the best for his new abode, Dunsmuir approached Warren Heywood Williams, an Oregon architect whose work he had admired.

Williams was no newcomer to Victoria, having designed the imposing Bank of British Columbia building at Government and Fort streets in 1885. He had likely never seen a Scottish baronial home, but he did his best to draw what he thought one would look like. Dunsmuir was impressed. The structure promised to reflect his status in a most satisfying way. Incorporating a mix of architectural styles, it looked more like a fairy-tale castle than a real one, with its tall, narrow outline, steep-pitched roof, and rounded entranceway. But its design was real enough, and Dunsmuir gave Williams permission to proceed.

Construction began in the fall of 1887. A few months later Williams died unexpectedly, the victim of pneumonia at only 43. His son, David L. Williams, and

Today, the grand entrance hall, with its trophies, wood panelling, 87-step staircase, and access to many of the castle's 39 rooms, provides visitors with an engaging and entertaining peek into its opulent past.

Arthur Smith, a colleague in his office, took over, and gradually Williams' vision took shape. Solid sandstone walls rose skywards. The red slate roof and tall, iron-braced chimneys could be seen from just about anywhere in the city. A high-pointed tower pierced the rarefied Rockland air.

Inside, an oak-panelled hall complete with massive fireplace and mounted rams' heads drew the eye toward the wide staircase—no elevators in Dunsmuir's day—leading up to the fourth-floor ballroom. Hand-painted and stencilled decorations graced the drawing room ceiling, mahogany gleamed in the library, and stained-glass windows added an elegant touch. There were sumptuous furnishings and ample space for Robert, Joan, and their three unmarried daughters. No fewer than 17 fireplaces helped keep the whole family warm.

Dunsmuir's castle was exactly what he wanted … but he didn't live to enjoy it. Taking to his bed in April 1889 with what seemed like a simple chill, Dunsmuir baffled his physicians by lapsing into a coma and dying six days later of liver disease. He was 63 years old.

A year later, when the castle was completed, Joan and her daughters moved in. She had come a long way from her humble beginnings as a coal master's wife in Nanaimo. But money had not, after all, brought her happiness. She had lost her husband of 52 years, and as the result of a bitter family feud, she was estranged from her only surviving son, James. The two were never reconciled. Even on her deathbed in 1908, Joan refused to see James, although by that time he lived at the end of her garden in Government House, as the queen's representative in British Columbia.

None of the surviving children—the Dunsmuirs had had two boys and eight girls—wanted Craigdarroch. Eventually the property was sold and subdivided. Since then the castle itself has served as a convalescent home for soldiers who served in the First World War, as Victoria College, as the Victoria School Board headquarters, and as the Victoria Conservatory of Music.

In 1959, threatened with demolition, the castle was saved by a group of concerned citizens and preserved as an historic landmark. Today it's a firm favourite with locals and visitors alike, the legacy of a man who built an empire out of coal mining in early British Columbia.

Chinese Public School
636 Fisgard Street

A prominent feature in Canada's oldest Chinatown, the 1909 Imperial Chinese School features an elegant tiled roof with upturned eaves and corners.

Among the fine buildings of Canada's oldest Chinatown, the Chinese Public School helps preserve the richness of the culture it represents.

The history of Victoria's Chinese community stretches back over a century and a half to 1858. When news of the Fraser River gold finds reached San Francisco, Chinese men were among the more than 20,000 European, Canadian, and American hopefuls who came to Vancouver Island. Some came from China. Others were successful merchants in San Francisco who moved lock, stock, and barrel to Victoria. Soon the Chinese business community rivalled the Hudson's Bay Company in providing goods and services.

A second influx of Chinese men occurred during construction of the Canadian Pacific Railway in the early 1880s. The railroad through the Canadian Rockies was built on the backs of Chinese workers. It was difficult and dangerous work, but there was a schedule to keep. Those who perished were quickly replaced. During those years, B.C. boosted Canada's Chinese population by almost 100 percent.

In the winters, when harsh weather

brought construction to a standstill, many Chinese workers came to Victoria. Unceremoniously laid off when the railroad was complete in 1885, they were left to their own devices. Most returned to Victoria where, thanks to the foresight of the local community, help was at hand in Victoria's Chinatown.

The early-day cheek-by-jowl wooden shanties had long since been replaced by two- and three-storey tenement buildings. Canada's oldest Chinatown was also the smallest, with its theatres, schools, churches, temples, shrines, gambling dens, and brothels contained within six short city blocks. The new arrivals from the mainland were housed side by side with Island-based workers—miners

and railway builders employed by coal magnate Robert Dunsmuir of Nanaimo.

Benevolent associations took care of Chinatown residents' welfare. In 1884 an umbrella organization—the Chinese Consolidated Benevolent Association (CCBA)—commissioned English-born architect John Teague to design its new headquarters on the 500 block of Fisgard Street.

Designer of churches, schools, Victoria's City Hall, and the Masonic Temple, Teague willingly tackled this new challenge. Within a year the new CCBA headquarters was complete. With its ground-floor stores, second-floor administration centre, and third-floor hospital and school, the building was well-equipped to

In the early 1900s the Lequn Free School was overwhelmed when a new "native-born students only" ruling excluded China-born children from Victoria's public schools. The new Imperial Chinese School provided classes in English and enabled Chinatown's young residents to attend white schools.

help newcomers adapt to this foreign environment with its strange customs and even stranger language.

As the city grew, so did Chinese fortunes. Victoria's citizens had Chinese house servants, purchased fresh fruit and vegetables from Chinese street peddlers, had their clothes cleaned at the Chinese laundry. Another major source of income for the Chinese was opium. No need for undercover operations—opium was a legal commodity in Canada. Any dealer with $250 to spare could walk into City Hall and buy a licence.

In 1889, according to Victoria's city directory, more than a dozen opium factories were processing almost 100,000 pounds of raw opium a year. It was shipped to Asia, smuggled to the United States, and—until the Dominion government declared its sale and possession illegal in 1908—consumed by Victoria's white citizens.

By 1910, little more than 50 years after the first Chinese men had stepped ashore, Victoria's Chinatown boasted 3,500 residents. More than 150 businesses catered to their needs, including a hospital and three schools. Largest and most important of these was the Chinese Imperial School, built by the CCBA in 1909 when non-English-speaking, Chinese-born students were barred from city education facilities. It was renamed the Chinese Public School after the collapse of China's ruling dynasty in 1912.

Designed by young architect David C. Frame, the school was situated half a block away from the CCBA building,

between the Gee Tuck Tong Benevolent Association and the Masonic Temple. Set back some 40 feet from the street, it featured an eclectic mix of architectural styles. The centred staircase leading to the school's entrance door ended under an ornate wooden second-floor balcony sporting Gothic trefoil fretwork. Upturned eaves and corners lined the pagoda-style tiled roof. Moorish windows and Italianate accents added to the mix, creating a look that was at once unique and spectacular.

Chinese children were educated separately from the whites for their first four school years, and this resolved the problem for a time. It resurfaced in the 1920s when racial bias once again reared its ugly head. An all-white school board called for total segregation of Chinese-speaking students. The CCBA retaliated by organizing a student strike. The result was partial segregation, which continued until after the Second World War—just another chapter in the long story of Victoria's Chinese community.

Today the Chinese Public School stands ready to celebrate its centenary, in the midst of a community that is almost 50 years older. Long since fully assimilated into the public education system, Chinese children still attend the Fisgard Street school. Each afternoon, when the regular school day is over, the great front doors are opened. Students fresh from their Western curriculum lessons climb the staircase to continue their education by keeping their Eastern language and culture alive.

City Hall
1 Centennial Square

John Teague's 1878 City Hall, on the corner of Douglas Street and Pandora Avenue, was much enhanced by later additions, including a huge clock from England. It is the oldest such municipal building in British Columbia.

It was January 1878, and Victoria was much changed from its early days. Instead of a stockaded settlement bounded by today's Government, Wharf, and Broughton streets and Bastion Square, the city had spread north, south, and east, with Government Street pushing its way north across Johnson Street.

Roderick Finlayson, who some 20 years earlier had been James Douglas's assistant at Fort Victoria, was now the city's mayor, and he was bound and determined that Victoria should have a proper city hall. The citizens objected, arguing that the sum suggested—$10,000—could be put to far better use. Others protested that the proposed city hall site—at the east end of the block bounded by Douglas, Pandora, Government, and Cormorant streets—was on the outskirts of town. But Finlayson prevailed, and the competition for plans was announced.

Winner of the contest was architect John Teague, a native of Cornwall, England, who came to Victoria in 1859 en route to the goldfields of the Cariboo. Settling here permanently in the early 1860s, Teague specialized in architecture and construction, setting up business on property owned by fellow Cornishman Thomas Trounce.

In 1874 the uninstructed (he had no formal architecture training), but some might think inspired, Teague decided to become a full-time architect. Before long his busy Trounce Alley office had produced plans for churches, hospitals, business establishments, and residences. Among the Teague-designed structures that survive to this day are Royal Jubilee and St. Joseph's hospitals, the Church of Our Lord at the foot of Blanshard Street, the Masonic Temple on Fisgard Street, and E.G. Prior's fine Pemberton Road home called The Priory.

Teague's design for Victoria's first purpose-built municipal headquarters, described by the *British Daily Colonist* as "Anglo-Italian," called for a brick structure with cement and stone accents. Apart from the council chamber, it was to house a police court, jail, surveyor's offices, committee rooms, and a museum room. By May 1878 the foundations had been laid. J. Huntington won the stone, brickwork, and plastering contract. J. Bennet was responsible for carpentry. D. Heal provided tinsmithing services, and painting and glazing was done by J. Sears.

Even as citizens once again objected to the cost, the structure rose before them. It was a rectangular block with a main entrance on Pandora Avenue. Dark red brick with white paint accents rose to the tin mansard roof with its tall chimneys.

In 1881 the building was enlarged at its southwest corner to house the fire department. The tin roof was extended over the new wing, and another arched doorway and windows were added to complement the existing structure. Ten years later the addition of a north wing, at a cost of $35,000, created the City Hall that we see today. John Pitcairn Elford won the contract to extend the building 80 feet along Douglas Street and 50 feet along Cormorant (now Centennial Square). The main entrance was moved from Pandora to the Douglas side, under the tall tower that now adorned the roof.

Former mayor Charles Redfern, who advertised his Government Street jewellery store with a huge clock high above its entrance, won the contract to install the city's civic timepieces. The clocks, ordered from Croydon, England, were a sight to behold. Four dials, each seven foot, six inches, in diameter, graced the square tower. The bell alone weighed more than 2,100 pounds. The clocks had to be wound once a week.

This was the last major addition until the mid-1960s, when the building was threatened with destruction, and Mayor Biggerstaff Wilson lobbied to save and improve it. The Centennial Square project included a $594,000 renovation of City Hall's interior and construction of an addition to its west end. The chimneys were removed, and the tin roof was replaced with one of galvanized iron. Architectural consultant Rod Clack worked with Wade, Stockdill, Armour & Partners and R.W. Siddall & Associates to complete the necessary work.

Today the City Hall tower is still visible from downtown, and the building stands as a symbol of the confidence that Roderick Finlayson had in Victoria over a century ago.

Today City Hall's Pandora Avenue entrance is graced by a bronze statue of Sir John A. Macdonald, who first visited the city in 1886.

Maynard Building
723 - 725 Pandora Avenue

In the 1890s the Maynards shared business premises with Richard's boot-making shop on the main floor and Hannah's photographic studio on the floor above.

Many of the black-and-white photos now stored in Victoria's archives were taken by Hannah and Richard Maynard, who arrived in this city well over a century ago.

The English-born Maynards emigrated to Bowmanville, Canada West (or Ontario, as it's known today), in 1852. In 1859, anxious to check out the gold-mining prospects on British Columbia's Fraser River, Richard left his shoemaking business in Hannah's capable hands and headed west.

He wrote home frequently. Letter after letter was filled with descriptions of beautiful scenery and tales of interesting people. Hannah, already fully occupied with running the business and bringing up the children, sensed an opportunity in all that Richard described. In his absence she had also been working in a photographer's shop. Now the enterprising Hannah decided to add another string to her bow—the study and practice of professional photography.

Buoyed by her news, Richard returned to Bowmanville only long enough

to gather up his boot-making tools, his wife, their four children, and Hannah's photographic equipment. They travelled down the east coast to Panama, crossed the Isthmus, and sailed on the *Sierra Nevada* up the west coast to Vancouver Island.

The Victoria that greeted their eyes in 1862 was showing all the signs of too-rapid growth. The population of the small Hudson's Bay settlement, which had been about 300 people some four years earlier, had swelled suddenly as thousands of gold miners from California passed through Victoria on their way to the Fraser River. Many had already returned from the mainland, some ecstatic, others empty-handed and despairing. But rich or poor, they had one thing in common: they were all fodder for Hannah's photographic appetite.

"Mrs. R. Maynard's Photographic Gallery" on Johnson Street, near Douglas, did not immediately gain the favourable attention of Victoria's citizens. Photography itself was only 20 years old at that time, and female photographers were unheard of, this being considered far too complicated a venture for the fairer sex. But Hannah wasn't going to let her cameras and her ideas go to waste. She taught Richard how to use the equipment and sent him out around the town to capture streetscapes and landscapes while she developed her studio business.

A trip on the *Princess Louise* around Vancouver Island convinced the Maynards that there was money to be made from photographs of Native people and their villages. In the late 1880s they journeyed to the Queen Charlotte Islands, taking all their equipment—complicated and cumbersome by today's standards—along.

Some years before, Hannah had moved her portrait studio to the floor above Richard's boot-making business, and now their different photographic strengths became apparent. Richard preferred places; Hannah's primary focus was people. She experimented with innovative techniques, such as double and triple exposures and montages. In the mid-1880s, after the tragic deaths from typhoid of daughters Lillian and Emma and daughter-in-law Adelaide, Hannah created a name for herself as a designer of what she called "living statuary." Her subjects, usually children, were often pictured standing on pedestals, draped and covered in white powder to look like stone statues. In 1892, when the Maynards moved to new business premises,

Hannah moderated her approach and reverted to her former style.

The purpose-built structure on Pandora Avenue was far more spacious and designed specifically for their individual endeavours. Looking at the front of the building, its double function could be easily discerned. On the lower level, a centred, recessed entrance led to Richard Maynard's shoe store. At the left side, a smaller arch-topped entrance door permitted access to Hannah's second-floor studio. Large windows on that level allowed the maximum amount of light to enter. A narrow centre door led onto a balcony where the Maynards could look out over the streets and buildings they had captured with their photographic lenses.

Five years after Richard died in 1907, Hannah retired, declaring quite rightly that she and her husband had, at one stage or another, probably photographed most of Victoria's citizens. She died in 1918 at the age of 84.

More than 15,000 of the Maynards' photographs, stored at the British Columbia Archives, document the growth of this province over more than five decades. At 733 Johnson Street, Maynard Court (formerly Grimm's Carriage Factory) bears their name. But the building at 723 Pandora Avenue is a physical reminder of the couple whose talents provided tantalizing glimpses of early British Columbia.

E.G. Prior Building
606 - 614 Johnson Street / 1401 Government Street

A distinctive corner arch topped the impressive 1888 commercial structure designed by L.B. Trimen to house E.G. Prior & Co.'s hardware business.

On the northeast corner of Government and Johnson streets, a plain building devoid of ornament or even much character gives no hint of its impressive appearance under the ownership of successful businessman and politician Edward Gawler Prior.

Born in Ripon, Yorkshire, the second son of a clergyman, Prior trained as a mining engineer before travelling to North America. Following a route that took him through Salt Lake City and on to San Francisco, he boarded the *Prince Alfred*, bound for Vancouver Island. When he arrived in Victoria in 1873, he was just 20 years old. The papers he carried in his pocket qualified him for a position with the Vancouver Coal Mining & Land Company, based in Nanaimo.

Devoid of its distinctive roofline, corner entrance, and exposed brick surface, E.G. Prior's building has lost much of its former grandeur.

Five years later Prior returned to Victoria and married Suzette, the youngest of John and Josette Work's eight girls. John Work, who had retired from the Hudson's Bay Company to manage Hillside Farm in the early 1850s, didn't survive to see his youngest daughter become Prior's bride. It was her brother David who walked her up the aisle of the iron church of St. John at Douglas and Fisgard streets.

The newlyweds lived for a short while on Fort Street, then moved to Nanaimo, where Prior was now based as inspector of mines for the province. After two years there he resigned and moved his family back to Victoria. They lived on Church Hill (today's Burdett Avenue) until Prior bought four acres of land just a few blocks south of Hillside, where Suzette had lived as a child. He commissioned architect John Teague to design a

home that would be large enough for his growing family and splendid enough for entertaining on a grand scale.

For Prior was rapidly moving up in the world. Realizing that his government post held little opportunity for advancement, he had returned to Victoria to enter into partnership with Alfred Fellows, hardware merchant in this town since 1859. Guided by Prior's entrepreneurial spirit and enthusiasm, what started out as a small Yates Street hardware store with one employee grew rapidly and prospered.

Prior continued to expand the business. When Alfred Fellows retired to England in 1883, Prior purchased his interest. Before long he had taken on three new partners, and the firm was incorporated as E.G. Prior & Company Ltd. in 1891. By that time almost 100 employees staffed the head office and warehouses in Victoria and the branch operations in Vancouver and Kamloops. The firm had gone international, with an office in London, England. Two catalogues—one for hardware, one for machinery—advertised its wares to mill, mine, and railway owners as well as to contractors and farmers throughout B.C.

By this time E.G. Prior & Co. had moved to grand new premises at Government and Johnson streets. Architect Leonard Buttress Trimen arrived in Victoria in the early 1880s, and in his brief but successful Victoria career (he died in 1891) he designed many prestigious residences. Dominating the northeast corner of the Government–Johnson intersection, the building he designed for Prior had a corner entrance, topped at roof level by an ornate arched structure. Including a later addition, the building took up most of the 600 block of Johnson, with ample frontage for horses and carts. As well as hardware items, the company sold carriages, wagons, and all kinds of farm machinery. A warehouse a few blocks north on Government, at Pembroke, provided storage space for goods imported from all over the world.

Prior became involved in provincial politics in 1886 and moved to the federal level in 1888, sitting in the House of Commons for fourteen years. "The Priory," as the family home in Rockland was called, had been the scene over those years of many a splendid social event. Sadly, it was also the place where, in 1897, Suzette died after a lengthy illness, leaving her husband with a son and three daughters. Two years later Prior married a second time, to Genevieve Wright.

In 1902 he returned to the B.C. legislature and succeeded James Dunsmuir as British Columbia's 11th premier. Barely six months into his term of office, a government contract awarded to his firm was deemed a conflict of interest, and Prior was asked to resign. He returned to his favourite pastimes—theatre, tennis, and local affairs—and continued an interest in the military that had started in 1874 and had seen him become colonel in command of the 5th Regiment Garrison Artillery, based in Victoria. By 1919 he was lieutenant-governor of B.C. Now he and Genevieve were residents of Government House, a short distance from their St. Charles Street home.

Prior was not to enjoy this latest success for long. Less than a year after his appointment, he was taken ill. He died a month later, at the age of 67, in Royal Jubilee Hospital. His firm was amalgamated with McLennan, McFeely & Co. and became known as "Mac & Mac & Prior." The company continued retail sales until 1961.

Recently granted heritage status, the building survives, much altered, to this day. The decorative roof elements are gone, and the original brick has been painted and covered over. If E.G. Prior were still around, he would barely recognize the building—or the town—where he first made his mark more than a century ago.

The New England Café, Victoria, B. C.

The Oldest on the Coast

New England Hotel

1312 Government Street

Building on the success of a popular restaurant on the same site, this 1892 40-room luxury hotel featured cast iron and distinctively styled tall, narrow windows.

Today on the fourth-floor façade of this lovely old building, you'll see the words "Established 1858." Indeed, it was at the very beginning of the Fraser River gold rush that George and Fritz Steitz founded an eatery here. They called it the New England Restaurant, and it quickly earned a reputation up and down the coast.

By 1864 the Steitz brothers were itching to get back to the goldfields. George had a notion to import camels for use as pack animals in the Cariboo. It seemed like a good idea, but the poor beasts' slow and cumbersome progress was totally unsuited to the task, and the venture failed. Steitz, sensing that in any

case his restaurant days were over, sold the New England to local businessmen Vogel and Weiler.

In 1876 Michael and Louis Young took it over. They opened a bakery and added a room "For Ladies and Families Exclusively," which served reasonably priced meals at all hours. They parlayed the business into such a success that by May 1892 they were ready to expand. The *Daily Colonist* of May 31 announced that the owners were moving out so the old 1858 structure could be demolished. Architect John Teague designed its replacement, and in October the paper covered the grand opening of the New England Hotel.

The four-storey brick building was an impressive sight, thanks to Teague's use of iron structural piers and tall, cast-iron, bay windows. Above the narrow fourth-floor windows, a false cornice topped the building. Inside, the 40 guest rooms were illuminated by electric light, and there were hot-water washrooms and bathrooms on each floor. The ceilings were high, and the full-length windows were hung with plush velvet drapes. The main-floor restaurant was complemented by a huge dining room and rooms for private parties. The basement housed extensive wine cellars and the bakery's great stone ovens.

The hotel became a firm favourite with visitors from far and wide. Many customers were farmers from out of town. In the days before automobiles, they would ready the rig and load up the wife and children. The Saturday morning ride into town took several hours. Leaving the horse and rig at a nearby livery, they would go to the New England Hotel and book rooms.

A leisurely four-course dinner might follow, then the men would retire to enjoy some sport—boxing was popular in those days—while the women shopped at stores that stayed open on Saturday nights till 11 p.m. to accommodate them. The next morning a legendary New England breakfast would fortify the family for the long journey home.

Toward the end of the 1890s a smallpox scare quarantined Victoria, and the city's hotels suffered tremendous losses. Somehow the New England Hotel survived, and a few years later it enjoyed an enviable reputation as one of the most desirable places to stay in Victoria. In the early 1900s it was at the height of its glory. Guests were appreciative, food was cheap, and business was as brisk as it had ever been in 45 years.

In 1915 Michael retired from the business and moved to Shawnigan Lake. In 1929 he moved back to Victoria, and in 1934, at the age of 91, he died at the home of his son, Louis.

The Depression had taken its toll, and a year after Michael's death the hotel was in debt. The furniture and fixtures were sold in 1935 for a total of $1,100. Subsequent owners and operators rented out its rooms, ran restaurants, leased space to a tattoo parlour, an antique shop, and a dance hall, among others. Eventually, in 1978, the current owner declared himself unable to come up with the $55,000 needed to comply with fire regulations. The upper-floor renters were forced to leave, and the space they vacated has been empty ever since.

The present owner, who ran Ivanhoe's Restaurant on the main floor for some years and has operated an ice-cream parlour there since 1982, would like to see the building in use and restored to its former glory. In the meantime it serves as a faded reminder of the once-vibrant business that started on this site a century and a half ago.

Theatres

The McPherson Playhouse, formerly one of a chain of vaudeville theatres opened by Greek immigrant Alexander Pantages, features an Edwardian-style auditorium complete with marbled columns, plaster cherubs, and mandolins.

In early Victoria a thriving and ever-changing network of theatres sought to entertain the local population. Most are long gone. Others serve a different purpose. Some are still with us today.

Long after the first homegrown attempts at entertainment in the assembly hall at Fort Victoria, and long before movies made their debut, live theatre was all the rage. The first facility to officially dedicate a space for performance was a downtown hostelry near the waterfront. In 1858 the Royal Hotel and Concert Rooms brought patrons flocking to the southeast corner of Wharf and Johnson streets. A year later, when a touring group called it the Royal Theatre, the space was already facing stiff competition.

Up on Government Street, an old fur-storage facility that had stood at the northeast corner of Fort Victoria since 1846 had been given a new lease on life. Judge Matthew Baillie Begbie bought the two huge barns that spanned the block between Government and Langley streets and converted them into the Victoria

Theatre. Patrons entered the 500-seat facility off Government, just south of Bastion Square (where the Bedford Regency Hotel stands today).

It was the first Victoria Theatre, but it wouldn't be the last. A quarter century later, another opened at the southwest corner of Douglas and View. The second Victoria Theatre was built at the prompting of the Victoria Theatre Company, led by chairman and local business tycoon Robert Dunsmuir. When it opened in 1885, the performance area was on the ground floor. The second-floor level was an extension of the Driard Hotel. In 1910, when David Spencer's department store was gutted by fire, he bought the Victoria Theatre and incorporated it into his shopping arcade (now The Bay Centre).

Three years later, in 1913, Victoria's third Royal Theatre—and first dedicated opera house—rose at 805 Broughton Street, on the corner of Blanshard. With its 1,574 seats, it was the city's pride and joy. But between the first Royal Theatre on Wharf Street and the third on Blanshard, there were many more theatres that came and went as half a century rolled by.

Touring groups being few and far between, Victorians were inspired to provide their own entertainment. The Victoria Amateur Dramatic Club drew its fair share of local talent, including the singing Franklin brothers, lawyer Henry Classon Courtney, and saloon-keeper Ben Griffin, whose Boomerang Inn stood ready to fortify the performers when their work was done.

In those early days it seemed there was a music hall on almost every downtown corner. They boasted names like Alhambra (southeast corner of Government and Yates), Concordia (southeast corner of Douglas and Yates), Delmonico (1300-block of Government), New Idea (southwest corner of Government and Johnson), and Trilby (southwest corner of Broad and Johnson).

As musical theatre and vaudeville came into their own, they were featured in theatres such as the 986-seat Pantages (now the McPherson Playhouse at 3 Centennial Square), where Eva Hart, one of Victoria's most popular singers and actresses, was a featured performer. Nearby, the 480-seat Savoy Theatre's floor was built on a steep incline so that those sitting at the back could see as well as those at the front, and there were no pillars to block anyone's view.

When moving pictures made their debut, many outlets were adapted to accommodate them, and new ones were created. In 1911 the Crystal Theatre on Broad Street boasted "the largest, best furnished and most comfortable picture theatre in the city," while the Majestic Theatre on lower Yates Street assured a "change of program three times a week" and proudly proclaimed, "We cater to ladies and children."

Talking pictures were next to appear on the scene and spawned movie theatres with names like the Dominion (which featured Victoria's first talking picture, Al Jolson's *Sonny Boy*), Capitol, Columbia, Princess (later renamed Playhouse, Plaza, then Haida), Atlas (later Coronet), Variety, and Totem. The Romano, at the corner of Government and Johnson—once Lawrence Goodacre's butcher shop, then an Army and Navy clothing store—was owned by the Quagliotti Brothers, who also owned the nearby Grand (later Empress, then Rio).

A few of these old-time entertainment centres remain as retail outlets. But for generations of theatre-lovers, live theatre is still the way to go. Today the Royal and McPherson draw huge crowds for local symphony, opera, and musical theatre performances, as well as big-name national and international artists. And at 805 Langham Court in Rockland, where the Victoria Theatre Guild first established a community theatre in an old barn in 1929, residents and visitors alike continue to enjoy professional-standard presentations in a proper theatre featuring talented local actors, directors, set designers, and production people.

In Victoria, a century and a half after the first live performance, the art of entertainment is alive and well.

Union Club

805 Gordon Street

San Francisco architect Loring P. Rexford designed the Union Club's new premises on Humboldt Street, seen here under construction in 1911–12.

In April 1879, more than three decades after the Hudson's Bay Company established a fort at the southern tip of Vancouver Island, two significant events occurred. Victorians learned that the overland railway—the carrot that coaxed them into Confederation—would not, after all, have its terminus in their town. And a group of businessmen in Victoria formed the first gentlemen's club west of Winnipeg, a place where they could discuss the pressing issues of the day as well as meet, greet, eat, imbibe, and enjoy the company of gentlemen like themselves.

Such clubs had an illustrious history dating back to the days when men of means would gather to exchange views and bemoan technological and other advances. At those clubs, members fraternized with others of their ilk— rich, older men who recognized those engaged in the army, navy, or diplomatic service, but wouldn't dream of colluding with people in the professions or men of commerce. Here in Victoria, however, it was the businessmen who brought the gentlemen's club to life and elected a lawyer, Sir Matthew Baillie Begbie, the so-called "hanging judge" who dispensed justice during the gold-rush days, as first

Viewed from what is now the Empress Hotel's entrance driveway, the new Union Club loomed large over neighbouring residences. Members may have missed the proximity to a bawdy house afforded by their previous premises.

president of the Union Club of British Columbia.

Cozily ensconced on the second floor of a building at Yates and Government streets, in premises recently vacated by the B.C. Mining Exchange, Victoria's answer to the gentlemen's social clubs of London took shape. By 1884 the club and its 150 members were casting about for more space. They purchased three blocks on the northwest corner of Courtenay and Douglas, and plans were drawn. In a city with still-unpaved streets and planked sidewalks, the new Union Club building was quite the grandest in the area when its members took possession in May 1885. In the years that followed, its virtues became even more apparent. Right across the street was one of Victoria's most popular bawdy houses, while at the other end of the same block, St. Andrew's Presbyterian Church lent an air of respectability.

Before long, however, troubles befell the facility. It was plagued by barking dogs, which club members tied to the iron railings outside, riddled with rats from the nearby livery stables, and infested with flies. What should have been a venerable institution became a hotbed of unrest. Members complained about the lack of heat and hot water, the misuse of toilet paper, and the odours that seemed to emanate from every nook and cranny.

A suggestion book was started so that annoyances could be recorded for all to see, and it soon became clear that fellow members were not excluded from this diary of discontent. One June 1907 entry suggests "that a certain gentleman … be ordered to discontinue using soup dishes and other plates for the purpose of feeding his dog." Another implores "that the present cooks be got rid of as soon as possible, as the cooking is abominable."

Yet another asserts "that the club flag is a disgrace. Ditto the Secretary's tie."

Space was provided in the book for the club's answer to each concern. One member waging a war against winged insects wrote, "Suggested that the windows of the lavatory be screened at once, as there is a plague of flies there." Answer: "Attended to. Flies can no longer use the lavatory."

Eventually, despite the fact that rat-chasing was considered by some members to be good sport—Saturday night rat hunts were a firm favourite—it became clear that other premises would have to be found. In 1910 land at the opposite end of the block was purchased, and club member Francis Rattenbury helped select design submissions. San Francisco native Loring P. Rexford designed an impressive Georgian Revival structure that echoed the style displayed in similar clubs across the country. Members took possession of their new building in 1912. Their original, rat-infested premises were demolished many years later.

More than 100 years after its creation, the Union Club finally agreed to admit women as full members. It was the end of an era, but the imposing structure at the corner of Gordon and Humboldt streets has managed to retain all the flavour of a gentlemen's club of long ago.

The Union Club (right) flanked the Belmont Building (centre), designed as a major hotel but destined to become an office block. Between the Belmont and the old post office (left, now the site of the Customs House), a streetcar leaves Government Street, headed south across the causeway.

Alexandra Club
716 Courtney Street

Its members scattered by First World War endeavours, the Alexandra Club premises served after 1922 as a guest house, then as the Hotel Windemere (pictured here in 1936) before being converted into office space.

In 1900, nearly two decades after the Union Club settled in a new building on the corner of Courtney and Douglas streets, Victoria's gentlewomen established a literary society and social club just a couple of blocks away. The Alexandra Club was named after the princess who was destined to become queen when Edward VII was crowned king of England the following year.

As of August 17, 1900, the club had attracted 56 enthusiastic ladies who were undeterred by the fact that the club entrance fee was five dollars, with a dollar-a-month subscription payable quarterly in advance. Charter members included such well-known figures as Joan, widow of Robert Dunsmuir, builder of Craigdarroch Castle; Laura, wife of Joan's son, Premier James Dunsmuir; Mary, sister of James Dunsmuir and wife of Harry Croft (after whom Crofton is named); and Jennie, wife of Dr. I.W. Powell (after whom Powell River is named). Also on the list were Sarah, wife of Charles Hayward, Victoria's first funeral director; Elizabeth, wife of Thomas Earle, MLA and owner of Victoria's first coffee and spice emporium; Genevieve, second wife of hardware business owner (and later lieutenant-governor) E.G. Prior; and Margaret (known as Daisy), wife of noted architect Samuel Maclure.

The club met in premises above Redfern's jewellery store on Government Street. Then, in 1909, club president Mary Croft suggested they build their own structure. The members incorporated the Alexandra Home Company and sold shares to raise funds. By September 1910 they had purchased land on Rae (now Courtney) Street and obtained a building permit for $51,200 (thanks to a generous donation from James Dunsmuir). Architect David C. Frame supervised construction.

The Alexandra Club, lauded in a *Daily Colonist* editorial as "significant of the new Victoria spirit," was completed in 1911 and was quite a sight, with its double-height second-floor windows and glazed French doors opening onto a balcony that afforded a magnificent panoramic view from Beacon Hill Park over to the Sooke hills. The front wood-panelled staircase and reception area were adorned with potted plants. On the upper levels, along with a large dining area and accommodations for club members, was an Empire-style ballroom. It was decorated in white and gold, lined with mirrors that made it look even larger than it really was, and featured hangings of "richest pale blue and gold brocade."

On May 9, 1911, Victoria's elite gathered for the Alexandra Club's inaugural ball. Musicians in the orchestra gallery played for dancers on the specially sprung floor below. The ladies' gowns blended with the abundant flowers in what was described by an onlooker as "a kaleidoscope of colour." Many other social events, receptions, weddings, and musical affairs graced the premises in the following years.

The Alexandra Club's second-floor ballroom, complete with sprung dance floor (pictured here around 1912), set the scene for many a "men by invitation only" affair.

The club prospered until rumours of a major conflict in Europe brought the festivities to a halt. Members joined the Red Cross, the Victorian Order of Nurses, and many other community efforts. During the First World War the club premises were used extensively for charitable work. Membership dwindled, and before long the club was in financial difficulties.

Eventually, in 1922, members gathered up the monogrammed china, silverware, and bed linen and moved to the Pemberton Building on Fort Street. In 1929 the club moved again, to the top of the Campbell Building on Douglas Street (later demolished to make way for the Royal Bank building). Eventually it disbanded. Meanwhile, its former home had become first the Alexandra Guest House, then the Windemere Hotel. From around 1950 to 1976 it served as the RCMP's division headquarters.

In 1977, with a $17-million provincial government building planned for most of the block bounded by Douglas, Blanshard, Broughton, and Courtney streets, a decision had to be made: Tear the old Alexandra Club down, or try to save it? Most of the structure was still serviceable and would provide ample space for government workers. Architects Hawthorn, Mansfield & Towers, and structural engineers Reed Jones Christofferson, recommended that the building be preserved. The rear one-third was demolished for safety reasons. The remainder was refurbished at a cost of $1.2 million, much to the delight of the original members' descendants, who recalled going there for dancing lessons as young girls.

In 1979 the venerable old building was integrated into the complex that housed the main branch of Victoria's public library. The second-floor front balcony is gone, but the huge fireplace in the ballroom remains. A peek under the carpet on the second floor reveals the original sprung floor. Close your eyes and listen carefully. You can almost hear the strains of the orchestra, playing while Victoria's elite danced the night away on Courtney Street.

Legislative Buildings
501 Belleville Street at Government Street

Almost a decade after the building's 1898 completion, the front lawn of the legislature was a popular focal point for parades and celebrations.

The stone legislature on Victoria's Inner Harbour—truly the jewel in British Columbia's crown—was the first waterfront building designed by a young Yorkshire architect over a century ago.

Forty years before the Legislative, or Parliament, Buildings rose majestically on the harbour's southern shore, the site was home to Victoria's first official seat of administration.

In 1858 Vancouver Island governor James Douglas realized that Fort Victoria was ill-equipped to handle the hordes of prospectors arriving from California en route to the Fraser River goldfields. He ordered the fort demolished, sold the land, and used the proceeds to pay for a new colonial administration centre located on the less-expensive south side of the harbour and—by fortunate coincidence—handy to the governor's home.

Douglas's architect of choice was Hermann Otto Tiedemann, who worked in surveyor-general Joseph Pemberton's department. Tiedemann, probably mindful of the ever-present danger of fire, designed a legislature comprising not one but five wood-frame administration buildings.

Looking south across the Inner Harbour from Wharf Street in the early 1900s, with the James Bay Bridge at left, the legislature stands boldly outlined against the cloud-covered peaks of the Olympic Peninsula mountains. The old CPR ticket office is at right in this picture.

It was the strangest set of structures Victorians had ever seen. *British Colonist* editor Amor De Cosmos heaped scorn on the hapless architect. They didn't look at all like government buildings, De Cosmos said. Squat and square, with up-tilted roofs, they looked like a cross between a pagoda and a birdcage! And "The Birdcages" they became. Complementing this notion, the road running along the east side of the buildings became Birdcage Walk, a name probably borrowed from St. James Park in London, England, where King Charles II once had an aviary.

The Birdcages served the province well for more than three decades, but in the early 1890s a new seat of power was proposed. The city had suffered in recent years. Relatively prosperous earlier times had given way to an era beset by economic depression, citizen unrest, and physical devastation from smallpox. The construction of a new legislature, argued its proponents, would provide work for the people and give a leg-up to the local economy. It would also show a confidence in the city's future that few at the time actually felt.

When it was advertised, the competition drew more than 60 entries from far and wide. The winning design, chosen by a government committee, was the work of Francis Mawson Rattenbury, a young architect from northern England who had only recently arrived in Vancouver. Rattenbury, 25 years old at the time, was hardly qualified for such an enterprise. The biggest building he had ever tackled, while he apprenticed in his uncle's firm, was a town hall in Yorkshire.

This didn't deter him. Fuelled by ambition, but concerned that he might be considered an outsider, he signed his drawings "B.C. Architect" and let

This view, taken from the CPR ticket office, shows the automobile-lined corner driveway entrance to the legislature and, atop the centre cupola, a three-metre gilded statue of Captain George Vancouver, after whom Vancouver Island is named.

his sweeping design speak for itself. It worked. The enthusiastic committee, much impressed by a vision that far exceeded its expectations, gave him the go-ahead. Rattenbury was ready. Confident of their approval, he had already taken up residence in Victoria in order to complete his plans.

Work began in earnest in 1893. It was an ambitious project, with some interesting stipulations. For example, the "short-stemmed T" design made it possible to erect the building between the two rows of Birdcages so that legislative proceedings could continue uninterrupted. The top of the T was a long frontage facing the harbour, stretching almost the length of a city block. In the centre of the block was the legislative hall and administrative offices. Smaller side blocks, connected by colonnades, housed the land registry and the printing office. In the middle of the centre block, a magnificent 415-foot dome on an octagonal base soared high above the area where the stem of the T joined the T-bar.

Local wood, bricks, slate, and stone were used for the interior, roof, and façade, but other hardwoods, marble, mosaics, and stained glass were imported from all over the world, along with art commissioned by European-trained artists and sculptors.

The structure, still incomplete by the time of Queen Victoria's Diamond Jubilee in 1897, was nevertheless illuminated by thousands of lights in celebration of the event.

It was a beautiful building all right, but the budget—$500,000—was well and truly blown. By the time it opened on February 10, 1898, Rattenbury's imposing structure had cost the province just under $925,000. After a suitable "pouting period," the public forgave him and brought him back in 1912 to design the south wing—the legislative library. It was completed in 1916.

The Birdcages, meanwhile, had not gone to waste. Demolished one by one, their bricks were used to pave the front carriageway. The original assembly hall, dating from 1859, survived to take on a new life as the department of mines and might be there to this day had it not been destroyed by fire in 1957.

Seven decades after Rattenbury's demise, the Legislative Buildings still dominate the Inner Harbour and are arguably the most photographed structure in the city. The public galleries are open when the legislature is in session, and tours of the interior offer an interesting insight into the province's past.

On the roof of the building,

standing atop the highest copper dome, gleams Victoria's own "golden boy," Captain George Vancouver. Turning his back on the ocean he loved so well, he gazes off into the distance toward the city that bears his name. Behind him, at the back of the building, larger-than-life-size likenesses of early navigators, explorers, and political figures stand guard, high up on the corners. On either side of the main Belleville Street entrance, Chief Justice Sir Matthew Baillie Begbie and Governor James Douglas gaze, stone-faced, upon the parades and protests that regularly take place on the steps at their feet.

Next time you wander past the Legislative Buildings, peek inside and remember the brash young Yorkshireman who burst onto the Victoria scene a century ago and designed the Inner Harbour's most impressive structures.

Otters and gulls grace the centre of the fountain at the rear of the Parliament Buildings, while bronze-cast creatures native to British Columbia—eagle, raven, wolf, and bear—decorate its rim.

Empress Hotel
721 Government Street

Looking northeast from the first legislative buildings in the late 1860s, a wooden bridge spanned James Bay. Four decades later the bay was dredged, drained, and levelled with landfill to form the foundation for the Empress Hotel.

It's hard to believe that schoolboys once paddled their canoes to school across the present-day site of the world-famous Empress Hotel.

Once upon a time, way back in the 1860s, a wooden bridge—the first of three—spanned Victoria's Inner Harbour. At its northern end was Government Street; at the southern end, the Legislative Buildings. Water from the harbour flowed under the bridge, its level fluctuating with the tides, covering and uncovering the mud flats that lay beneath. Many yards east of the bridge, the water lapped at the grassy bank behind the Church of Our Lord at the foot of Blanshard Street. This was the original James Bay.

As the decades passed, the residential area at the south end of the bridge took the name of that muddy pool. As the first humble dwellings along Humboldt Street were replaced by industrial ventures, the bay became an eyesore. It was filled with floating debris. The ebbing tide

Empress Hotel from Parliament Buildings, Victoria, B.C.

*By 1908 the Empress Hotel, newly landscaped and with its south and north wings
still to be added, was an imposing sight behind the causeway that replaced the
James Bay Bridge.*

revealed mud flats rank with rotting garbage and soiled with scum from a nearby soap factory. By the early 1890s the city was thoroughly embarrassed by this blot on its landscape. Canadian Pacific steamships were already tying up at the CPR dock in the Inner Harbour, and in the middle of it all, right where it mattered most, was this evil-smelling eyesore.

Local businessmen and politicians lobbied for a clean-up job and construction of a major hotel on the south side of James Bay, where the gardens of former governor James Douglas's home sloped down to the shore. Citizens approved a bylaw that authorized construction of a causeway and filling-in of the mud flats. Many schemes were proposed, but only one—for a hotel in grand CP style—was seriously considered.

Tempted by the promise of property, along with tax and water concessions,

CPR president Sir Thomas Shaughnessy agreed to investigate the possibility. Francis Rattenbury, architect of the recently constructed Legislative Buildings, sketched his vision of the harbour and included a magnificent hotel right behind the causeway. The drawings were received with enthusiasm. The city struck an agreement with CPR to build the hotel. And so, in the early years of the 1900s, work began.

First, the city bought the properties along Humboldt and cleaned up the street. Then a coffer dam was constructed at the site of the old James Bay Bridge, and over several months all the water was drained out. Next, just under 3,000 fifty-foot-long Douglas fir pilings were driven deep into the mud. Earth, sand, and gravel were trucked in from surrounding areas. Concrete was poured to form piers and a platform to serve as the foundation for the new structure.

Afternoon-tea guests lounged in the lush tropical surroundings of the Palm Court, bathed in the light from its multicoloured stained-glass dome.

The hotel's lounge (now the Tea Lobby), decorated in cream and green with rust-red accents, reflected the relative stiffness and formality of the Edwardian era.

Victorians watched with amazement as Rattenbury's design took shape, and by the end of January 1908 the hotel was open for business. It was named "The Empress" after Queen Victoria, who was also Empress of India.

In the dining room, fabric-covered walls complemented the rich wood of the pillars. The lounge was spacious, its large windows designed to let in plenty of light for reading and to afford a sweeping view of the Inner Harbour. Guest rooms were furnished with exquisite taste. Just off the main-floor lobby, the ladies' drawing room, with its unique arched, knobbed, and painted plaster ceiling, was a resting place for women to wait while their husbands registered them at the hotel desk. And outside, near the cream-coloured porte-cochere at the building's entrance, a special coach with solid brass mountings and leather-upholstered seats stood ready to take visitors to places of local interest.

The new hotel's 160 rooms were soon booked solid. By 1914, north and south wings had been built, along with a ballroom, a stained-glass-domed Palm Court, and a Reading and Writing Room

Rumours that the Empress was sinking were confirmed when it was discovered that one end of the original structure was settling more rapidly than the other. The subsidence was slowed by removing several metres of fill from around the main entrance. The new south wing, shown under construction behind the porte-cochère, was completed in 1912.

(later the Coronet Lounge, now the Bengal Lounge). The guest-room total was now 550.

The Empress was badly affected by the difficult Depression years and was also the victim of changing styles. Grand, expensive hotels were not as popular anymore. Forced to rethink its practices, at one point management allowed wealthy widows to rent their rooms for a dollar a day in order to keep the hotel occupied, and over the years undertook two major renovations.

In the 1960s, "Operation Teacup" attempted to modernize the hotel and give it a new lease on life. Teak furniture replaced the beautiful antiques. Broadloom covered the polished wood floors. The Humboldt Wing proudly proclaimed itself "The Empress Motor Lodge."

This carried the Empress through the next two decades, but then saner heads prevailed. In 1988 the teak was taken out; the antique furniture, fixtures, and fittings returned; and the broadloom was removed. This "Royal Restoration" brought the hotel back to its former glory.

Today the Fairmont Empress, as it's now called, stands proudly behind the Inner Harbour causeway, a tribute to its designer and to the forward-thinking citizens of Victoria who supported this remarkable structure 100 years ago.

Crystal Garden
701 Douglas Street

In return for tax and other concessions, the Canadian Pacific Railway agreed to finance a $300,000 amusement centre comprising a glass-roofed tropical pleasure garden complete with swimming pool. Crystal Garden, pictured from Humboldt Street looking south, was completed in 1925.

In the first years of the 20th century, Victoria was already enjoying a reputation as a tourist destination of choice. The city had recovered from the shock of the Canadian Pacific Railway's decision, in the early 1880s, to have its western terminus on the mainland. There had been compensations. Vancouver got the train, but Victoria got the steamship service. As the turn of the century came and went, vessels brought visitors from all over the world to Victoria's outer wharves, just north of Ogden Point. By 1908 they had

somewhere splendid to stay—the CPR's Empress Hotel.

Several proposals for use of the landfill area behind the hotel were considered and rejected in the early 1920s. Then, toward the end of 1923, public opinion was sought on a whole new concept—a combined swimming pool and amusement centre that would treat locals and tempt visitors. A referendum garnered solid support. The challenge now was to design an equally solid support for the building—and the 200,000-plus gallons of water its swimming pool would contain.

Canadian Pacific agreed to put up most of the money. The architect of choice was Francis Rattenbury, whose creative genius was already in evidence around the Inner Harbour. Rattenbury was joined in this latest venture by Percy L. James, who had worked on several large public facilities in England.

The foundation for the new facility was a 30-inch-thick concrete-and-steel platform. The iron-and-glass superstructure was constructed by a U.S. firm, with Victoria's own Luney Brothers carrying out the rest of the work. James supervised the project—and would later claim that he did the lion's share of the work and deserved a greater share of the commission.

By the early summer of 1925 the building was ready. That year the traditional May 24 Queen's Birthday celebrations were put on hold for one month so they could be combined with the official opening of the newly named Crystal Garden.

Victorians were suitably impressed by this latest addition to their city. Forty-five feet from the ground to the peak of its quarter-inch reinforced glass roof, it was 98 feet wide and almost 270 feet long. Above the 40-foot by 150-foot swimming pool there was an 18-foot-wide, 186-foot-long promenade. Huge maple-floored ballrooms anchored the building at either end. A third ballroom on the lower level doubled as a gymnasium and concert hall. There was permanent seating in the facility for 700 people, but it could accommodate 2,500 for special events.

The pool itself was a sight to behold. Its 250,000 gallons of seawater were

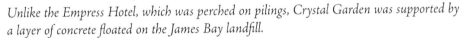

Unlike the Empress Hotel, which was perched on pilings, Crystal Garden was supported by a layer of concrete floated on the James Bay landfill.

pumped from the ocean at Dallas Road. Before it entered the pool, the water was warmed by heat generated by the Empress Hotel's laundry, which at that time stood immediately to the north.

To begin with, the pool was the city's pride and joy, but by the early 1930s the Great Depression was taking its toll and the public's enthusiasm had waned. In 1933 the CPR Hotels general manager, based in Winnipeg, chastised Victorians for not continuing to support the facility. Describing the pool as the largest and finest he had seen in his travels across Canada and through the U.S., he reminded citizens that his company had built it by popular demand and that the least they could do was use it. He assuaged concerns about contamination by ordering a series of tests proving that bacteria could not survive in chlorinated water. There was no excuse, he said, for not using and enjoying the Crystal Garden pool.

He was right—the water was safe. But what made it safe could also be harmful. Chlorine gas had been used in chemical warfare during the First World War. It was deadly in large doses and wreaked havoc in the bodies of those who managed to survive. The use of chlorine in public facilities was carefully regulated. Still, there was always the outside chance of accidents, and in the early summer of 1967, when the pool was already in need

The Crystal's upper balcony was the ideal place for relaxed conversation, bathed in light that filtered through a reinforced glass roof.

Crystal Garden, Victoria, B.C.

P. 1728

of substantial repairs, a dangerous incident occurred.

One Saturday morning in May, chlorine somehow escaped. Usually the heavy gas would stay close to the floor, but this time it dispersed quickly in the draught from doors opened to facilitate construction work. The alarm was quickly sounded. By the time staff had closed the building, 34 people—children, firefighters, and police officers—had been taken to hospital. Miraculously, all survived.

This near-fatal incident sealed the ailing facility's fate. The CPR had long since given up its interest in the structure, and the pool was permanently closed. By the mid-1970s it was a sorry sight. Rain leaking through the glass roof had corroded the steel supports and buckled the hardwood floors. The tiled bottom of the pool was full of garbage and broken glass from the roof.

Reluctant though the city was to spend the vast sums needed to bring it back to its former glory, it managed to survive in a different guise. Mention the Crystal Garden to anyone in Victoria today, and they will likely wax eloquent about the stunning tropical display that once delighted the senses beneath its glass-paned roof. Only the letters "CPR," still clearly visible above the Douglas Street entrance, hint at its amusement-palace origins in Victoria's early carefree days.

Crystal Garden boasted the largest heated indoor saltwater swimming pool in the British Empire and was billed as "the finest place of amusement on the Pacific Coast."

CPR Steamship Terminal

396 - 468 Belleville Street

C. P. R. Ticket Office and Landing Place, Victoria, B. C.

The first Canadian Steamship Ticket Office and Landing Stage was designed by Francis Rattenbury in 1904 to accommodate passengers sailing aboard the Canadian Pacific fleet.

Standing on the steps of the Legislative Buildings in the mid-1920s, architect Francis Rattenbury had good reason to be proud. To his right was the Empress Hotel and behind it, the Crystal Garden. To his left was the Canadian Pacific Railway Steamship Terminal—last of the major Inner Harbour structures that he designed.

Rattenbury had been the CPR's architect of choice since 1901, when he won the competition to design a new wing for the Hotel Vancouver and practically rebuild the rest of it. Shortly after, he refurbished the interior of the company's

luxury steamer *Princess Victoria*. Then he designed Victoria's Empress Hotel, as well as additions to other CPR-route hotels at Field and Banff.

By the time the CPR decided to establish a shipping terminal in Victoria, the city was a favoured port of long standing. In 1827 the Hudson's Bay Company had started a coastal steamship service when the *Cadboro* arrived on the Columbia River, where the HBC set up its western headquarters.

In 1836, when the side-wheeler *Beaver* steamed across the Atlantic to join

Commissioned by the CPR in 1924 to revisit his original design, Rattenbury sought to impress new arrivals with a colonnaded edifice complete with images of the Greek sea god Poseidon. Fellow architect Percy L. James was less than impressed when Rattenbury took all the glory for their shared design.

the fleet, the HBC was able to expand its service. The *Beaver* was used as a mobile trading post and transport ship, carrying furs, merchandise, passengers, and mail up and down the coast, calling at Fort Simpson on Chatham Sound, Fort Nisqually on Puget Sound, and Fort Victoria. The 1858 Fraser River gold rush spurred expansion of Island-to-mainland service.

In 1862 the HBC bought the side-wheeler *Enterprise* to compete with the many American vessels sailing up the coast from California. That same year, Captain William Irving of Victoria founded the Pioneer Line, a riverboat service. Twenty years later, when the transcontinental railway joining Canada's east and west coasts neared completion, Captain William's son John merged his line with the HBC's to form the Canadian

Pacific Navigation Company. The combined fleet of sidewheelers, sternwheelers, and steamships soon became a force to be reckoned with on the northwest coast.

Over the years the CPNC successfully competed with other companies, acquiring more ships toward the end of the 1890s in order to carry passengers and supplies to the goldfields of the Yukon. In 1899 CPNC purchased a property on Victoria's Belleville Street waterfront, near the Legislative Buildings. Two years later the CPR purchased controlling interest in the CPNC and quickly absorbed the Victoria company into its own operations.

The CPR's fleet of "Empress" ocean liners was augmented by steamships designed to develop tourist business along the scenic west coast. The high-speed *Princess Victoria*, purpose-built on the

Canadian Pacific steamships may no longer tie up there, but the CPR Terminal building, its Greek-temple style intact, remains an interesting focal point on Victoria's Inner Harbour.

Clyde in Scotland, arrived here in 1902 and began the famous "triangle service" between Victoria, Seattle, and Vancouver. Its speed—three hours and nine minutes from Victoria to Vancouver—was a credit to the builders. And the ship's luxurious accommodations, excellent cuisine, and superior service were enhanced by the vision of CPR architect Francis Rattenbury.

In 1905 the CPR purchased the Esquimalt & Nanaimo Railway and steamships. During the First World War the fleet was kept busy with troop and freight transport. The early 1920s saw the dawn of a new era for the CPR, as automobile traffic increased. Business on the company's first car ferry, *Motor Princess*, was brisk enough to inspire the CPR to knock down the wood-frame building on Belleville Street and build a new terminal in 1923.

Rattenbury again worked with Percy L. James, who was also heavily involved in the Crystal Garden project.

James later claimed that he did the bulk of the work, preparing drawings and specifications and then supervising construction. But it was Rattenbury who later declared self-importantly that the new terminal was "a handsome little building, as good as anything I have ever done."

The terminal was completed in 1924. At 122 feet by 54 feet, it was truly a monumental effort, resembling a Greek temple with massive columns and twin likenesses of the sea god Poseidon above the entrance. Its four storeys were constructed around a reinforced concrete frame, with enclosing walls of masonry and a finished surface of ground Newcastle Island stone mixed with white cement.

At basement level, on the wharf, were the port steward's office, pay office, and other offices, as well as a large lunchroom and kitchen. At ground level, street entrances on the south and east sides afforded access to a waiting room, agent's room, and ticket offices. A marble staircase from the Belleville Street entrance led to the first-floor offices of the B.C. Coast Service officials. The engineers' offices were on the second floor.

This was the heyday of the CPR's B.C. Coast Service. By 1963 the company no longer needed such a large building and transferred its Coast Service operations to Vancouver. By 1970 the upper floors of the building had been leased to a logging company and the lower floors to a wax museum.

By that time, Rattenbury was long gone. Once Victoria's darling, he had divorced, remarried, and left the city for good in 1930. Within five years he was dead—murdered by his second wife's younger lover and buried without fanfare in England. No gravestone, statue, or plaque tells us of his time here, but at key points around Victoria's Inner Harbour, his magnificent structures remain.

Likenesses of the Greek sea god Poseidon, a symbol of the building's original function, still grace the wall above its main entrance.

Queen's Printer
553 Superior Street

Formerly housed in the legislature's west wing, the Queen's Printer was established in a new building on the corner of Government and Superior streets in 1928.

On a busy intersection in James Bay, the King's (now Queen's) Printer opened for business 75 years ago.

Although the Queen's Printer has been on Superior Street since 1928, its function dates back to 1859, when Colonel Richard C. Moody of the Royal Engineers founded a capital city for the new colony of British Columbia at Queensborough (later New Westminster).

Working under instructions from Governor James Douglas, Moody's men established a printing office there. Captain E. Hammond King, appointed superintendent, supervised publication of the first *British Columbia Gazette*, a

Still dominating the same corner, the Queen's Printer building now features sophisticated equipment and on-line service, along with a main-floor display of primitive-looking printing tools that were once considered state of the art.

government paper, in September 1859, using a small hand-press and a few cases of type acquired by Moody from England. He was succeeded shortly afterward by Corporal Richard Wolfenden, who had apprenticed as a printer before joining the Royal Engineers.

In 1866 the mainland and Vancouver Island colonies were combined to form British Columbia, and in 1868 Victoria became the province's capital. Wolfenden, now a lieutenant-colonel, moved with his family to Vancouver Island to become superintendent of the printing office, located in one of the wooden administration buildings known as the Birdcages, which had stood on the Inner Harbour's south shore since 1859. In 1888 he was accorded the official title of Queen's Printer.

By 1897 new stone Parliament Buildings were taking shape on the same site. One by one the Birdcages bit the dust as their occupants found new homes in the grand new structure. The printing department was located in the west wing, which could house more up-to-date printing equipment to better serve the government's needs.

In 1901, when Queen Victoria died, Wolfenden became King's Printer, and two years later he was created Western Canada's first companion of the Imperial Service Order in recognition of his long and illustrious service.

In 1924 Saanich native Charles Banfield became King's Printer. Born to a Cornish father and a Welsh mother in 1877, Banfield had apprenticed as a printer in the office of H.G. Waterson after leaving school. Moving on to the *Daily Colonist* newspaper, he eventually finished his apprenticeship there in 1898 and went into business for himself. Later he was foreman for the Victoria Printing Company and worked in other printing houses before spending many more years back at the *Victoria Times*.

In 1928, four years after Banfield's appointment as King's Printer, the increasing demand for government printing services inspired a move to more spacious quarters. A corner site behind the Parliament Buildings was acquired, and the new, $156,000 cement-built King's Printer building quickly took shape.

The hand-fed printing presses were located on the first floor. The second floor housed printing and government office supplies. The third floor was home to the bindery. But soon after the building's official opening it became apparent that the printing operation would outgrow its new home before long. Thus began 60 years of modifications in an attempt to ease the space shortage problem.

In 1989 the Superior Street building underwent major renovations. By that time, with Elizabeth II on the British throne, King's Printer had become Queen's Printer once more, and Banfield was long gone. He had retired in 1946, four years later than usual, at the age of 69, after more than 22 years of service. Not content to rest on his laurels, he then entered civic politics and served two terms as alderman.

At age 76 he was still an active gardener and built a stone wall on the grounds of the Craigflower Road home he shared with his wife Effie and two children. However, in 1959, his incredible energy finally sapped by a lengthy illness, he died at Royal Jubilee Hospital. He was 82 years old.

The *British Columbia Gazette* is still published by today's Queen's Printer. Appearing once a week, on Thursdays, the *Gazette* is a valuable document that contains such items as notices to creditors; name changes; company registrations; notices of incorporation, amalgamation, dissolution; public tenders; and selected orders-in-council. Much of this information is also available on-line.

On the main floor of the Queen's Printer is a fascinating display of printing equipment that was once considered state of the art. The printing department now employs many more people than it did in Charles Banfield's day, and the equipment they work with is many times more sophisticated. But if Banfield were able to see it, he would be proud of the building that still provides the service he supervised there many decades ago.

James Bay Hotel

270 Government Street

In 1911 the red Spanish tiled James Bay Hotel was architect Charles Elwood Watkins' answer to the more expensive downtown hostelries. In the 1920s then-owner Major William Merston advertised room rates of $2.50 with a bath, $1.50 without. Dinner cost 75 cents.

Just a few blocks away from the Inner Harbour stands the building where Victoria's most famous daughter spent her final days.

Victoria's first—and once finest—residential area, James Bay is home to many bed-and-breakfast establishments and several hotels. But one hotel stands out among them all.

Many have stayed there, eaten in the restaurant, or played darts in the pub. But few know of its connection with the famous artist who once lived nearby.

When Emily Carr was born in 1871, James Bay was still the preferred place of residence for Victoria's elite. Separated by a wooden bridge from the dirt roads of downtown, surrounded on three sides by water and on the fourth side by a park, it maintained an air of quiet gentility. The Hudson's Bay Company's Beckley Farm, which once covered a large part of the peninsula, had long since been divided and subdivided until little trace of it remained.

Emily's father, Richard Carr, owned a business on Wharf Street. In 1863 he had moved his wife and two

In the 1960s, complaints from tourists disappointed to learn that the hotel was not actually situated on a bay occasioned a name change to "The Colonial Inn," but the original name was restored in 1973.

daughters to the peaceful community across the bridge, building a home near Beacon Hill Park. Three more girls—Emily was the youngest—and one boy were born in the Italianate villa that stands to this day at 217 Government Street (formerly Carr Street).

By the end of the 1800s most of Victoria's elite had moved to the Rockland area, and by the beginning of the 1900s James Bay had changed and grown. Gone were many of the mansions along the Inner Harbour's south shore where people like Governor James Douglas, department-store owner David Spencer, and coal-magnate Robert Dunsmuir once lived. James Bay had filled out and filled up, with homes of all shapes and sizes. On the west side, light industry and commercial ventures faced the Outer Harbour. On the east side, the wide track that would later become Katherine (now

Douglas) Street boasted homes, a college, and a couple of schools. It was near this track, in 1911, that the Parfitt brothers built their new hotel.

The five Parfitt brothers—James, Aaron, Fred, Mark, and Albert—who hailed from Cornwall, in England, had established themselves by rebuilding several business premises lost in two major downtown fires—one in 1907, the other in 1910—and quickly made their mark as contractors of note in the downtown core. Now they turned their attention to James Bay. Downtown hotels were expensive, so a well-designed, moderately priced hostelry was sure to succeed. But first they needed a good architect who could turn their vision into reality.

Charles Elwood Watkins was more than equal to the task. Born in Victoria and schooled both in his hometown and in Ontario, Watkins had apprenticed

Today, the James Bay Inn is the third-oldest hostelry—after the Dominion Hotel (1876) and the Fairmont Empress (1908)—still in operation in Victoria.

to Thomas Hooper at age 15 and worked with him on the Metropolitan United Church. His career advanced by leaps and bounds, and he and Hooper entered into partnership. The two designed the Victoria Public Library building at the corner of Yates and Blanshard streets, partially funded by the Andrew Carnegie Foundation, as well as the Bishop's Palace on View Street behind St. Andrew's Cathedral and several major structures in Chinatown. Watkins then struck out on his own and went on to design George Jay Elementary School and Begbie Hall, an administration building at Royal Jubilee Hospital. More schools were on the horizon. In the meantime, a hotel in James Bay would be a welcome change of pace.

Watkins' design called for reinforced concrete and brickwork, with the two upper storeys finished in stucco. Prominently situated at the corner of Government and Toronto streets, the front of the hotel afforded a view up Avalon Road toward Beacon Hill Park. An impressive staircase led to the hotel's entrance on the second-floor level. Its east side faced the park and enjoyed the sunrise; to the west,

the colours of the setting sun filled the evening sky.

Residents and out-of-town guests alike were impressed by the James Bay Hotel's red Spanish tile roof and spacious rooms, but were less enchanted by its name. It was buried in the middle of a residential community and did not, as its name implied, overlook a bay. This occasioned a brief change of name in the 1960s, to the Colonial Hotel.

By this time, Fred C. Smith, the facility's first lessee, had long since gone on his way. The hotel changed hands several times before it was bought in 1942 by a Benedictine order and turned into St. Mary's Priory Guest House.

Through what remained of the Second World War, the nursing sisters cared for residents and returning veterans. And it was here, less than a block from the house where she was born, that artist and author Emily Carr spent her final days. Recognized today as one of Canada's national treasures, Emily died in relative obscurity on March 2, 1945, at what is now the James Bay Inn.

St. Ann's Academy
835 Humboldt Street

This 1871 convent building was a far cry from the simple log cabin school–home occupied by the first Sisters of St. Ann, who arrived here in 1858. The old schoolhouse has since been relocated to Elliott Street Square, beside the Royal BC Museum.

St. Ann's Academy, nestling in the valley on the north side of Beacon Hill Park, is a lasting legacy to the Roman Catholic nuns who journeyed from Quebec in 1858.

In the spring of 1858 the first Fraser River-bound prospectors sailed from California at short notice, little prepared for what lay ahead. Not so the small group of women who followed them a few weeks later. The Sisters of St. Ann knew exactly what to expect, thanks to the man who had brought them here.

Bishop Modeste Demers had first visited the west coast 20 years earlier. Now his Quebec-based superiors wanted him to establish a Roman Catholic diocese for all British territory north of the 49th parallel and west of the Rockies. Demers approached the Sisters of St. Ann, a flourishing community of 45 religious women in St. Jacques (now Lachine), Quebec. When he outlined his plans and asked for a few people to help, all 45 volunteered. Demers chose four Sisters and another woman to accompany him to Vancouver Island.

After a two-month journey by train and steamer, the group reached Fort Victoria in June 1858. Their timing was perfect. Just over a month before they arrived, the settlement had seen the influx of hundreds of eager gold miners from California. Unbelievably, thousands more were still to come.

James Douglas, chief factor at the fort, gave Demers and his helpers a warm welcome. Douglas was supportive of Demers' dream of working with the Native people and encouraging their conversion to Christianity. But he was also aware that a growing number of local children needed a formal education. Demers, he decided, should concentrate his group's efforts in Victoria before venturing farther afield.

The four nuns—Sister Mary of the Sacred Heart, Sister Mary Lumena, Sister Mary Conception and Sister Mary Angèle—along with their helper, Marie Mainville, quickly assessed their surroundings. Singularly unimpressive at this early stage, "St. Ann's Victoria" was a crude 20- by 33-foot log cabin with broken windows and unlocked doors. It had been built in 1845 by an HBC employee. Its two rooms were separated by a partition and a chimney. There were doors and windows, but there was no heat, no light, and no ceiling.

In 1910 the Academy was completed with the addition of a new five-storey west wing.

Undaunted, the Sisters applied themselves to the task at hand. One side of the cabin was part sleeping quarters, part school. The other side was used for cooking. A total of 12 white, mixed-race, and Native children attended on the first school day, barely 48 hours after the Sisters had arrived. Within a year, 56 children were enrolled, and the focus had changed. St. Ann's was now a school for young ladies. Governor Douglas helped develop the curriculum and showed his support by sending three of his daughters there.

The schoolhouse was enlarged, and in September 1859 two more Sisters arrived from Quebec. Sister Mary Providence, the convent's Mother Superior, attempted to ease the cramped quarters by renting a building on Broad Street, soon replaced by a brick building on View Street.

In 1863 more reinforcements arrived. The following year the Sisters established a boarding school for Native girls in Quamichan, a 40-mile canoe trip from Victoria. Their dream of reaching out into distant communities was finally coming true.

Meanwhile, back in Victoria, plans were underway to centralize their activities. In September 1871, three months after British Columbia entered Confederation, Lieutenant-Governor Joseph Trutch laid the cornerstone for the new St. Ann's Convent. Sir James Douglas, Chief Justice Matthew Baillie Begbie, and many other political and church dignitaries attended. Sadly absent was Bishop Demers, who had died two months earlier at the age of 62.

This 1888 photo shows the rear (Collinson Street) entrance to St. Joseph's Hospital. Much enlarged and altered over the years but now restored, the original hospital building faces St. Ann's Academy across Humboldt Street, just as it did in days gone by.

Following Father Joseph Michaud's original plans, Montreal architect Charles Vereydhen supervised the building of the sand-finished red-brick central section of the four-storey building now known as St. Ann's Academy. In 1886 John Teague designed the "east block" and a relocated main entrance with an impressive curved double staircase and a pedimented gable pavilion. Father Michaud's 1858 wooden cathedral on the opposite side of Humboldt Street was put on skids, moved to the back of the convent, and encased in brick.

In 1876 Thomas Hooper designed St. Joseph's Hospital for the Sisters, on the opposite side of Humboldt Street, and in 1910 he completed St. Ann's by designing its third and final section—a west wing with a fifth storey under a mansard roof with dormer windows. By this time St. Ann's was the hub of activities that extended throughout British Columbia and into the Yukon.

Today a handful of downtown buildings and a cluster of white crosses in Ross Bay Cemetery tell the Sisters' story. St. Ann's Academy still stands, opposite St. Joseph's Hospital, on Humboldt Street. The original log cabin schoolhouse was moved to Elliot Street Square, behind the Royal British Columbia Museum. It is now the city's oldest building, a remarkable reminder of the brave women who gave Victoria's children the first formal schooling they had known.

St. Ann's Chapel, fashioned from California redwood in 1858 and recently restored, is Victoria's oldest religious building.

Close to 36,000 students from Canada, Central and South America, Mexico, and Hong Kong attended St. Ann's Academy before it closed in 1973. The building now houses provincial government offices.

Afterword

Victoria's growth from fur-trading post to provincial capital has been nothing short of remarkable. Starting out in 1843 as the first British-based venture on this island—indeed, on this coast—the former Hudson's Bay Company outpost was transformed in the late1850s into a way station for Fraser River gold miners. Subsequently burgeoning as a commercial centre, Victoria blossomed into the beautiful city we see today. Parks, gardens, and well-preserved heritage buildings combine to create a capital city that is the jewel in British Columbia's crown.

Bibliography

Source material at the British Columbia Archives and Records Service, City of Victoria Archives, Saanich Municipal Archives, Esquimalt Municipal Archives, Saanich Pioneer Society, Sooke Region Museum, and Greater Victoria Public Library was supplemented with information from the following books, as well as from the Victoria *Times Colonist* and newspapers of the News Group.

Akrigg, G.P.V., and Helen B. Akrigg. *British Columbia Chronicle 1847–1871: Gold and Colonists.* Vancouver, BC: Discovery Press, 1977.

Baskerville, Peter A. *Beyond the Island: An Illustrated History of Victoria.* Burlington, ON: Windsor Publications Ltd., 1986.

Bingham, Janet. *Samuel Maclure, Architect.* Ganges, BC: Horsdal and Schubart, 1985.

British Columbia from the Earliest Times to the Present, Biographical Vol. IV. Vancouver, BC: The S.J. Clarke Publishing Company, 1914.

Camas Historical Group. *Camas Chronicles of James Bay.* Victoria, BC: author, 1978.

Castle, Geoffrey (ed.). *Saanich: An Illustrated History.* Sidney, BC: Manning Press, 1989.

City of Victoria. *This Old House: An Inventory of Residential Heritage.* Victoria, BC: author, 1979.

———. *City of Victoria Downtown Heritage Registry.* Victoria, BC: author, 1996.

Crystal Gardens Preservation Society. *The Crystal Gardens: West Coast Pleasure Palace.* Victoria, BC: author, 1977.

Down, Edith E. *A Century of Service: A History of the Sisters of St. Ann and Their Contribution to Education in British Columbia, the Yukon and Alaska.* Victoria, BC: The Sisters of St. Ann, 1966.

Fawcett, Edgar. *Some Reminiscences of Old Victoria.* Toronto, ON: William Briggs, 1912.

Grant, Peter. Victoria: *A History in Photographs*. Canmore, AB: Altitude
 Publishing, 1995.

Green, Valerie. *No Ordinary People*. Victoria, BC: Beach Holme Publishers, 1992.

Humphreys, Danda. *On The Street Where You Live, Volume I: Pioneer Pathways
 of Early Victoria*. Surrey, BC: Heritage House, 1999.

———. *On The Street Where You Live, Volume II: Victoria's Early Roads and
 Railways*. Surrey, BC: Heritage House, 2000.

———. *On The Street Where You Live, Volume III: Sailors, Solicitors and
 Stargazers of Early Victoria*. Surrey, BC: Heritage House, 2001.

Kluckner, Michael. *Victoria: The Way It Was*. North Vancouver, BC: Whitecap
 Books, 1986.

Lai, David Chuenyan. *The Forbidden City Within Victoria*. Victoria, BC: Orca
 Books, 1991.

——— and Pamela Madoff. *Building and Rebuilding Harmony: The Gateway to
 Victoria's Chinatown*. Victoria, BC: University of Victoria, Western
 Geographical Series, 1997.

Ormsby, Margaret A. *British Columbia: A History*. Toronto, ON: Macmillan,
 1958.

Rayner, William. *British Columbia's Premiers in Profile*. Surrey, BC: Heritage
 House, 2000.

Reksten, Terry. Craigdarroch: *The Story of Dunsmuir Castle*. Victoria, BC: Orca
 Book Publishers, 1987.

———. *The Empress Hotel*. Vancouver/Toronto: Douglas and McIntyre, 1997.

Segger, Martin, and Douglas Franklin. *Exploring Victoria's Architecture*.
 Victoria, BC: Sono Nis Press, 1996.

Ward, Robin. *Echoes of Empire: Victoria and Its Remarkable Buildings*. Madeira
 Park, BC: Harbour Publishing, 1996.

Index

Photo Credits

B.C. Archives: 11 (F-06339), 23 top (PDP-02892), 24 (D-07225), 31 (F-09931), 33 (A-02604), 46 (A-02716), 52 (B-02225), 59 (F-07784), 63 (C-08993), 65 (B-02247), 72 (D-06418), 73 (F-02089). 76 (F-06593), 82(A-03407), 87 (I-20567), 100 (A-07737).

Christine Toller: Front cover (bottom right), 1, 13, 19, 23 bottom, 26, 27, 29, 32, 35, 39, 41, 44, 50, 53, 66, 70, 81, 92, 93, 95, 103,

Craigdarroch Castle: 55, 56 (Jeff Barber photo), 57 (Jeff Barber photo)

Edward Morris Store: 42

Heritage House: Front cover (postcard in centre), 48, 58, 62, 99, 104

Jim Munro: 43

John and Glenda Cheramy: Front cover (top three), back cover (lower left), 2, 8, 10, 14, 15, 18, 21, 28, 34, 37, 38, 40, 61, 68, 74, 75, 78, 79, 80, 83, 84, 85, 86, 88, 89, 90, 91, 97, 98, 101, 102

Queen's Printer: 94

Also available from Heritage House — Danda Humphreys' trilogy

On The Street Where You Live

VOLUMES ONE, TWO, AND THREE

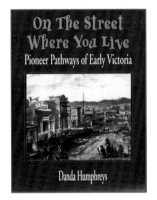 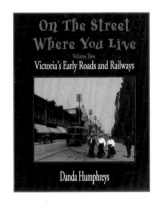 

1-895811-90-2   1-894384-09-1   1-894384-31-8
                                $34.95 each, hardcover

"The stories are full of snippets of the trials and tribulations, and of the triumphs, of lives lived at the edges of civilization … Early photographs are an integral part and are well presented with clear reproduction and helpful captions."
— George Newell, *B.C. Historical News*

"Humphreys' accounts go beyond standard textbook information and are enlivened with gossip and unusual details ... a good one to add to the Christmas list for friends living [in Victoria] and elsewhere, and for the wonderful peek into the past that it provides."
— Tracy O'Hara, *Times Colonist*

"A very readable book that uncovers a lot more than a mere printed date. The writing is clear and approachable ... A nice local history book which illuminates a significant period of British Columbia." — *B.C. Historical Federation*

"Students could write about streets in their own neighbourhoods using Humphreys' easy-to-read format as an example and a resource ... *On the Street Where You Live* provides a unique way to study history and create interest in the people who built our province."
— Anne Lansdell, *The Bookmark*, BCTLA

"These books are a joy to read." — Bob Griffin, *Discovery*

Originally from Cheshire, England, Danda has lived in Canada since 1972 and on the west coast since 1982. Her first career was nursing, and she then moved on to journalism. Over the years she has been an actor, broadcaster, public-relations person, and presentation skills coach.

Danda arrived in Victoria in late 1996 and was captivated by the city's short but unusual history. Curious about the street names, she started researching their historic origins and soon earned the unofficial title of "Victoria's Favourite Street Walker."

Danda's columns on street names appeared weekly in the Victoria Times Colonist for five years, and her first book, *On The Street Where You Live: Pioneer Pathways of Early Victoria*, was published in 1999. She has since written two more books in her "Streets" series. More than anything, Danda loves to tell stories. Residents and visitors alike enjoy her guided walks, talks, and tours through Victoria's historic downtown.